Antenna Handbook

Antenna Handbook

VOLUME I FUNDAMENTALS AND MATHEMATICAL TECHNIQUES

Edited by
Y. T. Lo
Electromagnetics Laboratory
Department of Electrical and Computer Engineering
University of Illinois–Urbana

S. W. Lee
Electromagnetics Laboratory
Department of Electrical and Computer Engineering
University of Illinois–Urbana

CHAPMAN & HALL
I(T)P An International Thomson Publishing Company

New York • Albany • Bonn • Boston • Cincinnati • Detroit • London • Madrid • Melbourne •
Mexico City • Pacific Grove • Paris • San Francisco • Singapore • Tokyo • Toronto • Washington

Copyright © 1993 by Van Nostrand Reinhold

This edition published by Chapman & Hall, New York, NY

Printed in the United States of America

For more information contact:

Chapman & Hall
115 Fifth Avenue
New York, NY 10003

Chapman & Hall
2-6 Boundary Row
London SE1 8HN
England

Thomas Nelson Australia
102 Dodds Street
South Melbourne, 3205
Victoria, Australia

Chapman & Hall GmbH
Postfach 100 263
D-69442 Weinheim
Germany

Nelson Canada
1120 Birchmount Road
Scarborough, Ontario
Canada M1K 5G4

International Thomson Publishing Asia
221 Henderson Road #05-10
Henderson Building
Singapore 0315

International Thomson Editores
Campos Eliseos 385, Piso 7
Col. Polanco
11560 Mexico D.F.
Mexico

International Thomson Publishing - Japan
Hirakawacho-cho Kyowa Building, 3F
1-2-1 Hirakawacho-cho
Chiyoda-ku, 102 Tokyo
Japan

All rights reserved. No part of this book covered by the copyright hereon may be reproduced or used in any form or by any means--graphic, electronic, or mechanical, including photocopying, recording, taping, or information storage and retrieval systems--without the written permission of the publisher.

2 3 4 5 6 7 8 9 XXX 01 00 99 98 97 96

Library of Congress Cataloging-in-Publication Data

The antenna handbook/edited by Y. T. Lo and S. W. Lee
        p.   Cm.
     Includes bibliographical references and indexes.
     Contents: v. 1. Fundamentals and mathematical techniques--v. 2. Antenna theory--v. 3. Applications--v. 4. Related topics.
     ISBN 0-442-01592-5 (v. 1).--ISBN 0-442-01593-3 (v.2).--ISBN 0-442-01594-1 (v. 3).--ISBN 0-442-01596-8 (v. 4)
     1. Antennas (Electronics) I. Lo, Y.T. II. Lee, S. W.
TK7871.6.A496 1993
621.382'4--dc20

93-6502
CIP

Visit Chapman & Hall on the Internet http://www.chaphall.com/chaphall.html

To order this or any other Chapman & Hall book, please contact **International Thomson Publishing, 7625 Empire Drive, Florence, KY 41042.** Phone (606) 525-6600 or 1-800-842-3636. Fax: (606) 525-7778. E-mail: order@chaphall.com.

For a complete listing of Chapman & Hall titles, send your request to **Chapman & Hall, Dept. BC, 115 Fifth Avenue, New York, NY 10003.**

Contents

Volume I FUNDAMENTALS AND MATHEMATICAL TECHNIQUES

1. **Basics**   1-3
   S. W. Lee

2. **Theorems and Formulas**   2-1
   S. W. Lee

3. **Techniques for Low-Frequency Problems**   3-1
   A. J. Poggio and E. K. Miller

4. **Techniques for High-Frequency Problems**   4-1
   P. H. Pathak

Appendices

A. Physical Constants, International Units, Conversion of Units, and Metric Prefixes   A-3
B. The Frequency Spectrum   B-1
C. Electromagnetic Properties of Materials   C-1
D. Vector Analysis   D-1
E. VSWR Versus Reflection Coefficient and Mismatch Loss   E-1
F. Decibels Versus Voltage and Power Ratios   F-1

Index   I-1

Preface

During the past decades, new demands for sophisticated space-age communication and remote sensing systems prompted a surge of R & D activities in the antenna field. There has been an awareness, in the professional community, of the need for a systematic and critical review of the progress made in those activities. This is evidenced by the sudden appearance of many excellent books on the subject after a long dormant period in the sixties and seventies. The goal of this book is to compile a reference to *complement* those books. We believe that this has been achieved to a great degree.

A book of this magnitude cannot be completed without difficulties. We are indebted to many for their dedication and patience and, in particular, to the forty-two contributing authors. Our first thanks go to Mr. Charlie Dresser and Dr. Edward C. Jordan, who initiated the project and persuaded us to make it a reality. After smooth sailing in the first period, the original sponsoring publisher had some unexpected financial problems which delayed its publication three years. In 1988, Van Nostrand Reinhold took over the publication tasks. There were many unsung heroes who devoted their talents to the perfection of the volume. In particular, Mr. Jack Davis spent many arduous hours editing the entire manuscript. Mr. Thomas R. Emrick redrew practically all of the figures with extraordinary precision and professionalism. Ms. Linda Venator, the last publication editor, tied up all of the loose ends at the final stage, including the preparation of the Index. Without their dedication and professionalism, the publication of this book would not have been possible.

Finally, we would like to express our appreciation to our teachers, students, and colleagues for their interest and comments. We are particularly indebted to Professor Edward C. Jordan and Professor George A. Deschamps for their encouragement and teaching, which have had a profound influence on our careers and on our ways of thinking about the matured field of electromagnetics and antennas.

This Preface was originally prepared for the first printing in 1988. Unfortunately, it was omitted at that time due to a change in the publication schedule. Since many readers questioned the lack of a Preface, we are pleased to include it here, and in all future printings.

Preface to the Second Printing

Since the publication of the first printing, we have received many constructive comments from the readers. The foremost was the bulkiness of a single volume for this massive book. The issue of dividing the book into multivolumes had been debated many times. Many users are interested in specific topics and not necessarily the entire book. To meet both needs, the publisher decided to reprint the book in multivolumes. We received this news with great joy, because we now have the opportunity to correct the typos and to insert the original Preface, which includes a heartfelt acknowledgment to all who contributed to this work.

We regret to announce the death of Professor Edward C. Jordan on October 18, 1991.

PART A

Fundamentals and Mathematical Techniques

Chapter 1

Basics

S. W. Lee
University of Illinois

CONTENTS

1. The Maxwell Equations and Time-Harmonic Fields	1-5
2. The Poynting Theorem	1-7
3. Boundary, Radiation, and Edge Conditions	1-8
4. Radiation from Sources	1-11
5. Plane Waves and Polarization	1-13
6. Antenna Near and Far Fields	1-16
7. Far-Field Representation	1-18
8. Calculation of the Far Field	1-21
From Given Currents 1-21	
From Given Fields over a Closed Surface 1-22	
9. Reference and Cross Polarizations	1-23
10. Antenna Gains	1-24
Incident Powers on an Antenna 1-24	
Three Gains in State $(\mathbf{k}, \mathbf{U})$ 1-25	
Three Gains in State $(\mathbf{k}, \mathbf{u})$ 1-25	
Peak Gains 1-26	
Gain-Related Terms 1-27	
11. Pattern Approximation by $(\cos\theta)^q$	1-28
12. TE and TM Field Representations	1-30
13. Plane-Wave Spectrum Representation	1-31
14. Periodic Structure	1-33
15. Rectangular Waveguide	1-37
16. Circular Waveguide	1-44
17. References	1-50

Shung-Wu Lee was born in Kiangsi, China. He received his BS degree in electrical engineering from Cheng Kung University in Tainan, Taiwan, in 1961. After one year of military service in Taiwan, he came to the United States in 1962. At the University of Illinois in Urbana, he received his MS and PhD degrees in electrical engineering and has been on the faculty since 1966. Currently he is a professor of electrical and computer engineering, and an associate director of the Electromagnetics Laboratory.

While on leave from the University of Illinois, Dr. Lee was with Hughes Aircraft Company, Fullerton, California, in 1969–1970, and with the Technical University at Eindhoven, The Netherlands, and the University of London, England, in 1973–74. Dr. Lee received several professional awards, including the 1968 Everitt Teaching Excellence Award from the University of Illinois, 1973 NSF NATO Senior Scientist Fellowship, 1977 Best Paper Award from IEEE Antennas and Propagation Society, and the 1985 Lockheed Million Dollar Award.

Dr. Lee has published more than 100 papers in technical journals on antennas and electromagnetic theory. He is the coauthor of a book on guided waves published by Macmillan in 1971, and a coeditor of this book. He is a Fellow of IEEE.

1. The Maxwell Equations and Time-Harmonic Fields

An electric field $\mathscr{E}$ is generally a function of spatial variable $\mathbf{r}$ and time variable t, and is denoted $\mathscr{E}(\mathbf{r},t)$. Unless specifically mentioned otherwise, the time variation is assumed to be sinusoidal:

$$\mathscr{E}(\mathbf{r},t) = \sqrt{2}\,\text{Re}\{\mathbf{E}(\mathbf{r})\,\epsilon^{+j\omega t}\} \qquad (1)$$

where $\omega = 2\pi f$ is the angular frequency in radians per second, and f is the frequency in hertz. We will work with the complex phasor $\mathbf{E}(\mathbf{r})$ throughout this book.*

In a continuous medium the behavior of the time-harmonic electromagnetic field is governed by the Maxwell equations. Using the International System of Units (SI units), the Maxwell equations for phasors are

$$\nabla \times \mathbf{E} = -j\omega \mathbf{B} - \mathbf{K} \qquad (2a)$$

$$\nabla \times \mathbf{H} = j\omega \mathbf{D} + \mathbf{J} \qquad (2b)$$

$$\nabla \cdot \mathbf{B} = \varrho_m \qquad (2c)$$

$$\nabla \cdot \mathbf{D} = \varrho \qquad (2d)$$

where

$\mathbf{E}$ = electric field in volts per meter (V/m)

$\mathbf{H}$ = magnetic field in amperes per meter (A/m)

$\mathbf{D}$ = electric flux density in coulombs per square meter (C/m^2)

$\mathbf{B}$ = magnetic flux density in teslas (T) or in webers per square meter (Wb/m^2)

$\mathbf{J}$ = electric current density in amperes per square meter, (A/m^2)

$\mathbf{K}$ = magnetic current density in volts per square meter (V/m^2)

ϱ = electric charge density in coulombs per cubic meter (C/m^3)

---

*Most antenna terms and definitions used in this chapter are consistent with those in *IEEE Standard Definitions of Terms for Antennas*, IEEE Std. 145–1983, appeared in *IEEE Trans. Antennas Propag.*, vol. AP-31, no. 8, November 1983.

ϱ_m = magnetic charge density in webers per cubic meter (Wb/m³)

From (2) we can deduce the equations of continuity:

$$\nabla \cdot \mathbf{J} + j\omega\varrho = 0 \tag{3a}$$

$$\nabla \cdot \mathbf{K} + j\omega\varrho_m = 0 \tag{3b}$$

which state the conservation of charges.

In an isotropic medium the constitutive relation is

$$\mathbf{D} = \epsilon\mathbf{E}, \quad \mathbf{B} = \mu\mathbf{H}, \quad \mathbf{J} = \sigma\mathbf{E} \tag{4}$$

The conductivity σ is given in siemens per meter (S/m) or in mhos per meter (℧/m) and is generally real, while the other two parameters, the dielectric constant (permittivity) ϵ and the permeability μ, are complex:

$$\epsilon = \epsilon' - j\epsilon'' \quad \text{in farads per meter (F/m)}, \tag{5a}$$

$$\mu = \mu' - j\mu'' \quad \text{in henrys per meter (H/m)} \tag{5b}$$

where ϵ', ϵ'', μ', and μ'' are positive real. The factors $(\epsilon''/\epsilon', \mu''/\mu')$ are called (*dielectric, magnetic*) *loss tangents*, respectively. They become vanishingly small if the medium is lossless. The *wave number* k in (meter)$^{-1}$ is defined by

$$k = k' - jk'' = \omega\sqrt{\mu\epsilon(1 - j\sigma/\omega\epsilon)} \tag{6}$$

where k' and k'' are positive real. The wavelength in meters is

$$\lambda = \frac{2\pi}{k'} \tag{7}$$

The intrinsic wave impedance Z in ohms and admittance Y in siemens of the medium are

$$Z = \frac{1}{Y} = \frac{\sqrt{\mu}}{\sqrt{\epsilon(1 - j\sigma/\omega\epsilon)}} \tag{8}$$

Here the square roots should be taken such that, if $a = \sqrt{b}$, Re $\{a\} \geq 0$ and Im $\{a\} \leq 0$. It follows that the phase angle of the complex number Z is between $-\pi/4$ and $+\pi/4$.

Free space or a vacuum is an isotropic medium in which*

---

*The approximate value of ϵ_0 in (9a) is obtained from $\epsilon_0 = \mu_0^{-1}c^{-2}$ with the speed of light $c \cong 3 \times 10^8$ m/s. If the more exact value $c \cong 2.997\,925 \times 10^8$ m/s is used, then $\epsilon_0 \cong 8.854\,185 \times 10^{-12}$ F/m.

Basics

$$\epsilon = \epsilon_0 \cong \frac{1}{36\pi \times 10^9} \quad \text{(F/m)} \tag{9a}$$

$$\mu = \mu_0 = 4\pi \times 10^{-7} \quad \text{(H/m)} \tag{9b}$$

$$\sigma = 0 \quad \text{(S/m)} \tag{9c}$$

For other isotropic media it is convenient to introduce the dimensionless ratios

$$\epsilon_r = \epsilon/\epsilon_0, \qquad \mu_r = \mu/\mu_r \tag{10}$$

which are known as the *relative dielectric constant* and *relative permeability*, respectively. In a lossless medium the wave number k and wavelength λ are given explicitly by

$$\begin{aligned}
k &= 0.20944 \times f \times \sqrt{\mu_r \epsilon_r} \quad \text{(cm}^{-1}) \\
&= 0.53198 \times f \times \sqrt{\mu_r \epsilon_r} \quad \text{(in}^{-1}) \\
\lambda &= \frac{30}{f \times \sqrt{\mu_r \epsilon_r}} \quad \text{(cm)} \\
&= \frac{11.81102}{f \times \sqrt{\mu_r \epsilon_r}} \quad \text{(in)}
\end{aligned}$$

in which f is given in gigahertz.

2. The Poynting Theorem

Consider a volume V enclosed by the boundary surface $S = \partial V$ in an isotropic medium characterized by constitutive parameters (ϵ, μ, σ). In general, (ϵ, μ) are functions of ω (dispersive medium). The complex power in watts supplied by sources inside V is

$$P_s = -\iiint_V (\mathbf{E} \cdot \mathbf{J}^* + \mathbf{H}^* \cdot \mathbf{K}) \, dv \tag{11}$$

[As described in (1) we use rms values for phasors. Therefore the factor 1/2 does not appear in (11).] The time-averaged power supplied by sources inside V is equal to the real part of P_s. The complex power leaving V across S is

$$P = \iint_S \mathbf{E} \times \mathbf{H}^* \cdot d\mathbf{s} \tag{12}$$

The direction of $d\mathbf{s}$ is in the outward normal direction of surface S (pointing away from V). The time-averaged power dissipated (conduction loss) inside V is

$$P_c = \iiint_V \sigma |\mathbf{E}|^2 \, dv \tag{13}$$

The time-averaged energy in joules stored in electric and magnetic fields inside V is

$$\overline{W}_e = \frac{1}{2} \iiint_V \overline{\epsilon} |\mathbf{E}|^2 \, dv \tag{14a}$$

$$\overline{W}_m = \frac{1}{2} \iiint_V \overline{\mu} |\mathbf{H}|^2 \, dv \tag{14b}$$

where

$$\overline{\epsilon} = \epsilon + \omega \frac{d\epsilon}{d\omega} \tag{14c}$$

$$\overline{\mu} = \mu + \omega \frac{d\mu}{d\omega} \tag{14d}$$

For the special case in which (ϵ, μ) are not functions of ω (nondispersive medium) the quantities in (14) reduce to

$$(\overline{\epsilon}, \overline{\mu}) \to (\epsilon, \mu) \tag{15a}$$

$$(\overline{W}_e, \overline{W}_m) \to (W_e, W_m) \tag{15b}$$

where

$$W_e = \frac{1}{2} \iiint_V \epsilon |\mathbf{E}|^2 \, dv \tag{15c}$$

$$W_m = \frac{1}{2} \iiint_V \mu |\mathbf{H}|^2 \, dv \tag{15d}$$

Note that in a dispersive medium (W_e, W_m) do not represent time-averaged energy stored in the electromagnetic field, as (ϵ, μ) may be negative. The stored energies are given by $(\overline{W}_e, \overline{W}_m)$, which are always positive. The Poynting theorem for the time-harmonic fields is

$$P_s = P + P_c + j\omega(W_m - W_e) \tag{16}$$

which holds for both dispersive and nondispersive media.

3. Boundary, Radiation, and Edge Conditions

At a point at the interface of two media we define a surface normal $\hat{n}_{21}$ pointing into medium 1 (Fig. 1). Then the fields in the two media at this point are related by boundary conditions:

$$\hat{n}_{21} \times (\mathbf{E}_1 - \mathbf{E}_2) = -\mathbf{K}_s \tag{17a}$$

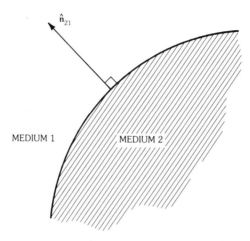

Fig. 1. Interface between two media.

$$\hat{n}_{21} \times (\mathbf{H}_1 - \mathbf{H}_2) = \mathbf{J}_s \qquad (17b)$$

$$\hat{n}_{21} \cdot (\mathbf{D}_1 - \mathbf{D}_2) = \varrho_s \qquad (17c)$$

$$\hat{n}_{21} \cdot (\mathbf{B}_1 - \mathbf{B}_2) = \varrho_{ms} \qquad (17d)$$

where $(\mathbf{J}_s, \varrho_s; \mathbf{K}_s, \varrho_{ms})$ are possible surface current and charge densities at the interface.

In an unbounded space the solution of the Maxwell equations may not be unique. To make it unique we impose an additional constraint to the Maxwell equations, namely, the radiation condition. For definiteness let us assume that all sources are contained in a finite region near $r = 0$. If the medium in space is lossy, the radiation condition requires that the fields vanish as $r \to \infty$. If the medium is lossless and isotropic, it requires that all field components have a phase progressively outward, and have an amplitude that decreases at least as rapidly as r^{-1}.

In problems where geometrical singularities (edges or tips) of perfect conductors are present, the solutions of the Maxwell equations must satisfy yet another constraint, namely, the edge condition [1]. It requires that the electric and magnetic energies stored in any finite neighborhood of a geometrical singularity be finite; that is,

$$\int_V |\mathbf{E}|^2 \, dv \to 0, \qquad \int_V |\mathbf{H}|^2 \, dv \to 0 \qquad (18)$$

as the volume V contracts to the geometrical singularity. For the conducting wedge in Fig. 2 the upper bounds of the fields near an edge point O are, as $\varrho \to 0$,

$$E_\varrho, E_\phi, H_\varrho, H_\phi = O(\varrho^{(\beta-\pi)/(2\pi-\beta)}) \qquad (19a)$$

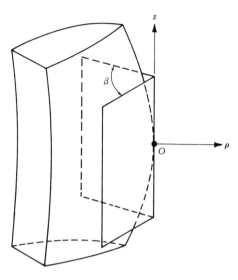

Fig. 2. A conducting wedge with interior wedge angle β and tangent planes at point O.

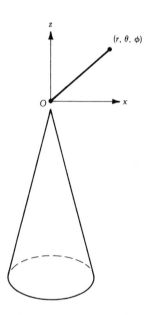

Fig. 3. A conducting cone with tip O.

$$E_z = O(\varrho^{\pi/(2\pi-\beta)}) \tag{19b}$$

$$H_z = O(1) \tag{19c}$$

where β is the interior wedge angle at point O. For the conducting cone in Fig. 3 all field components must grow slower than $r^{-3/2}$ as $r \to 0$.

4. Radiation from Sources

A source described by current densities $(\mathbf{J},\mathbf{K})$ radiates in an unbounded homogeneous space characterized by constitutive parameters (ϵ,μ). See Fig. 4. The radiation field at an arbitrary observation point $\mathbf{r}$ can be calculated from two vector potentials $(\mathbf{A},\mathbf{F})$ by the relations [2]

$$\mathbf{E}(\mathbf{r}) = -\nabla \times \mathbf{F} + \frac{1}{j\omega\epsilon}(\nabla \times \nabla \times \mathbf{A} - \mathbf{J}) \qquad (20a)$$

$$\mathbf{H}(\mathbf{r}) = \nabla \times \mathbf{A} + \frac{1}{j\omega\mu}(\nabla \times \nabla \times \mathbf{F} - \mathbf{K}) \qquad (20b)$$

where

$$\mathbf{A}(\mathbf{r}) = \iiint_V \mathbf{J}(\mathbf{r}')g(\mathbf{r}-\mathbf{r}')\,dv'$$

$$\mathbf{F}(\mathbf{r}) = \iiint_V \mathbf{K}(\mathbf{r}')g(\mathbf{r}-\mathbf{r}')\,dv'$$

$$g(\mathbf{r}-\mathbf{r}') = \frac{e^{-jk|\mathbf{r}-\mathbf{r}'|}}{4\pi|\mathbf{r}-\mathbf{r}'|}$$

$$k = \omega\sqrt{\mu\epsilon} = 2\pi/\lambda$$

The integrals above are over the source region V, and $\mathbf{r}'$ is a typical point in V.

Now consider an important special case of the radiation problem sketched in Fig. 4: the source region V is situated near the origin O, whereas the observation point $\mathbf{r} = (r,\theta,\phi)$ is far away $(kr \gg 1)$. Then Fig. 4 becomes Fig. 5, and the

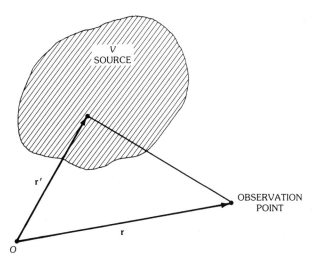

Fig. 4. Calculation of radiation field at an observation point $\mathbf{r}$.

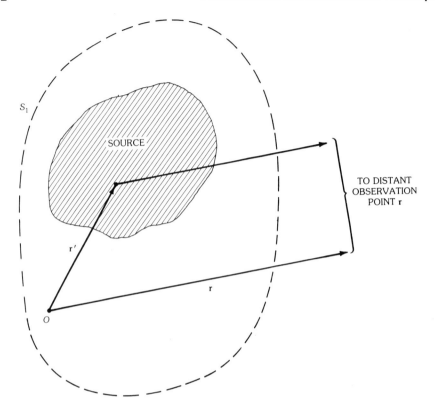

Fig. 5. Calculation of radiation field with **r** in the far zone and S_1 an arbitrary surface enclosing the source.

formulas in (20) can be simplified. Neglecting terms of order r^{-2} and higher in amplitude and phase, (20) is reduced to, as $r \to \infty$, the asymptotic equalities

$$E_r, H_r \sim 0 \tag{21a}$$

$$E_\theta \sim ZH_\phi \sim -jkZA_\theta - jkF_\phi \tag{21b}$$

$$E_\phi \sim -ZH_\theta \sim -jkZA_\phi + jkF_\theta \tag{21c}$$

where

$$\begin{Bmatrix} \mathbf{A(r)} \\ \mathbf{F(r)} \end{Bmatrix} = \frac{e^{-jkr}}{4\pi r} \iiint_V \begin{Bmatrix} \mathbf{J(r')} \\ \mathbf{K(r')} \end{Bmatrix} e^{j\mathbf{k}\cdot\mathbf{r'}} \, dv'$$

$\mathbf{k} = k\hat{\mathbf{r}} =$ wave vector

$Z = \sqrt{\mu/\epsilon}$

$\mathbf{r'} =$ source point with coordinates (x', y', z'), (ϱ', ϕ', z'), or (r', θ', ϕ')

Basics

$$\begin{aligned}\mathbf{k}\cdot\mathbf{r}' &= k[(x'\cos\phi + y'\sin\phi)\sin\theta + z'\cos\theta]\\ &= k[\varrho'\sin\theta\cos(\phi' - \phi) + z'\cos\theta]\\ &= kr'[\sin\theta'\sin\theta\cos(\phi' - \phi) + \cos\theta'\cos\theta]\end{aligned}$$

We note the following characteristics of the far field (21) of a finite source: (*a*) the field is an outgoing spherical wave, (*b*) both **E** and **H** are transverse to the direction of propagation, and (*c*) **E** and **H** satisfy the impedance relation $\mathbf{E} \sim Z\mathbf{H} \times \hat{\mathbf{r}}$.

5. Plane Waves and Polarization

The far field in (21) is locally a plane wave. A plane wave in an isotropic homogeneous medium has the following representation:

$$\begin{Bmatrix}\mathbf{E}(\mathbf{r})\\ \mathbf{H}(\mathbf{r})\end{Bmatrix} = C\begin{Bmatrix}\sqrt{Z}\mathbf{u}\\ \sqrt{Y}\hat{\mathbf{k}}\times\mathbf{u}\end{Bmatrix}e^{-j\mathbf{k}\cdot\mathbf{r}} \qquad (22)$$

Here $\mathbf{k} = k\hat{\mathbf{k}}$, with spherical coordinates (k, θ, ϕ) describing the direction of propagation (Fig. 6). The term $Z = Y^{-1} = \sqrt{\mu/\epsilon}$ is the wave impedance of the medium. The complex vector **u** satisfies two conditions:

(*a*) $\mathbf{u}\cdot\mathbf{u}^* = 1$ (unitary)   (23a)

(*b*) $\mathbf{u}\cdot\mathbf{k} = 0$ (transverse to direction of propagation)   (23b)

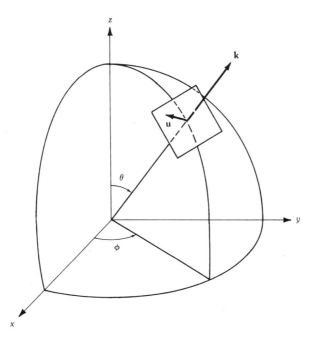

Fig. 6. A plane wave propagating in the direction $\mathbf{k} = (k, \theta, \phi)$ and with the polarization **u**.

The amplitude C in (18) is, in general, a complex number and has the unit (watt)$^{1/2}$ (meter)$^{-1}$. The power density of the plane wave is $|C|^2$ W/m^2.

The unitary vector $\mathbf{u}$ describes the polarization of the plane wave. It is defined only within a phase angle. Thus $\mathbf{u}$ and $\mathbf{u}e^{j\delta}$ describe the same polarization. To study $\mathbf{u}$ in further detail it is convenient to fix the direction of propagation in the z direction ($\mathbf{k} = k\hat{\mathbf{z}}$). Then (22) becomes

$$\left\{\begin{array}{l}\mathbf{E}(\mathbf{r})\\ \mathbf{H}(\mathbf{r})\end{array}\right\} = C \left\{\begin{array}{l}\sqrt{Z}\,\mathbf{u}\\ \sqrt{Y}\,\hat{\mathbf{z}} \times \mathbf{u}\end{array}\right\} e^{-jkz} \qquad (24)$$

Because of (23), a general expression for $\mathbf{u}$ has the form

$$\mathbf{u} = (\hat{\mathbf{x}}a_x + \hat{\mathbf{y}}a_y) \qquad (25)$$

Here a_x and a_y are two complex constants

$$a_x = |a_x|\,e^{j\delta_x}, \qquad a_y = |a_y|\,e^{j\delta_y} \qquad (26a)$$

subject to the constraint

$$\sqrt{|a_x|^2 + |a_y|^2} = 1 \qquad (26b)$$

In the time domain the electric-field vector at a reference plane (say $z = 0$) derived from (24) is

$$\mathscr{E}(z = 0, t) = \sqrt{2}\,\text{Re}\{\mathbf{E}(z = 0)e^{j\omega t}\} = \sqrt{2}\,C\sqrt{Z}\,\mathbf{U}(t) \qquad (27a)$$

Here the real vector

$$\mathbf{U}(t) = \text{Re}\{\mathbf{u}e^{j\omega t}\} \qquad (27b)$$

is a vector rotating in the xy plane at a speed such that it sweeps through a constant area per unit time. The locus of the extremity of the rotating vector is an ellipse, called the *polarization ellipse*. Depending on its shape, there are three cases to be considered:

(*a*) *Linear polarization.* If the x component and y component of $\mathbf{u}$ are in phase ($\delta_x = \delta_y$), the polarization ellipse degenerates into a straight-line segment. Thus, $\mathbf{U}(t)$ points to a constant direction in space (Fig. 7a), which makes an angle

$$\alpha = \tan^{-1}(|a_y|/|a_x|), \qquad 0 \leq \alpha \leq \pi/2 \qquad (28)$$

with the x axis. Apart from an unimportant phase factor, $\mathbf{u}$ in (25) is reduced to

$$\mathbf{u} = \hat{\mathbf{x}}\cos\alpha + \hat{\mathbf{y}}\sin\alpha \qquad (29)$$

which describes a linear polarization.

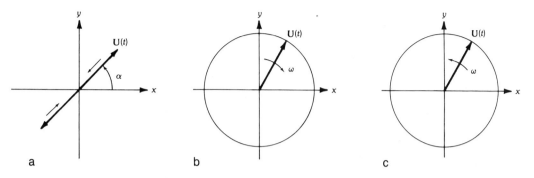

Fig. 7. Rotating vector U(*t*) for linear and circular polarizations. (*a*) Linear. (*b*) Left-hand circular. (*c*) Right-hand circular.

(*b*) *Circular polarization.* If the *x* component and *y* component of **u** are equal in magnitude ($|a_x| = |a_y|$) and 90° out of phase ($\delta_y - \delta_x = \pm\pi/2$), then the polarization ellipse becomes a circle (Figs. 7b and 7c). Apart from an unimportant phase factor, **u** in (25) is reduced to one of the following two unitary vectors:

$$\mathbf{L} = (\hat{\mathbf{x}} + j\hat{\mathbf{y}})/\sqrt{2} \quad \text{(LHCP)} \tag{30a}$$

$$\mathbf{R} = (\hat{\mathbf{x}} - j\hat{\mathbf{y}})/\sqrt{2} \quad \text{(RHCP)} \tag{30b}$$

Vector **L** (vector **R**) represents a vector in the time domain which rotates in the left-hand (right-hand) sense with respect to the direction of propagation, and is called a left-hand (right-hand) circular polarization vector.

(*c*) *Elliptical polarization.* For general values of a_x and a_y, **u** in (25) represents an elliptical polarization. The extremity of the time-domain vector **U**(*t*) traces out an ellipse (Fig. 8). An alternative presentation of (25) is obtained by decomposing **u** into two circular polarizations, viz.,

$$\mathbf{u} = \hat{\mathbf{x}}a_x + \hat{\mathbf{y}}a_y \tag{31a}$$
$$= \mathbf{L}a_L + \mathbf{R}a_R$$

where

$$a_L = |a_L|e^{j\delta_L} = \mathbf{u}\cdot\mathbf{L}^* = \frac{1}{\sqrt{2}}(a_x - ja_y) \tag{31b}$$

$$a_R = |a_R|e^{j\delta_R} = \mathbf{u}\cdot\mathbf{R}^* = \frac{1}{\sqrt{2}}(a_x + ja_y) \tag{31c}$$

The polarization ellipse in Fig. 8 is characterized by three parameters described as follows. (*a*) The axis ratio *AR* is defined by the ratio of maximum length and minimum length of **U**(*t*), or the ratio of the semimajor and semiminor axes of the polarization ellipse. It may be calculated from

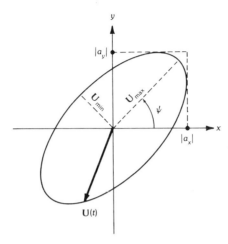

Fig. 8. Elliptical polarization: the tip of the rotating vector $\mathbf{U}(t)$ traces out an ellipse.

$$AR = \frac{U_{\max}}{U_{\min}} = \left| \frac{|a_L| + |a_R|}{|a_L| - |a_R|} \right| \quad (32)$$

For a linear polarization, $AR = \infty$. For a circular polarization, $AR = 1$. (b) The tilt angle ψ is measured from the x axis to a semimajor axis of the ellipse. It may be calculated from

$$\psi = \frac{1}{2}(\delta_R - \delta_L) + m\pi, \quad m = 0, \pm 1, \pm 2, \ldots \quad (33)$$

Customarily, we choose the integer m so that ψ falls in the range $0 \leq \psi < \pi$. (c) The sense of rotation of $\mathbf{U}(t)$ is determined by the comparison of $|a_L|$ and $|a_R|$. If $|a_L| > |a_R|$, $\mathbf{u}$ in (31a) represents a left-hand elliptical polarization. If $|a_L| = |a_R|$, $\mathbf{u}$ represents a linear polarization. If $|a_L| < |a_R|$, $\mathbf{u}$ represents a right-hand elliptical polarization.

The plane wave in (22) is characterized by its propagation direction $\mathbf{k}$ and its polarization $\mathbf{u}$. We say that the plane wave is in a state $(\mathbf{k}, \mathbf{u})$. In Chapter 2 we will use the time-reversed field of (22), namely, the plane wave propagates in the opposite direction as the original one, and has the *same* polarization. This time-reversed plane wave is in the state $(-\mathbf{k}, \mathbf{u}^*)$.

6. Antenna Near and Far Fields

The radiation field from a transmitting antenna can be roughly decomposed into two components: (a) radiating field in which the complex Poynting vector $\mathbf{E} \times \mathbf{H}^*$ is real, and $(\mathbf{E}, \mathbf{H})$ decay as r^{-1}, where r is the distance from the antenna; and (b) reactive field in which $\mathbf{E} \times \mathbf{H}^*$ is imaginary, and $(\mathbf{E}, \mathbf{H})$ decay more rapidly than r^{-1}. These two field components dominate in different regions in the space around the antenna. Based on this we can divide the space into three regions:

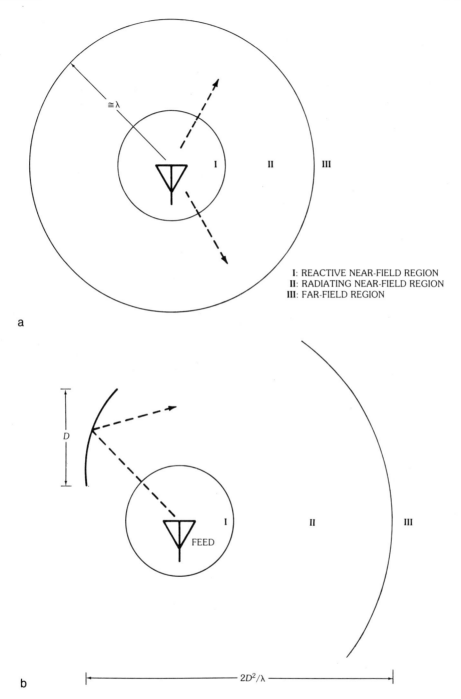

Fig. 9. Near and far fields. (*a*) Electrically small antennas. (*b*) Electrically large reflector.

(a) *Reactive near-field region.* This region is the space immediately surrounding the antenna in which the reactive field dominates the radiating field. For most electrically small antennas (Fig. 9a), the outer limit is on the order of a few wavelengths or less.

(b) *Radiating near-field region.* Beyond the immediate neighborhood of the antenna the radiating field begins to dominate. However, in this region, it is not dominant to the extent that the relative angular distribution of the field is independent of r, and the usual antenna radiation pattern can be defined. This region is sometimes referred to as the *Fresnel region* in analogy to optical terminology. For electrically large antennas (Fig. 9b) a commonly used criterion to define the outer boundary of the radiating near-field region is

$$r = 2D^2/\lambda \qquad (34)$$

where D is the largest dimension of the antenna, and λ is the wavelength.

(c) *Far-field region.* Beyond the reactive/radiating near-field region the reactive field becomes negligible. To a good approximation the field radiated by the antenna can be represented by the radiating field component alone, and the relative angular field distribution is independent of r. This region is sometimes known as the *Fraunhofer region* in analogy to optical terminology.

7. Far-Field Representation

Consider an antenna radiating in an unbounded isotropic homogeneous medium characterized by constitutive parameters (ϵ, μ). Its radiation field in the far-field region is a spherical wave that can be represented by*

$$\begin{Bmatrix} \mathbf{E}(\mathbf{r}) \\ \mathbf{H}(\mathbf{r}) \end{Bmatrix} \sim \begin{Bmatrix} \sqrt{Z}\mathbf{A}(\mathbf{k}) \\ \sqrt{Y}\hat{\mathbf{k}} \times \mathbf{A}(\mathbf{k}) \end{Bmatrix} \frac{e^{-jkr}}{r}, \qquad r \to \infty \qquad (35)$$

Here $\mathbf{r}$ is the observation point with spherical coordinates (r, θ, ϕ). The origin O of the coordinates is in the vicinity of the antenna (Fig. 10). The wave vector

$$\mathbf{k} = \hat{\mathbf{k}}k = \hat{\mathbf{r}}k \qquad (36)$$

is in the same direction as $\mathbf{r}$. Thus, we can use either $\hat{\mathbf{k}}$ or $\hat{\mathbf{r}}$ to describe the observation direction. The intrinsic wave impedance Z and admittance Y of the medium are defined by

$$Z = \frac{1}{Y} = \sqrt{\frac{\mu}{\epsilon}} \qquad (37)$$

If the medium is free space, then Z has the numerical value

---

*Do not confuse amplitude $\mathbf{A}(\mathbf{k})$ with vector potential $\mathbf{A}(\mathbf{r})$ used in (20). The present representation of the far field was described in G. A. Deschamps [3].

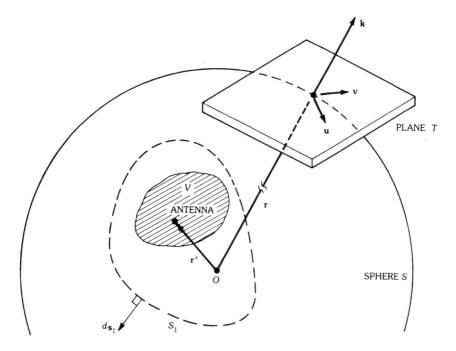

Fig. 10. An antenna located inside V radiates in unbounded space, with observation point **r** on a large sphere S, whose tangent plane at **r** is T.

$$Z_0 = Y_0^{-1} \cong 120\pi \quad \text{ohms} \tag{38}$$

The amplitude vector $\mathbf{A}(\mathbf{k})$ in (35) is generally complex, and has the unit (watt)$^{1/2}$. We will discuss it in detail below.

Referring to Fig. 10, we draw a sphere S, called the *radiation sphere*, passing through the observation point **r** and centered at O. A property of $\mathbf{A}(\mathbf{k})$ is that it is transverse to **k**. Thus **A** lies in the tangent plane T of sphere S. Therefore the set of **A**s forms a four-dimensional manifold (the tangent bundle of S). We decompose **A** into two orthogonal components:

$$\mathbf{A}(\mathbf{k}) = \mathbf{u}A(\mathbf{k},\mathbf{u}) + \mathbf{v}A(\mathbf{k},\mathbf{v}) \tag{39}$$

Here $(\mathbf{u},\mathbf{v})$ satisfy the following restrictions:

$$|\mathbf{u}|^2 = \mathbf{u}\cdot\mathbf{u}^* = |\mathbf{v}|^2 = 1 \tag{40a}$$

$$\mathbf{u}^*\cdot\mathbf{v} = \mathbf{u}\cdot\mathbf{v}^* = 0 \tag{40b}$$

$$\mathbf{u}\cdot\mathbf{k} = \mathbf{v}\cdot\mathbf{k} = 0 \tag{40c}$$

Unitary vectors $(\mathbf{u},\mathbf{v})$ describe the polarization of the radiated field (*cf.* Section 5). The scalar $A(\mathbf{k},\mathbf{u})$ is the component of $\mathbf{A}(\mathbf{k})$ in the direction **k** and with polarization **u**, or the component in the state $(\mathbf{k},\mathbf{u})$. It is calculated by

$$A(\mathbf{k}, \mathbf{u}) = \mathbf{A}(\mathbf{k}) \cdot \mathbf{u}^* \tag{41}$$

A similar formula applies to $A(\mathbf{k}, \mathbf{v})$. As an example, we choose $\mathbf{u} = \hat{\boldsymbol{\theta}}$ and $\mathbf{v} = \hat{\boldsymbol{\phi}}$. Then (39) becomes

$$\mathbf{A}(\mathbf{k}) = \hat{\boldsymbol{\theta}} A_\theta(\mathbf{k}) + \hat{\boldsymbol{\phi}} A_\phi(\mathbf{k}) \tag{42}$$

where we used the more common notations (A_θ, A_ϕ) for $(A(\mathbf{k}, \hat{\boldsymbol{\theta}}), A(\mathbf{k}, \hat{\boldsymbol{\phi}}))$. Other examples of $(\mathbf{u}, \mathbf{v})$ are

(a) linear polarization:

$$\mathbf{u} = \hat{\boldsymbol{\theta}} \cos \phi - \hat{\boldsymbol{\phi}} \sin \phi, \quad \mathbf{v} = \hat{\boldsymbol{\theta}} \sin \phi + \hat{\boldsymbol{\phi}} \cos \phi \tag{43a}$$

(b) circular polarization:

$$\mathbf{u} = (\hat{\boldsymbol{\theta}} + j\hat{\boldsymbol{\phi}})/\sqrt{2}, \quad \mathbf{v} = (\hat{\boldsymbol{\theta}} - j\hat{\boldsymbol{\phi}})/\sqrt{2} \tag{43b}$$

In (43a), $\mathbf{u}$ is along the projection of $\hat{\mathbf{x}}$, and $\mathbf{v}$ is along the projection of $\hat{\mathbf{y}}$ onto the tangent plane T in Fig. 10. In (43b) $\mathbf{u}$ represents the left-hand circular polarization, and $\mathbf{v}$ represents the right-hand circular polarization.

For the far field in (35) the radiation intensity in watts per steradian (power per solid angle) in the state $(\mathbf{k}, \mathbf{u})$ is defined by

$$I(\mathbf{k}, \mathbf{u}) = (\mathbf{A} \cdot \mathbf{u}^*)(\mathbf{A} \cdot \mathbf{u}^*)^* = |A(\mathbf{k}, \mathbf{u})|^2 \tag{44}$$

A plot of $I(\mathbf{k}, \mathbf{u})$ as a function of observation direction $\hat{\mathbf{k}} = (\theta, \phi)$ is known as the antenna pattern. A typical pattern is shown in Fig. 11, where $I(\mathbf{k}, \hat{\boldsymbol{\phi}})$ is plotted as a function of θ for a fixed $\phi = \phi_0$. The maximum intensity for all observation directions occurs at a direction $\mathbf{k} = \mathbf{k}_0 = (k, \theta_0, \phi_0)$. The lobe in this direction is called the *main beam*, while all other lobes are called *side lobes*. The vertical axis of Fig. 11 shows the decibel (dB) value of the normalized intensity defined by

$$10 \log_{10} \left[\frac{I(\mathbf{k}, \mathbf{u})}{I(\mathbf{k}_0, \mathbf{u})} \right]$$

Thus the main beam is at the 0-dB level. The highest side lobe in Fig. 11 is -10 dB. It is customary to say that the side lobe level in $\phi = \phi_0$ cut is 10 dB down from the main beam. The width of the main beam at -3 dB level is called the *half-power beamwidth*, and in the cut $\phi = \phi_0$ is given by $\theta_2 - \theta_1$ in radians or in degrees.

For the far field in (35) the radiation intensity in all (both) polarizations is

$$|\mathbf{A}(\mathbf{k})|^2 = |A(\mathbf{k}, \mathbf{u})|^2 + |A(\mathbf{k}, \mathbf{v})|^2 \tag{45}$$

Alternatively, we can rewrite (39) as

$$\mathbf{A}(\mathbf{k}) = |A(k)|\mathbf{U} \tag{46}$$

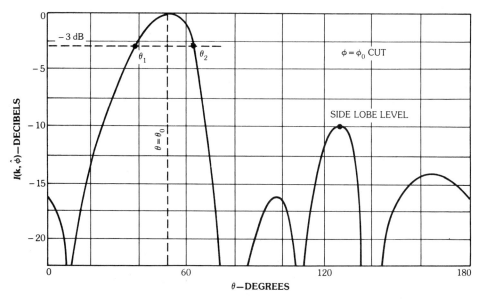

Fig. 11. An antenna pattern: the normalized intensity in the polarization $\hat{\phi}$ versus observation polar angle θ for a fixed $\phi = \phi_0$.

where the unitary vector **U** is given by

$$\mathbf{U} = \frac{1}{|\mathbf{A}(\mathbf{k})|}[\mathbf{u}A(\mathbf{k},\mathbf{u}) + \mathbf{v}A(\mathbf{k},\mathbf{v})] \qquad (47)$$

Then the polarization of the far field in the direction **k** is described by a single unitary vector **U**.

In summary, an antenna far field can be represented by the spherical wave in (35). To describe amplitude vector $\mathbf{A}(\mathbf{k})$ we need to specify an observation direction **k**, and a polarization **u**, or a state $(\mathbf{k}, \mathbf{u})$. The component of $\mathbf{A}(\mathbf{k})$ in this state is $A(\mathbf{k}, \mathbf{u})$ in (41), and the radiation intensity is $I(\mathbf{k}, \mathbf{u})$ in (44). There are two degrees of freedom in the polarization of $\mathbf{A}(\mathbf{k})$. We customarily describe them by two orthogonal unitary vectors $(\mathbf{u}, \mathbf{v})$ in the manner stated in (39). A common choice is $(\mathbf{u} = \hat{\theta}, \mathbf{v} = \hat{\phi})$; another one is (**u** = reference polarization, **v** = cross polarization), which will be described in Section 9.

8. Calculation of the Far Field

We will look at the calculation of the far field (*a*) from given currents and (*b*) from given fields over a closed surface.

From Given Currents

The antenna far field is represented by the expression in (35). We will now consider its calculation. For a given transmitting antenna we may replace it by a mathematical current source $(\mathbf{J}, \mathbf{K})$ which radiates in the unbounded medium after the antenna structure is removed. The far field radiated by $(\mathbf{J}, \mathbf{K})$ may be calculated from (21) or an equivalent formula given below. For a given state $(\mathbf{k}, \mathbf{u})$ the component of the amplitude vector $\mathbf{A}(\mathbf{k})$ in this state is defined in (41) and is given by (see Fig. 10)

$$A(\mathbf{k},\mathbf{u}) = \frac{k}{j4\pi} \iiint_V [\sqrt{Z}\mathbf{J}(\mathbf{r}')\cdot\mathbf{u}^* + \sqrt{Y}\mathbf{K}(\mathbf{r}')\cdot(\hat{\mathbf{k}} \times \mathbf{u}^*)] e^{j\mathbf{k}\cdot\mathbf{r}'} dv' \qquad (48)$$

where the integration is over a typical source point $\mathbf{r}'$ located inside the source region V. Explicit forms of $\mathbf{k}\cdot\mathbf{r}'$ are given in (21). If $\mathbf{u} = \hat{\boldsymbol{\theta}}$ (for calculating the A_θ component), the two dot products in (48) become

$$\mathbf{J}\cdot\mathbf{u}^* = (J_x \cos\phi + J_y \sin\phi)\cos\theta - J_z \sin\theta \qquad (49a)$$

$$\mathbf{K}\cdot(\hat{\mathbf{k}} \times \mathbf{u}^*) = -K_x \sin\phi + K_y \cos\phi \qquad (49b)$$

If $\mathbf{u} = \hat{\boldsymbol{\phi}}$ (for calculating the A_ϕ component), the dot products in (48) become

$$\mathbf{J}\cdot\mathbf{u}^* = -J_x \sin\phi + J_y \cos\phi \qquad (50a)$$

$$\mathbf{K}\cdot(\hat{\mathbf{k}} \times \mathbf{u}^*) = (K_x \cos\phi + K_y \sin\phi)(-\cos\theta) + K_z \sin\theta \qquad (50b)$$

From Given Fields over a Closed Surface

Referring to the radiation problem in Fig. 10, let us denote the (exact) radiation field from the antenna over a closed surface S_1 enclosing the antenna by $(\mathbf{E}_1, \mathbf{H}_1)$. If $(\mathbf{E}_1, \mathbf{H}_1)$ are given, we can calculate the far field (35) from them. The component of the amplitude vector $\mathbf{A}(\mathbf{k})$ in the state $(\mathbf{k}, \mathbf{u})$ is calculated from

$$A(\mathbf{k},\mathbf{u}) = \frac{k}{j4\pi} \iint_{S_1} e^{j\mathbf{k}\cdot\mathbf{r}'_1}[\sqrt{Y}(\hat{\mathbf{k}} \times \mathbf{u}^*) \times \mathbf{E}_1(\mathbf{r}'_1) - \sqrt{Z}\mathbf{u}^* \times \mathbf{H}_1(\mathbf{r}'_1)]\cdot d\mathbf{s}'_1 \qquad (51)$$

where the integration is over the closed surface S_1, and $\mathbf{r}'_1$ is a typical point on S_1, and the differential surface $d\mathbf{s}'_1$ is in the direction of outward normal of S_1. For the special case that S_1 is the infinite plane $z = 0$ (Fig. 12), the two spherical components of $\mathbf{A}(\mathbf{k})$ deduced from (51) are

$$\begin{Bmatrix} A_\theta(\mathbf{k}) \\ A_\phi(\mathbf{k}) \end{Bmatrix} = \frac{k}{j4\pi} \iint_{-\infty}^{\infty} \begin{Bmatrix} B_\theta \\ B_\phi \end{Bmatrix} e^{jk\sin\theta(x'\cos\phi + y'\sin\phi)} dx' dy' \qquad (52)$$

where

$$B_\theta = \sqrt{Y}(-1)(E_x \cos\phi + E_y \sin\phi) + \sqrt{Z}(H_x \sin\phi - H_y \cos\phi)\cos\theta \qquad (53a)$$

$$B_\phi = \sqrt{Y}(E_x \sin\phi - E_y \cos\phi)\cos\theta + \sqrt{Z}(H_x \cos\phi + H_y \sin\phi) \qquad (53b)$$

Basics

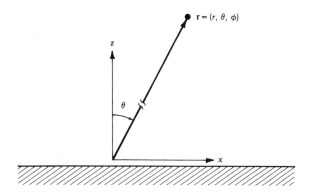

Fig. 12. Calculation of far field at **r** from a given field distribution over $z = 0$ plane.

In (53), (E_x, E_y, H_x, H_y) are the components of $(\mathbf{E}_1, \mathbf{H}_1)$ evaluated at a typical source point $(x', y', z' = 0)$ on surface S_1. The formula in (51) is related to that in (48) by Huygens' principle, which is discussed in Chapter 2.

9. Reference and Cross Polarizations

Consider a transmitting antenna with its far field given by (35). The amplitude vector $\mathbf{A}(\mathbf{k})$ can be expressed in terms of two polarization vectors $(\mathbf{u}, \mathbf{v})$ in the manner described in (39). Among many possible choices for $(\mathbf{u}, \mathbf{v})$, we often select a particular set $(\mathbf{u}_R, \mathbf{u}_C)$ and call them

$\mathbf{u}_R$ = polarization vector for the reference polarization

$\mathbf{u}_C$ = polarization vector for the cross polarization

(Alternative names for reference polarization are copolarization and principal polarization.) The unitary vectors $(\mathbf{u}_R, \mathbf{u}_C)$ are, in general, functions of observation direction $\mathbf{k}$, and satisfy the constraints in (40).

There is no standard way to define $(\mathbf{u}_R, \mathbf{u}_C)$. This difficulty may be traced to the geometrical property that there is no unique way to transport a tangent vector at one point on the sphere S in Fig. 10 to another point on S. Consequently, if $(\mathbf{u}_R, \mathbf{u}_C)$ are defined in one direction, say the antenna main beam direction, there is no unique way to extend into other directions.

Nevertheless, a popular definition for $(\mathbf{u}_R, \mathbf{u}_C)$ is given by Ludwig [4]. His definition is useful for a class of antennas in which:

(a) the antenna main beam is nearly in the z direction, and
(b) the antenna current (or equivalent current) has the form

$$\mathbf{J}(\mathbf{r}) = J(\mathbf{r})(\hat{\mathbf{x}} a e^{j\psi} + \hat{\mathbf{y}} b) \tag{54}$$

where (a, b, ψ) are real (independent of $\mathbf{r}$), and $a^2 + b^2 = 1$.

Then Ludwig's definition is

$$\mathbf{u}_R = \hat{\boldsymbol{\theta}}(ae^{j\psi}\cos\phi + b\sin\phi) + \hat{\boldsymbol{\phi}}(-ae^{j\psi}\sin\phi + b\cos\phi) \quad (55a)$$

$$\mathbf{u}_C = \hat{\boldsymbol{\theta}}(ae^{-j\psi}\sin\phi - b\cos\phi) + \hat{\boldsymbol{\phi}}(ae^{-j\psi}\cos\phi + b\sin\phi) \quad (55b)$$

For a linear wire antenna along the x direction we have ($a = 1, b = 0, \psi = 0$) and

$$\mathbf{u}_R = \hat{\boldsymbol{\theta}}\cos\phi - \hat{\boldsymbol{\phi}}\sin\phi, \quad \mathbf{u}_C = \hat{\boldsymbol{\theta}}\sin\phi + \hat{\boldsymbol{\phi}}\cos\phi$$

For a cross dipole with $J_x : J_y = 1 : (-j)$, we have, apart from an unimportant phase factor,

$$\mathbf{u}_R = \frac{1}{\sqrt{2}}(\hat{\boldsymbol{\theta}} - j\hat{\boldsymbol{\phi}}), \quad \mathbf{u}_C = \frac{1}{\sqrt{2}}(\hat{\boldsymbol{\theta}} + j\hat{\boldsymbol{\phi}})$$

which describe circular polarizations as expected.

10. Antenna Gains

There are several different terms relating to the gain of a transmitting antenna, depending on the incident power normalization, the direction of observation, and polarization of its far field.

Incident Powers on an Antenna

A transmitting antenna is connected to a generator through a transmission line (or a waveguide) as sketched in Fig. 13. The (time-averaged) power in watts incident from the generator to the antenna is denoted by P_1. Due to the mismatch between the transmission line and the antenna, only P_2 is accepted by the antenna, where

$$P_2 = (1 - |\Gamma|^2)P_1 \quad (56)$$

and $|\Gamma|^2$ is the power reflection coefficient. If the antenna is lossless, P_2 is radiated

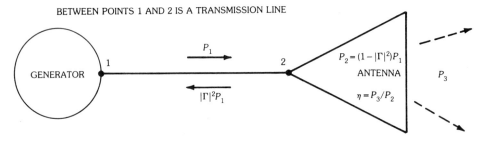

Fig. 13. The power incident to the antenna is P_1, the power accepted by the antenna is P_2, and that radiated is P_3.

Basics

into the space. If the antenna is lossy, the radiated power P_3 is only a portion of P_2. The ratio

$$\eta = P_3/P_2 \tag{57}$$

is called the *radiation efficiency*.

Three Gains in State (k, U)

As indicated in (39) the antenna radiates in two orthogonal polarizations $(\mathbf{u}, \mathbf{v})$. Alternatively, its polarization can be represented by (within a phase factor) a single unitary vector $\mathbf{U}$ in the manner described in (46) and (47). Then the radiation intensity in watts per steradian in state $(\mathbf{k}, \mathbf{U})$ is

$$I(\mathbf{k}) = |\mathbf{A}(\mathbf{k})|^2 = |A(\mathbf{k}, \mathbf{u})|^2 + |A(\mathbf{k}, \mathbf{v})|^2 \tag{58}$$

which includes radiation in all (both) polarizations. The radiated power P_3 by the antenna in space is

$$P_3 = \int_0^\pi d\theta \int_0^{2\pi} d\phi \, [I(\mathbf{k}) \sin \theta] \tag{59}$$

The general definition of a gain (dimensionless) in the direction $\mathbf{k}$ and for all polarizations is

$$G(\mathbf{k}) = \frac{4\pi}{P_n} I(\mathbf{k}) = \frac{4\pi}{P_n} |\mathbf{A}(\mathbf{k})|^2 = 4\pi \frac{\text{intensity of the antenna in direction } \mathbf{k}}{\text{a reference power}} \tag{60}$$

where P_n is a reference power. Depending on P_n, there are three commonly used gains:

(a) realized gain $G_1(\mathbf{k})$ if $P_n = P_1 = $ power incident at the antenna,
(b) gain $G_2(\mathbf{k})$ if $P_n = P_2 = $ power accepted by the antenna,
(c) directivity $D(\mathbf{k})$ if $P_n = P_3 = $ power radiated by the antenna.

The relations among the three gains are

$$G_1(\mathbf{k}) = (1 - |\Gamma|^2) G_2(\mathbf{k}) = \eta(1 - |\Gamma|^2) D(\mathbf{k}) \tag{61}$$

The three gains are graphically illustrated in the lower half of Fig. 14.

Three Gains in State (k, u)

The antenna has a polarization $\mathbf{U}$ as defined in (47). In defining gains we may specify a preferred polarization $\mathbf{u}$. Then the radiation intensity in state $(\mathbf{k}, \mathbf{u})$ is

$$I(\mathbf{k}, \mathbf{u}) = |A(\mathbf{k}, \mathbf{u})|^2 = |\mathbf{A}(\mathbf{k}) \cdot \mathbf{u}^*|^2 \tag{62}$$

in watts per steradian. The ratio

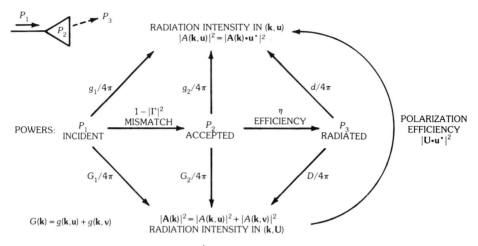

Fig. 14. Six definitions of gain ($a \xrightarrow{t} b$ means $at = b$). (*Courtesy G. A. Deschamps*)

$$p = \frac{I(\mathbf{k},\mathbf{u})}{I(\mathbf{k})} = |\mathbf{U}\cdot\mathbf{u}^*|^2 \tag{63}$$

is called the *polarization efficiency in state* $(\mathbf{k},\mathbf{u})$ *for an antenna that has polarization* $\mathbf{U}$. With respect to state $(\mathbf{k},\mathbf{u})$ we define a gain

$$g(\mathbf{k},\mathbf{u}) = \frac{4\pi}{P_n} I(\mathbf{k},\mathbf{u}) = \frac{4\pi}{P_n}|\mathbf{A}(\mathbf{k})\cdot\mathbf{u}^*|^2 = pG(\mathbf{k}) \tag{64}$$

$$= 4\pi \frac{\text{intensity of the antenna in state } (\mathbf{k},\mathbf{u})}{\text{a reference power}}$$

Again, depending on the reference power P_n, there are three gains g_1, g_2, and d, which are analogous to the cases associated with (60). These three gains are illustrated in the upper half of Fig. 14.

The gain defined in (64) is called the *partial gain* for a specific polarization $\mathbf{u}$. The (total) gain $G(\mathbf{k})$ in (60) is the sum of partial gains for any two orthogonal polarizations:

$$G(\mathbf{k}) = g(\mathbf{k},\mathbf{u}) + g(\mathbf{k},\mathbf{v})$$

which follows from (58).

Peak Gains

All of the six gains depend on the observation direction $\hat{\mathbf{k}} = (\theta,\phi)$. In applications we often use the maximum of a gain as a function of $\hat{\mathbf{k}}$. Thus, corresponding to (60), we define a *peak gain* for all polarizations by

$$G = \max G(\mathbf{k}) = \frac{4\pi}{P_n} \max |\mathbf{A}(\mathbf{k})|^2 \tag{65}$$

Basics

In a similar manner we define a peak gain for a specific polarization **u**. When a gain is given without a specified observation direction, it is customarily assumed to be the peak gain.

Gain-Related Terms

When defining antenna terms related to gain, we can use any one of the six definitions of gain. If the radiation in all polarizations is of interest, we use $G(\mathbf{k})$ in (60). If the radiation in a specific polarization is of interest, we use $g(\mathbf{k},\mathbf{u})$ in (62). With this understanding we use $G(\mathbf{k})$ to represent a typical gain in the discussion below.

An isotropic radiator is a hypothetical lossless antenna having equal radiation intensity in all directions, i.e., $I(\mathbf{k})$ = a constant. If an input power P_n were fed to an isotropic radiator, its radiation intensity in watts per steradian would be

$$I_{\text{iso}}(\mathbf{k}) = \frac{P_n}{4\pi} \tag{66}$$

Here P_n can be any one of three powers explained at the beginning of this section. In terms of (66) the gain definition in (60) has the following interpretation:

$$G(\mathbf{k}) = \frac{I(\mathbf{k})}{I_{\text{iso}}(\mathbf{k})}$$

$$= \frac{\text{intensity of the antenna in direction } \mathbf{k}}{\text{intensity of an isotropic radiator fed by the same power}} \tag{67}$$

We often express the dimensionless G by its decibel (dB) value: $10\log_{10} G$. Sometimes we write dB as dBi, where the letter "i" emphasizes that the gain is over an isotropic radiator.

Consider an antenna with gain $G(\mathbf{k})$ fed by an input power P_n. Its radiation intensity $I(\mathbf{k})$ would be the same as that for an isotropic radiator if the latter were fed with an input power given in watts by

$$\text{EIRP} = P_n G(\mathbf{k}) \tag{68}$$

where EIRP stands for equivalent (effective) isotropically radiated power. We often express EIRP by $10\log_{10} P_n G$ in decibels referred to 1 W (dBW).

When an antenna is used for receiving, a figure of merit of the antenna is defined by

$$\frac{G}{T_a} = \frac{\text{peak gain of the antenna}}{\text{noise temperature of the antenna}} \tag{69}$$

T_a, which is usually given in kelvins (K), is discussed in Chapter 2.

11. Pattern Approximation by $(\cos\theta)^q$

The far-field patterns of many aperture-type antennas have two characteristics: (a) A single major lobe exists in the forward half-space $z > 0$ and the lobe maximum is in the $+z$ direction. (b) The radiation in the backward half-space $z < 0$ is negligible. For these types of antennas their far fields can be approximated by simple analytical functions described in this section. For an x-polarized antenna, such as the flanged waveguide shown in Fig. 15, we have (for $\theta \leq \pi/2$, $0 \leq \phi \leq 2\pi$, and $r \to \infty$)

$$\mathbf{E}(\mathbf{r}) = A_0[\hat{\boldsymbol{\theta}}\, C_E(\theta)\cos\phi - \hat{\boldsymbol{\phi}}\, C_H(\theta)\sin\phi]\sqrt{Z}\,\frac{e^{-jkr}}{r} \tag{70a}$$

For a y-polarized antenna we have

$$\mathbf{E}(\mathbf{r}) = A_0[\hat{\boldsymbol{\theta}}\, C_E(\theta)\sin\phi + \hat{\boldsymbol{\phi}}\, C_H(\theta)\cos\phi]\sqrt{Z}\,\frac{e^{-jkr}}{r} \tag{70b}$$

Here A_0 is a complex constant in (watts)$^{1/2}$, and

$$C_E(\theta) = (\cos\theta)^{q_E} = E\text{-plane pattern} \tag{71a}$$

$$C_H(\theta) = (\cos\theta)^{q_H} = H\text{-plane pattern} \tag{71b}$$

The shape of the pattern is controlled by indices (q_E, q_H). For a given linearly polarized antenna pattern we can approximate it by (70), with (q_E, q_H) determined by matching the given pattern and (70) at two selected directions. A proper superposition of (70a) and (70b) gives a circularly polarized far field, namely,

$$\mathbf{E}(\mathbf{r}) = A_0 e^{j\tau\phi}[\hat{\boldsymbol{\theta}}\, C_E(\theta) + \hat{\boldsymbol{\phi}}\, j\tau\, C_H(\theta)]\sqrt{Z}\,\frac{e^{-jkr}}{r} \tag{72}$$

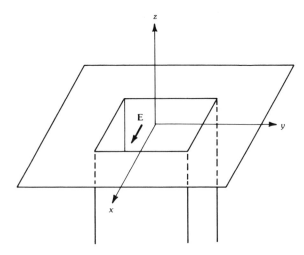

Fig. 15. Radiation from a flanged rectangular waveguide.

where $\tau = +1$ for left-handed circular polarization, and $\tau = -1$ for right-handed circular polarization. Note that (72) represents a perfectly circularly polarized wave only in the main beam direction ($\theta = 0$). Away from this direction it is generally elliptically polarized (unless $q_E = q_H$).

The directivity of the pattern in (70) or (72) can be calculated from (60) and (59). The result for the peak directivity in the main beam direction is

$$D(\theta = 0) = \frac{2(2q_E + 1)(2q_H + 1)}{q_E + q_H + 1} \tag{73}$$

which is valid for both linear and circular polarizations. It should be noted that (73) applies only if the antenna has no radiation in the backward half-space ($z < 0$ in Fig. 15). For an antenna which has a symmetrical pattern in the forward and backward half-spaces, the directivity is one half of the value calculated from (73). For example, an electric dipole oriented in the x direction has a pattern described by (70a), with $q_E = 1$ and $q_H = 0$ for all $\theta \leq \pi$. Its directivity is 3/2, one half of the value given by (73).

The formula in (73) can be expressed in a different form. The half-power beamwidths (θ_E, θ_H) in the (E, H) planes are related to (q_E, q_H) by

$$\left(\cos \frac{\theta_n}{2}\right)^{q_n} = \frac{1}{\sqrt{2}}, \quad \text{for } n = E, H \tag{74}$$

The use of (74) in (73) leads to

$$D(\theta = 0) = \frac{4}{\dfrac{1}{1 + 0.3/|\log \cos(\theta_E/2)|} + \dfrac{1}{1 + 0.3/|\log \cos(\theta_H/2)|}} \tag{75}$$

where the log is of base 10. If θ_E and θ_H are much less than one radian, (75) is simplified to read

$$D(\theta = 0) \cong \frac{36\,300}{(\theta_E^2 + \theta_H^2)/2} \tag{76}$$

where θ_E and θ_H are expressed in degrees.

Another directivity formula, which is similar to (76), is based on the approximation that $D \cong 4\pi/\Omega$, where Ω is the solid angle extended by the main beam at the half-power level. Using a further approximation that $\Omega \cong \theta_E \theta_H$, we obtain

$$D \cong \frac{41\,300}{\theta_E \theta_H} \tag{77}$$

where θ_E and θ_H are expressed in degrees. Formulas similar to (76) and (77) exist in the literature, and the coefficients in the numerator usually fall in the range from 27 900 to 41 300 (see [5, 6]).

12. TE and TM Field Representations

Consider an isotropic, homogeneous medium characterized by the constitutive parameters (ϵ, μ). In a source-free region of the medium, the field can be decomposed into two components: TE (transverse electric) field and TM (transverse magnetic) field with respect to the z direction. For the TE component the fields are derived from an electric vector potential $\mathbf{F(r)}$ by

$$\mathbf{E} = -\nabla \times \mathbf{F} \tag{78a}$$

$$\mathbf{H} = \frac{k}{jZ}\left[\mathbf{F} + \frac{1}{k^2}\nabla(\nabla\cdot\mathbf{F})\right] \tag{78b}$$

where

$$\mathbf{F} = \hat{\mathbf{z}}\psi(\mathbf{r}) \tag{79a}$$

$$(\nabla^2 + k^2)\psi(\mathbf{r}) = 0 \tag{79b}$$

$$k = \omega\sqrt{\mu\epsilon}, \quad Z = \sqrt{\mu/\epsilon} \tag{79c}$$

For the TM field the corresponding equations are

$$\mathbf{E} = -jkZ\left[\mathbf{A} + \frac{1}{k^2}\nabla(\nabla\cdot\mathbf{A})\right] \tag{80a}$$

$$\mathbf{H} = \nabla \times \mathbf{A} \tag{80b}$$

where

$$\mathbf{A} = \hat{\mathbf{z}}\frac{1}{Z}\overline{\psi}(\mathbf{r}) \tag{81a}$$

$$(\nabla^2 + k^2)\overline{\psi}(\mathbf{r}) = 0 \tag{81b}$$

We introduce Z^{-1} in (81a) so that ψ and $\overline{\psi}$ have the same dimensions in volts. In rectangular coordinates, (78) and (80) may be written explicitly:

Transverse Electric

$$\begin{aligned}
E_x &= -\frac{\partial \psi}{\partial y} & H_x &= \frac{1}{jkZ}\frac{\partial^2 \psi}{\partial x \partial z} \\
E_y &= \frac{\partial \psi}{\partial x} & H_y &= \frac{1}{jkZ}\frac{\partial^2 \psi}{\partial y \partial z} \\
E_z &= 0 & H_z &= \frac{1}{jkZ}\left(\frac{\partial^2}{\partial z^2} + k^2\right)\psi
\end{aligned} \tag{82}$$

Transverse Magnetic

$$E_x = \frac{1}{jk}\frac{\partial^2 \overline{\psi}}{\partial x \partial z} \qquad H_x = \frac{1}{Z}\frac{\partial \overline{\psi}}{\partial y}$$

$$E_y = \frac{1}{jk}\frac{\partial^2 \overline{\psi}}{\partial y \partial z} \qquad H_y = -\frac{1}{Z}\frac{\partial \overline{\psi}}{\partial x} \qquad (83)$$

$$E_z = \frac{1}{jk}\left(\frac{\partial^2}{\partial z^2} + k^2\right)\overline{\psi} \qquad H_z = 0$$

In cylindrical coordinates, (78) and (80) may be written explicitly:

Transverse Electric

$$E_\varrho = -\frac{1}{\varrho}\frac{\partial \psi}{\partial \phi} \qquad H_\varrho = \frac{1}{jkZ}\frac{\partial^2 \psi}{\partial \varrho \partial z}$$

$$E_\phi = \frac{\partial \psi}{\partial \varrho} \qquad H_\phi = \frac{1}{jkZ\varrho}\frac{\partial^2 \psi}{\partial \phi \partial z} \qquad (84)$$

$$E_z = 0 \qquad H_z = \frac{1}{jkZ}\left(\frac{\partial^2}{\partial z^2} + k^2\right)\psi$$

Transverse Magnetic

$$E_\varrho = \frac{1}{jk}\frac{\partial^2 \overline{\psi}}{\partial \varrho \partial z} \qquad H_\varrho = \frac{1}{Z\varrho}\frac{\partial \overline{\psi}}{\partial \phi}$$

$$E_\phi = \frac{1}{jk\varrho}\frac{\partial^2 \overline{\psi}}{\partial \phi \partial z} \qquad H_\phi = -\frac{1}{Z}\frac{\partial \overline{\psi}}{\partial \varrho} \qquad (85)$$

$$E_z = \frac{1}{jk}\left(\frac{\partial^2}{\partial z^2} + k^2\right)\overline{\psi} \qquad H_z = 0$$

Two remarks are in order. First, in a conducting medium with conductivity σ all the formulas in this section remain valid if ϵ is replaced by $\epsilon(1 - j\sigma/\omega\epsilon)$. Second, to decompose the field into TE and TM with respect to the x direction, we make the following changes in (82) and (83): $x \to y$, $y \to z$, and $z \to x$.

13. Plane-Wave Spectrum Representation

A general solution of the wave equations in (79b) or (81b) is

$$\begin{Bmatrix}\psi(r)\\ \overline{\psi}(r)\end{Bmatrix} = \int_{-\infty}^{\infty} dk_x \int_{-\infty}^{\infty} dk_y \begin{Bmatrix}\Psi(k_x, k_y)\\ \overline{\Psi}(k_x, k_y)\end{Bmatrix} e^{-j(k_x x + k_y y + k_z z)} \qquad (86)$$

Here (k_x, k_y) are two integration variables, and

$$k_z = \sqrt{k^2 - (k_x^2 + k_y^2)} \qquad (87a)$$

where $k = \omega\sqrt{\mu\epsilon}$. The square root in (87a) is taken such that

$$\text{Re}\{k_z\} \geqq 0 \quad \text{and} \quad \text{Im}\{k_z\} \leqq 0 \qquad (87b)$$

The integral in (86) represents a superposition of plane waves and constitutes the plane-wave spectrum representation (or Fourier transform representation) of fields in a source-free region.

The amplitudes of the plane waves are related to potentials via the inverse Fourier theorem, namely,

$$\Psi(k_x, k_y) = \left(\frac{1}{2\pi}\right)^2 e^{jk_z z_0} \int_{-\infty}^{\infty} dx_0 \int_{-\infty}^{\infty} dy_0\, \psi(x_0, y_0, z_0)\, e^{+j(k_x x_0 + k_y y_0)} \qquad (88)$$

Thus $\Psi(k_x, k_y)$ can be determined from the values of $\psi(\mathbf{r})$ over a reference plane P_0, on which a typical point is $(x_0, y_0, z_0) = $ a constant (Fig. 16).

The explicit field components of $(\mathbf{E}, \mathbf{H})$ in the plane-wave representation can be obtained by substituting (86) into (82) and (83). Including both TE and TM components, the $\mathbf{E}$ field is

$$\begin{Bmatrix} E_x(\mathbf{r}) \\ E_y(\mathbf{r}) \\ E_z(\mathbf{r}) \end{Bmatrix} = \int_{-\infty}^{\infty} dk_x \int_{-\infty}^{\infty} dk_y\, e^{-j(k_x x + k_y y + k_z z)}$$

$$\times \left[\begin{Bmatrix} jk_y \\ -jk_x \\ 0 \end{Bmatrix} \Psi(k_x, k_y) + \begin{Bmatrix} -k_x k_z \\ -k_y k_z \\ k_x^2 + k_y^2 \end{Bmatrix} \frac{1}{jk} \overline{\Psi}(k_x, k_y) \right] \qquad (89)$$

A similar formula exists for the $\mathbf{H}$ field. We note that any rectangular component of $(\mathbf{E}, \mathbf{H})$ has the same form as $\psi(\mathbf{r})$ given in (86). In the following we will give several formulas related to the integral of $\psi(\mathbf{r})$. These formulas apply equally well to any rectangular components of $(\mathbf{E}, \mathbf{H})$.

(*a*) *Far field due to finite sources.* If the sources that produce the field in (86) are confined in a finite region near $\mathbf{r} = 0$, then amplitude $\Psi(k_x, k_y)$ is an entire function of three independent complex variables k_x, k_y, and k_z. In the far-field region the integral in (88) can be asymptotically evaluated, with the result

$$\psi(\mathbf{r}) \sim \frac{e^{-jkr}}{r} (2\pi j\, k \cos\theta)\, \Psi(k_x = k \sin\theta \cos\phi,\; k_y = k \sin\theta \sin\phi) \qquad (90)$$

valid for $r \to \infty$. Here (r, θ, ϕ) are the spherical coordinates of the observation point $\mathbf{r}$.

(*b*) *Huygens-Fresnel principle.* Substituting (88) into (86) and making use of (90), we obtain another far-field formula valid for $R \to \infty$ (see Fig. 16):

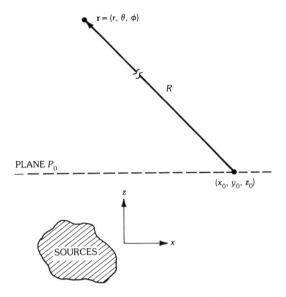

Fig. 16. Far field at **r** is calculated from the field over a plane P_0.

$$\psi(\mathbf{r}) \sim \left(\frac{jk}{2\pi}\right) \int_{-\infty}^{\infty} dx_0 \int_{-\infty}^{\infty} dy_0 \, \psi(x_0, y_0, z_0) \frac{z - z_0}{R} \frac{e^{-jkR}}{R} \quad (91a)$$

where

$$R = \sqrt{(x - x_0)^2 + (y - y_0)^2 + (z - z_0)^2} \quad (91b)$$

Due to finite sources near plane P_0, the factor $(z - z_0)/R$ in (91a) is close to unity. Apart from this factor, (91a) agrees with a corresponding formula derived by using the Huygens-Fresnel principle in optics [7].

(c) *Field with azimuthal symmetry.* If the field is independent of ϕ, the representation in (86) can be rewritten as

$$\psi(\varrho, z) = 2\pi \int_0^{\infty} dk_\varrho [k_\varrho J_0(k_\varrho \varrho) \Psi(k_\varrho)] e^{-j\sqrt{k^2 - k_\varrho^2} \, z} \quad (92)$$

The representation in (92) is valid for any finite ϱ (including $\varrho = 0$). In a region which is unbounded in the ϱ direction and excludes $\varrho = 0$, however, we must replace $J_0(\cdot)$ in (92) by $H_0^{(2)}(\cdot)$ on account of the radiation condition.

14. Periodic Structure

Consider an infinitely large periodic structure of thickness τ situated in an unbounded, isotropic, homogeneous space as sketched in Fig. 17. In terms of geometry and material the structure is repetitive from cell to cell, with its periodic lattice specified by three parameters (a, b, Ω). The structure is illuminated by an

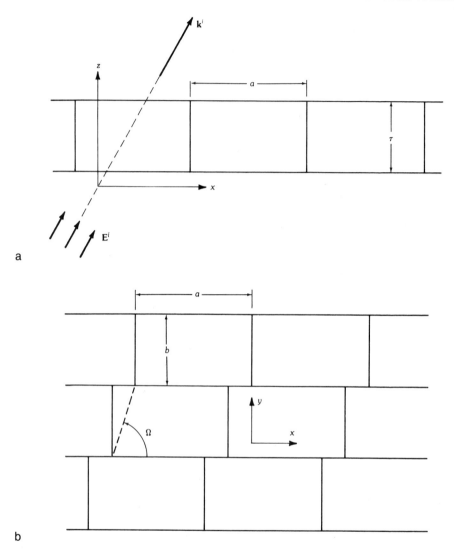

Fig. 17. A periodic structure illuminated by an incident plane wave. (*a*) Side view. (*b*) Top view.

incident plane wave described in (22). The plane wave propagates in the direction $\mathbf{k} = \mathbf{k}^i$, where

$$\mathbf{k}^i = k(\hat{\mathbf{x}} \sin\theta_0 \cos\phi_0 + \hat{\mathbf{y}} \sin\theta_0 \sin\phi_0 + \hat{\mathbf{z}} \cos\theta_0) \qquad (93)$$

The problem at hand is to find a field representation for the scattered field in the unbounded space (outside of the periodic slab).

Due to the periodic nature of the structure and the uniformity of the incident field, the scattered field can be expressed in terms of Floquet space harmonics. For the (p,q)th space harmonics, its transverse field variation is given by

Basics

$$Q_{pq}(x, y) = \exp[-j(u_{pq}x + v_{pq}y)], \qquad p, q = 0, \pm 1, \pm 2, \ldots \qquad (94a)$$

where

$$u_{pq} = (2p\pi/a) + k \sin\theta_0 \cos\phi_0 \qquad (94b)$$

$$v_{pq} = (-2p\pi/a)\cot\Omega + (2q\pi/b) + k \sin\theta_0 \sin\phi_0 \qquad (94c)$$

Its z variation is $\exp(\pm j\gamma_{pq}z)$, where the propagation constant is

$$\gamma_{pq} = \sqrt{k^2 - (u_{pq}^2 + v_{pq}^2)} \qquad (95)$$

The square root in (95) is taken such that

$$\operatorname{Re}\{\gamma_{pq}\} \geq 0 \quad \text{and} \quad \operatorname{Im}\{\gamma_{pq}\} \leq 0 \qquad (96)$$

In a lossless medium (with real k), only a finite number of $\{\gamma_{pq}\}$s are real, corresponding to propagating space harmonics. The remaining $\{\gamma_{pq}\}$s are negative imaginary, corresponding to attenuating space harmonics. In phased array terminology, the dominant space harmonic with $(p=0, q=0)$ is called the *main beam*, whereas all others are *grating lobes*. In most applications we wish to choose an array lattice so that only the main beam is propagating and all grating lobes are attenuating. To achieve this for the incident polar angle in the range $0 \leq \theta_0 \leq \theta_1$ and for $0 \leq \phi_0 \leq 2\pi$, the lattice parameters must be such that

$$1 + \sin\theta_1 < \Gamma \qquad (97a)$$

where

$$\Gamma = \min\left\{\frac{\lambda}{a\sin\Omega}, \frac{\lambda}{b}, \left[\left(\frac{\lambda}{a}\right)^2 + \left(\frac{\lambda}{b} - \frac{\lambda}{a}\cot\Omega\right)^2\right]^{1/2}\right\} \qquad (97b)$$

In other words, when (97a) is satisfied, only γ_{00} in (95) is real in the range $0 \leq \theta_0 \leq \theta_1$ and $0 \leq \phi_0 \leq 2\pi$. For several commonly used lattices, (97b) can be simplified to the following:

(a) For a rectangular lattice ($\Omega = 90°$),

$$\Gamma = \min\left\{\frac{\lambda}{a}, \frac{\lambda}{b}\right\} \qquad (98a)$$

(b) For an exact triangular lattice ($b = \sqrt{3}a/2$, $\Omega = 60°$),

$$\Gamma = \frac{2\lambda}{\sqrt{3}\,a} \qquad (98b)$$

(c) For an isosceles triangular lattice ($\cot\Omega = 0.5a/b$),

$$\Gamma = \min\left\{\frac{\lambda \csc \Omega}{a}, \frac{\lambda}{b}\right\} \quad (98c)$$

which reduces to $\Gamma = \sqrt{5}\lambda/(2a)$ if $a = b$ ($\Omega = 63.435°$).

The scattered field in the unbounded space can be decomposed into TE$_z$ and TM$_z$ components in the manner described in Section 12. The two potentials are expressed in terms of Floquet space harmonics:

$$\left\{\begin{array}{l}\psi(\mathbf{r})\\ \overline{\psi}(\mathbf{r})\end{array}\right\} = \sum_{p=-\infty}^{\infty}\sum_{q=-\infty}^{\infty}\left[\left\{\begin{array}{l}C_{pq}\\ \overline{C}_{pq}\end{array}\right\}e^{-j\gamma_{pq}z} + \left\{\begin{array}{l}D_{pq}\\ \overline{D}_{pq}\end{array}\right\}e^{+j\gamma_{pq}z}\right]Q_{pq}(x,y) \quad (99)$$

The fields $(\mathbf{E}, \mathbf{H})$ can be calculated from (99), (82), and (83). Retaining only the component traveling in the $+z$ direction, we have

$$\left\{\begin{array}{l}\mathbf{E}_t\\ E_z\\ \mathbf{H}_t\\ H_z\end{array}\right\} = \sum_p\sum_q\left[\left\{\begin{array}{c}\boldsymbol{\alpha}_{pq}\\ 0\\ (\gamma_{pq}/kZ)\boldsymbol{\beta}_{pq}\\ -w_{pq}^2/kZ\end{array}\right\}C_{pq} + \left\{\begin{array}{c}(\gamma_{pq}/k)\boldsymbol{\beta}_{pq}\\ -w_{pq}^2/k\\ -Z^{-1}\boldsymbol{\alpha}_{pq}\\ 0\end{array}\right\}\overline{C}_{pq}\right]$$

$$\times [jQ_{pq}(x,y)e^{-j\gamma_{pq}z}] \quad (100)$$

where

$$\boldsymbol{\alpha}_{pq} = (\hat{\mathbf{x}}v_{pq} - \hat{\mathbf{y}}u_{pq})$$
$$\boldsymbol{\beta}_{pq} = (\hat{\mathbf{x}}u_{pq} + \hat{\mathbf{y}}v_{pq}) = \hat{\mathbf{z}} \times \hat{\boldsymbol{\alpha}}_{pq}$$
$$w_{pq}^2 = u_{pq}^2 + v_{pq}^2$$
$$Z = \sqrt{\mu/\epsilon}$$

The field component traveling in the $-z$ direction associated with coefficients $(D_{pq}, \overline{D}_{pq})$ is also given by (100) after changing the sign of γ_{pq}. The time-averaged power carried by the total field in the $+z$ direction in each unit cell $a \times b$ is defined by

$$\overline{P} = \iint_{\text{a cell}} \text{Re}\{(\mathbf{E} \times \mathbf{H}^*)\cdot\hat{\mathbf{z}}\}\,dx\,dy \quad (101a)$$

If the medium is lossless, we have

$$\overline{P} = \frac{ab}{kZ}\sum_p\sum_q^{\text{prop.}} w_{pq}^2\gamma_{pq}\{|C_{pq}|^2 + |\overline{C}_{pq}|^2 - |D_{pq}|^2 - |\overline{D}_{pq}|^2\} \quad (101b)$$

in which the summation includes only propagating space harmonics with real γ_{pq}s. The constant coefficients C_{pq} in (99) are to be determined by matching

Basics

boundary conditions at $z = 0$ and $z = \tau$. This can be done only after a more detailed description of the periodic slab is given. Of particular interest is the dominant space harmonic with $(p = 0, q = 0)$. The fields with coefficients $(C_{00}, \overline{C}_{00})$ are the (perpendicular, parallel) components of a plane wave propagating in the direction of $\mathbf{k}^i$:

$$\mathbf{E}(\mathbf{r}) = (jk \sin \theta_0)[(-\hat{\boldsymbol{\phi}}^i)C_{00} + \hat{\boldsymbol{\theta}}^i \overline{C}_{00}] e^{-j\mathbf{k}^i \cdot \mathbf{r}} + \cdots \tag{102a}$$

$$\mathbf{H}(\mathbf{r}) = \frac{1}{Z}(jk \sin \theta_0)[\hat{\boldsymbol{\theta}}^i C_{00} + \hat{\boldsymbol{\phi}}^i \overline{C}_{00}] e^{-j\mathbf{k}^i \cdot \mathbf{r}} + \cdots \tag{102b}$$

where $\mathbf{k}^i$ is given in (93), and

$$\hat{\boldsymbol{\theta}}^i = (\hat{\mathbf{x}} \cos \phi_0 + \hat{\mathbf{y}} \sin \phi_0) \cos \theta_0 - \hat{\mathbf{z}} \sin \theta_0 = \hat{\boldsymbol{\phi}}^i \times \hat{\mathbf{k}}^i \tag{102c}$$

$$\hat{\boldsymbol{\phi}}^i = -\hat{\mathbf{x}} \sin \phi_0 + \hat{\mathbf{y}} \cos \phi_0 = \hat{\mathbf{k}}^i \times \hat{\boldsymbol{\theta}}^i \tag{102d}$$

The fields with coefficients $(D_{00}, \overline{D}_{00})$ are the (perpendicular, parallel) components of a plane wave propagating in the direction of $\mathbf{k}^r$, where $\mathbf{k}^r = \mathbf{k}^i - 2\hat{\mathbf{z}}(\hat{\mathbf{z}} \cdot \mathbf{k}^i)$ is the reflected wave vector. By "perpendicular" component we mean that the $\mathbf{E}$ vector of the field is linearly polarized and is perpendicular to the plane of incidence defined by $\hat{\mathbf{z}}$ and $\mathbf{k}^i$.

15. Rectangular Waveguide

Consider a metallic rectangular waveguide filled with an isotropic homogeneous medium (Fig. 18). The field in the waveguide can be decomposed into TE_z and TM_z. They are derivable from two potential functions (for the component traveling in the $+z$ direction):

$$\psi = \sum_{m=0}^{\infty} \sum_{n=0}^{\infty}{}' C_{mn} \cos\left(\frac{m\pi}{a}x\right) \cos\left(\frac{n\pi}{b}y\right) e^{-j\gamma_{mn}z} \tag{103a}$$

$$\overline{\psi} = \sum_{m=1}^{\infty} \sum_{n=1}^{\infty} \overline{C}_{mn} \sin\left(\frac{m\pi}{a}x\right) \sin\left(\frac{n\pi}{b}y\right) e^{-j\gamma_{mn}z} \tag{103b}$$

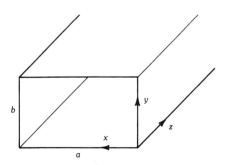

Fig. 18. A rectangular waveguide.

In (103a) the prime signifies that the term with $m = n = 0$ is omitted in the summation. The propagation constants γ_{mn} for either TE$_{mn}$ or TM$_{mn}$ are given by

$$\gamma_{mn} = \sqrt{k^2 - (m\pi/a)^2 - (n\pi/b)^2} \qquad (104)$$

The square-root convention stated in (96) should be observed. The field components for the (m,n)th mode calculated from (82) and (83) are

$$\mathbf{E}_t = C_{mn}\boldsymbol{\alpha}_{mn} + \overline{C}_{mn}\boldsymbol{\beta}_{mn} \qquad (105a)$$

$$E_z = \frac{1}{jk}\left[\left(\frac{m\pi}{a}\right)^2 + \left(\frac{n\pi}{b}\right)^2\right]\psi \qquad (105b)$$

$$\mathbf{H}_t = C_{mn}\frac{\gamma_{mn}}{kZ}\hat{\mathbf{z}} \times \boldsymbol{\alpha}_{mn} + \overline{C}_{mn}\frac{k}{\gamma_{mn}Z}\hat{\mathbf{z}} \times \boldsymbol{\beta}_{mn} \qquad (105c)$$

$$H_z = \frac{1}{jkZ}\left[\left(\frac{m\pi}{a}\right)^2 + \left(\frac{n\pi}{b}\right)^2\right]\psi \qquad (105d)$$

where

$$\boldsymbol{\alpha}_{mn} = \left[\hat{\mathbf{x}}\left(\frac{n\pi}{b}\right)\cos\left(\frac{m\pi}{a}x\right)\sin\left(\frac{n\pi}{b}y\right)\right.$$
$$\left. - \hat{\mathbf{y}}\left(\frac{m\pi}{a}\right)\sin\left(\frac{m\pi}{a}x\right)\cos\left(\frac{n\pi}{b}y\right)\right]e^{-j\gamma_{mn}z} \qquad (106a)$$

$$\boldsymbol{\beta}_{mn} = \left(-\frac{\gamma_{mn}}{k}\right)\left[\hat{\mathbf{x}}\left(\frac{m\pi}{a}\right)\cos\left(\frac{m\pi}{a}x\right)\sin\left(\frac{n\pi}{b}y\right)\right.$$
$$\left. + \hat{\mathbf{y}}\left(\frac{n\pi}{b}\right)\sin\left(\frac{m\pi}{a}x\right)\cos\left(\frac{n\pi}{b}y\right)\right]e^{-j\gamma_{mn}z} \qquad (106b)$$

In particular, the nonzero field components of the dominant TE$_{10}$ mode are

$$\text{TE}_{10}: \quad E_y = -\left(\frac{\pi}{a}\right)C_{10}\sin\left(\frac{\pi}{a}x\right)e^{-j\gamma_{10}z} \qquad (107a)$$

$$H_x = \left(\frac{\pi}{a}\right)\left(\frac{\gamma_{10}}{kZ}\right)C_{10}\sin\left(\frac{\pi}{a}x\right)e^{-j\gamma_{10}z} \qquad (107b)$$

$$H_z = \frac{1}{jkZ}\left(\frac{\pi}{a}\right)^2 C_{10}\cos\left(\frac{\pi}{a}x\right)e^{-j\gamma_{10}z} \qquad (107c)$$

For the field component traveling in the $-z$ direction we change the sign of γ_{mn} in (103) through (107). The transverse field variations of several lower-order modes are sketched in Fig. 19.

If the medium in the waveguide is lossless, the cutoff frequency of TE$_{mn}$ or TM$_{mn}$ is

Basics

$$f_c = \frac{1}{2\sqrt{\epsilon\mu}} \sqrt{\left(\frac{m}{a}\right)^2 + \left(\frac{n}{b}\right)^2} \tag{108}$$

which is plotted in Fig. 20 for several lower-order modes. The propagation constant of TE_{mn} or TM_{mn} is

$$\gamma_{mn} = \begin{cases} k\sqrt{1 - (f_c/f)^2} = 2\pi/\lambda_g & \text{if } f > f_c \\ -jk\sqrt{(f_c/f)^2 - 1} & \text{if } f < f_c \end{cases} \tag{109}$$

where λ_g is the waveguide wavelength. The time-averaged power carried by the field in (105) is

$$\bar{P} = \frac{ab}{4Z} \sum_m \sum_n^{\text{prop.}} \Delta_m \Delta_n \frac{\gamma_{mn}}{k} \left[\left(\frac{m\pi}{a}\right)^2 + \left(\frac{n\pi}{b}\right)^2\right](|\bar{C}_{mn}|^2 + |C_{mn}|^2) \tag{110}$$

when the summation includes only propagating modes with real γ_{mn}s, and $\Delta_m = 2$ if $m = 0$, and $\Delta_m = 1$ if $m \neq 0$.

There are two common sources of waveguide loss. One is the dielectric loss in the medium inside the waveguide: $k = k' - jk''$. The other is the conduction loss due to the finite conductivity σ of the metallic waveguide walls. When both losses are small, the propagation constant of a propagating mode sufficiently above cutoff is approximately given by

$$\gamma_{mn} = \beta - j\alpha, \quad \beta \gg \alpha \tag{111}$$

where

$$\beta = \sqrt{(k')^2 - (m\pi/a)^2 - (n\pi/b)^2} \tag{112}$$

$$\alpha = \alpha_d + \alpha_c \tag{113}$$

$$\alpha_d = k'k''/\beta \tag{114}$$

with β in radians per meter and α in nepers per meter. The attenuation constant due to conduction loss is approximately given by

$$TE_{mo}: \quad \alpha_c = v\left(\frac{1}{b}u^2 + \frac{1}{2a}\right) \tag{115a}$$

$$TE_{on}: \quad \alpha_c = v\left(\frac{1}{a}u^2 + \frac{1}{2b}\right) \tag{115b}$$

$$TE_{mm}: \quad \alpha_c = v\left[\left(\frac{1}{a} + \frac{1}{b}\right)u^2 + (1 - u^2)\frac{m^2b + n^2a}{m^2b^2 + n^2a^2}\right] \tag{115c}$$

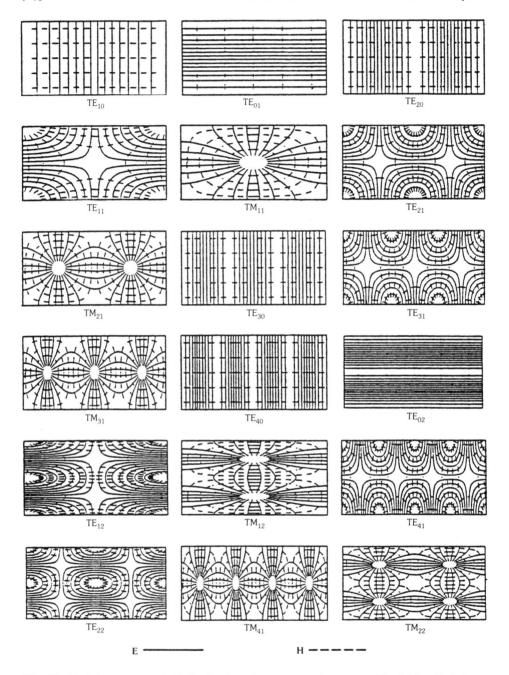

Fig. 19. Modal transverse field distributions in a rectangular waveguide. (*After C. S. Lee, S. W. Lee, and S. L. Chuang [8]*, © *1985 IEEE*)

Basics

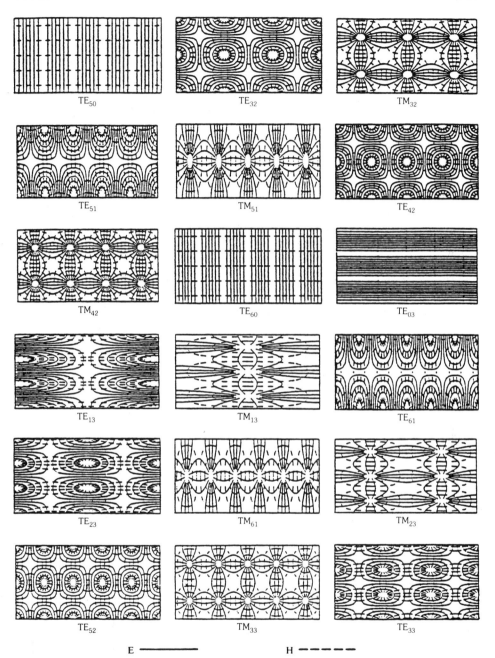

Fig. 19, *continued.*

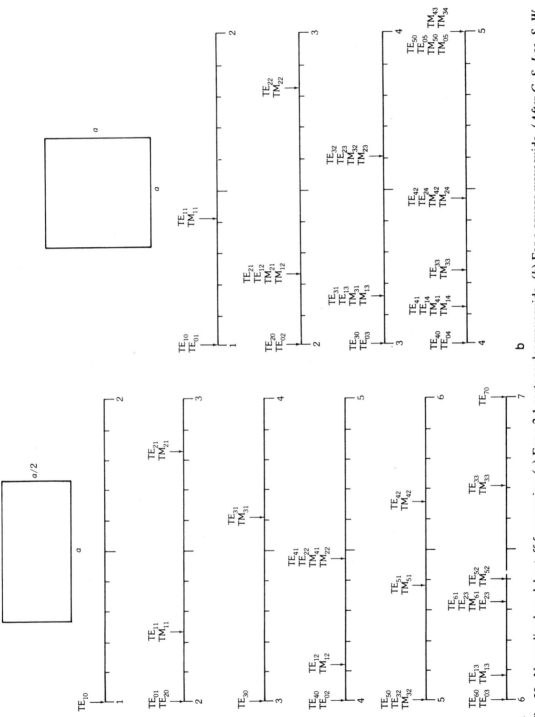

Fig. 20. Normalized modal cutoff frequencies. (*a*) For a 2:1 rectangular waveguide. (*b*) For a square waveguide. (*After C. S. Lee, S. W. Lee, and S. L. Chuang [8], © 1985 IEEE*)

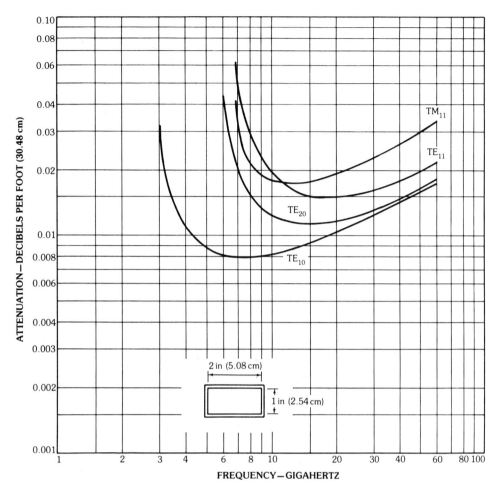

Fig. 21. Attenuation versus frequency curves for lower-order modes in a typical rectangular brass waveguide. (*After Jordan and Balmain [9], © 1968 Prentice-Hall, Inc.; reprinted by permission of Prentice-Hall, Inc., Englewood Cliffs, NJ.*)

$$\text{TM}_{mn}: \quad \alpha_c = v\left[\frac{m^2 b^3 + n^2 a^3}{ab(m^2 b^2 + n^2 a^2)}\right] \tag{115d}$$

where

$$u = f_c/f \tag{116a}$$

$$v = \frac{2R_s}{Z\sqrt{1 - u^2}} \tag{116b}$$

$$R_s = (\omega\mu/2\sigma)^{1/2} = \text{skin-effect surface resistance of waveguide wall} \tag{116c}$$

The attenuation of TE_{mn} or TM_{mn} is $8.686\,\alpha$ dB/m. An example is given in Fig. 21 for a typical air-filled, rectangular, brass waveguide.

16. Circular Waveguide

Consider a metallic circular waveguide filled with an isotropic homogeneous medium (Fig. 22). The field in the waveguide can be decomposed into TE_z and TM_z, which are derivable from two potential functions (for the component traveling in the $+z$ direction):

$$\psi = \sum_m \sum_n \begin{Bmatrix} C^v_{mn} \\ C^h_{mn} \end{Bmatrix} J_m(\xi'_{mn}\varrho/a) \begin{Bmatrix} \cos m\phi \\ \sin m\phi \end{Bmatrix} e^{-j\gamma_{mn}z} \qquad (117a)$$

$$\overline{\psi} = \sum_m \sum_n \begin{Bmatrix} \overline{C}^v_{mn} \\ \overline{C}^h_{mn} \end{Bmatrix} J_m(\xi_{mn}\varrho/a) \begin{Bmatrix} \sin m\phi \\ \cos m\phi \end{Bmatrix} e^{-j\overline{\gamma}_{mn}z} \qquad (117b)$$

Here the superscripts v and h denote vertical and horizontal modes, respectively. For example, C^v_{mn} is the modal coefficient of the TEV_{mn} mode, and (ξ_{mn}, ξ'_{mn}) are the nth roots of $(J_m(x), J'_m(x))$, respectively. Their lower-order values are tabulated in Table 1. The propagation constants in (117) are given by

$$TE_{mn}: \quad \gamma_{mn} = \sqrt{k^2 - (\xi'_{mn}/a)^2} \qquad (118a)$$

$$TM_{mn}: \quad \overline{\gamma}_{mn} = \sqrt{k^2 - (\xi_{mn}/a)^2} \qquad (118b)$$

The square-root convention stated in (96) should be observed. The field components of TE_{mn} calculated from (117a) and (84) are

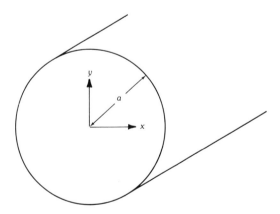

Fig. 22. A circular waveguide.

Basics

Table 1. Zeros of Bessel Functions*

Number	Transverse Electric		Transverse Magnetic	
	m, n	ξ'_{mn}	m, n	ξ_{mn}
1	1, 1	1.841 18	0, 1	2.404 83
2	2, 1	3.054 24	1, 1	3.831 71
3	0, 1	3.831 71	2, 1	5.135 62
4	3, 1	4.201 19	0, 2	5.520 08
5	4, 1	5.317 55	3, 1	6.380 16
6	1, 2	5.331 44	1, 2	7.015 59
7	5, 1	6.415 62	4, 1	7.588 34
8	2, 2	6.706 13	2, 2	8.417 24
9	0, 2	7.015 59	0, 3	8.653 73
10	6, 1	7.501 27	5, 1	8.771 48
11	3, 2	8.015 24	3, 2	9.761 02
12	7, 1	8.577 84	6, 1	9.936 11

*The nth zeros of $[J_m(x), J'_m(x)]$ are (ξ_{mn}, ξ'_{mn}), respectively.

$$\text{TE}_{mn}: \quad \mathbf{E} = C^v_{mn}\boldsymbol{\alpha}^v_{mn} + C^h_{mn}\boldsymbol{\alpha}^h_{mn} \tag{119a}$$

$$\mathbf{H}_t = \frac{\gamma_{mn}}{kZ}\hat{\mathbf{z}} \times \mathbf{E} \tag{119b}$$

$$H_z = \frac{1}{jkZ}\left(\frac{\xi'_{mn}}{a}\right)^2 \psi \tag{119c}$$

where

$$\begin{Bmatrix}\boldsymbol{\alpha}^v_{mn}\\ \boldsymbol{\alpha}^h_{mn}\end{Bmatrix} = \left[\hat{\boldsymbol{\varrho}}\frac{m}{\varrho}J_m(\xi'_{mn}\varrho/a)\begin{Bmatrix}\sin m\phi\\ -\cos m\phi\end{Bmatrix}\right.$$
$$\left. + \hat{\boldsymbol{\phi}}\left(\frac{\xi'_{mn}}{a}\right)J'_m(\xi'_{mn}\varrho/a)\begin{Bmatrix}\cos m\phi\\ \sin m\phi\end{Bmatrix}\right]e^{-j\gamma_{mn}z} \tag{119d}$$

The field components of TM_{mn} calculated from (117b) and (85) are

$$\text{TM}_{mn}: \quad \mathbf{E}_t = (\overline{C}^v_{mn}\boldsymbol{\beta}^v_{mn} + \overline{C}^h_{mn}\boldsymbol{\beta}^h_{mn})(\overline{\gamma}_{mn}/k) \tag{120a}$$

$$\mathbf{E}_z = \frac{1}{jk}\left(\frac{\xi_{mn}}{a}\right)^2 \overline{\psi} \tag{120b}$$

$$\mathbf{H} = \frac{k}{\overline{\gamma}_{mn}Z}\hat{\mathbf{z}} \times \mathbf{E}_t \tag{120c}$$

where

$$\begin{Bmatrix} \beta_{mn}^v \\ \beta_{mn}^h \end{Bmatrix} = \left[-\hat{\varrho} \left(\frac{\xi_{mn}}{a} \right) J'_m(\xi_{mn}\varrho/a) \begin{Bmatrix} \sin m\phi \\ \cos m\phi \end{Bmatrix} \right.$$

$$\left. + \hat{\phi} \frac{m}{\varrho} J_m(\xi_{mn}\varrho/a) \begin{Bmatrix} -\cos m\phi \\ \sin m\phi \end{Bmatrix} \right] e^{-j\gamma_{mn}z} \qquad (120d)$$

In particular, the nonzero field components of the dominant TE_{11} are

$$TE_{11}: \quad E_\varrho = \begin{Bmatrix} C_{11}^v \\ C_{11}^h \end{Bmatrix} \frac{1}{\varrho} J_1(\xi'_{11}\varrho/a) \begin{Bmatrix} \sin\phi \\ -\cos\phi \end{Bmatrix} e^{-j\gamma_{11}z}$$

$$E_\phi = \begin{Bmatrix} C_{11}^v \\ C_{11}^h \end{Bmatrix} \frac{\xi'_{11}}{a} J'_1(\xi'_{11}\varrho/a) \begin{Bmatrix} \cos\phi \\ \sin\phi \end{Bmatrix} e^{-j\gamma_{11}z}$$

$$H_\varrho = -\frac{\gamma_{11}}{kZ} E_\phi$$

$$H_\phi = \frac{\gamma_{11}}{kZ} E_\varrho$$

$$H_z = \frac{1}{jkZ} \left\{ \frac{\xi'_{11}}{a} \right\}^2 J_1(\xi'_{11}\varrho/a) \begin{Bmatrix} \cos\phi \\ \sin\phi \end{Bmatrix} e^{-j\gamma_{11}z} \qquad (121)$$

The transverse field variations of several lower-order modes are sketched in Fig. 23.

If the medium in the guide is lossless, the modal cutoff frequencies are

$$TE_{mn}: \quad f_c = \frac{\xi'_{mn}}{2\pi a \sqrt{\epsilon\mu}} \qquad (122a)$$

$$TM_{mn}: \quad f_c = \frac{\xi_{mn}}{2\pi a \sqrt{\epsilon\mu}} \qquad (122b)$$

which are plotted in Fig. 24 for several lower-order modes. In terms of f_c, the propagation constants $(\gamma_{mn}, \overline{\gamma}_{mn})$ can be again written in the form of (109). The time-averaged power carried by the modal fields is

$$\overline{P} = \frac{\pi}{2Z} \sum_m^{\text{prop.}} \sum_n \Delta_m \frac{1}{k} \{\gamma_{mn}[(\xi'_{mn})^2 - m^2][J_m(\xi'_{mn})]^2(|C_{mn}^v|^2 + |C_{mn}^h|^2)$$

$$+ \overline{\gamma}_{mn}[\xi_{mn}J_{m+1}(\xi_{mn})]^2[|\overline{C}_{mn}^v|^2 + |\overline{C}_{mn}^h|^2]\} \qquad (123)$$

Fig. 23. Transverse modal field distributions for a circular waveguide. (*After C. S. Lee, S. W. Lee, and S. L. Chuang [8], © 1985 IEEE*)

Fig. 23, *continued.*

Basics

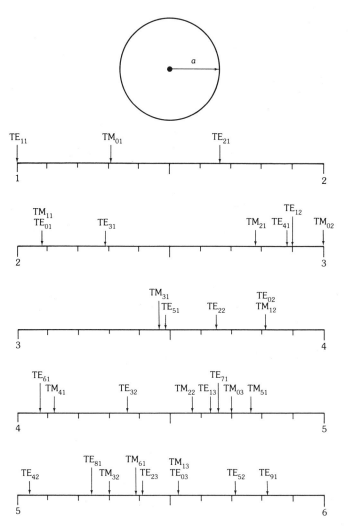

Fig. 24. Normalized modal cutoff frequencies for a circular waveguide. (*After C. S. Lee, S. W. Lee, and S. L. Chuang [8], © 1985 IEEE*)

where the summation includes only propagating modes with real γ_{mn} or $\bar{\gamma}_{mn}$s, and $\Delta_m = 2$ if $m = 0$, and $\Delta_m = 1$ if $m \neq 0$.

The loss in a circular waveguide is similar to that in a rectangular waveguide. Thus the formulas in (111), (113), and (114) still hold. The attenuation constant due to the conduction loss in a circular waveguide is

$$\text{TE}_{mn}: \quad \alpha_c = \frac{v}{2a}\left[\frac{m^2}{(\xi'_{mn})^2 - m^2} + u^2\right] \tag{124a}$$

$$\text{TM}_{mn}: \quad \alpha_c = \frac{v}{2a} \tag{124b}$$

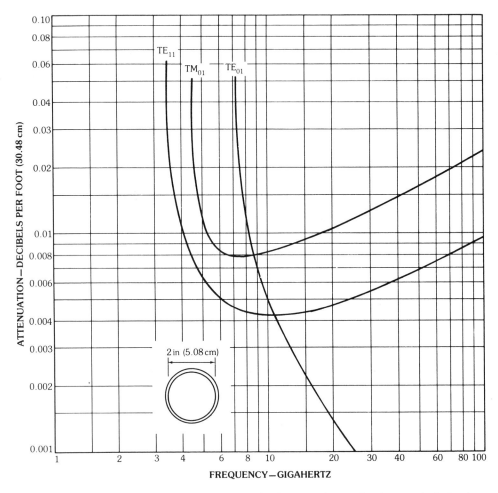

Fig. 25. Attenuation versus frequency curves for lower-order modes in a typical circular brass waveguide. (*After Jordan and Balmain [9]*, © *1968 Prentice-Hall Inc.; reprinted by permission of Prentice-Hall, Inc., Englewood Cliffs, NJ.*)

where u and v are defined in (116), and f_c is given in (122). Note that the conduction loss of TE_{0n} decreases without limit as $f \to \infty$. An example of the conduction loss in a typical air-filled circular brass waveguide is shown in Fig. 25.

17. References

[1] D. S. Jones, *The Theory of Electromagnetism*, New York: Macmillan Co., 1964, pp. 566–569.
[2] R. F. Harrington, *Time-Harmonic Electromagnetic Fields*, New York: McGraw-Hill Book Co., 1961, pp. 99–100.
[3] G. A. Deschamps, "I. Le principe de réciprocité en électromagnétisme. II. Application du principe de réciprocité aux antennes et aux guides d'ondes," *Revue du CETHEDEC* (Paris) N° 8-4°, pp. 71–101, 1966.

[4] A. C. Ludwig, "The definition of cross polarization," *IEEE Trans. Antennas Propag.*, vol. AP-21, pp. 116–119, 1973.

[5] R. J. Stegen, "The gain-beamwidth product of an antenna," *IEEE Trans. Antennas Propag.*, vol. AP-21, pp. 505–506, 1964.

[6] C. T. Tai and C. S. Pereira, "An approximate formula for calculating the directivity of an antenna," *IEEE Trans. Antennas Propag.*, vol. AP-24, pp. 235–236, 1976.

[7] M. Born and E. Wolf, *Principles of Optics*, 5th ed., New York: Pergamon Press, 1975, p. 436.

[8] C. S. Lee, S. W. Lee, and S. L. Chuang, "Plot of modal field distribution in rectangular and circular waveguides," *IEEE Trans. Microwave Theory Tech.*, vol. MTT-33, pp. 271–274, 1985.

[9] E. C. Jordan and K. G. Balmain, *Electromagnetic Waves and Radiating Systems*, 2nd ed., Englewood Cliffs: Prentice-Hall, 1968, p. 271.

Chapter 2

Theorems and Formulas

S. W. Lee
University of Illinois

CONTENTS

1. Duality — 2-5
2. Green's Function in an Unbounded Space — 2-6
   Scalar Wave Equation 2-6
   Vector Wave Equation 2-7
   The Electric Field 2-8
   Integration Involving the R^{-3} Singularity 2-9
   Explicit Expressions for C_{mn} 2-11
3. Image Theory — 2-11
4. The Babinet Principle — 2-13
   Scattering Problems 2-13
   An Impedance Problem 2-15
5. Reciprocity — 2-16
   Bra-Ket Notation 2-16
   The Lorentz Relation 2-16
   The Reciprocity Theorem 2-18
6. Huygens' Principle — 2-18
7. The Kirchhoff Approximation — 2-19
8. Scattering by an Obstacle — 2-21
   Bistatic Cross Section 2-22
   Radar Cross Section 2-23
   Reciprocity 2-24
   Scattering Cross Section 2-24
   Physical Optics Approximation for Scattering by a Conductor 2-25
   Two Other Versions of the Physical Optics Approximation 2-26
9. The Antenna as a One-Port Device — 2-27
10. Three Ideal Sources for Transmitting Antennas — 2-29
11. Three Ideal Meters for Receiving Antennas — 2-30

Shung-Wu Lee was born in Kiangsi, China. He received his BS degree in electrical engineering from Cheng Kung University in Tainan, Taiwan, in 1961. After one year of military service in Taiwan, he came to the United States in 1962. At the University of Illinois in Urbana, he received his MS and PhD degrees in electrical engineering and has been on the faculty since 1966. Currently he is a professor of electrical and computer engineering, and an associate director of the Electromagnetics Laboratory.

While on leave from the University of Illinois, Dr. Lee was with Hughes Aircraft Company, Fullerton, California, in 1969–1970, and with the Technical University at Eindhoven, The Netherlands, and the University of London, England, in 1973–74. Dr. Lee received several professional awards, including the 1968 Everitt Teaching Excellence Award from the University of Illinois, 1973 NSF NATO Senior Scientist Fellowship, 1977 Best Paper Award from IEEE Antennas and Propagation Society, and the 1985 Lockheed Million Dollar Award.

Dr. Lee has published more than 100 papers in technical journals on antennas and electromagnetic theory. He is the coauthor of a book on guided waves published by Macmillan in 1971, and a coeditor of this book. He is a Fellow of IEEE.

12. Reciprocity between Antenna Transmitting and Receiving — 2-32
    Reciprocity Involving General Incident Fields   2-32
    Reciprocity Involving Plane Waves   2-33
    Antenna Effective Length   2-34
    Transmitting Field in Terms of Effective Length   2-34
    Receiving Cross Section   2-35
13. The Radar Equation and Friis Transmission Formula — 2-36
14. Noise Temperature of an Antenna — 2-39
    Antenna Noise Temperature   2-39
    Brightness Temperature of an Emitter   2-40
    Calculation of Antenna Noise Temperature   2-41
    Noise Power at the Receiver's Terminal   2-42
15. References — 2-43

1. Duality

Consider the two radiation problems sketched in Fig. 1. In problem 1, the source $(\mathbf{J}_1, \mathbf{K}_1)$ radiates in medium described by $\epsilon_1(\mathbf{r})$ and $\mu_1(\mathbf{r})$, and in the presence of two typical scatterers: a perfect electric conductor $(PEC)_1$, and a perfect magnetic conductor $(PMC)_1$. The total (radiation) field in space is $(\mathbf{E}_1, \mathbf{H}_1)$. A similar description holds for problem 2. Now, these two problems are "dual" if*

$$\mathbf{J}_1(\mathbf{r}) \to \mathbf{K}_2(\mathbf{r}) \tag{1a}$$

$$\mathbf{K}_1(\mathbf{r}) \to -\mathbf{J}_2(\mathbf{r}) \tag{1b}$$

$$\epsilon_1(\mathbf{r}) \to \mu_2(\mathbf{r}) \tag{1c}$$

$$\mu_1(\mathbf{r}) \to \epsilon_2(\mathbf{r}) \tag{1d}$$

$$(PEC)_1 \to (PMC)_2 \tag{1e}$$

$$(PMC)_1 \to (PEC)_2 \tag{1f}$$

For example, (1e) indicates that the PEC in problem 1 is replaced by a PMC in problem 2. A consequence of (1) is

$$\mathbf{E}_1(\mathbf{r}) \to \mathbf{H}_2(\mathbf{r}) \tag{2a}$$

$$\mathbf{H}_1(\mathbf{r}) \to -\mathbf{E}_2(\mathbf{r}) \tag{2b}$$

Thus, knowing $(\mathbf{E}_1, \mathbf{H}_1)$, we can immediately write down, with the help of (2), the field solution $(\mathbf{E}_2, \mathbf{H}_2)$ of the dual problem. In other words, dual sources and media imply dual fields.

In some dual scattering problems the sources are not explicitly specified. We are given instead the incident field $(\mathbf{E}_1^i, \mathbf{H}_1^i)$ in problem 1 and field $(\mathbf{E}_2^i, \mathbf{H}_2^i)$ in problem 2. Then, these two problems remain dual provided that (1a) and (1b) are replaced by

$$\mathbf{E}_1^i(\mathbf{r}) \to \mathbf{H}_2^i(\mathbf{r}) \tag{3a}$$

$$\mathbf{H}_1^i(\mathbf{r}) \to -\mathbf{E}_2^i(\mathbf{r}) \tag{3b}$$

---

*The symbol $A \to B$ means replacing A in problem 1 by B in problem 2.

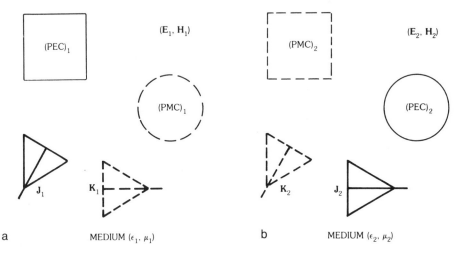

Fig. 1. Dual radiation problems. (*a*) Problem 1. (*b*) Problem 2.

After this replacement the satisfaction of (1) again implies the validity of (2). The fields in (2) can be either total fields or the scattered fields.

2. Green's Function in an Unbounded Space

A Green's function is the field due to a point source described by a delta function. Once it is known, the field due to an arbitrary source can be calculated by a convolution integral involving the source distribution and the Green's function.

Scalar Wave Equation

A Green's function $G(\mathbf{r}, \mathbf{r}')$ involves two points:

$$\mathbf{r} = (x_1, x_2, x_3) = \text{observation point} \tag{4a}$$

$$\mathbf{r}' = (x_1', x_2', x_3') = \text{source point} \tag{4b}$$

In an unbounded isotropic medium, G depends only the distance

$$R = |\mathbf{r} - \mathbf{r}'| = \sqrt{(x_1 - x_1')^2 + (x_2 - x_2')^2 + (x_3 - x_3')^2} \tag{4c}$$

Hence we often write $G(\mathbf{r}, \mathbf{r}')$ as $G(R)$.

In a three-dimensional space filled with a homogeneous isotropic medium, the scalar wave equation with a point source is

$$(\nabla^2 + k^2)g(R) = -\delta(\mathbf{r} - \mathbf{r}') \tag{5}$$

where $\delta(\cdot)$ is the delta function. The solution of (5) subject to the radiation condition is [1]

Theorems and Formulas

$$g(R) = \frac{e^{-jkR}}{4\pi R} \tag{6}$$

For a two-dimensional case the corresponding Green's function is

$$\text{2D:} \quad g(R_2) = \frac{1}{4j} H_0^{(2)}(kR_2) \tag{7}$$

$$\sim \frac{e^{-j(kR_2 + \pi/4)}}{2\sqrt{2\pi kR_2}}, \quad kR_2 \gg 1 \tag{8}$$

where $R_2 = \sqrt{(x_1 - x_1')^2 + (x_2 - x_2')^2}$. For a one-dimensional case it is

$$\text{1D:} \quad g(R_1) = \frac{1}{2jk} e^{-jkR_1} \tag{9}$$

where $R_1 = |x_1 - x_1'|$.

It is important to realize that Green's function $g(R)$ is singular or has discontinuous derivatives when the observation point is at the source point ($\mathbf{r} = \mathbf{r}'$). We must properly interpret their meanings when we perform integrations/differentiations involving $g(R)$. In the following manipulations we consider $g(R)$ a generalized function. We manipulate $g(R)$ formally and explain its singular terms later.

Vector Wave Equation

A vector wave equation with a point source oriented in the x_n direction reads

$$(\nabla \times \nabla \times -k^2)\mathbf{G}^{(n)}(R) = \hat{\mathbf{x}}_n \delta(\mathbf{r} - \mathbf{r}'), \quad \text{for } n = 1, 2, 3 \tag{10}$$

The solution of (10) is

$$\mathbf{G}^{(n)} = \left[1 + \frac{1}{k^2}\nabla\nabla\cdot\right][\hat{\mathbf{x}}_n g(R)], \quad \text{for } n = 1, 2, 3 \tag{11}$$

Using the dyadic notation we may rewrite (10) and (11) as

$$(\nabla \times \nabla \times -k^2)\bar{\bar{\mathbf{G}}}(R) = \bar{\bar{\mathbf{I}}}\delta(\mathbf{r} - \mathbf{r}') \tag{12}$$

$$\bar{\bar{\mathbf{G}}}(R) = \left(1 + \frac{1}{k^2}\nabla\nabla\cdot\right)\bar{\bar{\mathbf{I}}}g(R) \tag{13a}$$

$$= \left(\bar{\bar{\mathbf{I}}} + \frac{1}{k^2}\nabla\nabla\right)g(R) \tag{13b}$$

where $\bar{\bar{\mathbf{I}}}$ is the unit dyad defined by

$$\bar{\bar{I}} = \hat{x}_1\hat{x}_1 + \hat{x}_2\hat{x}_2 + \hat{x}_3\hat{x}_3 \tag{14}$$

Since $g(R)$ is symmetrical in $\mathbf{r}$ and $\mathbf{r}'$, an alternative expression of (13b) reads

$$\bar{\bar{G}}(R) = (\bar{\bar{I}} + \frac{1}{k^2}\nabla'\nabla')g(R) \tag{15}$$

Note the difference in (13b) and (15):

$$\nabla = \hat{x}_1\frac{\partial}{\partial x_1} + \hat{x}_2\frac{\partial}{\partial x_2} + \hat{x}_3\frac{\partial}{\partial x_3} \tag{16}$$

$$\nabla' = \hat{x}_1\frac{\partial}{\partial x'_1} + \hat{x}_2\frac{\partial}{\partial x'_2} + \hat{x}_3\frac{\partial}{\partial x'_3} \tag{17}$$

In matrix notation, (15) may be explicitly written as

$$\bar{\bar{G}}(R) = \begin{bmatrix} 1 + \frac{1}{k^2}\left(\frac{\partial}{\partial x'_1}\right)^2 & \frac{1}{k^2}\frac{\partial^2}{\partial x'_1 \partial x'_2} & \frac{1}{k^2}\frac{\partial^2}{\partial x'_1 \partial x'_3} \\ \frac{1}{k^2}\frac{\partial^2}{\partial x'_2 \partial x'_1} & 1 + \frac{1}{k^2}\left(\frac{\partial}{\partial x'_2}\right)^2 & \frac{1}{k^2}\frac{\partial^2}{\partial x'_2 \partial x'_3} \\ \frac{1}{k^2}\frac{\partial^2}{\partial x'_3 \partial x'_1} & \frac{1}{k^2}\frac{\partial^2}{\partial x'_3 \partial x'_2} & 1 + \frac{1}{k^2}\left(\frac{\partial}{\partial x'_3}\right)^2 \end{bmatrix} \frac{e^{-jkR}}{4\pi R} \tag{18}$$

The differentiations in (18) formally yield the following results:

$$\frac{\partial^2 g}{\partial x_m'^2} = k^2\left[-\cos^2\theta_m + \frac{j}{kR}\left(1 - \frac{j}{kR}\right)(3\cos^2\theta_m - 1)\right]g(R),$$
$$m = 1, 2, 3 \tag{19a}$$

$$\frac{\partial^2 g}{\partial x'_m \partial x'_n} = k^2\cos\theta_m\cos\theta_n\left[-1 + \frac{3j}{kR}\left(1 - \frac{j}{kR}\right)\right]g(R),$$
$$m \neq n \tag{19b}$$

where

$$\cos\theta_n = (x'_n - x_n)/R, \quad \text{for } n = 1, 2, 3 \tag{19c}$$

Note that $\bar{\bar{G}}(R)$ contains the R^{-3} singularity.

The Electric Field

In an unbounded, isotropic, homogeneous medium the electric field $\mathbf{E}$ due to a current $\mathbf{J}$ satisfies the following wave equation:

Theorems and Formulas

$$(\nabla^2 + k^2)\left(\mathbf{E} + \frac{1}{j\omega\epsilon}\mathbf{J}\right) = \frac{-1}{j\omega\epsilon}\nabla \times \nabla \times \mathbf{J} \tag{20}$$

For a given $\mathbf{J}$, we wish to calculate $\mathbf{E}(\mathbf{r})$ by using the Green's functions defined above. We shall give three formulas. The first formula is derived from (5) and (20), namely,

$$\mathbf{E}(\mathbf{r}) = \frac{1}{j\omega\epsilon}\int_V [\nabla' \times \nabla' \times \mathbf{J}(\mathbf{r}')]g(R)\,dv' - \frac{1}{j\omega\epsilon}\mathbf{J}(\mathbf{r}) \tag{21}$$

where V is the support of $\mathbf{J}$ (outside volume V, the current $\mathbf{J}$ is identically zero). The second formula makes use of $\bar{\bar{G}}$ defined in (13b), namely,

$$\mathbf{E}(\mathbf{r}) = (-j\omega\mu)\left(\bar{\bar{I}} + \frac{1}{k^2}\nabla\nabla\right)\int_V \mathbf{J}(\mathbf{r}')g(R)\,dv' \tag{22}$$

The third formula makes use of $\bar{\bar{G}}$ defined in (15), namely,

$$\mathbf{E}(\mathbf{r}) = (-j\omega\mu)\int_V \left[\left(\bar{\bar{I}} + \frac{1}{k^2}\nabla'\nabla'\right)g(R)\right]\mathbf{J}(\mathbf{r}')\,dv' \tag{23}$$

Note the difference between (22) and (23). In (22) the differentiation with respect to the observation point coordinates (x_1, x_2, x_3) takes place after the integration over the source is performed, whereas in (23) the differentiation with respect to the source point coordinates (x_1', x_2', x_3') takes place before. If the integral is to be evaluated numerically, (23) is preferred to (22). The problem with (23) is how to integrate over the R^{-3} singularity. This is discussed next.

Integration Involving the R^{-3} Singularity

In evaluating the integral in (23), we encounter the following typical integral:

$$I_{mn}(\mathbf{r}) = \int_V J(\mathbf{r}')\frac{\partial^2 g}{\partial x_m' \partial x_n'}\,dv', \quad \text{for } m, n = 1, 2, 3 \tag{24}$$

As shown in (19) the second derivatives of g have an R^{-3} singularity at $R = 0$ (when observation point $\mathbf{r}$ is at a source point $\mathbf{r}'$). This singularity is not generally integrable. Thus (24) may be a divergent integral. We shall regularize (24) so that $I_{mn}(\mathbf{r})$ is a well-defined function of $\mathbf{r}$ for all $\mathbf{r}$. The result of a particular regularization is [2]

$$I_{mn}(\mathbf{r}) = A_{mn} + B_{mn} + C_{mn}, \quad \text{for } m, n = 1, 2, 3 \tag{25a}$$

where

$$A_{mn} = \int_{V-V_\epsilon} J(\mathbf{r}')\frac{\partial^2 g}{\partial x_m' \partial x_n'}\,dv' \tag{25b}$$

$$B_{mn} = \int_{V_\epsilon} \left[J(\mathbf{r}') \frac{\partial^2 g}{\partial x'_m \partial x'_n} - J(\mathbf{r}) \frac{\partial^2 g_0}{\partial x'_m \partial x'_n} \right] dv' \qquad (25c)$$

$$C_{mn} = J(\mathbf{r}) \left(\frac{-1}{4\pi} \right) \int_{\partial V_\epsilon} \frac{(\hat{\mathbf{x}}_m \cdot \hat{\mathbf{N}})(\hat{\mathbf{x}}_n \cdot \hat{\mathbf{R}})}{R^2} ds' \qquad (25d)$$

The various notations in (25) are explained below. Volume V is the support of current $J(\mathbf{r}')$ (see Fig. 2). Volume V_ϵ is an arbitrary volume inside V and contains the observation point $\mathbf{r}$. We emphasize that:

(a) V_ϵ need not be small, and
(b) the value of I_{mn} is independent of the choice of V_ϵ.

Term A_{mn} in (25b) is convergent because the region V_ϵ is excluded from the domain of integration. The static Green's function g_0 in (25c) is given by

$$g_0(R) = \frac{1}{4\pi R} \qquad (26)$$

which is obtained by setting $k = 0$ in the (dynamic) Green's function in (6). Note that g and g_0 have the same R^{-1} singularity at $R = 0$. This fact ensures the convergence of term B_{mn} in (25c). In term C_{mn} in (25d), there is a surface integral over the boundary surface of V_ϵ, which is denoted by ∂V_ϵ (Fig. 3). In this figure $\hat{\mathbf{N}}$ is the unit-outward normal of ∂V_ϵ at a point $\mathbf{r}'$, and $\hat{\mathbf{R}}$ is the unit vector along $\mathbf{R} = \mathbf{r}' - \mathbf{r}$. In summary, we regularize the divergent integral (24) so that it is defined by (25a). With this particular regularization it can be shown that I_{mn} is identical with

$$I_{mn}(\mathbf{r}) = \frac{\partial^2}{\partial x_m \partial x_n} \int_V J(\mathbf{r}') g(R) \, dv', \qquad \text{for } m, n = 1, 2, 3 \qquad (27)$$

In other words, the convergent integral in (27) can be alternatively evaluated via (25). This establishes the equivalence of two Green's functions in (13b) and (15) when the latter is regularized in the manner described in (25).

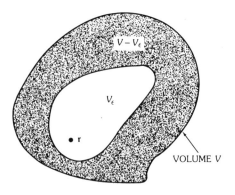

Fig. 2. Arbitrary volume V_ϵ is inside V and contains observation point $\mathbf{r}$.

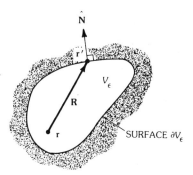

Fig. 3. Symbols for the surface integral in (25d): **r** is the observation point and **r'** is the integration point on surface ∂V_ϵ.

Explicit Expressions for C_{mn}

The term C_{mn} defined in (25d) can be explicitly evaluated for several special V_ϵs [3].

(a) V_ϵ is a sphere of radius a with the observation point **r** at an arbitrary point inside V_ϵ:

$$C_{mn} = -\frac{1}{3} J(\mathbf{r}) \delta_{mn} \tag{28}$$

(b) V_ϵ is a cube with center at **r** but is arbitrarily oriented, i.e., the faces of the cube need not be orthogonal to the x_1, x_2, x_3 axes:

$$C_{mn} = -\frac{1}{3} J(\mathbf{r}) \delta_{mn} \tag{29}$$

(c) V_ϵ is a cylinder with center at **r**, radius a, and height $2b$, and its axis coincides with the x_3 axis:

$$C_{11} = C_{22} = -\frac{b}{2\sqrt{a^2 + b^2}} J(\mathbf{r}) \tag{30}$$

$$C_{33} = \left(\frac{b}{\sqrt{a^2 + b^2}} - 1\right) J(\mathbf{r}) \tag{31}$$

$$C_{mn} = 0, \quad \text{if } m \neq n \tag{32}$$

3. Image Theory

Consider the radiation problem sketched in Fig. 4a with source $(\mathbf{J}_1, \mathbf{K}_1)$ and scatterer Σ_1, both situated above an infinitely large, planar, perfect electric conductor at $z = 0$. To calculate the field in the upper half-space $z > 0$, we may replace the configuration in Fig. 4a by that in Fig. 4b. In Fig. 4b the conductor is removed, scatterer Σ_2 is the mirror image of Σ_1, and source $(\mathbf{J}_2, \mathbf{K}_2)$ is given by:

$$J_{1x}(x, y, z) = -J_{2x}(x, y, -z) \tag{33a}$$

$$J_{1y}(x, y, z) = -J_{2y}(x, y, -z) \tag{33b}$$

$$J_{1z}(x, y, z) = +J_{2z}(x, y, -z) \tag{33c}$$

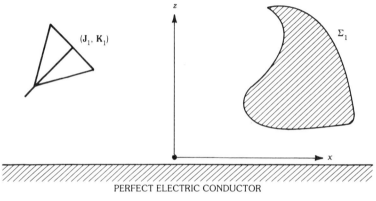

a

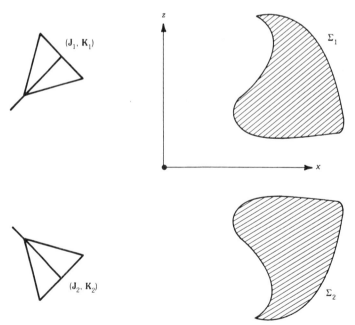

b

Fig. 4. By the image theory the problem in (a) may be replaced by that in (b) for calculating fields in the upper half-space $z > 0$. (a) Given radiation problem. (b) Calculating field in $z > 0$.

Theorems and Formulas

$$K_{1x}(x,y,z) = +K_{2x}(x,y,-z) \tag{34a}$$

$$K_{1y}(x,y,z) = +K_{2y}(x,y,-z) \tag{34b}$$

$$K_{1z}(x,y,z) = -K_{2z}(x,y,-z) \tag{34c}$$

The above image theory can be also applied if sources $(\mathbf{J}_1, \mathbf{J}_2)$ in (33) are replaced by incident fields $(\mathbf{E}_1^i, \mathbf{E}_2^i)$, and $(\mathbf{K}_1, \mathbf{K}_2)$ in (34) by $(\mathbf{H}_1^i, \mathbf{H}_2^i)$.

Now if the perfect electric conductor in Fig. 4a is replaced by a perfect magnetic conductor, the relations in (33) and (34) must be replaced by:

$$J_{1x}(x,y,z) = +J_{2x}(x,y,-z) \tag{35a}$$

$$J_{1y}(x,y,z) = +J_{2y}(x,y,-z) \tag{35b}$$

$$J_{1z}(x,y,z) = -J_{2z}(x,y,-z) \tag{35c}$$

$$K_{1x}(x,y,z) = -K_{2x}(x,y,-z) \tag{36a}$$

$$K_{1y}(x,y,z) = -K_{2y}(x,y,-z) \tag{36b}$$

$$K_{1z}(x,y,z) = +K_{2z}(x,y,-z) \tag{36c}$$

4. The Babinet Principle

The Babinet principle relates the field solutions of two problems with complementary configurations and dual sources.

Scattering Problems

Let us first consider the scattering version of the Babinet principle involving the following two scattering problems:

(a) As sketched in Fig. 5a, an infinitely large, thin, perfectly conducting plane Σ_1 with apertures lies in the plane $z = 0$. It is illuminated by an incident field, from $z < 0$, described by

$$\mathbf{E}_1^i = \mathbf{E}_0, \quad \mathbf{H}_1^i = \mathbf{H}_0 \tag{37a}$$

We express the total electric field by

$$\mathbf{E}_1^{\text{total}} = \begin{cases} \mathbf{E}_1^i + \mathbf{E}_1^s & \text{for } z > 0 \\ \mathbf{E}_1^i + \mathbf{E}_1^r + \mathbf{E}_1^d & \text{for } z < 0 \end{cases} \tag{37b}$$

Here $\mathbf{E}_1^r$ is the reflected field from an infinite conducting plane (without aperture) at $z = 0$ due to the incidence of (37a), and $\mathbf{E}_1^d$ is the diffracted

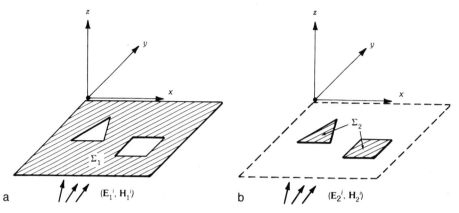

Fig. 5. Two scattering problems related by the Babinet principle. (*a*) Perfectly conducting plane Σ_1 is infinitely large. (*b*) Planes Σ_2 are complementary to Σ_1.

field. (If apertures were absent, $\mathbf{E}_1^d$ would be zero.) A similar expression holds for the total magnetic field $\mathbf{H}_1^{\text{total}}$.

(*b*) Complementary to the infinite plane Σ_1 in Fig. 5a, we have two thin, perfectly conducting planes sketched in Fig. 5b. They are denoted by Σ_2. Let Σ_2 be illuminated by an incident field from $z < 0$ described by

$$\mathbf{E}_2^i = -Z\mathbf{H}_0, \qquad \mathbf{H}_2^i = Y\mathbf{E}_0 \tag{38a}$$

where $Z = Y^{-1} = \sqrt{\mu/\epsilon}$. We express the total electric field everywhere by

$$\mathbf{E}_2^{\text{total}} = \mathbf{E}_2^i + \mathbf{E}_2^s, \qquad \text{for all } z \tag{38b}$$

A similar expression holds for $\mathbf{H}_2^{\text{total}}$.

In the transmitted half-space the Babinet principle states that

$$\left.\begin{array}{r}\mathbf{E}_1^{\text{total}} + Z\mathbf{H}_2^{\text{total}} - \mathbf{E}_0 = 0\\ \mathbf{H}_1^{\text{total}} - Y\mathbf{E}_2^{\text{total}} - \mathbf{H}_0 = 0\end{array}\right\} \text{ for } z > 0 \tag{39a}$$

when stating in terms of total fields; or

$$\left.\begin{array}{r}\mathbf{E}_1^s + Z\mathbf{H}_2^s + \mathbf{E}_0 = 0\\ \mathbf{H}_1^s - Y\mathbf{E}_2^s + \mathbf{H}_0 = 0\end{array}\right\} \text{ for } z > 0 \tag{39b}$$

when stating in terms of scattered fields. In the reflected half-space the Babinet principle states that

$$\left.\begin{array}{r}\mathbf{E}_1^{\text{total}} - Z\mathbf{H}_2^{\text{total}} - \mathbf{E}_1^r = 0\\ \mathbf{H}_1^{\text{total}} + Y\mathbf{E}_2^{\text{total}} - \mathbf{H}_1^r = 0\end{array}\right\} \text{ for } z < 0 \tag{40a}$$

Theorems and Formulas

when stating in terms of total fields; or

$$\left.\begin{array}{l}\mathbf{E}_1^d - Z\mathbf{H}_2^s = 0 \\ \mathbf{H}_1^d + Y\mathbf{E}_2^s = 0\end{array}\right\} \quad \text{for } z < 0 \quad (40b)$$

when stating in terms of scattered/diffracted fields. In either problem (*a*) or (*b*), the following relations hold for the field in the aperture (nonmetal part) at the $z = 0$ plane:

$$E_z^{\text{total}} = E_z^i \quad (41)$$

$$H_x^{\text{total}} = H_x^i, \quad H_y^{\text{total}} = H_y^i \quad (42)$$

This is due to the fact that the induced surface current on the conducting screen Σ_1 or Σ_2 produces symmetric (E_x^s, E_y^s, H_z^s), and antisymmetric (E_z^s, H_x^s, H_y^s) on the two sides of the $z = 0$ plane.

An Impedance Problem

Consider two complementary planar antennas: the slot antenna excited by a voltage source across the small gap ab in Fig. 6a, and the strip antenna excited by a voltage source across the small gap cd in Fig. 6b. Their impedances are related by

$$Z_{\text{slot}} Z_{\text{strip}} = \frac{1}{4} Z^2 \quad (43)$$

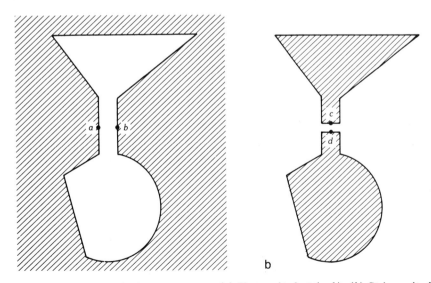

Fig. 6. Two complementary planar antennas. (*a*) Slot excited at (a, b). (*b*) Strip excited at (c, d).

where $Z = \sqrt{\mu/\epsilon}$ is the intrinsic wave impedance of the unbounded medium where both antennas are situated.

5. Reciprocity

The reciprocity theorem is useful in formulating problems as well as checking final answers. It "removes" the explicit dependence of media from a radiation or scattering problem.

Bra-Ket Notation

A source A generally consists of an electric current described by its density $\mathbf{J}_a(\mathbf{r})$ and a magnetic current described by its density $\mathbf{K}_a(\mathbf{r})$. In a given environment the field produced by A is $(\mathbf{E}_a, \mathbf{H}_a)$, denoted by F_a. Over a volume V the "reaction" between a source A and a field $F_b = (\mathbf{E}_b, \mathbf{H}_b)$ is defined by [4]

$$\langle AVF_b \rangle = \int_V (\mathbf{J}_a \cdot \mathbf{E}_b - \mathbf{K}_a \cdot \mathbf{H}_b) \, dv \tag{44}$$

which has the unit of watt. Through a surface S (not necessarily closed) the *cross-flux* of two fields F_a and F_b is defined by [4]

$$\langle F_a S F_b \rangle = \int_S (\mathbf{E}_a \times \mathbf{H}_b - \mathbf{E}_b \times \mathbf{H}_a) \cdot \hat{\mathbf{N}} \, ds \tag{45}$$

which again has the unit of watt. The surface S in (45) is oriented, meaning a unit normal $\hat{\mathbf{N}}$ is specified.

The Lorentz Relation

For a volume V, its boundary is a closed surface S. To emphasize this relation we use the notation $S = \partial V$. We choose the normal $\hat{\mathbf{N}}$ of S pointing outward (away from V). Consider two fields F_a and F_b produced by sources A and B, respectively (Fig. 7). The Lorentz relation reads

$$\langle F_a S F_b \rangle = \langle AVF_b \rangle - \langle BVF_a \rangle \tag{46a}$$

or, more explicitly,

$$\int_S (\mathbf{E}_a \times \mathbf{H}_b - \mathbf{E}_b \times \mathbf{H}_a) \cdot \hat{\mathbf{N}} \, ds = \int_V (\mathbf{J}_a \cdot \mathbf{E}_b - \mathbf{K}_a \cdot \mathbf{H}_b) \, dv$$
$$- \int_V (\mathbf{J}_b \cdot \mathbf{E}_a - \mathbf{K}_b \cdot \mathbf{H}_a) \, dv \tag{46b}$$

The relation in (46) holds under rather general conditions, namely:

(a) Sources A and/or B may be inside or outside V.
(b) Fields F_a and F_b may be produced in different media provided that these two media coincide inside V.

Theorems and Formulas

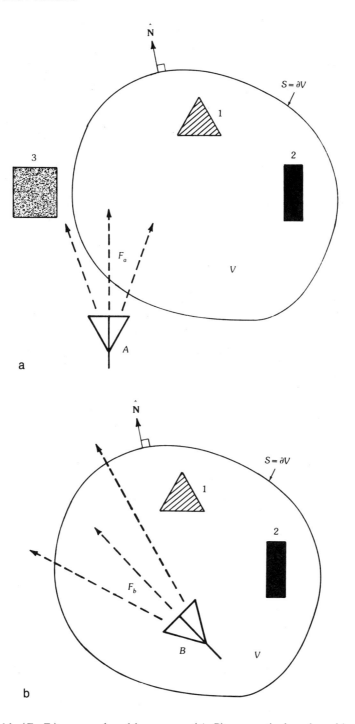

Fig. 7. Fields (F_a, F_b) are produced by sources (A, B), respectively, where blocks $(1, 2, 3)$ are scatterers and the medium in (a) is identical with the medium in (b) only inside V, not necessarily outside. (a) Field produced by A. (b) Field produced by B.

(c) The media are isotropic, but may be inhomogeneous with scatterers presented.

Several useful properties of the cross-flux term in (46) are listed below:

(a) $\langle F_a S F_b \rangle = -\langle F_b S F_a \rangle$
(b) $\langle F_a S F_b \rangle = \langle F_a S_1 F_b \rangle$ if there is no source between the closed surfaces S and S_1. Thus we can deform surface S to S_1 without crossing sources.
(c) Over the closed surface S, if an impedance boundary condition of the form

$$\mathbf{E}_{\tan} = Z\hat{\mathbf{N}} \times \mathbf{H}_{\tan} \tag{47a}$$

is satisfied for both fields F_a and F_b, then

$$\langle F_a S F_b \rangle = 0 \tag{47b}$$

A more general version of (47a) is

$$\mathbf{E}_{\tan} = \bar{\bar{Z}} \mathbf{H}_{\tan} \tag{47c}$$

where $\bar{\bar{Z}}$ is a 2×2 matrix. The relation (47b) holds if the trace of $\bar{\bar{Z}}$ is zero.

We also mention that $\langle AVF_b \rangle = 0$ if source A is outside V.

The Reciprocity Theorem

A useful application of (47) is stated below. Let volume V be an infinitely large spherical volume enclosing sources A and B, and be denoted by V_∞. Over the surface $S = \partial V_\infty$ both fields are locally plane waves and therefore satisfy (47a). The use of (47b) in (46) leads to the reciprocity theorem:

$$\langle AV_\infty F_b \rangle = \langle BV_\infty F_a \rangle \tag{48a}$$

or, more explicitly,

$$\int_{V_\infty} (\mathbf{J}_a \cdot \mathbf{E}_b - \mathbf{K}_a \cdot \mathbf{H}_b) \, dv = \int_{V_\infty} (\mathbf{J}_b \cdot \mathbf{E}_a - \mathbf{K}_b \cdot \mathbf{H}_a) \, dv \tag{48b}$$

Reciprocity holds for isotropic media, homogeneous or inhomogeneous, but it generally does not apply to an anisotropic medium, such as a magnetoplasma.

6. Huygens' Principle

Let $F = (\mathbf{E}, \mathbf{H})$ be the field produced by source $A = (\mathbf{J}, \mathbf{K})$ in medium a (Fig. 8a). Over an arbitrary closed surface S enclosing A, we define a Huygens' source described by surface current densities as

$$\left.\begin{aligned}\mathbf{J}_s &= \hat{\mathbf{N}} \times \mathbf{H} \\ \mathbf{K}_s &= \mathbf{E} \times \hat{\mathbf{N}}\end{aligned}\right\} \text{ over } S \tag{49}$$

Theorems and Formulas 2-19

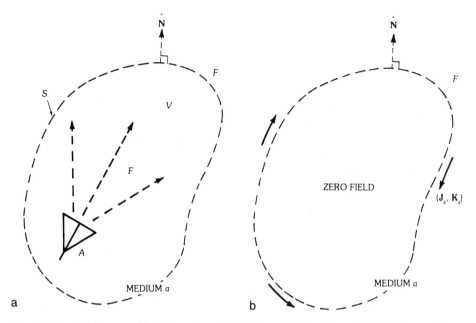

Fig. 8. Huygens' source $(\mathbf{J}_s, \mathbf{K}_s)$ produces field F outside V and zero field inside V. (*a*) Field F produced by source. (*b*) Huygens' source.

It may be shown that, when radiating in medium a, the Huygens' source produces

(*a*) zero field inside V, and
(*b*) the original field F outside V.

Thus, for observation points outside V, source A and its Huygens' source are equivalent. From $(\mathbf{J}_s, \mathbf{K}_s)$ in (49) we may calculate its exact radiation field by using (20) in Chapter 1, or its far field by using (21) or (48) in Chapter 1.

The Huygens' source is not unique. In addition to the one given in (49), an alternative Huygens' source is

$$\left. \begin{array}{l} \mathbf{J}'_s = \hat{\mathbf{N}} \times \mathbf{H}' \\ \mathbf{K}'_s = \mathbf{E}' \times \hat{\mathbf{N}} \end{array} \right\} \text{ over } S \qquad (50)$$

Here $(\mathbf{E}', \mathbf{H}')$ is the field produced by source A in medium b, which coincides with medium a inside V and may be different from medium a outside V (Fig. 9). The source $(\mathbf{J}'_s, \mathbf{K}'_s)$ radiating in medium a (not b) produces the original field F outside V.

7. The Kirchhoff Approximation

Consider an infinitely large perfectly conducting screen Σ_1, with a finite aperture Σ_2 (Fig. 10a). For a given source $A = (\mathbf{J}, \mathbf{K})$, the problem is to find the transmitted field $F = (\mathbf{E}, \mathbf{H})$ in the upper half-space $z > 0$. If the dimension of Σ_2 is

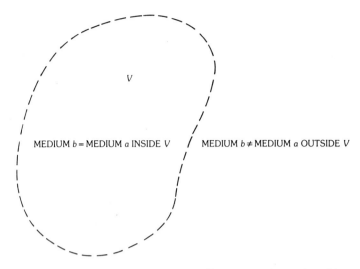

Fig. 9. In calculating the Huygens' source, medium *a* may be replaced by medium *b*.

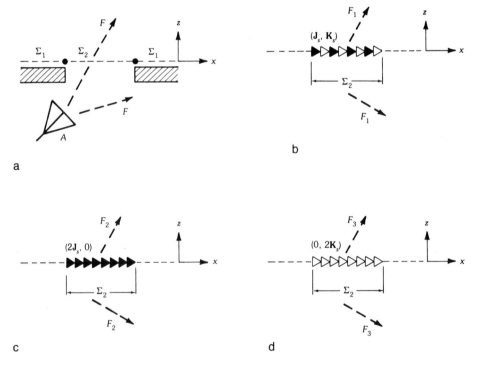

Fig. 10. Three versions of the Kirchhoff approximation for the field in the upper half-space $z > 0$. (*a*) Radiation problem. (*b*) Approximating electric and magnetic currents. (*c*) Approximating electric current alone. (*d*) Approximating magnetic current alone.

Theorems and Formulas

large in terms of wavelength, the field F may be approximately calculated by using one of the following three versions of Kirchhoff's approximations:

(a) *Electric and magnetic currents* (Fig. 10b). We replace source A by an equivalent surface source distributed over the aperture Σ_2:

$$\left.\begin{array}{l}\mathbf{J}_s = \hat{\mathbf{N}} \times \mathbf{H}^i \\ \mathbf{K}_s = \mathbf{E}^i \times \hat{\mathbf{N}}\end{array}\right\} \quad \text{over } \Sigma_2 \qquad (51)$$

Here $\hat{\mathbf{N}} = \hat{\mathbf{z}}$ is the outward normal of Σ_2. Field $(\mathbf{E}^i, \mathbf{H}^i)$ is that produced by source A in the absence of screen Σ_1. The field produced by $(\mathbf{J}_s, \mathbf{K}_s)$ in the absence of Σ_1 is F_1, which is an approximation of F in the upper half-space $z > 0$. Note that F_1 is not a good approximation of F for $z < 0$.

(b) *Electric current alone* (Fig. 10c). The equivalent source in (51) consists of both electric and magnetic surface currents. An alternative equivalent source is $(2\mathbf{J}_s, 0)$, which doubles the electric current in (51) and contains no magnetic current. The field produced by $(2\mathbf{J}s, 0)$ in the absence of Σ_1 is F_2, which is the second approximation of F for $z > 0$.

(c) *Magnetic current alone* (Fig. 10d). Another equivalent source is $(0, 2\mathbf{K}_s)$, which doubles the magnetic current in (51) and contains no electric current. The field produced by it in the absence of Σ_1 is F_3, which is the third approximation of F for $z > 0$.

Version (a) above is sometimes known as the Stratton-Chu formula. Field F_1 is the average of F_2 and F_3. Kirchhoff's approximation has been applied to transmission through aperture problems in which the aperture is not necessarily in a planar screen, such as radiation from an open-ended waveguide.

8. Scattering by an Obstacle

Consider the scattering problem sketched in Fig. 11. The obstacle (scatterer) is illuminated by an incident plane wave F^i given by

$$\begin{Bmatrix}\mathbf{E}^i(\mathbf{r}) \\ \mathbf{H}^i(\mathbf{r})\end{Bmatrix} = C \begin{Bmatrix}\sqrt{Z}\,\mathbf{u}_1 \\ \sqrt{Y}\,\hat{\mathbf{k}}_1 \times \mathbf{u}_1\end{Bmatrix} e^{-j\mathbf{k}_1 \cdot \mathbf{r}} \qquad (52)$$

where

$\mathbf{k}_1$ = wave vector with magnitude $k = \omega\sqrt{\mu\epsilon}$ and pointing in the direction of propagation of F^i

$Z = Y^{-1} = \sqrt{\mu/\epsilon}$

$\mathbf{u}_1$ = a unitary vector which describes the polarization of F^i and is orthogonal to $\mathbf{k}_1$

C = amplitude of F^i in $(\text{watt})^{1/2}\,(\text{meter})^{-1}$

We say that F^i is in state $(\mathbf{k}_1, \mathbf{u}_1)$. In the presence of the obstacle the total field

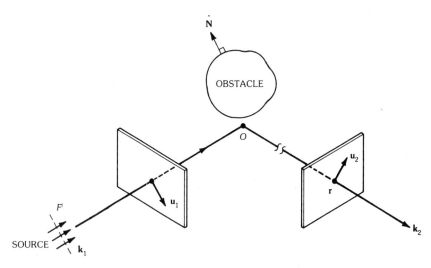

Fig. 11. We are interested in the scattered far field at **r** with state $(\mathbf{k}_2, \mathbf{u}_2)$ when an obstacle is illuminated by an incident plane wave F^i with state $(\mathbf{k}_1, \mathbf{u}_1)$.

everywhere is the sum of F^i and a scattered field F. At a far-field observation point **r**, field F is represented by a spherical wave

$$\begin{Bmatrix} \mathbf{E}(\mathbf{r}) \\ \mathbf{H}(\mathbf{r}) \end{Bmatrix} \sim \begin{Bmatrix} \sqrt{Z}\mathbf{A}(\mathbf{k}_2) \\ \sqrt{Y}\hat{\mathbf{k}}_2 \times \mathbf{A}(\mathbf{k}_2) \end{Bmatrix} \frac{e^{-jkr}}{r}, \quad r \to \infty \quad (53)$$

The distance r is measured from a reference point O in the vicinity of the obstacle. Vector $\mathbf{k}_2$ has a magnitude equal to k and is in the direction of $O\mathbf{r}$. The amplitude vector $\mathbf{A}(\mathbf{k}_2)$ given in (watt)$^{1/2}$ may be decomposed into two orthogonal components, as discussed in Section 7, Chapter 1. Let us concentrate on a particular component of $\mathbf{A}(\mathbf{k}_2)$ with polarization $\mathbf{u}_2$, namely, $\mathbf{A} \cdot \mathbf{u}_2^*$. We introduce the notation (after G. A. Deschamps)

$$A(2,1) = \mathbf{A}(\mathbf{k}_2) \cdot \mathbf{u}_2^* \quad (54)$$

which represents the scattering amplitude in state 2 (direction $\mathbf{k}_2$ and polarization $\mathbf{u}_2$) due to an incident plane wave in state 1. Using this notation, we will introduce some definitions and theorems for the scattering problem in Fig. 11.

Bistatic Cross Section

The bistatic cross section (BCS) in square meters from state 1 to state 2 is defined by

$$\text{BCS} = \frac{4\pi |A(2,1)|^2}{|C|^2} = \frac{4\pi |\mathbf{A}(\mathbf{k}_2) \cdot \mathbf{u}_2^*|^2}{|C|^2} \quad (55a)$$

Theorems and Formulas

$$= \frac{4\pi \text{ (intensity of scattered field in state 2)}}{\text{power density of incident plane wave in state 1}} \tag{55b}$$

Radar Cross Section

The plane wave in (52) is in state 1 described by $(\mathbf{k}_1, \mathbf{u}_1)$. Its time-reversed counterpart is in state 1' described by $(-\mathbf{k}_1, \mathbf{u}_1^*)$. Note that these two fields have the *same* polarization.* The radar cross section (RCS) in square meters is a special case of the bistatic cross section BCS(2, 1) with state 2 equal to state 1', namely,

$$\text{RCS} = \frac{4\pi |A(1', 1)|^2}{|C|^2} = \frac{4\pi |A(-\mathbf{k}_1) \cdot \mathbf{u}_1|^2}{|C|^2} \tag{56a}$$

$$= \frac{4\pi \text{ (intensity of scattered field in state 1')}}{\text{power density of incident plane wave in state 1}} \tag{56b}$$

We will consider two examples of RCS. (*a*) For a smooth conductor whose dimension is large in terms of wavelength and is illuminated by a linearly polarized incident field, its RCS is approximately independent of polarization and is given by

$$\text{RCS} \cong \pi |R_1 R_2| \tag{57a}$$

where R_1 and R_2 are two principal radii of curvature of the conducting body at the specular point P (Fig. 12). The point P is determined by the relation in which the surface normal $\hat{\mathbf{N}}$ is in the opposite direction of $\mathbf{k}_1$. We assume that there is only one such specular point. If the conductor is a sphere of radius a, the use of (57a) leads to RCS $\cong \pi a^2$, a well-known result. (*b*) For a rectangular conducting plate of dimension $a \times b$ in the $z = 0$ plane, the RCS is again approximately independent of polarization and is given by

$$\text{RCS} \cong \frac{1}{\pi}\left(kab \cos\theta_0 \frac{\sin\alpha}{\alpha} \frac{\sin\beta}{\beta}\right)^2 \tag{57b}$$

where

(θ_0, ϕ_0) = spherical angles of back-scattered wave vector $\mathbf{k}_2$ or $-\mathbf{k}_1$

$\alpha = ka \sin\theta_0 \cos\phi_0$

$\beta = kb \sin\theta_0 \sin\phi_0$

The expression in (57b) is approximately valid when dimensions (a, b) are large in terms of wavelength, and incident angle θ_0 is not too large (near normal incidence).

The RCS defined in (56) is proportional to the scattered intensity in the same polarization as the incident one. It is also known as the RCS for the copolarization. We can define an RCS in square meters for the cross polarization by

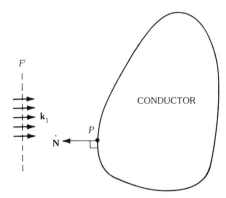

Fig. 12. The radar cross section (RCS) of a large, smooth conductor.

$$\text{RCS for cross polarization} = \frac{4\pi |\mathbf{A}(-\mathbf{k}_1)\cdot\mathbf{v}_1|^2}{|C|^2} \quad (58a)$$

Here $\mathbf{v}_1$ is the unitary vector describing the cross polarization satisfying

$$\mathbf{v}_1\cdot\mathbf{k}_1 = 0, \quad \mathbf{v}_1\cdot\mathbf{v}_1^* = 1, \quad \mathbf{v}_1\cdot\mathbf{u}_1^* = 0 \quad (58b)$$

A common choice for the copolarization and cross polarization is $\mathbf{u}_1$ describing vertical polarization and $\mathbf{v}_1$ horizontal polarization, or vice versa. This results in four values for RCS, with two of them being equal.

Reciprocity

In terms of the notation in (54), the reciprocity for the scattering amplitude may be stated as

$$A(2,1) = A(1',2') \quad (59)$$

The amplitude $A(2,1)$ is explained in Fig. 11, and amplitude $A(1',2')$ in Fig. 13.

Scattering Cross Section

For the scattered field in (53), the scattering cross section (SCS) in square meters is defined by

$$\text{SCS} = |C|^{-2} \int_{4\pi} |\mathbf{A}|^2 \, d\Omega \quad (60a)$$

$$= \frac{\text{total scattered power in all directions and all polarizations}}{\text{power density of incident plane wave in state 1}} \quad (60b)$$

Clearly, SCS is a function of the incident state 1. It can be shown that SCS of a lossless obstacle is related to the forward scattering amplitude by

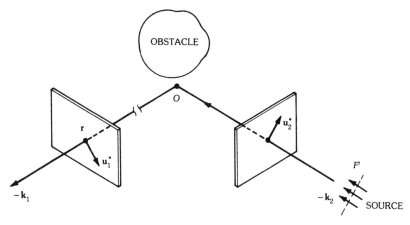

Fig. 13. Reciprocity applies to the above situation and that in Fig. 11.

$$\text{SCS} = \frac{4\pi}{k}\text{Im}\{A(1,1)\} = \frac{4\pi}{k}\text{Im}\{\mathbf{A}(\mathbf{k}_1)\cdot\mathbf{u}_1^*\} \tag{61}$$

which is known as the *scattering cross section theorem*.

Physical Optics Approximation for Scattering by a Conductor

Consider the scattering problem in Fig. 11, with the incident field described by (52) and the scattered field by (53). If the obstacle in Fig. 11 is a perfect conductor, the scattered field is due to the radiation of a surface electric current $\mathbf{J}_s$ on the very surface of the conductor. At high frequencies we may use the so-called physical optics approximation for $\mathbf{J}_s$, namely,

$$\mathbf{J}_s(\mathbf{r}) \cong \begin{cases} 2\hat{\mathbf{N}} \times \mathbf{H}^i & \text{for } \mathbf{r} \text{ in the lit region} \\ 0 & \text{for } \mathbf{r} \text{ in the shadow} \end{cases} \tag{62}$$

For a convex obstacle the lit region is defined by relation $\mathbf{k}_1\cdot\hat{\mathbf{N}} < 0$, and shadow region by $\mathbf{k}_1\cdot\hat{\mathbf{N}} > 0$. The scattered field defined in (53) is the radiation field of $\mathbf{J}_s$ in the unbounded space. The scattering amplitude in state $(\mathbf{k}_2, \mathbf{u}_2)$ may be calculated from (62) above and (48) in Chapter 1. The final result is

$$\mathbf{A}(\mathbf{k}_2)\cdot\mathbf{u}_2^* \cong \frac{jkC}{2\pi}\left[\mathbf{u}_2^* \times (\hat{\mathbf{k}}_1 \times \mathbf{u}_1)\right]\cdot\left[\iint_{\text{lit}} \hat{\mathbf{N}} e^{j(\mathbf{k}_2-\mathbf{k}_1)\cdot\mathbf{r}'}\,ds'\right] \tag{63a}$$

where the surface integration is over the source point $\mathbf{r}'$ in the lit portion of the conductor surface. For backscattering, $\mathbf{k}_2 = -\mathbf{k}_1$, and (63a) is reduced to

$$\mathbf{A}(-\mathbf{k}_1)\cdot\mathbf{u}_2^* \cong \frac{jkC}{2\pi}\left[(\mathbf{u}_2^*\cdot\mathbf{u}_1)\hat{\mathbf{k}}_1\right]\cdot\left[\iint_{\text{lit}} \hat{\mathbf{N}} e^{-j2\mathbf{k}_1\cdot\mathbf{r}'}\,ds'\right] \tag{63b}$$

For backscattering in the *same* polarization as the incident one, we set $\mathbf{u}_2 = \mathbf{u}_1^*$ in (63b) and obtain

$$\mathbf{A}(-\mathbf{k}_1)\cdot(\mathbf{u}_1^*)^* \cong \frac{jkC}{2\pi}\left[(\mathbf{u}_1\cdot\mathbf{u}_1)\,\hat{\mathbf{k}}_1\right]\cdot\left[\iint_{\text{lit}} \hat{\mathbf{N}}\,e^{-j2\mathbf{k}_1\cdot\mathbf{r}'}\,ds'\right] \qquad (63c)$$

If the incident field is linearly polarized, $\mathbf{u}_1$ is real and $\mathbf{u}_1\cdot\mathbf{u}_1 = 1$ (see Section 5, Chapter 1). If the incident field is circularly polarized, we have $\mathbf{u}_1\cdot\mathbf{u}_1 = 0$, implying that the backscattered field calculated from (63c) is zero in the copolarization, and all of the backscattered field is in the cross polarization. At high frequencies we may evaluate the integral in (63a) by the stationary phase point method [5]. The leading term of this evaluation gives precisely the geometrical optics field that can be directly calculated by ray techniques.

Two Other Versions of the Physical Optics Approximation

Consider the scattering problem sketched in Fig. 14. An electrically large perfect conductor is illuminated by an incident field $(\mathbf{E}^i, \mathbf{H}^i)$ from a source at point A. The scattered field $F(B)$ at B is to be found. According to the physical optics approximation described above, we may approximate $F(B)$ by $F_1(B)$. Here $F_1(B)$ is the field produced by a surface current $\mathbf{J}_s$ on the conductor's surface radiating in the free space, and $\mathbf{J}_s$ is given by

$$\mathbf{J}_s(\mathbf{r}) = \begin{cases} 2\hat{\mathbf{N}} \times \mathbf{H}^i & \text{for } \mathbf{r} \text{ in the surface portion visible from } A \\ 0 & \text{elsewhere on the conductor's surface} \end{cases} \qquad (64)$$

The surface portion with nonzero $\mathbf{J}_s$ is indicated by a zigzag line in Fig. 14a. The second version of the physical optics approximation is to approximate $F(B)$ by $F_2(B)$. The latter is produced by a magnetic surface current $\mathbf{K}_s$ radiating in the free space, and $\mathbf{K}_s$ is given by

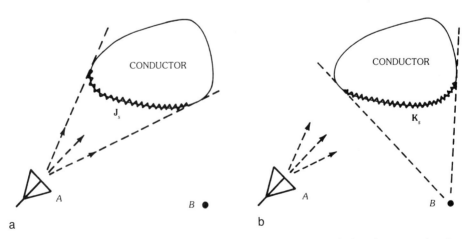

Fig. 14. Scattering by a large conductor: two versions of the physical optics approximation. (a) Using electric surface current $\mathbf{J}_s$. (b) Using magnetic surface current $\mathbf{K}_s$.

Theorems and Formulas

$$\mathbf{K}_s(\mathbf{r}) = \begin{cases} 2\mathbf{E}^i \times \hat{\mathbf{N}} & \text{for } \mathbf{r} \text{ in the surface portion visible from } B \\ 0 & \text{elsewhere on the conductor's surface} \end{cases} \quad (65)$$

The surface portion with nonzero $\mathbf{K}_s$ is indicated by a zigzag line in Fig. 14b. Note that $\mathbf{J}_s$ and $\mathbf{K}_s$ are distributed over different portions of the conductor's surface. A third version of the physical optics approximation is to approximate $F(B)$ by the average of $F_1(B)$ and $F_2(B)$. When the positions of the source and the observation points are interchanged, the reciprocity holds in the third version, but not in the first two. It is also interesting to mention that the three versions of physical optics are the counterparts of the three Kirchhoff approximations discussed in Section 7.

9. The Antenna as a One-Port Device

An antenna is usually fed through a transmission line or a waveguide, as the two examples sketched in Fig. 15. We assume that the transmission line (waveguide) supports only one propagating mode. The total field along the line (inside the waveguide) is a superposition of two traveling waves, one in the $+z$ direction and one in the $-z$ direction:

$$\mathbf{E}(\mathbf{r}) = \mathbf{e}(x,y)[ae^{-j\beta z} + be^{+j\beta z}] \quad (66a)$$

$$\mathbf{H}(\mathbf{r}) = \mathbf{h}(x,y)[ae^{-j\beta z} - be^{+j\beta z}] \quad (66b)$$

The reference plane $z = 0$ can be arbitrarily chosen, and need not be exactly at the antenna–transmission-line junction. Here β is the propagation constant of the dominant mode, and (a, b) are wave amplitudes in (watts)$^{1/2}$. Transverse field variations are described by $(\mathbf{e}, \mathbf{h})$, whose units are $(\Omega^{1/2}\text{m}^{-1}, \Omega^{-1/2}\text{m}^{-1})$, respectively, and satisfy the normalization condition

$$\iint (\mathbf{e} \times \mathbf{h}^*) \cdot \hat{\mathbf{z}} \, dx \, dy = 1 \quad (67)$$

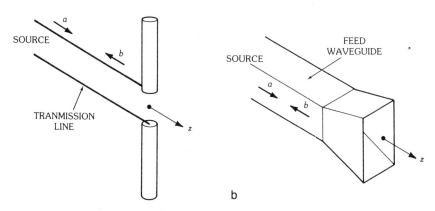

Fig. 15. Antennas are fed via transmission lines or waveguides. (a) Dipole. (b) Horn.

where the integration is over the infinite plane transverse to the transmission line, or the transverse cross section of the waveguide. The ratio

$$\Gamma = b/a \tag{68}$$

is called the E-field (voltage) reflection coefficient at a reference plane $z = 0$. The H-field (current) reflection coefficient is $-\Gamma$.

The above description of the field in the waveguide is from the wave viewpoint. Alternatively, it may be described from the circuit viewpoint. To do so, we are forced to introduce a so-called characteristic impedance $Z_c = Y_c^{-1}$, and then define the (modal) voltage and current by

$$V^+ = \sqrt{Z_c}\,a, \qquad I^+ = \sqrt{Y_c}\,a \tag{69}$$

where the superscript $+$ signifies a wave traveling in the $(+z)$ direction. We emphasize that in a general waveguide there is no obvious (unique) way to define Z_c because the ratio of transverse electric and magnetic fields varies from point to point. As an example, for a rectangular-feed waveguide having cross section $(c \times d)$ with $c > d$, we have

$$\mathbf{e}(x,y) = \hat{\mathbf{y}}\left(Z\frac{k}{\beta}\frac{2}{cd}\right)^{1/2}\cos\left(\frac{\pi}{c}x\right) \tag{70a}$$

$$\mathbf{h}(x,y) = (Zk/\beta)^{-1}\hat{\mathbf{z}} \times \mathbf{e} \tag{70b}$$

where

$$Z = Y^{-1} = (\mu/\epsilon)^{1/2}$$
$$\beta = k[1 - (\pi/kc)^2]^{1/2} \tag{71}$$

As for the characteristic impedance, there are at least four commonly used definitions, namely,

$$Z_{c1} = Zk/\beta \tag{72a}$$

$$Z_{c2} = 2(d/c)Z_{c1} \tag{72b}$$

$$Z_{c3} = (\pi^2/8)(d/c)Z_{c1} \tag{72c}$$

$$Z_{c4} = (\pi/2)(d/c)Z_{c1} \tag{72d}$$

Thus there are at least four different ways to define the (modal) voltage and current in the rectangular waveguide, corresponding to the four choices of Z_c in (72).

Once a choice of Z_c is made, we find the (total) voltage and current at the reference plane $z = 0$ are given by

Theorems and Formulas

$$V = V^+ + V^- = \sqrt{Z_c}(a + b) \tag{73}$$

$$I = I^+ + I^- = \sqrt{Y_c}(a - b) \tag{74}$$

The input impedance of the antenna is defined by

$$Z_{in} = \frac{1}{Y_{in}} = \frac{V}{I} \tag{75}$$

The relations among Γ, Z_{in}, and Y_{in} are

$$\frac{Z_{in}}{Z_c} = \frac{Y_c}{Y_{in}} = \frac{1 + \Gamma}{1 - \Gamma} \tag{76}$$

$$\Gamma = \frac{(Z_{in}/Z_c) - 1}{(Z_{in}/Z_c) + 1} = \frac{1 - (Y_{in}/Y_c)}{1 + (Y_{in}/Y_c)} \tag{77}$$

It is clear from (76) that antenna impedance Z_{in} also depends on the choice of Z_c.

The power transmitted (radiated) from the source into the free space via the antenna is given by

$$P_t = \text{Re}\left\{\iint (\mathbf{E} \times \mathbf{H}^*)\cdot\hat{z}\,dx\,dy\right\} \tag{78a}$$

$$= |a|^2 - |b|^2 = |a|^2(1 - |\Gamma|^2) \tag{78b}$$

$$= \text{Re}\{VI^*\} \tag{79}$$

The reflected power back toward the source (due to the mismatch of Z_c and Z_{in}) is $|b|^2$.

10. Three Ideal Sources for Transmitting Antennas

As discussed in the previous section, an antenna as a one-port device can be described by either wave amplitudes (a, b), or by circuit parameters (V, I). When the antenna is used for transmitting, a source is connected to the transmission line (or feed waveguide). Following Deschamps [6], we will introduce three principal ideal sources:

(a) unit amplitude source (defined by $a = 1$ (watt)$^{1/2}$)
(b) unit voltage source ($V = 1$ volt)
(c) unit current source ($I = 1$ ampere)

Their graphical representation is shown in Fig. 16. Note that the dot in each source indicates the propagation direction of the incident wave leaving the dot, the positive voltage terminal at the dot, or the current out of the dot. The term "ideal" is used because those sources have special internal impedance Z_s, viz.,

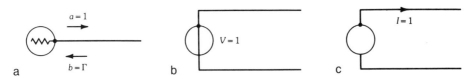

Fig. 16. Three ideal sources for transmitting antennas.

(a) $Z_s = Z_c$ for the amplitude source, in which Z_c is the characteristic impedance of the transmission line. Thus the source is matched to the transmission line, and there is no reflection at the source–transmission-line junction. (However, the mismatch at the antenna–transmission-line junction may still exist; Z_c may or may not be equal to Z_{in}.)

(b) $Z_s = 0$ for the voltage source so that there is no internal voltage drop within the source.

(c) $Z_s \to \infty$ for the current source so that the total source current enters the transmission line.

In Fig. 16 we use the zigzag line, straight line, and gap inside the circles to indicate the source impedances.

For a transmitting antenna any one of the three ideal sources may be used as the excitation. For each case, we list in Table 1 the internal feed waveguide quantities (V, I, a, b) and the external radiation quantities (radiated field F and its power P_t). Here F stands for vector fields $(\mathbf{E}, \mathbf{H})$. The subscripts $(1, 2, 3)$ are used to identify excitations due to a unit (amplitude, voltage, current) source. For example, a_1 is the value of a when the excitation is a unit amplitude source, and F_3 is the value of F when the excitation is a unit current source. The three fields (F_1, F_2, F_3) in Table 1 are related by

$$F_2 = \frac{\sqrt{Y_c}}{1+\Gamma} \frac{V_2}{a_1} F_1 = \frac{1}{2}\sqrt{Y_c}\left(1 + \frac{Y_{in}}{Y_c}\right) \frac{V_2}{a_1} F_1 \tag{80}$$

$$F_3 = \frac{\sqrt{Z_c}}{1-\Gamma} \frac{I_3}{a_1} F_1 = \frac{1}{2}\sqrt{Z_c}\left(1 + \frac{Z_{in}}{Z_c}\right) \frac{I_3}{a_1} F_1 \tag{81}$$

where $a_1 = 1$ (watt)$^{1/2}$, $V_2 = 1$ volt, and $I_3 = 1$ ampere.

11. Three Ideal Meters for Receiving Antennas

When an antenna is used for receiving, the source is replaced by a receiver or, for our present purpose, a meter. Again, following Deschamps we introduce three principal ideal meters:

(a) Amplitude meter, which measures the incoming traveling-wave amplitude b at a reference plane $z = 0$, and has internal source impedance $Z_s = Z_c$ such that the meter is matched to the feed waveguide. (The antenna impedance Z_{in} may or may not be matched to Z_c.) Note that b is a complex number, including both magnitude and phase information.

Theorems and Formulas

Table 1. Relationships for Transmitting Antennas

Source \ Quantity	Amplitude	Voltage	Current
V (volt)	$\sqrt{Z_c}(1+\Gamma)a_1$	$V_2 = 1$	$Z_{in}I_3$
I (ampere)	$\sqrt{Y_c}(1-\Gamma)a_1$	$Y_{in}V_2$	$I_3 = 1$
a (watt)$^{1/2}$	$a_1 = 1$	$\frac{1}{2}\sqrt{Y_c}\left(1+\frac{Y_{in}}{Y_c}\right)V_2$	$\frac{1}{2}\sqrt{Z_c}\left(\frac{Z_{in}}{Z_c}+1\right)I_3$
b (watt)$^{1/2}$	Γa_1	$\frac{1}{2}\sqrt{Y_c}\left(1-\frac{Y_{in}}{Y_c}\right)V_2$	$\frac{1}{2}\sqrt{Z_c}\left(\frac{Z_{in}}{Z_c}-1\right)I_3$
P_t (watt)	$(1-|\Gamma|^2)a_1^2$	$V_2^2 \operatorname{Re} Y_{in}$	$I_3^2 \operatorname{Re} Z_{in}$
Field	F_1	F_2	F_3
Source impedance	Z_c (matched)	0 (short)	∞ (open)

(b) Voltmeter, which measures the voltage V at a reference plane $z = 0$, and has an infinite internal source impedance ($Z_s \to \infty$) such that V is the open-circuit voltage.

(c) Ammeter, which measures the current I at a reference plane $z = 0$, and has a zero internal source impedance ($Z_s = 0$) such that I is the short-circuit current.

The graphical representation of meters is given in Fig. 17. Note our convention that the circles represent sources, while squares represent meters (compare Figs. 16 and 17).

When the receiving antenna is illuminated by an incident field we may connect any of the three ideal meters in Fig. 17 to its feed waveguide, corresponding to matched-load, open-circuit, and short-circuit situations. Table 2 lists the internal feed waveguide quantities (V, I, a, b). The subscripts (4, 5, 6) are used to identify the use of an (amplitude meter, voltmeter, ammeter). For the same incident field the three situations are related by

$$V_5 = \sqrt{Z_c}\frac{2b_4}{1-\Gamma} = b_4\sqrt{Z_c}\left(1+\frac{Z_{in}}{Z_c}\right) \tag{82}$$

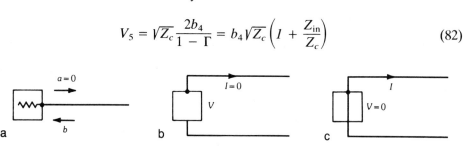

Fig. 17. Three ideal meters for receiving antennas. (*a*) Amplitude. (*b*) Voltage. (*c*) Current.

Table 2. Relationships for Receiving Antennas

Quantity \ Meter	Amplitude	Voltmeter	Ammeter
V (volt)	$\sqrt{Z_c}\, b_4$	V_5	0
I (ampere)	$-\sqrt{Y_c}\, b_4$	0	I_6
a (watt)$^{1/2}$	0	$\frac{1}{2}\sqrt{Y_c}\, V_5$	$\frac{1}{2}\sqrt{Z_c}\, I_6$
b (watt)$^{1/2}$	b_4	$\frac{1}{2}\sqrt{Y_c}\, V_5$	$-\frac{1}{2}\sqrt{Z_c}\, I_6$
Source impedance	Z_c (matched)	∞ (open)	0 (short)

$$I_6 = \sqrt{Y_c}\frac{(-2b_4)}{1+\Gamma} = -b_4\sqrt{Y_c}\left(1+\frac{Y_{in}}{Y_c}\right) \tag{83}$$

Commonly, V_5 is known as the *open-circuit voltage*, and I_6 as the *short-circuit current* of a receiving antenna.

12. Reciprocity between Antenna Transmitting and Receiving

The reciprocity theorem in a circuit is well known. Its application to antennas, however, is not simple for the reason explained below. If viewed from the transmission line, the antenna looks like a circuit element whose transmitting and receiving properties are describable by two (complex) numbers: (a, b) or (V, I), as discussed in Section 9. Outside the antenna in the free-space region, either the radiated field of the transmitting antenna or the incident field on the receiving antenna is more complex. They are vector fields characterized by polarization and spatial variation, which are not describable by circuit quantities. Hence the reciprocity for an antenna cannot be simply stated by the usual exchange of sources and meters. In this section we will give two reciprocity relations for antennas.

Reciprocity Involving General Incident Fields

Consider the transmitting situation in Fig. 18a, where the antenna is excited by a unit amplitude source with $a_1 = 1$ (watt)$^{1/2}$. The radiated field is $F_1 = (\mathbf{E}_1, \mathbf{H}_1)$. In the receiving situation (Fig. 18b) the same antenna is connected to an amplitude meter with matched impedance (so that $a_4 = 0$), and is illuminated by an incident field $F_4 = (\mathbf{E}_4, \mathbf{H}_4)$. Then a reciprocity states [6]

$$b_4 = \frac{1}{2a_1}\int_S (\mathbf{E}_4 \times \mathbf{H}_1 - \mathbf{E}_1 \times \mathbf{H}_4)\cdot\hat{\mathbf{N}}\, dS \tag{84}$$

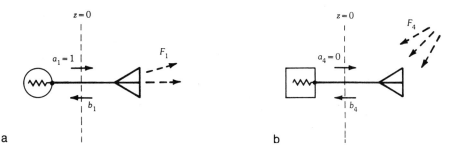

Fig. 18. An antenna in transmitting excited with a unit-amplitude source, and in receiving connected with an amplitude meter. (*a*) Transmitting. (*b*) Receiving.

$$= \frac{1}{2a_1} \langle F_4 S F_1 \rangle$$

Here S is an arbitrary closed surface which encloses the antenna but excludes the source of F_4. Its outward unit normal is $\hat{N}$. It is emphasized that F_4 is the incident field that would exist in the absence of the antenna in Fig. 18b, and does not include the scattered field F_4' from the antenna. It can be shown, however, that (84) remains valid if F_4 is replaced by $F_4 + F_4'$, because the cross flux $\langle F_4' S F_1 \rangle = 0$.

Reciprocity Involving Plane Waves

In many applications we are interested in a special case of Fig. 18, namely, the radiated field F_1 in the transmitting situation is known in the far-field zone (Fig. 19a), and the incident field F_4 in the receiving situation is a plane wave (Fig. 19b). We express the radiated field F_1 by

$$F_1 = \begin{Bmatrix} \mathbf{E}_1(\mathbf{r}) \\ \mathbf{H}_1(\mathbf{r}) \end{Bmatrix} \sim \begin{Bmatrix} \sqrt{Z}\mathbf{A}(\mathbf{k}) \\ \sqrt{Y}\hat{\mathbf{k}} \times \mathbf{A}(\mathbf{k}) \end{Bmatrix} \frac{e^{-jkr}}{r}, \qquad r \to \infty \qquad (85)$$

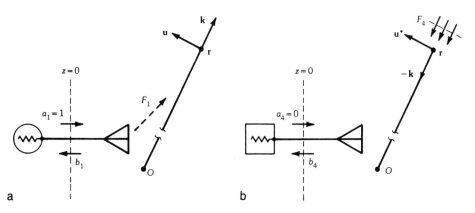

Fig. 19. Same antenna as Fig. 18 except incident field F_4 is a plane wave. (*a*) Transmitting. (*b*) Receiving.

where $Z = Y^{-1} = (\mu/\epsilon)^{1/2}$. We represent the incident field F_4 by

$$F_4 = \begin{Bmatrix} \mathbf{E}_4(\mathbf{r}) \\ \mathbf{H}_4(\mathbf{r}) \end{Bmatrix} = C \begin{Bmatrix} \sqrt{Z}\mathbf{u}^* \\ -\sqrt{Y}\hat{\mathbf{k}} \times \mathbf{u}^* \end{Bmatrix} e^{j\mathbf{k}\cdot\mathbf{r}} \tag{86}$$

whose amplitude is C in (watt)$^{1/2}$/meter, propagation vector $(-\mathbf{k})$, and polarization vector $\mathbf{u}^*$. Then a reciprocity relation states [6]

$$b_4 = -j\lambda \left(\frac{C}{a_1}\right) [\mathbf{A}(\mathbf{k}) \cdot \mathbf{u}^*] \tag{87}$$

where $\lambda = 2\pi/k$ is the free-space wavelength. If the open-circuit voltage and short-circuit current in the receiving situation are of interest, we may use (87) or (84) in conjunction with (82) and (83). The interpretation of (87) is as follows: The received amplitude b_4 of an antenna under the matched condition due to the incidence of a plane wave in state $(-\mathbf{k}, \mathbf{u}^*)$ is proportional to the antenna radiation far-field amplitude in state $(\mathbf{k}, \mathbf{u})$. [Remember that $(-\mathbf{k}, \mathbf{u}^*)$ and $(\mathbf{k}, \mathbf{u})$ have the same polarization.]

Antenna Effective Length

Consider the receiving situation in Fig. 19b, where the antenna is connected to a matched meter and is illuminated by an incident plane wave F_4 defined in (86). The effective length $\mathbf{h}$ is defined by the relation

$$V_5 = \mathbf{h} \cdot \mathbf{E}_4(\mathbf{r} = 0) \tag{88}$$

where V_5 is the open-circuit voltage of the transmission line at a reference plane $z = 0$, and $\mathbf{E}_4(\mathbf{r} = 0)$ is the electric field at a reference point O of the incident plane wave given in (86). From (88), (87), and (82) we conclude that

$$\mathbf{h} = \sqrt{\frac{Z_c}{Z}} \left(1 + \frac{Z_{\text{in}}}{Z_c}\right)\left(-j\frac{\lambda}{a_1}\right) \mathbf{A}(\mathbf{k}) \tag{89}$$

where $Z = (\mu/\epsilon)^{1/2}$. Clearly $\mathbf{h}$ is a complex vector in meters. The relation in (89) relates the effective length $\mathbf{h}$ and the far-field amplitude $\mathbf{A}(\mathbf{k})$ of an antenna.

Transmitting Field in Terms of Effective Length

Excited by a unit amplitude source $(a = 1)$, the transmitted far field $\mathbf{E}_1$ is given in (85). Now if the same antenna is excited by a unit current source $(I = 1)$, the corresponding transmitted far field $\mathbf{E}_3$ is then given by

$$\mathbf{E}_3(\mathbf{r}) \sim \frac{1}{2}\sqrt{Z_c Z}\left(1 + \frac{Z_{\text{in}}}{Z_c}\right) \mathbf{A}(\mathbf{k}) \frac{e^{-jkr}}{r}, \quad r \to \infty \tag{90}$$

where we have used the formula (81) in relating $\mathbf{E}_1$ and $\mathbf{E}_3$. Replacing $\mathbf{A}$ by $\mathbf{h}$ in accordance with (89), we rewrite (90) as

Theorems and Formulas

$$E_3(r) \sim \frac{jZ}{2\lambda} \mathbf{h} \frac{e^{-jkr}}{r} \tag{91}$$

which may be used as an alternative definition of the effective length $\mathbf{h}$.

Receiving Cross Section

Consider the receiving situation in Fig. 19b, where the antenna is connected to a matched amplitude meter. The incident plane wave is given in (86), with amplitude C and state $(-\mathbf{k}, \mathbf{u}^*)$. The received wave amplitude is b_4, given in (87). Then we define the receiving cross section (effective area) σ in state $(\mathbf{k}, \mathbf{u})$ of the antenna by

$$\sigma(\mathbf{k}, \mathbf{u}) = \left|\frac{b_4}{C}\right|^2$$

$$= \frac{\text{received power of the antenna under matched condition}}{\text{power density of incident plane wave in state } (-\mathbf{k}, \mathbf{u}^*)} \tag{92}$$

in square meters. Making use of (87), we have

$$\sigma(\mathbf{k}, \mathbf{u}) = \lambda^2 \left|\frac{1}{a_1} \mathbf{A}(\mathbf{k}) \cdot \mathbf{u}^*\right|^2 \tag{93}$$

A partial gain of the antenna in state $(\mathbf{k}, \mathbf{u})$ is defined by (Section 10, Chapter 1)

$$g_1(\mathbf{k}, \mathbf{u}) = 4\pi \left|\frac{1}{a_1} \mathbf{A}(\mathbf{k}) \cdot \mathbf{u}^*\right|^2 \tag{94}$$

From (93) and (94) we have

$$g_1(\mathbf{k}, \mathbf{u}) = \frac{4\pi}{\lambda^2} \sigma(\mathbf{k}, \mathbf{u}) \tag{95}$$

where $\sigma_\lambda = \lambda^2/4\pi$ is sometimes known as the *receiving cross section of the fictitious isotropic radiator*. As discussed in Section 10, Chapter 1, the partial gain g_1 in a preferred polarization $\mathbf{u}$ is related to the (total) gain G_1 in all (both) polarizations by

$$g_1(\mathbf{k}, \mathbf{u}) = G_1(\mathbf{k}) |\mathbf{u}^* \cdot \mathbf{U}|^2 \tag{96}$$

where $\mathbf{U}$ describes the polarization of the antenna in direction $\mathbf{k}$. Substituting (96) in (95) gives

$$\sigma(\mathbf{k}, \mathbf{u}) = \frac{\lambda^2}{4\pi} |\mathbf{u}^* \cdot \mathbf{U}|^2 G_1(\mathbf{k}) \tag{97}$$

which relates the receiving cross section and the gain of an antenna. The factor

$$p = |\mathbf{u}^* \cdot \mathbf{U}|^2 \tag{98}$$

is called the *polarization efficiency* (or *polarization mismatch factor*). We emphasize that unitary vector $\mathbf{u}^*$ (not $\mathbf{u}$) describes the polarization of the incoming plane wave in direction $(-\mathbf{k})$, while $\mathbf{U}$ describes the polarization of the receiving antenna in the outgoing direction $(+\mathbf{k})$. For example, let $\mathbf{k} = \hat{\mathbf{z}}$ and $\mathbf{U} = (\hat{\mathbf{x}} - j\hat{\mathbf{y}})/\sqrt{2}$ for a right-hand circularly polarized antenna. If the incoming plane wave is also right-hand circularly polarized, we have $\mathbf{u}^* = (\hat{\mathbf{x}} + j\hat{\mathbf{y}})/\sqrt{2}$ and $p = 1$.

13. The Radar Equation and Friis Transmission Formula

The configuration of a bistatic radar is sketched in Fig. 20. The obstacle (target) in the vicinity of point 0 is illuminated by an incident wave from the transmitting antenna at point 1. A part of the scattered energy is received by a receiving antenna at point 2. Both distances R_1 and R_2 are large in terms of wavelength so that the scatterer is in the far-field zones of the antennas. The problem at hand is to determine the power ratio P_2/P_1, where

P_1 = power incident from the generator to the transmitting antenna,

P_2 = power received by the receiver via the receiving antenna.

To this end we need to define the antennas and the obstacle more precisely, as below.

(*a*) In the vicinity of point 0, the radiated field of the transmitting antenna is in

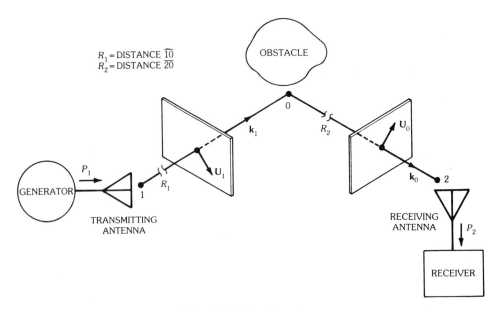

Fig. 20. A bistatic radar.

Theorems and Formulas

the direction $\mathbf{k}_1$ and has a polarization described by a unitary vector $\mathbf{U}_1$. In other words, the state of the transmitting antenna at point 0 is $(\mathbf{k}_1, \mathbf{U}_1)$. The gain of the transmitting antenna in the direction $\mathbf{k}_1$ is $G_1(\mathbf{k}_1)$, which is related to the directivity $D_1(\mathbf{k}_1)$ by

$$G_1(\mathbf{k}_1) = \eta_1(1 - |\Gamma|^2)D_1(\mathbf{k}_1) \tag{99}$$

Here η_1 is the transmitting antenna efficiency accounting for the conductor and dielectric losses, and Γ_1 is the reflection coefficient accounting for the impedance mismatch between the transmitting antenna and its feed (Section 10, Chapter 1).

(b) If the receiving antenna were used for transmitting (Fig. 21), the state of its radiated field at point 0 would be $(\mathbf{k}_2, \mathbf{U}_2)$, and its gain would be $G_2(\mathbf{k}_2)$. A relation similar to (99) holds for G_2 and D_2.

(c) Due to the illumination from the transmitting antenna the obstacle produces a scattered field at point 2, with state $(\mathbf{k}_0 = -\mathbf{k}_2, \mathbf{U}_0)$. Its bistatic cross section from state $(\mathbf{k}_1, \mathbf{U}_1)$ to state $(\mathbf{k}_0, \mathbf{U}_0)$ is denoted by the BCS.

With the above description we may now calculate various power quantities. The power density W_0 of the radiated field at point 0 from the transmitting antenna is

$$W_0 = \frac{P_1}{4\pi R_1^2} G_1(\mathbf{k}_1) \tag{100}$$

in watts per square meter. The power density W_2 of the scattered field at point 2 from the obstacle is

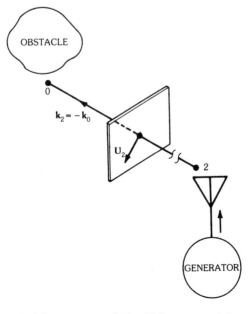

Fig. 21. The receiving antenna of Fig. 20 is now used for transmitting.

$$W_2 = W_0 \frac{\text{BCS}}{4\pi R_2^2} \tag{101}$$

The received power P_2 by the receiver at point 2 via the receiving antenna may be calculated from (92) and (97) with the result

$$P_2 = W_2 \frac{\lambda^2}{4\pi} |\mathbf{U}_0 \cdot \mathbf{U}_2|^2 G_2(\mathbf{k}_2) \tag{102}$$

Combining (100) through (102), we have the power ratio for the bistatic radar in Fig. 20, namely,

$$\frac{P_2}{P_1} = \frac{\text{BCS}}{4\pi} \left(\frac{\lambda}{4\pi R_1 R_2} \right)^2 |\mathbf{U}_0 \cdot \mathbf{U}_2|^2 G_1(\mathbf{k}_1) G_2(\mathbf{k}_2) \tag{103}$$

which is known as the *radar (range) equation*.

Next, we will consider a special case of (103). Let us remove the obstacle in Fig. 21, and study the direct power transmission from point 1 to point 2 (Fig. 22). Making use of the following relations:

$$\mathbf{k}_2 = -\mathbf{k}_1, \qquad \mathbf{U}_0 = \mathbf{U}_1, \qquad R_1 = R$$

$$W_0 = W_2, \qquad \text{BCS} = 4\pi R_2^2$$

then (103) becomes the Friis transmission formula for the far-field transmission between two antennas, namely,

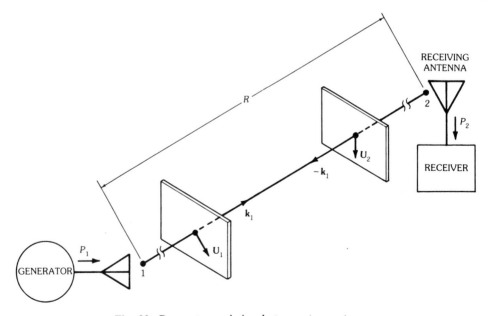

Fig. 22. Power transmission between two antennas.

Theorems and Formulas

$$\frac{P_2}{P_1} = \left(\frac{\lambda}{4\pi R}\right)^2 |\mathbf{U}_1 \cdot \mathbf{U}_2|^2 G_1(\mathbf{k}_1) G_2(-\mathbf{k}_1) \tag{104}$$

The factor $(\lambda/4\pi R)^2$ is called the *free-space loss factor*, and it accounts for the loss due to the spherical spread of the transmitted field. The polarization efficiency

$$p = |\mathbf{U}_1 \cdot \mathbf{U}_2|^2 \tag{105}$$

accounts for the loss due to polarization mismatch between the transmitting and the receiving antennas. If both antennas have the same polarization, $\mathbf{U}_1 = \mathbf{U}_2^*$ and $p = 1$. (The conjugate on $\mathbf{U}_2^*$ is due to the fact that $\mathbf{U}_2$ refers to a propagation direction opposite that of $\mathbf{U}_1$.) Note the symmetry between 1 and 2 in the right-hand side of (104), implying that the same formula applies if the roles of the receiving and transmitting antennas are interchanged in Fig. 22.

14. Noise Temperature of an Antenna

For a high-resolution antenna a crucial factor that limits its ability to detect a weak signal is the antenna noise, which is the subject of the present section.

Antenna Noise Temperature

The receiving antenna sketched in Fig. 23 receives the desired signal as well as noise. We denote the available noise power at terminal Σ by P_a in watts. It is a common practice to express P_a in terms of an (effective) antenna noise temperature T_a in kelvins via the relation

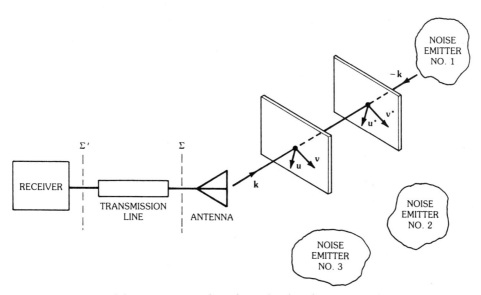

Fig. 23. A receiving antenna receives thermal noises from external noise emitters.

$$T_a = P_a/(k'\Delta f) \qquad (106)$$

where

k' = Boltzmann's constant = 1.38×10^{-23} J/K

Δf = bandwidth of the antenna receiving system in hertz

The interpretation of (106) is that antenna noise power P_a is equal to that of a matched resistor whose physical temperature is T_a. Hereafter we will use P_a and T_a interchangeably. Clearly, T_a depends on the antenna receiving characteristics and external noise emitter.

Brightness Temperature of an Emitter

Every object with its physical temperature above the absolute zero (0 K) is an emitter of thermal energy in the form of electromagnetic waves. Within a narrow frequency band the amount of energy radiated in direction $\mathbf{k}$ and polarization $\mathbf{u}$ or simply in state $(\mathbf{k}, \mathbf{u})$ is proportional to a parameter called the *brightness temperature* $T_b(\mathbf{k}, \mathbf{u})$. The latter is related to the physical temperature T_p of the emitter via the relation

$$T_b(\mathbf{k}, \mathbf{u}) = \varepsilon(\mathbf{k}, \mathbf{u}) T_p \qquad (107)$$

where the dimensionless parameter $\varepsilon(\mathbf{k}, \mathbf{u})$ is called the *emissivity in state* $(\mathbf{k}, \mathbf{u})$ of the emitter. To determine $\varepsilon(\mathbf{k}, \mathbf{u})$ we illuminate the emitter by an incident plane wave in state $(-\mathbf{k}, \mathbf{u}^*)$. See Fig. 24. Then

$$\varepsilon(\mathbf{k}, \mathbf{u}) = \frac{\text{incident power absorbed by emitter}}{\text{incident power intercepted by emitter}}$$

For a large emitter an approximate formula for calculating its emissivity is

$$\varepsilon(\mathbf{k}, \mathbf{u}) \cong 1 - |\Gamma(\mathbf{k}, \mathbf{u})|^2 \qquad (108)$$

Here $\Gamma(\mathbf{k}, \mathbf{u})$ is the reflection coefficient at the surface of the emitter. Since there are two independent polarizations we can define two independent (partial) brightness temperatures $T_b(\mathbf{k}, \mathbf{u})$ and $T_b(\mathbf{k}, \mathbf{v})$ in a given direction $\mathbf{k}$, for each choice of orthogonal, unitary, polarization vectors $(\mathbf{u}, \mathbf{v})$. A common choice of $(\mathbf{u}, \mathbf{v})$ is

$$\mathbf{u} = \text{horizontal polarization}, \quad \mathbf{v} = \text{vertical polarization} \qquad (109)$$

The sum of the two (partial) brightness temperatures gives the total brightness temperature

$$T_b(\mathbf{k}) = T_b(\mathbf{k}, \mathbf{u}) + T_b(\mathbf{k}, \mathbf{v}) \qquad (110)$$

which is proportional to the total energy radiated by the emitter in direction $\mathbf{k}$ in

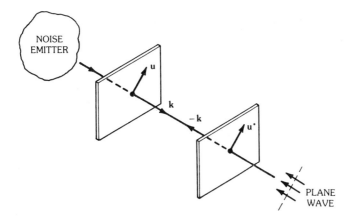

Fig. 24. To determine the emissivity $\epsilon(\mathbf{k}, \mathbf{u})$ of a noise emitter we illuminate the emitter by an incident plane wave of state $(-\mathbf{k}, \mathbf{u}^*)$.

both polarizations. A blackbody absorbs all the incoming energy impinging on it (a perfect absorber). Its emissivity is unity (a perfect emitter), and $\tfrac{1}{2} T_b(\mathbf{k}) = T_b(\mathbf{k}, \mathbf{u}) = T_b(\mathbf{k}, \mathbf{v}) = T_p$ for any direction $\mathbf{k}$.

Calculation of Antenna Noise Temperature

The power received by a receiving antenna can be traced to three sources: the desired signal, interference from other coherent radiators, and incoherent noise from noise emitters. Fig. 25 shows some important noise emitters in the free-space environment. Let us concentrate on a typical noise emitter (No. 1 in Fig. 23). Its contribution to the antenna noise temperature can be calculated from the following formula:

$$T_a = \frac{1}{4\pi} \int_0^\pi \sin\theta \, d\theta \int_0^{2\pi} d\phi \, [T_b(-\mathbf{k}, \mathbf{u}^*) d(\mathbf{k}, \mathbf{u}) + T_b(-\mathbf{k}, \mathbf{v}^*) d(\mathbf{k}, \mathbf{v})] \quad (111)$$

where

$(\mathbf{u}, \mathbf{v})$ = unitary vectors describing two orthogonal polarizations

$d(\mathbf{k}, \mathbf{u})$ = (partial) directivity of the antenna in state $(\mathbf{k}, \mathbf{u})$

$T_b(-\mathbf{k}, \mathbf{u}^*)$ = (partial) brightness temperature of the emitter in state $(-\mathbf{k}, \mathbf{u}^*)$. Note that states $(\mathbf{k}, \mathbf{u})$ and $(-\mathbf{k}, \mathbf{u}^*)$ have the opposite directions but the same polarization

If there is more than one noise emitter in space, the superposition principle applies to the calculation of T_a. This is so because noise emitters are incoherent and the superposition of powers (temperatures) is permissible. For an idealized omnidirectional antenna, which radiates equally in both polarizations, we have $d(\mathbf{k}, \mathbf{u}) = d(\mathbf{k}, \mathbf{v}) = 1/2$, and

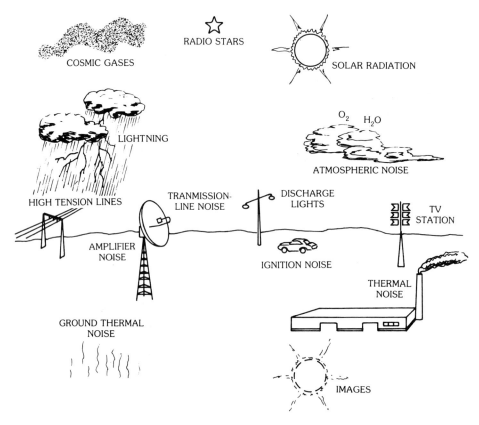

Fig. 25. Important noise emitters for a receiving antenna. (*Courtesy Y. T. Lo*)

$$T_a = T_{a0} = \frac{1}{8\pi} \int_0^\pi \sin\theta \, d\theta \int_0^{2\pi} d\phi \, [T_b(-\mathbf{k}, \mathbf{u}^*) + T_b(-\mathbf{k}, \mathbf{v}^*)] \qquad (112)$$

Fig. 26 presents some typical values of T_{a0} of an omnidirectional antenna, together with the noise temperature of a typical receiver.

Noise Power at the Receiver's Terminal

Corresponding to the antenna noise temperature T_a in (111), the available noise power at terminal Σ in Fig. 23 is given by P_a according to (106). This incoming power propagates through the transmission line, and arrives at the receiver's terminal Σ' with its value in watts equal to

$$P'_a = k' \Delta f [\eta T_a + (1 - \eta) T_0] \qquad (113)$$

where

η = power transmission efficiency between terminals Σ and Σ'
T_0 = physical temperature of the transmission line

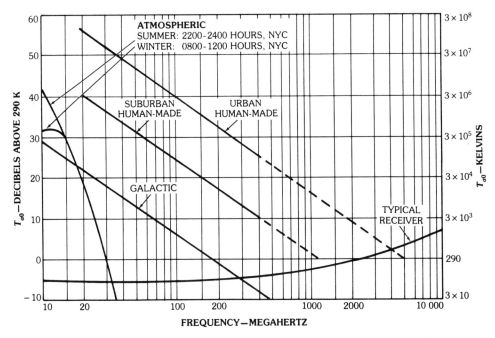

Fig. 26. Median values of average antenna noise temperature for an omnidirectional antenna near the earth's surface. (*After Sams [7]*, © *1975 Howard W. Sams & Company, Indianapolis; reprinted with permission*)

The power transmission efficiency is defined by

$$\eta = \frac{\text{output power at terminal } \Sigma'}{\text{input power at terminal } \Sigma} \tag{114}$$

It includes the power loss due to the mismatches at Σ and Σ', and conductor/dielectric losses of the transmission line.

15. References

[1] C. T. Tai, *Dyadic Green's Functions in Electromagnetic Theory*, Scranton: Intext Educational, 1971.
[2] S. W. Lee, J. Boersma, C. L. Law, and G. A. Deschamps, "Singularity in Green's function and its numerical evaluation," *IEEE Trans. Antennas Propag.*, vol. AP-28, pp. 311–317, 1980.
[3] A. D. Yaghjian, "Electric dyadic Green's functions in the source region," *Proc. IEEE*, vol. 68, pp. 248–263, 1980.
[4] G. A. Deschamps, "Scattering diagrams in electromagnetic theory," in *Electromagnetic Theory and Antennas*, Part I, ed. by E. C. Jordan, New York: Pergamon, 1963, pp. 235–251.
[5] R. D. Kodis, "A note on the theory of scattering from an irregular surface," *IEEE Trans. Antennas Propag.*, vol. AP-14, pp. 77–82, 1966.

[6] G. A. Deschamps, "I. Le principe de réciprocité en électromagnétisme. II. Application du principe de réciprocité aux antennes et aux guides d'ondes," *Revue du CETHEDEC* (Paris) N° 8–4°, pp. 71–101, 1966.
[7] Howard W. Sams & Co., *Reference Data for Radio Engineers*, 6th ed., Indianapolis: Howard W. Sams & Co., 1975, p. 29-2.

Chapter 3

Techniques for Low-Frequency Problems

A. J. Poggio
Lawrence Livermore National Laboratory

E. K. Miller
Rockwell International Science Center

CONTENTS

Introduction to Low-Frequency Techniques	3-5
Part 1. Selected Analytical Issues for Antenna Engineering	3-5
1. Theory	3-6
The Electromagnetic Field Equations for Antenna Analysis 3-6	
Integral Representations for Far Fields 3-8	
Duality 3-9	
Radiated Power 3-11	
Directive Gain, Directivity, Efficiency, and Gain 3-12	
2. Characteristics of Some Classical Antennas	3-13
The Electric Dipole Antenna 3-13	
The Sinusoidal Current Distribution 3-14	
The Traveling-Wave Antenna 3-16	
The Small Loop Antenna 3-16	
The Perfect Ground Plane 3-18	
The Rectangular-Aperture Antenna 3-21	
The Biconical Antenna 3-24	
An Antenna Reference Chart 3-26	
Imperfect Grounds 3-27	
Arrays 3-33	
3. Integral Equations in Antenna Analysis	3-36
Perfectly Conducting Wires and Bodies 3-36	

Andrew J. Poggio was born in New York City in 1941. He received a PhD in electrical engineering from the University of Illinois in 1969. His doctoral research was directed toward the numerical solution of integral equations for dipole and slot antennas. As a research associate at MBAssociates he performed research in numerical techniques and mathematical modeling in electromagnetic wave theory and was instrumental in developing computer methods for solving boundary value problems associated with electromagnetic scatterers and antennas. At the Cornell Aeronautical Laboratory he was a research engineer studying the propagation of electromagnetic waves in random media. Now a staff member of the Lawrence Livermore National Laboratory, he is at present involved in microwave and millimeter-wave analytical and experimental programs and leads the Electromagnetic Diagnostic Systems Group and the Microwave Engineering Facility.

 Edmund K. Miller was born in Milwaukee in 1935. He received a PhD in electrical engineering from the University of Michigan in 1965. At the Radiation Laboratory, and later at the High Altitude Research Laboratory, of the University of Michigan, he conducted experimental and analytical research in plasma–electromagnetic-wave interaction. In 1968 he joined MBAssociates, where he worked on integral-equation methods for modeling antennas in both the time and frequency domains. His involvement in numerical techniques continued at the Lawrence Livermore National Laboratory, which he joined in 1971 and where he served as leader of the Engineering Research Division and the Nuclear Energy Systems Division until 1985. He was Regent's Distinguished Professor of Electrical and Computer Engineering at the University of Kansas at Lawrence from 1985 to 1987 and is now Manager of Electromagnetics at Rockwell International Science Center. He has lectured widely on computer methods in electromagnetics and has taught numerous short courses on the subject. Dr. Miller has published more than 60 articles on numerical methods, their electromagnetic applications, signal processing, graphics, and related topics.

    Thin-Wire Integral Equations   3-44
    Integral Equations for Solid Bodies   3-46
    The Imperfectly Conducting Ground   3-47
    Integral Equations in the Time Domain   3-50
Part 2. Numerical Issues Involved in Integral Equations for
  Antenna Analysis   3-52
1. Introduction   3-52
2. Preliminary Discussion   3-52
3. Numerical Implementation   3-55
    The General Idea   3-55
    Frequency-Domain Method of Moments   3-57
    Time-Domain Method of Moments   3-62
    The N-Port Analogy   3-64
    Comparison of Frequency- and Time-Domain Approaches   3-65
    Benefits of Symmetry   3-65
4. Computation   3-68
    Modeling Errors   3-69
    Limitations   3-69
5. Validation   3-72
    Experimental Validation   3-72
    Analytical Validation   3-76
    Numerical Validation   3-76
6. A Guided Tour of Some Codes and Their Features   3-80
    The General Versus the Specific   3-81
    Generic Characteristics   3-81
    The Importance of the Details   3-86
    A Code Catalog   3-87
    A Closer Examination of Two Specific Codes   3-87
    Application Guidelines   3-87
7. References   3-96

Introduction to Low-Frequency Techniques

In this chapter we survey various techniques that can be used to evaluate antennas in the regime where their physical size is a maximum of a few wavelengths in extent. There are two basic approaches besides experimentation that can be considered for such problems: analytical and numerical.

Analytical procedures are the older of the two, representing the only practical way to design and characterize new antennas prior to the advent of the digital computer. These techniques are discussed in Part 1 of this chapter, beginning with the problem of finding the fields of a prescribed current for various geometries. The presentation follows a sequence of problems of increasing complexity, including expressions for radiation resistance and near and far fields. Also considered are a variety of tools that can be used to simplify or extend analytical techniques, based for example on image theory, duality, and the like.

A common denominator of these analytical techniques, with the exception of the few problems that can be solved using boundary-value formulations, is that of assuming a current distribution. That approach, adequate for a surprising range of problems, can be extended in applicability by using variational techniques. But for really general problems more elaborate numerical computations are necessary. Part 1 concludes with presentation of several integral equations suitable for such numerical modeling, specialized to the wire geometries which make up the majority of low-frequency problems.

The emphasis of Part 2 of the chapter is on numerical techniques. It extends the treatment of Part 1 to situations where the current distribution cannot be assumed, but must be solved as part of a generalized boundary-value problem. A discussion of the issues needing consideration in developing and applying numerical techniques is given. These include the formulation, numerical implementation, computation, and validation. Both frequency-domain and time-domain techniques are considered. Numerous tables are used to summarize salient points, and a number of existing computer codes are outlined to give more concrete examples of what can be accomplished.

Part 1
Selected Analytical Issues for Antenna Engineering

In this part on analysis we will cover the electromagnetic field equations used in antenna analysis, some antenna characteristics, and integral equations in antenna analysis.

1. Theory

This section treats the basic theory of electromagnetic fields as applied to the characterization of antennas.

The Electromagnetic Field Equations for Antenna Analysis

The fundamental equations for electromagnetics can be written in their time-dependent form or specialized to time-harmonic behavior as in Chapter 1, (1) through (3b). For the purposes of this chapter and the definition of antenna behavior in the frequency domain, we will confine our attention to the time-harmonic Maxwell equations given by

$$\nabla \times \mathbf{E} = -j\omega\mu\mathbf{H} - \mathbf{K} \quad (1)$$

$$\nabla \times \mathbf{H} = j\omega\varepsilon\mathbf{E} + \mathbf{J} \quad (2)$$

$$\nabla \cdot \mathbf{E} = \varrho/\epsilon \quad (3)$$

$$\nabla \cdot \mathbf{H} = \varrho_m/\mu \quad (4)$$

with the continuity equations

$$\nabla \cdot \mathbf{J} = -j\omega\varrho \quad (5)$$

$$\nabla \cdot \mathbf{K} = j\omega\varrho_m \quad (6)$$

The solution of the coupled differential equations for $\mathbf{E}$ and $\mathbf{H}$ when driven by the forcing functions $\mathbf{J}$ and $\mathbf{K}$ has been the subject of numerous books and is beyond the scope of this work. Rather, we will employ the expressions in Chapter 1—(20a) and (20b)—which are written in terms of vector potentials [1], viz., the magnetic vector potential, $\mathbf{A}(\mathbf{r})$, and the electric vector potential, $\mathbf{F}(\mathbf{r})$. These expressions are presented in Chart 1. Also shown in that chart are alternate representations obtained from direct integration of the field equations [2, 3, 4, 5, 6, 7]. In the equations, ∇' implies operations in the source or primed coordinates. The equations in Chart 1 are the general representations for the electromagnetic fields due to volumetric distributions of electric and magnetic sources in an unbounded homogeneous medium.

The Chart 1 representations are most useful when evaluating the fields due to assumed (or approximate) current distributions. While this may often be adequate, the current distributions on complex structures can be difficult to accurately predict or approximate. In such cases it may be possible to solve Maxwell's equations directly subject to appropriate boundary conditions and driving conditions for the fields and the induced surface currents. This approach is particularly useful when the antenna geometry conforms to a separable coordinate system, but since the class of antennas fitting separable coordinate systems is limited, the direct solution of Maxwell's equations (or the corresponding wave equation) has also been limited. An alternative to solving Maxwell's equations is to cast the field equations into

Chart 1. Field Equation Pairs

E Field	H Field
$\mathbf{E}(\mathbf{r}) = -\nabla \times \mathbf{F} + \dfrac{1}{j\omega\epsilon}(\nabla \times \nabla \times \mathbf{A} - \mathbf{J})$ (7a)	$\mathbf{H}(\mathbf{r}) = \nabla \times \mathbf{A} + \dfrac{1}{j\omega\mu}(\nabla \times \nabla \times \mathbf{F} - \mathbf{K})$ (8a)
$\mathbf{E}(\mathbf{r}) = -\nabla \times \mathbf{F} + \dfrac{1}{j\omega\epsilon}(\nabla\nabla\cdot + k^2)\mathbf{A}$ (7b)	$\mathbf{H}(\mathbf{r}) = \nabla \times \mathbf{A} + \dfrac{1}{j\omega\mu}(\nabla\nabla\cdot + k^2)\mathbf{F}$ (8b)

$$\mathbf{A}(\mathbf{r}) = \frac{1}{4\pi}\int_V \mathbf{J}(\mathbf{r}')\varphi(\mathbf{r},\mathbf{r}')\,dv' \quad (9)$$

$$\mathbf{F}(\mathbf{r}) = \frac{1}{4\pi}\int_V \mathbf{K}(\mathbf{r}')\varphi(\mathbf{r},\mathbf{r}')\,dv' \quad (10)$$

$$\varphi(\mathbf{r},\mathbf{r}') = \frac{e^{-jk|\mathbf{r}-\mathbf{r}'|}}{|\mathbf{r}-\mathbf{r}'|}$$

$$\mathbf{E}(\mathbf{r}) = -\frac{1}{4\pi}\int_V \left(j\omega\mu\mathbf{J}\varphi + \mathbf{K}\times\nabla'\varphi + \frac{\nabla'\cdot\mathbf{J}}{j\omega\epsilon}\nabla'\varphi\right)dv' \quad (11)$$

$$\mathbf{H}(\mathbf{r}) = \frac{1}{4\pi}\int_V \left(-j\omega\epsilon\varphi + \mathbf{J}\times\nabla'\varphi + \frac{\nabla'\cdot\mathbf{K}}{j\omega\mu}\nabla'\varphi\right)dv' \quad (12)$$

$$\mathbf{E}(\mathbf{r}) = -\frac{1}{4\pi}\int_V (j\omega\mu\mathbf{J}\varphi + \mathbf{K}\times\nabla'\varphi)\,dv' + \frac{1}{4\pi j\omega\epsilon}\nabla\int_V \nabla'\cdot\mathbf{J}\varphi\,dv'$$

$$\mathbf{H}(\mathbf{r}) = \frac{1}{4\pi}\int_V (-j\omega\epsilon\mathbf{K}\varphi + \mathbf{J}\times\nabla'\varphi)\,dv' + \frac{1}{4\pi j\omega\mu}\nabla\int_V \nabla'\cdot\mathbf{K}\varphi\,dv'$$

$$\mathbf{E}(\mathbf{r}) = \frac{1}{4\pi j\omega\epsilon}\int_V (\mathbf{J}\cdot(k^2\bar{\bar{\mathbf{I}}} + \nabla\nabla)\varphi - j\omega\epsilon\mathbf{K}\times\nabla'\varphi)\,dv'$$

$$\mathbf{H}(\mathbf{r}') = \frac{1}{4\pi j\omega\mu}\int_V (\mathbf{K}\cdot(k^2\bar{\bar{\mathbf{I}}} + \nabla\nabla)\varphi + j\omega\mu\mathbf{J}\times\nabla'\varphi)\,dv'$$

$$\bar{\bar{\mathbf{I}}} = \text{unit dyad}$$

integral equations in which the induced source distributions in the form of **J** and **K** on the conducting surface are the unknowns driven by the exciting source. The radiated fields are then computed using equations of the form given in Chart 1. This particular approach, which has found significant usage in recent times, will be discussed in a later section.

The complexity of the solution processes discussed above is justified for general geometries since the current distributions or the fields which result from the computations are consistent, i.e., they satisfy Maxwell's equations with the appropriate boundary and driving conditions. As a result a great deal of confidence can be placed in the results of the computation. On the other hand, for geometries where the induced distributions are reasonably well known so that confidence in the results can be maintained, the fields can be directly evaluated. Fig. 1 attempts to capture the essence of the above discussion.

Integral Representations for Far Fields

The evaluation of fields can progress directly once the current distribution over the radiating structure is known. Using (7) through (10) in Chart 1, it is rather straightforward though sometimes demanding to perform this computation. The integral representations are a complete and precise description of the field based on

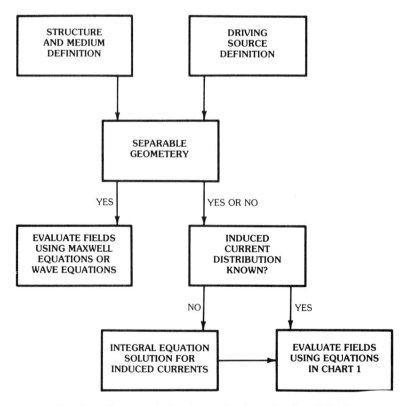

Fig. 1. A flow graph for the evaluation of radiated fields.

Techniques for Low-Frequency Problems

the source distributions. In keeping with the discussion of Chapter 1, Section 6, it is often convenient to compute only the portion of the complete field which dominates over a particular region of space. In the following we focus our attention on the far field since the development of expressions for the near-zone fields is most conveniently pursued for specific structures.

The fields at large distances from source distributions can be written in simplified forms by investigation of the vector potential in (9) and (10). In the far zone $|\mathbf{r}| \gg |\mathbf{r}'|$ and $|kr| \gg 1$, so that

$$|\mathbf{r} - \mathbf{r}'| = \to r - \mathbf{r} \cdot \mathbf{r}'/r = r - \hat{\mathbf{r}} \cdot \mathbf{r}' \text{ and}$$

$$\varphi(\mathbf{r}, \mathbf{r}') = \frac{e^{-jk|\mathbf{r}-\mathbf{r}'|}}{|\mathbf{r} - \mathbf{r}'|} \to \frac{e^{-jkr}}{r} e^{jk\hat{\mathbf{r}} \cdot \mathbf{r}'} \qquad (13)$$

$$\nabla'\varphi = \nabla' \frac{e^{-jk|\mathbf{r}-\mathbf{r}'|}}{|\mathbf{r} - \mathbf{r}'|} \to \hat{\mathbf{r}} jk \frac{e^{-jkr}}{r} e^{jk\hat{\mathbf{r}} \cdot \mathbf{r}'}$$

Hence the far-zone representations of the equations in Chart 1 can be written by using (13) and letting the ∇ operator be expressed as $\nabla \to -jk\hat{\mathbf{r}}$. The equations for the far-zone fields (14 through 21) are presented in Chart 2. Note that the equations are related through the far-zone plane-wave relation

$$\mathbf{H}(\mathbf{r}) = \frac{1}{Z_0} \hat{\mathbf{r}} \times \mathbf{E}(\mathbf{r}), \qquad Z_0 = \sqrt{\mu/\epsilon} \qquad (22)$$

and furthermore that the far-zone fields satisfy the conditions imposed in Chapter 1, Section 6, namely, that $\mathbf{E} \times \mathbf{H}^*$ be real and that the fields decay as r^{-1}.

Duality

The similarity in the form of Maxwell's equations for $\mathbf{E}(\mathbf{r})$ and $\mathbf{H}(\mathbf{r})$ allows them to be referred to as dual equations. Two examples of this duality are captured in Chart 3, where the indicated changes will lead to a set of Maxwell's equations for the dual case which are identical with those for the original [1]. Thus, given the fields that satisfy Maxwell's equations with the sources $[\mathbf{J}(\mathbf{r}), \mathbf{K}(\mathbf{r})]$ and constitutive parameters (ϵ, μ), the solution for the fields satisfying Maxwell's equations with the sources $[\mathbf{K}'(\mathbf{r}) = \mathbf{J}(\mathbf{r}), \mathbf{J}'(\mathbf{r}) = -\mathbf{K}'(\mathbf{r})]$ and constitutive parameters $(\epsilon' = \mu, \mu' = \epsilon)$ will be given by $[\mathbf{E}'(\mathbf{r}) = -\mathbf{H}(\mathbf{r}), \mathbf{H}'(\mathbf{r}) = \mathbf{E}(\mathbf{r})]$. Note that in this case we have dealt with dual problems having "dual" conditions achieved by interchanging μ and ϵ. Also shown in Chart 3 is a case where the medium remains unchanged. Thus the solution to a problem in the dual space can be related to a solution of a problem in the original space with the same constitutive parameters by the definitions in example 2 of Chart 3. The usefulness of the "duality" approach will be illustrated later for the small loop antenna.

It should be noted that the described duality relationships, while being the most widely used, are not exhaustive.

Chart 2. Some Convenient Expressions for Far-Zone Fields

E Field	H Field
	$\mathbf{F} = \dfrac{1}{4\pi}\dfrac{e^{-jkr}}{r}\displaystyle\int_V \mathbf{K}(\mathbf{r'})\, e^{jk\hat{\mathbf{r}}\cdot\mathbf{r'}}\,dv' \quad (14)$
	$\mathbf{A} = \dfrac{1}{4\pi}\dfrac{e^{-jkr}}{r}\displaystyle\int_V \mathbf{J}(\mathbf{r'})\, e^{jk\hat{\mathbf{r}}\cdot\mathbf{r'}}\,dv' \quad (15)$
$\mathbf{E}(\mathbf{r}) = jk\hat{\mathbf{r}} \times \mathbf{F} + j\omega\mu\hat{\mathbf{r}} \times \hat{\mathbf{r}} \times \mathbf{A} \quad (16)$	$\mathbf{H}(\mathbf{r}) = -jk\hat{\mathbf{r}} \times \mathbf{A} + j\omega\epsilon\hat{\mathbf{r}} \times \hat{\mathbf{r}} \times \mathbf{F} \quad (17)$
$\mathbf{E}(\mathbf{r}) = -\dfrac{1}{4\pi}\dfrac{e^{-jkr}}{r}\displaystyle\int_V (j\omega\mu\mathbf{J} + jk\mathbf{K}\times\hat{\mathbf{r}} + \sqrt{\mu/\epsilon}\,\nabla'\cdot\mathbf{J}\hat{\mathbf{r}})\, e^{jk\hat{\mathbf{r}}\cdot\mathbf{r'}}\,dv' \quad (18)$	$\mathbf{H}(\mathbf{r}) = \dfrac{1}{4\pi}\dfrac{e^{-jkr}}{r}\displaystyle\int_V (-j\omega\epsilon\mathbf{K} + jk\mathbf{J}\times\hat{\mathbf{r}} - \sqrt{\epsilon/\mu}\,\nabla'\cdot\mathbf{K}\hat{\mathbf{r}})\, e^{jk\hat{\mathbf{r}}\cdot\mathbf{r'}}\,dv' \quad (19)$
$\mathbf{E}(\mathbf{r}) = \dfrac{j\omega\mu\, e^{-jkr}}{4\pi\, r}\displaystyle\int_V [(\hat{\mathbf{r}}\cdot\mathbf{J})\hat{\mathbf{r}} - \mathbf{J} - \sqrt{\epsilon/\mu}\,\mathbf{K}\times\hat{\mathbf{r}}]\, e^{jk\hat{\mathbf{r}}\cdot\mathbf{r'}}\,dv' \quad (20)$	$\mathbf{H}(\mathbf{r}) = \dfrac{j\omega\epsilon\, e^{-jkr}}{4\pi\, r}\displaystyle\int_V [(\hat{\mathbf{r}}\cdot\mathbf{K})\hat{\mathbf{r}} + \mathbf{K} - \sqrt{\mu/\epsilon}\,\mathbf{J}\times\hat{\mathbf{r}}]\, e^{jk\hat{\mathbf{r}}\cdot\mathbf{r'}}\,dv' \quad (21)$

Chart 3. Examples of Duality Relationships

	Initial Quantities	Dual Quantities
Example 1	$E(r)$ $H(r)$ $J(r)$ $K(r)$ $\mu(r)$ $\epsilon(r)$	$H(r)$ $-E(r)$ $K(r)$ $-J(r)$ $\epsilon(r)$ $\mu(r)$
Example 2	$E(r)$ $H(r)$ $J(r)$ $K(r)$ $\mu(r)$ $\epsilon(r)$	$\sqrt{\mu/\epsilon}\, H(r)$ $-\sqrt{\epsilon/\mu}\, E(r)$ $\sqrt{\epsilon/\mu}\, K(r)$ $-\sqrt{\mu/\epsilon}\, J(r)$ $\mu(r)$ $\epsilon(r)$

Radiated Power

The time-average power exiting a region of space or equivalently crossing the bounding surface can be found by integrating the complex Poynting vector (power flux density) over the surface. Thus

$$P = \text{Re}\left\{ \oiint \mathbf{E} \times \mathbf{H}^* \cdot d\mathbf{s} \right\} \tag{23}$$

where $d\mathbf{s}$ has a unit normal in the outward direction and Re denotes the real part.

In a similar manner the time-average power supplied by sources is given by

$$P_{\text{sources}} = \text{Re}\left\{ \int_V (\mathbf{E} \cdot \mathbf{J}^* + \mathbf{H}^* \cdot \mathbf{K}) \, dv \right\} \tag{24}$$

The power P_r radiated by an antenna is defined with (23) evaluated using the far-zone fields. Similarly, it can be evaluated using (23) and then taking its limiting value as the observation point $\mathbf{r}$ approaches infinity.

The time-average power balance for an antenna is given by

$$P_r + P_{\text{losses}} = P_{\text{sources}} \tag{25}$$

Thus the power supplied by sources is either radiated or dissipated through loss mechanisms in the antenna or the medium.

For a lossless antenna composed of perfect conductors, the losses vanish so that the input power is equal to the radiated power. Over a small antenna source region, which we can treat in an analogous network manner,

$$\text{Re}\left\{ \int_{\substack{\text{source} \\ \text{region}}} \mathbf{E} \cdot \mathbf{J}^* \, dv \right\} = \text{Re}\{VI^*\} \tag{26}$$

Thus the antenna time-average input power can lead to the evaluation of antenna radiation resistance R_r:

$$\text{Re}\{VI^*\} = \frac{1}{R_{in}}|I_{in}|^2 \tag{27a}$$

and

$$R_r = \frac{1}{|I_{in}|^2}\text{Re}\left\{\oiint (\mathbf{E} \times \mathbf{H}^*) \cdot d\mathbf{s}\right\} \tag{27b}$$

where I_{in} is the current at input terminals.

Directive Gain, Directivity, Efficiency, and Gain

A number of useful quantities are defined for the characterization of antennas. These are dealt with in Chapter 1 and in several texts [3, 8, 9, 10, 11, 12, 13].

The *directive gain* of an antenna $G_D(\theta, \varphi)$ is defined as the ratio of real power flux density in the far zone in a given direction (θ, φ) to its average value over the entire radiation sphere. Thus

$$G_D(\theta, \varphi) = \frac{\text{Re}\{(\mathbf{E} \times \mathbf{H}^*) \cdot \hat{\mathbf{r}}\}}{\lim_{r \to \infty}(1/4\pi r^2)\oiint (\mathbf{E} \times \mathbf{H}^*) \cdot d\mathbf{s}} = \frac{\text{Re}\{(\mathbf{E} \times \mathbf{H}^*) \cdot \hat{\mathbf{r}}\}}{P_r/4\pi r^2} \tag{28}$$

The quantity $P_r/4\pi r^2$ is equivalent to the power density of an isotropic antenna radiating the power P_r. The *directivity* D of an antenna is defined as the maximum value of the directive gain or simply

$$D = \max\{G_D(\theta, \varphi)\} \tag{29}$$

For lossy antennas, i.e., antennas for which the total input power at the antenna terminals is not radiated, additional quantities of interest are defined, namely, radiation efficiency η and gain $G(\theta, \varphi)$. The ratio of power P_r radiated to total input power P_{in} is the *radiation efficiency* η. Thus η can be related to the radiation resistance through the radiated power and, through the total input power P_{in}, to the total input resistance composed of the radiation resistance and resistances accounting for the other losses R_L [9]. Thus

$$\eta = \frac{P_r}{P_{in}} = \frac{R_r}{R_r + R_L} \tag{30}$$

The *gain* of an antenna $G(\theta, \varphi)$ is defined in a manner similar to directive gain but with the inclusion of losses. It is the ratio of the real power flux density in the far zone in a given direction to the power flux density which would be obtained if the input power were radiated isotropically. Thus

Techniques for Low-Frequency Problems

$$G(\theta, \varphi) = \frac{\text{Re}\{(\mathbf{E} \times \mathbf{H}^*) \cdot \hat{\mathbf{r}}\}}{P_{\text{in}}/4\pi r^2} = \eta G_D(\theta, \varphi) \tag{31}$$

The descriptions above can be related to the concepts in Fig. 14 in Chapter 1 by the substitution of P_{in} for P_1 and P_r for P_3. The consequences of input impedance mismatch to a feed line have not been included but can be readily taken into account as in Chapter 1 and Fig. 14 of Chapter 1.

2. Characteristics of Some Classical Antennas

In the following we will tabulate the characteristics of several simple antennas which are computed using the equations established in the preceding section. The examples illustrate how, with a certain amount of approximation (e.g., an assumed current distribution), one can obtain realistic estimates of antenna characteristics. In subsequent sections we will deal with more complicated structures and more precise techniques.

The Electric Dipole Antenna

The linear dipole antenna has received treatment in a very large number of works on antennas. Here we will consider the dipole antenna in a progression of models, from the very short to the capacitively loaded and then to the somewhat longer dipole, for each of which we can estimate a realistic current distribution.

The Point Source—The electric point-current source (sometimes called a Hertzian source), while possessing little realism, permits an individual to conceptualize the radiation due to electric currents. In a coordinate system as shown in Chart 4 the point source is defined as

$$\mathbf{J}(\mathbf{r}) = \hat{\mathbf{z}} \delta(\mathbf{r}) \tag{32}$$

where $\delta(\mathbf{r})$ is the Dirac delta function. The utility of this current source is that all other sources can be constructed as a superposition of these infinitesimal elements. Using (7), (8), (9), (16), (17), and (23) we can construct the relevant characteristics for the electric point-current source given in Chart 4. From the expressions for $\mathbf{E}(\mathbf{r})$ and $\mathbf{H}(\mathbf{r})$ one can identify [3] the far-zone terms behaving as $1/r$ and the near-zone terms composed of the "static" portion behaving as $1/r^3$, similar to that of a static dipole distribution, and the "induction" field behaving as $1/r^2$.

The Short Dipole Antenna—The short electric dipole antenna [3, 8, 11, 13, 14, 15] is of interest since it represents the behavior of dipole antennas at low frequencies where $L/\lambda \ll 1$. Two electric current distributions are of interest for the short electric dipole driven at its center. These distributions and the relevant antenna characteristics are presented in Chart 5. The first case, possessing a current distribution similar to the unit current element, i.e., a constant current over its length, can be constructed in reality by capacitively loading a short dipole with end caps. The second is the triangular distribution with maximum current at its center

Chart 4. Characteristics for the Electric Point-Current Source

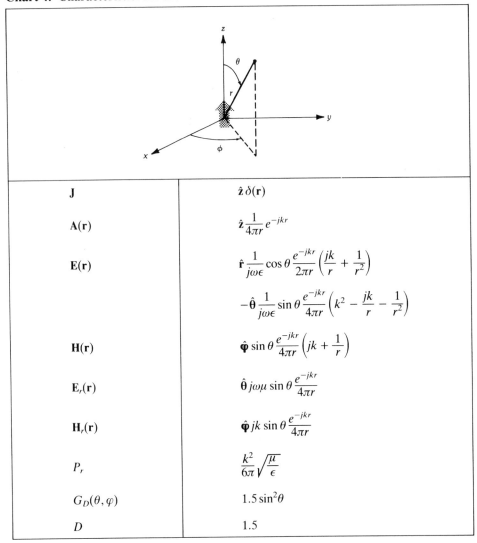

J	$\hat{z}\,\delta(\mathbf{r})$
A(r)	$\hat{z}\dfrac{1}{4\pi r}e^{-jkr}$
E(r)	$\hat{\mathbf{r}}\dfrac{1}{j\omega\epsilon}\cos\theta\,\dfrac{e^{-jkr}}{2\pi r}\left(\dfrac{jk}{r}+\dfrac{1}{r^2}\right)$
	$-\hat{\boldsymbol{\theta}}\dfrac{1}{j\omega\epsilon}\sin\theta\,\dfrac{e^{-jkr}}{4\pi r}\left(k^2-\dfrac{jk}{r}-\dfrac{1}{r^2}\right)$
H(r)	$\hat{\boldsymbol{\varphi}}\sin\theta\,\dfrac{e^{-jkr}}{4\pi r}\left(jk+\dfrac{1}{r}\right)$
$E_r(\mathbf{r})$	$\hat{\boldsymbol{\theta}}\,j\omega\mu\sin\theta\,\dfrac{e^{-jkr}}{4\pi r}$
$H_r(\mathbf{r})$	$\hat{\boldsymbol{\varphi}}\,jk\sin\theta\,\dfrac{e^{-jkr}}{4\pi r}$
P_r	$\dfrac{k^2}{6\pi}\sqrt{\dfrac{\mu}{\epsilon}}$
$G_D(\theta,\varphi)$	$1.5\sin^2\theta$
D	1.5

and zero current at its ends. This current realistically approximates the current on a very short center-driven dipole.

The Sinusoidal Current Distribution

Current distribution measurements have indicated that for thin cylindrical antennas of diameter less than approximately $\lambda/100$, the current distribution can be well approximated by a sinusoid [3, 9, 10, 11, 12, 14, 15]. Theoretical analyses have supported these observations. Thus the sinusoidal current distribution whose characteristics are presented in Chart 6 occupies an important place in antenna

Chart 5. Characteristics for the Short Electric Dipole Antenna

	Constant Current	Triangular Current
$I(z)$	$I_0, \quad -L/2 \leqq z \leqq L/2$	$I_0\left(1 - \dfrac{|z|}{L/2}\right), \quad -L/2 \leqq z \leqq L/2$
$A(z)$	$(I_0 L) \dfrac{e^{-jkr}}{4\pi r} \dfrac{\sin[(kL/2)\cos\theta]}{(kL/2)\cos\theta}$	$\dfrac{(I_0 L)}{2} \dfrac{e^{-jkr}}{4\pi r} \dfrac{\sin^2[(kL/4)\cos\theta]}{[(kL/4)\cos\theta]^2}$
$\mathbf{E}_r(\mathbf{r})$	$\hat{\boldsymbol{\theta}} j\omega\mu (I_0 L) \sin\theta \dfrac{e^{-jkr}}{4\pi r} \dfrac{\sin[(kL/2)\cos\theta]}{(kL/2)\cos\theta}$	$\hat{\boldsymbol{\theta}} j\omega\mu \dfrac{(I_0 L)}{2} \sin\theta \dfrac{e^{-jkr}}{4\pi r} \dfrac{\sin^2[(kL/4)\cos\theta]}{[(kL/4)\cos\theta]^2}$
$\mathbf{H}_r(\mathbf{r})$	$\hat{\boldsymbol{\phi}} jk (I_0 L) \sin\theta \dfrac{e^{-jkr}}{4\pi r} \dfrac{\sin[(kL/2)\cos\theta]}{(kL/2)\cos\theta}$	$\hat{\boldsymbol{\phi}} jk \dfrac{(I_0 L)}{2} \sin\theta \dfrac{e^{-jkr}}{4\pi r} \dfrac{\sin^2[(kL/4)\cos\theta]}{[(kL/4)\cos\theta]^2}$
P_r	$\dfrac{k^2}{6\pi}\sqrt{\dfrac{\mu}{\epsilon}}(I_0 L)^2 = \dfrac{2\pi}{3}\sqrt{\dfrac{\mu}{\epsilon}} I_0^2 \left(\dfrac{L}{\lambda}\right)^2$	$\dfrac{k^2}{6\pi}\sqrt{\dfrac{\mu}{\epsilon}}\left(\dfrac{I_0 L}{2}\right)^2 = \dfrac{\pi}{6}\sqrt{\dfrac{\mu}{\epsilon}} I_0^2 \left(\dfrac{L}{\lambda}\right)^2$
$G_D(\theta, \phi)$	$1.5 \sin^2\theta$	$1.5 \sin^2\theta$
D	1.5	1.5
R_r	$\dfrac{2\pi}{3}\sqrt{\dfrac{\mu}{\epsilon}}\left(\dfrac{L}{\lambda}\right)^2$	$\dfrac{\pi}{6}\sqrt{\dfrac{\mu}{\epsilon}}\left(\dfrac{L}{\lambda}\right)^2$

Far zone applies to $\mathbf{E}_r(\mathbf{r})$ and $\mathbf{H}_r(\mathbf{r})$.

*$kL \ll 1$ so that $\sin(kL\cos\theta)/kL\cos\theta \to 1$.

analysis. In Chart 6, results are presented for an arbitrary-length dipole and for a half-wavelength dipole ($L = \lambda/2$).

The Traveling-Wave Antenna

The traveling-wave antenna [8, 10, 13, 15, 16] is defined as an antenna whose conductors support a current distribution described by a simple traveling wave. In comparison to the dipole antenna, which supports two counter-directed traveling waves giving rise to the sinusoidal (standing-wave) distribution, the traveling-wave antenna supports a single wave. A simple example of a traveling-wave antenna and its attendant characteristics are presented in Chart 7. For this antenna the current distribution is given by

$$I(z) = I_0 e^{-jkz}, \qquad 0 \leq z \leq L$$

where I_0 is its amplitude and k the free-space wave number.

The important feature of the traveling-wave antenna is that the maximum radiation is neither end-fire ($\theta = 0$) nor broadside ($\theta = \pi/2$). Rather it occurs at an intermediate angle and creates a single major conical lobe independent of antenna length and tilted in the direction of wave travel with minor lobes determined by length. An approximation for the angle of the main lobe from the end-fire direction [13, 16] is given by $\theta_m = \cos^{-1}[1 - 0.371/(L/\lambda)]$.

The Small Loop Antenna

The small loop antenna [8, 9, 10, 11, 13, 14, 15] of radius a, carrying a uniform current I, can be analyzed using the techniques presented previously. The characteristics for the small loop are presented in Chart 8.

The striking similarity between the expressions in Chart 8 for the small loop and those for the short dipole in Chart 4 is immediately evident. Clearly, the electric current I on a dipole of length L generates fields which are obviously related to those of a loop of radius a carrying an electric current I. Since the fields radiated by such a loop are identical with those radiated by a magnetic dipole [1] if

$$mL = j\omega\mu I(\pi a^2)$$

where m is the magnetic current, we can obtain the fields due to a small current loop from those of a short electric dipole using duality. The procedure is as follows:

(a) The original problem is a linear electric current I of length L in a medium with constitutive parameters ϵ and μ. Obtain the solution to this problem.

(b) By the duality of Chart 3, construct the solution of the problem of a linear magnetic current m and length L in a medium with constitutive parameters ϵ and μ. This is obtained from the solution in (a) above by setting $I_1 = \sqrt{\epsilon/\mu}\, m = jkI(\pi a^2)/L$, $\mathbf{E}_1(\mathbf{r}) = \sqrt{\mu/\epsilon}\,\mathbf{H}_2(\mathbf{r})$, and $\mathbf{H}_1(\mathbf{r}) = -\sqrt{\epsilon/\mu}\,\mathbf{E}_2(\mathbf{r})$. Note that the magnetic current is spatially orthogonal to the plane of the loop.

(c) Replace mL by $j\omega\mu I(\pi a^2)$ to provide the solution for the small loop radiating element.

Techniques for Low-Frequency Problems

Chart 6. Characteristics for a Dipole with Sinusoidal Current Distribution

[Diagram: sinusoidal current distribution $I(z)$ on a dipole from $-L/2$ to $L/2$ along z-axis]

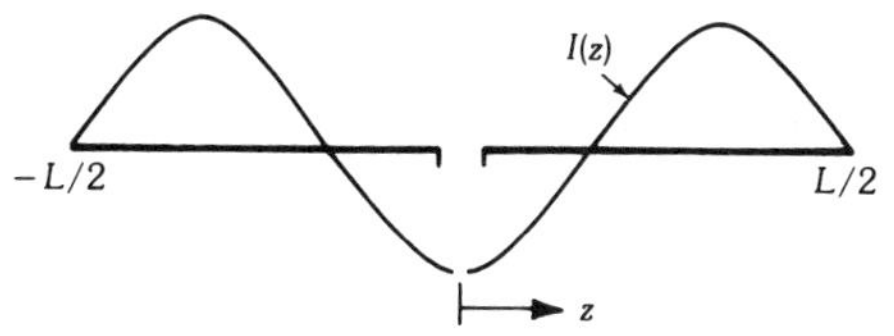

$I(z)$	$I_0 \sin k_0(L/2 - |z|)$, $\quad -L/2 \leq z \leq L/2$
	Arbitrary-Length Dipole
$A_z(\mathbf{r})$	$I_0 \dfrac{e^{-jkr}}{2\pi r} \dfrac{\cos[(kL/2)\cos\theta] - \cos(kL/2)}{k \sin^2\theta}$
$\mathbf{E}_r(\mathbf{r})$	$\hat{\boldsymbol{\theta}} j \sqrt{\dfrac{\mu}{\epsilon}} I_0 \dfrac{e^{-jkr}}{2\pi r} \dfrac{\cos[(kL/2)\cos\theta] - \cos(kL/2)}{\sin\theta}$ far zone
$\mathbf{H}_r(\mathbf{r})$	$\hat{\boldsymbol{\phi}} j I_0 \dfrac{e^{-jkr}}{2\pi r} \dfrac{\cos[(kL/2)\cos\theta] - \cos(kL/2)}{\sin\theta}$
$P_r[11,12]$	$\sqrt{\dfrac{\mu}{\epsilon}} I_0^2 \dfrac{1}{2\pi} \Big\{ \sin kL \, [Si(kL) - \tfrac{1}{2}Si(2kL)]$ $\quad + (1 + \cos kL)[\ln(kL\gamma) - Ci(kL)]$ $\quad - \dfrac{\cos kL}{2}[\ln(2kL\gamma) - Ci(2kL)] \Big\}$
	where $\quad Si(x) = \displaystyle\int_0^x \dfrac{\sin x}{x} dx, \quad Ci(x) = -\displaystyle\int_x^\infty \dfrac{\cos x}{x} dx, \quad \ln\gamma = 0.5772$
	Half-Wavelength Dipole ($L = \lambda/2$)
$A_z(\mathbf{r})$	$I_0 \dfrac{e^{-jkr}}{2\pi r} \dfrac{\cos[(\pi/2)\cos\theta]}{k \sin^2\theta}$
$\mathbf{E}_r(\mathbf{r})$	$\hat{\boldsymbol{\theta}} j \sqrt{\dfrac{\mu}{\epsilon}} I_0 \dfrac{e^{-jkr}}{2\pi r} \dfrac{\cos[(\pi/2)\cos\theta]}{\sin\theta}$
P_r	$\sqrt{\dfrac{\mu}{\epsilon}} I_0^2 \dfrac{1}{4\pi} [\ln(2\pi\gamma) - Ci(2\pi)] = 0.194 \sqrt{\mu/\epsilon}\, I_0^2$
R_r	$\sqrt{\dfrac{\mu}{\epsilon}} \dfrac{1}{4\pi} [\ln(2\pi\gamma) - Ci(2\pi)] = 0.194\sqrt{\mu/\epsilon}\,(R_r = 73\,\Omega$ in free space$)$
$G_D(\theta,\phi)$	$\dfrac{4\cos^2[(\pi/2)\cos\theta]}{\sin^2\theta\,[\ln(2\pi\gamma) - Ci(2\pi)]} = 1.644 \dfrac{\cos^2[(\pi/2)\cos\theta]}{\sin^2\theta}$
D	$\dfrac{4}{\ln(2\pi\gamma) - Ci(2\pi)} = 1.644$

Chart 7. Characteristics for the Traveling-Wave Antenna

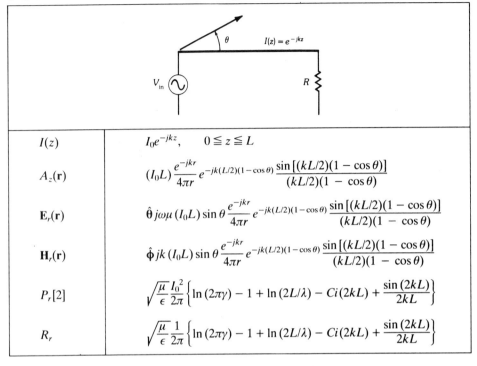

$I(z)$	$I_0 e^{-jkz}, \quad 0 \leq z \leq L$
$A_z(\mathbf{r})$	$(I_0 L)\dfrac{e^{-jkr}}{4\pi r} e^{-jk(L/2)(1-\cos\theta)} \dfrac{\sin[(kL/2)(1-\cos\theta)]}{(kL/2)(1-\cos\theta)}$
$\mathbf{E}_r(\mathbf{r})$	$\hat{\boldsymbol{\theta}} j\omega\mu (I_0 L) \sin\theta \dfrac{e^{-jkr}}{4\pi r} e^{-jk(L/2)(1-\cos\theta)} \dfrac{\sin[(kL/2)(1-\cos\theta)]}{(kL/2)(1-\cos\theta)}$
$\mathbf{H}_r(\mathbf{r})$	$\hat{\boldsymbol{\phi}} jk (I_0 L) \sin\theta \dfrac{e^{-jkr}}{4\pi r} e^{-jk(L/2)(1-\cos\theta)} \dfrac{\sin[(kL/2)(1-\cos\theta)]}{(kL/2)(1-\cos\theta)}$
P_r [2]	$\sqrt{\dfrac{\mu}{\epsilon}} \dfrac{I_0^2}{2\pi} \left\{ \ln(2\pi\gamma) - 1 + \ln(2L/\lambda) - Ci(2kL) + \dfrac{\sin(2kL)}{2kL} \right\}$
R_r	$\sqrt{\dfrac{\mu}{\epsilon}} \dfrac{1}{2\pi} \left\{ \ln(2\pi\gamma) - 1 + \ln(2L/\lambda) - Ci(2kL) + \dfrac{\sin(2kL)}{2kL} \right\}$

The Perfect Ground Plane

The perfect ground plane, i.e., the planar, perfectly conducting surface with conductivity $\sigma \to \infty$, is often found in the environment of antenna structures. Since the previous characterizations have been for antennas in homogeneous media with real ϵ and μ, the introduction of conducting bodies such as infinite planes will require a modification.

The boundary conditions at any perfectly conducting surface require that the tangential electric field satisfy, with $\hat{\mathbf{n}}(\mathbf{r})$ the outward normal to the surface and $\mathbf{r}$ on the surface,

$$\hat{\mathbf{n}}(\mathbf{r}) \times \mathbf{E}(\mathbf{r}) = 0 \qquad (33)$$

and that the magnetic field satisfy

$$\hat{\mathbf{n}}(\mathbf{r}) \cdot \mathbf{H}(\mathbf{r}) = 0 \qquad (34)$$

The exclusion of fields from the interior of the surface leads to the generation of surface charge and current densities given by

$$\hat{\mathbf{n}}(\mathbf{r}) \cdot \mathbf{E}(\mathbf{r}) = \varrho_s(\mathbf{r})/\epsilon$$
$$\hat{\mathbf{n}}(\mathbf{r}) \times \mathbf{H}(\mathbf{r}) = \mathbf{J}_s(\mathbf{r}) \qquad (35)$$

Chart 8. Characteristics of the Small Loop Antenna

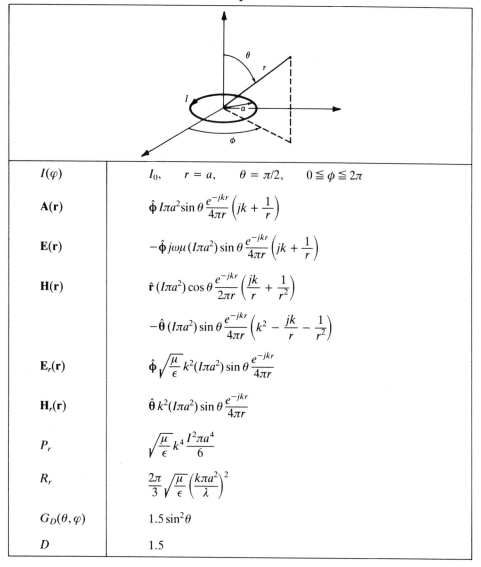

$I(\varphi)$	$I_0, \quad r = a, \quad \theta = \pi/2, \quad 0 \leq \varphi \leq 2\pi$
$\mathbf{A}(\mathbf{r})$	$\hat{\boldsymbol{\varphi}} \, I\pi a^2 \sin\theta \dfrac{e^{-jkr}}{4\pi r} \left(jk + \dfrac{1}{r}\right)$
$\mathbf{E}(\mathbf{r})$	$-\hat{\boldsymbol{\varphi}} \, j\omega\mu (I\pi a^2) \sin\theta \dfrac{e^{-jkr}}{4\pi r} \left(jk + \dfrac{1}{r}\right)$
$\mathbf{H}(\mathbf{r})$	$\hat{\mathbf{r}} (I\pi a^2) \cos\theta \dfrac{e^{-jkr}}{2\pi r} \left(\dfrac{jk}{r} + \dfrac{1}{r^2}\right)$
	$-\hat{\boldsymbol{\theta}} (I\pi a^2) \sin\theta \dfrac{e^{-jkr}}{4\pi r} \left(k^2 - \dfrac{jk}{r} - \dfrac{1}{r^2}\right)$
$\mathbf{E}_r(\mathbf{r})$	$\hat{\boldsymbol{\varphi}} \sqrt{\dfrac{\mu}{\epsilon}} k^2 (I\pi a^2) \sin\theta \dfrac{e^{-jkr}}{4\pi r}$
$\mathbf{H}_r(\mathbf{r})$	$\hat{\boldsymbol{\theta}} \, k^2 (I\pi a^2) \sin\theta \dfrac{e^{-jkr}}{4\pi r}$
P_r	$\sqrt{\dfrac{\mu}{\epsilon}} k^4 \dfrac{I^2 \pi a^4}{6}$
R_r	$\dfrac{2\pi}{3} \sqrt{\dfrac{\mu}{\epsilon}} \left(\dfrac{k\pi a^2}{\lambda}\right)^2$
$G_D(\theta,\varphi)$	$1.5 \sin^2\theta$
D	1.5

An arbitrarily oriented antenna located above a perfectly conducting ground plane must generate a vanishing electric field at the surface in order to satisfy the boundary condition on the electric field. Such a condition can be satisfied by creating an equivalent problem where the ground plane is removed and an image source is introduced to produce, when combined with the original source, a vanishing tangential electric field at the location of the plane. The appropriate images for electric and magnetic currents are shown in Fig. 2. A close inspection will show that the tangential electric and normal magnetic fields vanish at the plane surface and that the fields are identical in the upper half-space.

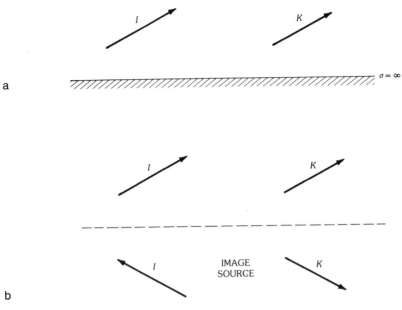

Fig. 2. Equivalent source distributions for sources above perfectly conducting ground planes. (*a*) Original problem. (*b*) Equivalent problem.

The theory of images can be applied to configurations other than infinite planes. In fact, a source in the presence of any surface composed of intersecting planes can be represented in terms of the source and multiple images [1,8]. This concept applies to both exterior and interior (e.g., waveguide) problems.

An example of the use of images is provided by the monopole antenna driven against a perfectly conducting ground plane. The fields radiated by the monopole into the upper half-space are identical with those of the dipole. The radiated power is contained only in the upper half-space and thus is only half of that for the dipole antenna. Hence we can immediately infer that the input voltage to the monopole need only be one-half that of the dipole to produce the same fields in the upper half-space. Thus

$$P_r = \sqrt{\frac{\mu}{\epsilon}} I_0^2 \frac{1}{4\pi} \{[\sin(kL)][Si(kL) - \tfrac{1}{2} Si(2kL)]$$

$$+ [1 - \cos(kL)][\ln(kL\gamma) - Ci(kL)] - [\cos(kL)][\ln(2kL\gamma) - Ci(2kL)]\}$$

The input resistance is given by $R_r = P_r/I_0^2$, and for a quarter-wavelength monopole

$$R_r = \sqrt{\frac{\mu}{\epsilon}} \frac{1}{8\pi} \{\ln(2\pi\gamma) - Ci(2\pi)\} = 36.5 \ \Omega$$

The Rectangular-Aperture Antenna

The rectangular-aperture antenna [1, 3, 8, 9, 10, 11, 13, 15], shown in Fig. 3, has many features in common with the electric dipole. In the following we will illustrate the use of equivalence and images to provide a convenient mechanism for evaluating its fields. In each case the electric-field distribution in the aperture will be assumed to be known. Then the fields produced by the aperture can be obtained using equivalence, images, and the field representations presented earlier.

For an assumed field distribution in the aperture in a perfectly conducting plane, the pictorial representation of Fig. 4a is appropriate. Using equivalence, the situation in Fig. 4b applies and the fields in the half-space $z > 0$ are identical with those in the original problem but zero for $z < 0$. In this figure $\mathbf{J}_a = \hat{\mathbf{z}} \times \mathbf{H}$ and $\mathbf{K}_a = -\hat{\mathbf{z}} \times \mathbf{E}$. As a result of the vanishing fields for $z < 0$, the perfectly conducting plane is completed through the aperture as shown in Fig. 4c. Using images, the plane can be removed and the situation in Fig. 4d holds. Note that the fields are the correct fields for the half-space $z > 0$ and not for $z < 0$.

The fields radiated by the aperture distribution can be evaluated using equations in Charts 1 and 2 with only a magnetic current source or an equivalent electric vector potential. In the following we tabulate the expressions necessary to construct the radiation fields and certain characteristics for some commonly encountered aperture distributions defined in general as

$$\mathbf{E}^a(x, y, z) = \hat{\mathbf{y}} E^a(x, y) \delta(z) \qquad -L_x/2 < x < L_x/2, \quad -L_y/2 < y < L_y/2 \quad (36)$$

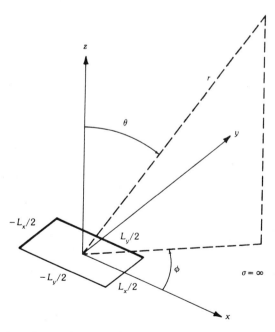

Fig. 3. The rectangular aperture or slot antenna in a perfectly conducting plane.

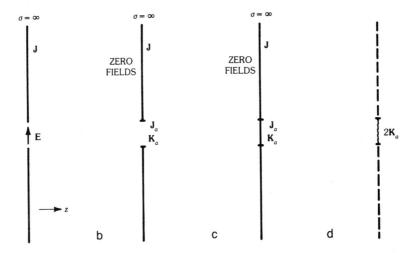

Fig. 4. Equivalent source distributions and their environments for fields in the right half-space. (*a*) Aperture field distribution in a perfectly conducting plane. (*b*) Using equivalence with (*a*). (*c*) Perfectly conducting plane is completed through aperture. (*d*) Using images to (*c*).

In the far zone the electric vector potential is given by

$$\mathbf{F} = \hat{\mathbf{x}} \frac{1}{2\pi} \frac{e^{-jkr}}{r} \int_{-\infty}^{\infty} dz'\, \delta(z') \int_{-L_y/2}^{L_y/2} dy'\, e^{jky' \sin\theta \sin\varphi} \int_{-L_x/2}^{L_x/2} dx'\, E^a(x', y')\, e^{jkx' \sin\theta \cos\varphi} \tag{37}$$

and the radiated fields by

$$\mathbf{E}_r(\mathbf{r}) = jk\hat{\mathbf{r}} \times \mathbf{F} = jk|\mathbf{F}|\hat{\mathbf{e}}(\theta, \varphi) \tag{38}$$

$$\mathbf{H}_r(\mathbf{r}) = \sqrt{\epsilon/\mu}\,\hat{\mathbf{r}} \times \mathbf{E}_r(\mathbf{r}) = j\omega\epsilon|\mathbf{F}|\hat{\mathbf{h}}(\theta, \varphi) \tag{39}$$

where

$$\hat{\mathbf{e}}(\theta, \varphi) = \hat{\boldsymbol{\varphi}} \cos\theta \cos\varphi + \hat{\boldsymbol{\theta}} \sin\varphi$$
$$\hat{\mathbf{h}}(\theta, \varphi) = \hat{\mathbf{r}} \times \hat{\mathbf{e}}(\theta, \varphi) = \hat{\boldsymbol{\varphi}} \sin\varphi - \hat{\boldsymbol{\theta}} \cos\theta \cos\varphi$$

Since the far-zone fields are simply related to the far-zone electric vector potential, only the electric vector potential will be presented in the following charts.

In Chart 9 the far-zone electric vector potential is presented for various lengths and widths of rectangular apertures with uniform excitation. Also presented are some radiation characteristics for the short, thin-slot antenna excited by a uniform transverse electric field. The results presented for the short, thin slot are consistent with the relationship between the impedances of complementary structures [13], i.e.,

Techniques for Low-Frequency Problems

Chart 9. Far-Zone Characteristics for Apertures with Uniform Aperture Distribution

---

Uniform Aperture Distribution

$$\mathbf{E}^a(x,y,z) = \hat{\mathbf{y}} E_0 \delta(z), \quad -L_x/2 < x < L_x/2, \quad -L_y/2 < y < L_y/2$$

Aribtrary Length: L_x, L_y

$$\mathbf{F}(\mathbf{r}) = \hat{\mathbf{x}} \frac{e^{-jkr}}{2\pi r} E_0 L_x L_y \frac{\sin[(kL_x/2)\sin\theta\cos\varphi]}{(kL_x/2)\sin\theta\cos\varphi} \frac{\sin[(kL_y/2)\sin\theta\sin\varphi]}{(kL_y/2)\sin\theta\sin\varphi}$$

Thin Slot: $kL_y \ll 1$, L_x Arbitrary

$$E_0 L_y \to V_0$$

$$\mathbf{F}(\mathbf{r}) = \hat{\mathbf{x}} \frac{e^{-jkr}}{2\pi r} V_0 L_x \frac{\sin[(kL_x/2)\sin\theta\cos\varphi]}{(kL_x/2)\sin\theta\cos\varphi}$$

Short, Thin Slot: $kL_x \ll 1$, $kL_y \ll 1$

$$E_0 L_y \to V_0$$

$$\mathbf{F}(\mathbf{r}) = \hat{\mathbf{x}} \frac{e^{-jkr}}{2\pi r} V_0 L_x$$

$$\mathbf{E}(\mathbf{r}) = jk V_0 L_x \frac{e^{-jkr}}{2\pi r} \hat{\mathbf{e}}(\theta,\varphi)$$

$$\mathbf{H}(\mathbf{r}) = j\omega\epsilon V_0 L_x \frac{e^{-jkr}}{2\pi r} \hat{\mathbf{h}}(\theta,\varphi)$$

$$P_r = \frac{8\pi}{3} \sqrt{\frac{\epsilon}{\mu}} V_0 \left(\frac{L_x}{\lambda}\right)^2$$

$$R_r = V_0^2/P_r = \left[\frac{8\pi}{3}\sqrt{\frac{\epsilon}{\mu}}\left(\frac{L_x}{\lambda}\right)^2\right]^{-1}$$

$$G_D(\theta,\varphi) = \frac{3}{2}(1 - \cos^2\varphi \sin^2\theta), \quad D = 1.5$$

$$Z_S Z_D = \frac{1}{4}\left(\frac{\mu}{\epsilon}\right) \tag{40}$$

where

Z_S = the slot impedance (Chart 9)

Z_D = the dipole impedance (Chart 5)

The far-zone representations for the electric vector potentials for triangular and sinusoidal aperture distributions are presented in Chart 10.

Chart 10. Far-Zone Vector Potentials for Triangular and Sinusoidal Aperture Distributions

Triangular Aperture Distribution

$$\mathbf{E}^a(x,y,z) = \hat{y} E_0 (1 - 2|x|/L_x)\delta(z), \quad -L_x/2 < x < L_x/2, \quad -L_y/2 < y < L_y/2$$

$$\mathbf{F}(\mathbf{r}) = \hat{x} \frac{e^{-jkr}}{4\pi r} E_0 L_x L_y \frac{\sin[(kL_y/2)\sin\theta\sin\varphi]}{(kL_y/2)\sin\theta\sin\varphi} \frac{\sin^2[(kL_x/2)\sin\theta\cos\varphi]}{[(kL_x/4)\sin\theta\cos\varphi]^2}$$

$$\mathbf{E}_r(\mathbf{r}) = jk|\mathbf{F}|\hat{e}(\theta,\varphi), \quad \mathbf{H}_r(\mathbf{r}) = j\omega\epsilon|\mathbf{F}|\hat{h}(\theta,\varphi)$$

Sinusoidal Aperture Distribution

$$\mathbf{E}^a(x,y,z) = \hat{y} E_0 \sin[k(L_x/2 - |x|)], \quad -L_x/2 < x < L_x/2, \quad -L_y/2 < x < L_y/2$$

$$\mathbf{F}(\mathbf{r}) = \hat{x} \frac{e^{-jkr}}{2\pi r} E_0 L_x L_y \frac{\sin[(kL_y/2)\sin\theta\sin\varphi]}{(kL_y/2)\sin\theta\sin\varphi} \frac{\cos[(kL_x/2)\sin\theta\sin\varphi] - \cos(kL_x/2)}{(kL_x/2)(1 - \sin^2\theta\sin^2\varphi)}$$

$$\mathbf{E}_r(\mathbf{r}) = jk|\mathbf{F}|\hat{e}(\theta,\varphi), \quad \mathbf{H}_r(\mathbf{r}) = j\omega\epsilon|\mathbf{F}|\hat{h}(\theta,\varphi)$$

The Biconical Antenna

The solution for antenna characteristics is simplified when the antenna surface coincides with coordinate surfaces. While there are only a few such coordinate systems and furthermore only one of finite dimensionality, the spheroidal, the method of separation of variables has been used to study some specific antennas. By the "perturbation" of the separable geometries, such as the spheroidal and conical shapes, other shapes such as the cylindrical antenna have been investigated.

The infinite biconical antenna shown in Fig. 5 has been studied extensively [4, 5, 8, 9, 11, 13, 14, 15, 17]. Such a structure coincides with spherical coordinate surfaces and serves as a guide for spherical waves. By solving the wave equation in the spherical coordinate system, the components of the assumed TEM wave are

$$H_\varphi = \frac{1}{r\sin\theta} H_0 e^{-jkr} \tag{41}$$

$$E_\theta = \sqrt{\mu/\epsilon}\, H_\varphi \tag{42}$$

The voltage and current for the biconical transmission line are

$$V(r) = \int_{\theta_c}^{\pi-\theta_c} E_\theta r\, d\theta = 2\sqrt{\mu/\epsilon}\, H_0 e^{-jkr} \ln[\cot(\theta_c/2)] \tag{43}$$

$$I(r) = \int_0^{2\pi} H_\varphi r\sin\theta\, d\varphi = 2\pi H_0 e^{-jkr} \tag{44}$$

The characteristic impedance of the biconical transmission line is then

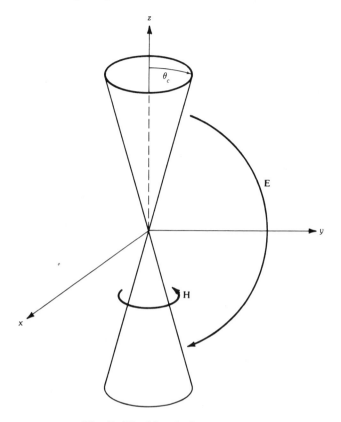

Fig. 5. The biconical antenna.

$$Z_0 = \frac{V(r)}{I(r)} = \frac{\sqrt{\mu/\epsilon}}{\pi} \ln\left[\cot\left(\theta_c/2\right)\right] \qquad (45)$$

The radiated power is

$$P_r = 4\pi\sqrt{\mu/\epsilon}\, H_0^2 \ln\left[\cot\left(\theta_c/2\right)\right] \qquad (46)$$

and the directive gain is

$$G_D = \frac{1}{\sin^2\theta \ln\left[\cot\left(\theta_c/2\right)\right]} \qquad (47)$$

Note that because of the TEM fields and the structure's infinite length there is no frequency dependence exhibited in the preceding quantities.

A realistic biconical antenna is not infinite in length as required by the model created above. For the practical antenna of finite length ($r = L$), the space is divided into two regions. One is the modal or TEM region containing the bounded wave descriptions ($r < L$), and the other is the radiation region ($r > L$). In the modal region outgoing and reflected TEM and higher-order modes can exist, while in the radiation region only the outgoing higher-order modes exist.

Using the transmission line analogy, Schelkunoff [4] has computed an input impedance

$$Z_{in} = Z_0 \frac{Z_L + jZ_0 \tan(kL)}{Z_0 + jZ_L \tan(kL)} \tag{48}$$

where Z_L is a load impedance at the end of the biconical line representing the transition region of the antenna defined as

$$Z_L = Z_0^2/Z_m \tag{49}$$

where $Z_m = R_m + jX_m$. Then, for the thin cone, with $Ci(x)$ and $Si(x)$ defined in Chart 6,

$$\begin{aligned} R_m &= 60Ci(2kL) + 30[0.577 + \ln(kL) - 2Ci(kL) + Ci(4kL)]\cos(2kL) \\ &\quad + 30[Si(4kL) - 2Si(2kL)]\sin(2kL) \\ X_m &= 60Si(2kL) + 30[Ci(4kL) - \ln(kL) - 0.577]\sin(2kL) \\ &\quad - 30Si(4kL)\cos(2kL) \end{aligned} \tag{50}$$

A detailed analysis of the radiation properties of conical structures has been pursued and presented in many works [4, 5, 11, 17, 18, 19, 20]. A formal solution is presented in "Electromagnetic radiation from conical structures" by J. R. Wait in [11] and by Schelkunoff and Friis in [5], where expansions for the fields in the two regions are obtained and a formalism for determining the unknown coefficients is established. The limiting cases of a spherical antenna ($\theta_c \to \pi/2$) and the thin-wire antenna ($\theta_c \to 0$) are considered as well.

Other antennas which have received theoretical attention because of their coincidence with coordinate surfaces are the spherical [8, 16] and spheroidal [2, 4, 8, 15] antennas.

An Antenna Reference Chart

A number of works have characterized many antenna systems. The work represented in Jasik [10], *Antenna Engineering Handbook*, is monumental and extremely useful. Here we will augment the approximate analysis methods useful for simple antennas presented earlier with an antenna reference chart useful for qualitative and approximate quantitative analysis [21].

The antenna reference chart is duplicated with minor corrections in Fig. 6. The chart includes the antenna name, some physical characteristics such as size, a diagram with coordinate system, its resistance at the lowest resonance frequency, the half-power bandwidth in percent, the antenna gain in decibels relative to a half-wavelength dipole and isotropic source, the polarization characteristic of the antenna, and the principal-plane radiation patterns characterizing the antenna. In

Techniques for Low-Frequency Problems

the text, mention is made of another valuable reference on antenna characteristics which has, as well, a handy section of tabulated antenna characteristics [22].

Imperfect Grounds

The introduction of a ground plane, such as the earth, with realistic electrical parameters in the vicinity of an antenna can modify the antenna characteristics. The current distribution is affected through near-field interactions between the antenna and ground, and the radiated field is modified by the altered antenna currents and the ground reflection of the radiation field. The former effect, where near-field interactions perturb the current distribution, is considered in a following section. The latter effect can be included in the solution process by using plane-wave reflection techniques where the radiated fields are constructed from a direct and a ground-reflected wave [9, 13, 23].

The use of image and reflection coefficients is rather straightforward, being only somewhat more involved than that for the perfect ground plane as exhibited previously. The approach uses the image as induced in a perfect ground plane but with a modified strength which is proportional to the appropriate Fresnel plane-wave reflection coefficient. The relevant geometry for an electric current source and ground with parameters ϵ, μ_0, and σ is shown in Fig. 7.

The reflection of the incident **E** and **H** fields at the interface will depend on the polarization of the field with respect to the plane of incidence (the plane containing the surface normal and the propagation vector for the wave). Thus the Fresnel reflection coefficients, which are strictly true for an infinite plane-wave field, will exhibit this dependency on polarization.

The two cases of interest are illustrated in Fig. 8, where the wave with **E(r)** in the plane of incidence is termed vertically polarized and **E(r)** normal to the plane of incidence as horizontally polarized.

The Fresnel reflection coefficients for the vertically and horizontally polarized waves are

$$R_v = \frac{\epsilon' \cos\theta - \sqrt{\epsilon' - \sin^2\theta}}{\epsilon' \cos\theta + \sqrt{\epsilon' - \sin^2\theta}} \tag{51}$$

and

$$R_h = \frac{\cos\theta - \sqrt{\epsilon' - \sin^2\theta}}{\cos\theta + \sqrt{\epsilon' - \sin^2\theta}} \tag{52}$$

where

$$\cos\theta = -\hat{\mathbf{k}}\cdot\hat{\mathbf{z}} = \hat{\mathbf{k}}_R\cdot\hat{\mathbf{z}}$$
$$\epsilon = \frac{\epsilon}{\epsilon_0} - j\frac{\sigma}{\omega\epsilon_0} \tag{53}$$

The determination of the radiated field $\mathbf{E}(\theta, \varphi)$ for an arbitrarily oriented antenna over a finitely conducting ground can be readily computed. First, the

Type	Configuration	Impedance: Resistive at f_r, $R\,(\Omega)$	−3-dB Bandwidth Percent	Gain: dB above Isotropic	Gain: dB above Dipole	Polarization	Pattern Number
Isotropic radiator (theoretical)		—	—	0	−2.14	none	A
Small dipole $L > \lambda/2$		—	—	1.74	−0.4	h	B
Thin dipole $L = \lambda/2$ $L/D = 276$		60	34	2.14	0	h	B
Thick dipole $L = \lambda/2$ $L/D = 51$		49	55	2.14	0	h	B
Cylindrical dipole $L = \lambda/2$ $L/D = 10$		37	100	2.14	0	h	B
Cylindrical dipole $L = \lambda$ $L/D = 9.6$		150	130	3.64	1.5	h	B

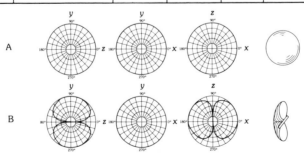

Fig. 6. An antenna chart. (*After Salati [21]*)

Techniques for Low-Frequency Problems

Type	Configuration	Impedance: Resistive at f_r, $R\,(\Omega)$	−3-dB Bandwidth Percent	Gain: dB above Isotropic	Gain: dB above Dipole	Polarization	Pattern Number
Folded dipole $L = \lambda/4$ $L/d = 13$		6000	5	1.64	−0.5	h	B
Folded dipole $L = \lambda/2$ $L/d = 25.5$		300	45	2.14	0	h	B
Biconical $L = \lambda/2$		72	100	2.14	0	h	B
Biconical $L = \lambda$		350	200	2.14	0	h	B
Turnstile $L = \lambda/2$ $L/d = 25.5$		150	50	−0.86	−3	h	C
Folded dipole over reflecting sheet $L = \lambda/2$ $L/d = 25.5$ $\lambda/8$ above sheet		150	20	7.14	5	h	D

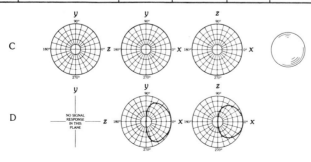

Fig. 6, *continued.*

Type	Configuration	Impedance: Resistive at f_r, $R(\Omega)$	−3-dB Bandwidth Percent	Gain: dB above Isotropic	Gain: dB above Dipole	Polarization	Pattern Number
Dipole over small ground plane $L = \lambda/4$ $L/D = 53$ $\ell = 2\lambda$		28	40	2.14	0	v	B
Folded unipole over small ground plane $L = \lambda/4$ $L/D = 53$ $\ell = 2\lambda$ $L/d = 13$		150	45	2.14	0	v	B
Coaxial dipole $L = \lambda/4$ $L/D = 40$		50	16	2.14	0	v	B
Biconical coaxial dipole $L = \lambda/2$ $d = \lambda/8$ $D = 3\lambda/8$		72	200	2.14	0	v	B
Disc-cone or rod disc-cone $L = \lambda/4$ $\ell = \lambda$		50	300	2.14	0	v	B
Biconical horn $L = 9\lambda/2$ $D = 14\lambda$		20	25	14.14	12	v	B

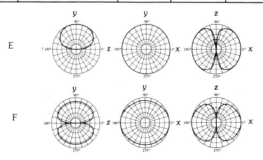

Fig. 6, *continued.*

Techniques for Low-Frequency Problems

Type	Configuration	Impedance: Resistive at f_r, R (Ω)	−3-dB Bandwidth Percent	Gain: dB above Isotropic	Gain: dB above Dipole	Polarization	Pattern Number
Slot in large ground plane $L = \lambda/2$ $l/d = 29$		350	70	2.14	0	h	E
Vertical full-wave loop $D = \lambda/\pi$ $D/d = 36$		45	13	3.14	1	h	F
Helical over reflector screen, tube 6λ long coiled into 6 turns $\lambda/4$ apart		130	200	10.14	8	circ.	G
Rhombic $L = 9\lambda$ $l = 9\lambda/2$		600	100	16.74	14.5	h	H
Parabolic with folded dipole feed ($\lambda/2$) $D = 5\lambda/2$		300	30	14.74	12.5	h	H
Horn, coaxial feed $L = 3\lambda$ $l = 3\lambda$		50	35	15.14	13	h	H

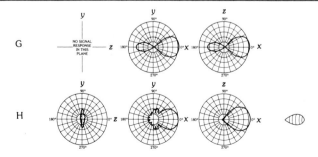

Fig. 6, *continued.*

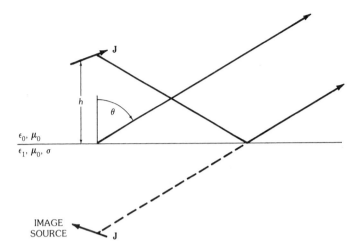

Fig. 7. Images for imperfect ground analysis.

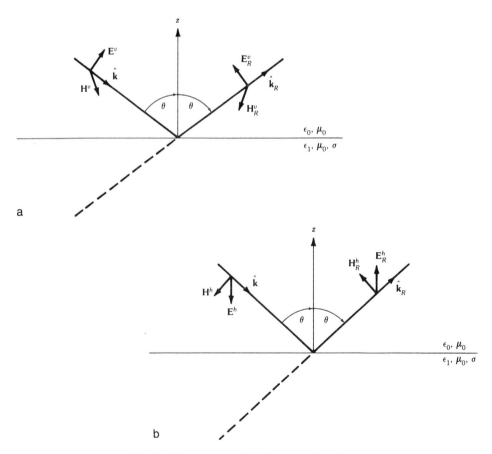

Fig. 8. Plane-wave reflection at an interface.

radiated field $\mathbf{E}_r(\theta, \varphi)$ due to the given antenna is computed for the case when it is located in an infinite homogeneous medium at its original location. Then the reflected field $\mathbf{F}_R(\theta, \varphi)$ is calculated from

$$\mathbf{E}_R(\theta, \varphi) = R_v \mathbf{E}_I(\theta, \varphi) + (R_h - R_v)[\mathbf{E}_I(\theta, \varphi) \cdot \hat{\mathbf{p}}]\hat{\mathbf{p}} \quad (54)$$

where $\mathbf{E}_I(\theta, \varphi)$ is the field due to the image of the original source in a perfect ground plane and $\hat{\mathbf{p}}$ is the unit vector normal to the plane of incidence ($\mathbf{p} = \hat{\mathbf{z}} \times \hat{\mathbf{k}}$). Finally, the total field $\mathbf{E}(\theta, \varphi)$ is the sum of the two contributions:

$$\mathbf{E}(\theta, \varphi) = \mathbf{E}_r(\theta, \varphi) + \mathbf{E}_R(\theta, \varphi) \quad (55)$$

Chart 11 illustrates the steps and the results of a procedure for evaluating the fields of a vertical and horizontal short dipole carrying a constant current I_0. The procedure for analyzing an arbitrary antenna, though more involved, can proceed in a similar manner.

The Fresnel reflection coefficients are strictly valid only for plane waves and are therefore not rigorously valid for antennas near ground planes [13, 23, 24, 25].

Arrays

The subject of antenna arrays has been extensively documented and will be discussed in Chapters 13, 14, and 17. Previous work is described in the writings of Bach and Hansen [26], Kraus [14], Jasik [10], Stutzman and Thiele [13], Weeks [9], and Ma [27]. Here we will merely show the relationship of some aspects of array analysis to previously discussed subjects in this chapter.

In the following we will present expressions which are useful for expressing the fields due to an array of identical antenna elements, each having an identical current distribution differing by, at most, a complex scaling constant. The similarity of current distribution implies identical orientation of the elements. The field due to an electric current source $\mathbf{J}(\mathbf{r})$ can be evaluated using the equations in Charts 1 and 2.

Some elements in an antenna array are shown in Fig. 9. Using element number 1 as the reference element in an N-element array we establish our coordinate system definitions so that a linear shift of the nth element will cause it to be coincident with element 1. In this case such a shift between coincident points is $\mathbf{r}_n - \mathbf{r}_1$ or $\mathbf{r}'_n - \mathbf{r}'_1$. The relationship between the element current distributions is given by

$$J_r(\mathbf{r}'_n) = a_n \mathbf{J}_1(\mathbf{r}'_1) \quad (56)$$

The magnetic vector potential in the far zone for the nth element is

$$\mathbf{A}_n(\mathbf{r}) = a_n e^{j k \hat{\mathbf{r}} \cdot (\mathbf{r}_n - \mathbf{r}_1)} \frac{1}{4\pi} \frac{e^{-jkr}}{r} \int_V \mathbf{J}_1(\mathbf{r}') e^{jk\hat{\mathbf{r}} \cdot \mathbf{r}'} dv' \quad (57)$$

or equivalently,

Chart 11. Evaluation of Field Components for a Short Dipole of Constant Current in Presence of a Finitely Conducting Ground

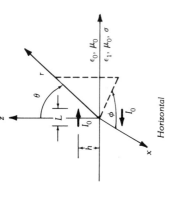

Vertical

$$\mathbf{E}_r(\theta, \varphi) = \hat{\boldsymbol{\theta}} j\omega\mu_0 I_0 L \sin\theta \frac{e^{-jk(r-h\cos\theta)}}{4\pi r}$$

Step 1. The field, due to a primary source in free space, is evaluated:

$$A_y = I_0 L \frac{e^{-jk(r-h\cos\theta)}}{4\pi r}$$

$$\mathbf{E}_r(\theta, \varphi) = j\omega\mu_0 \hat{\mathbf{r}} \times \mathbf{A}$$

$$= j\omega\mu_0 I_0 L \frac{e^{-jk(r-h\cos\theta)}}{4\pi r} (-\cos\varphi \, \hat{\boldsymbol{\varphi}} - \cos\theta \sin\varphi \, \hat{\boldsymbol{\theta}})$$

Step 2. The field, due to the image source, is evaluated:

$$\mathbf{E}_I(\theta, \varphi) = \hat{\boldsymbol{\theta}} j\omega\mu_0 I_0 L \sin\theta \frac{e^{-jk(r+h\cos\theta)}}{4\pi r}$$

$$\mathbf{E}_I(\theta, \varphi) = j\omega\mu_0 I_0 L \frac{e^{-jk(r+h\cos\theta)}}{4\pi r} (\cos\varphi \, \hat{\boldsymbol{\varphi}} + \cos\theta \sin\varphi \, \hat{\boldsymbol{\theta}})$$

Step 3. Evaluate the ground-reflected field using (54):

$$\mathbf{E}_R(\theta, \varphi) = R_v \mathbf{E}_I(\theta, \varphi)$$

$$\mathbf{E}_R(\theta, \varphi) = R_v \mathbf{E}_I(\theta, \varphi) + (R_h - R_v)(\mathbf{E}_I(\theta, \varphi) \cdot \hat{\boldsymbol{\varphi}})\hat{\boldsymbol{\varphi}}$$

$$= j\omega\mu_0 I_0 L \frac{e^{-jk(r+h\cos\theta)}}{4\pi r} (-R_h \cos\varphi \, \hat{\boldsymbol{\varphi}} + R_v \cos\theta \sin\varphi \, \hat{\boldsymbol{\theta}})$$

Step 4. Evaluate the total field using (55):

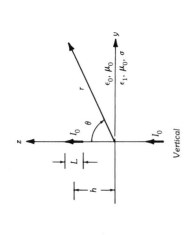

Horizontal

$$\mathbf{E}(\theta, \varphi) = \hat{\boldsymbol{\theta}} j\omega\mu_0 I_0 L \sin\theta \frac{e^{-jkr}}{4\pi r} (e^{jkh\cos\theta} + R_v e^{-jkh\cos\theta})$$

$$\mathbf{E}(\theta, \varphi) = j\omega\mu_0 I_0 L \frac{e^{-jkr}}{4\pi r} [-\hat{\boldsymbol{\varphi}} \cos\theta (e^{jkh\cos\theta} + R_h e^{-jkh\cos\theta})$$

$$- \hat{\boldsymbol{\theta}} \cos\theta \sin\varphi (e^{jkh\cos\theta} - R_v e^{-jkh\cos\theta})]$$

Techniques for Low-Frequency Problems

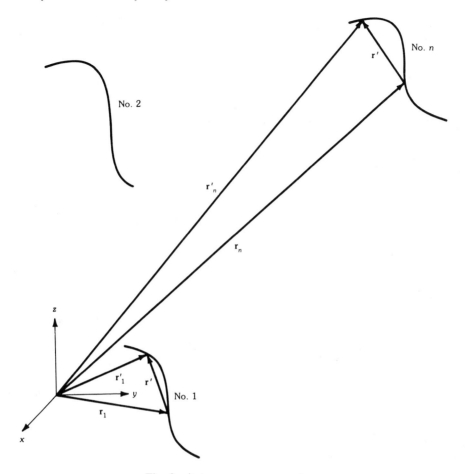

Fig. 9. Antenna array geometry.

$$\mathbf{A}_n(\mathbf{r}) = a_n e^{jk\hat{\mathbf{r}}\cdot(\mathbf{r}_n - \mathbf{r}_1)} \mathbf{A}_1(\mathbf{r}) \tag{58}$$

The far-zone magnetic vector potential for the array is then

$$\mathbf{A}(\mathbf{r}) = \mathbf{A}_1(\mathbf{r}) \sum_{n=1}^{N} a_n e^{jk\hat{\mathbf{r}}\cdot(\mathbf{r}_n - \mathbf{r}_1)} \tag{59}$$

where the summation, denoted by $f(\theta, \varphi)$, is referred to as the *array factor*:

$$f(\theta, \varphi) = \sum_{n=1}^{N} a_n e^{jk\hat{\mathbf{r}}\cdot(\mathbf{r}_n - \mathbf{r}_1)} \tag{60}$$

This factorization separates the element contribution or element pattern $\mathbf{A}_1(\mathbf{r})$ from the array pattern $f(\theta, \varphi)$, which depends only on the relative source strengths and locations. This factoring is involved in a procedure referred to as *pattern multiplication*.

The electric and magnetic fields in the far zone can then be calculated using (16) and (17) in Chart 2:

$$\mathbf{E}(\mathbf{r}) = j\omega\mu \, f(\theta,\varphi) \, \hat{\mathbf{r}} \times \hat{\mathbf{r}} \times \mathbf{A}(\mathbf{r})$$
$$\mathbf{H}(\mathbf{r}) = -jk \, f(\theta,\varphi) \, \hat{\mathbf{r}} \times \mathbf{A}(\mathbf{r}) \tag{61}$$

Chart 12 presents the radiation fields for N-element arrays of uniform current (I_0) and equally phased dipoles of spacing d. The computations have been performed using the preceding equations for two cases, namely, arrays of collinear elements and parallel elements. The procedure used in deriving the directive gain for short elements ($kL \ll 1$) follows that in [26]. Results obtained using the expressions for directive gain at broadside ($\theta = \pi/2$, $\varphi = \pi/2$) in decibels relative to a single isotropic source are presented versus element spacing d in Fig. 10 for a collinear array and in Fig. 11 for the parallel array. The number of elements N is a parameter.

Some reference data for practical arrays is also presented in Fig. 12 [21].

In the previous discussion the current distribution on each element was assumed to be known and unperturbed by the presence of the other elements in the array. For adequate interelement spacing this is a valid assumption, but for close spacing the interaction of the elements can cause perturbations of the elemental current distributions and, of course, the input impedance for the elements. The rigorous treatment of these effects requires a consistent treatment of the boundary-value problems, as discussed later in this chapter.

3. Integral Equations in Antenna Analysis

In previous discussions the current distributions on radiating elements were assumed to be known. This, however, is generally not the case since the precise description of the current distribution on a metallic structure such as an antenna in the presence of an exciting source, such as a voltage generator at its terminals, involves the solution of a complicated boundary-value problem. In the following we will describe the integral equations which can be solved for the unknown source distributions induced by specified excitations, and in Part 2 we will consider the numerical solution of these equations and some associated issues.

Perfectly Conducting Wires and Bodies

Here we will focus our attention on radiating structures composed of perfect electric conductors over which the boundary conditions given by (33), (34), and (35) must hold. Furthermore, we will devote our attention mainly to wire structures, with conducting bodies touched on briefly.

In order to facilitate ensuing discussions concerning integral equations, questions concerning validity of each specific equation, the existence or uniqueness of solutions, and various features of the limiting process which reduce the integral representations for radiated fields to integral equations for unknown source dis-

Chart 12. Radiation Fields for Vertical Collinear and Parallel Arrays of Short Dipoles Supporting Equal Amplitude and Equal Phase Currents

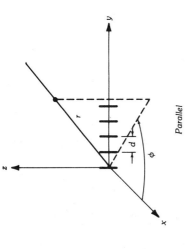

Collinear

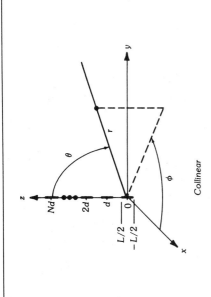

Parallel

Elemental Vector Potential

$$A_1(\mathbf{r}) = \hat{\mathbf{z}} \frac{e^{-jkr}}{4\pi r}(I_0 L) \frac{\sin[(kL/2)\cos\theta]}{(kL/2)\cos\theta}$$

Array Factor ($\alpha_n = 1$)

$$f(\theta,\varphi) = \sum_{n=1}^{N} e^{jk(n-1)d\cos\theta} = e^{jk(N-1)d/2]\cos\theta} \frac{\sin[(kNd/2)\cos\theta]}{\sin[(kd/2)\cos\theta]}$$

Far-Zone Field

$$\mathbf{E}_r(\mathbf{r}) = \hat{\boldsymbol{\theta}} j\omega\mu(I_0 L)\sin\theta \frac{e^{-jkr}}{4\pi r} \frac{\sin[(kL/2)\cos\theta]}{(kL/2)\cos\theta} \frac{\sin[(kNd/2)\cos\theta]}{\sin[(kd/2)\cos\theta]}$$

$$\times e^{j[k(N-1)d/2]\cos\theta}$$

Directive Gain

$$G_D(\theta,\varphi) =$$

$$\frac{\sin^2\theta\sin^2[(kNd/2)\cos\theta]\sin^2[(kd/2)\cos\theta]}{2N/3 + 4\sum\limits_{m=1}^{N-1}(N-m)[(\sin mkd)/(mkd)]^3 - (\cos mkd)/(mkd)^2]}$$

Elemental Vector Potential

$$A_1(\mathbf{r}) = \hat{\mathbf{z}} \frac{e^{-jkr}}{4\pi r}(I_0 L) \frac{\sin[(kL/2)\cos\theta]}{(kL/2)\cos\theta}$$

Array Factor ($\alpha_n = 1$)

$$f(\theta,\varphi) = e^{j[k(N-1)d/2]\sin\theta\sin\varphi} \frac{\sin[(kNd/2)\sin\theta\sin\varphi]}{\sin[(kd/2)\sin\theta\sin\varphi]}$$

Far-Zone Field

$$\mathbf{E}_r(\mathbf{r}) = \hat{\boldsymbol{\theta}} j\omega\mu(I_0 L)\sin\theta \frac{e^{-jkr}}{4\pi r} \frac{\sin[(kL/2)\cos\theta]}{(kL/2)\cos\theta} \frac{\sin[(kNd/2)\sin\theta\sin\varphi]}{\sin[(kd/2)\sin\theta\sin\varphi]}$$

$$\times e^{jk(N-1)d/2]\sin\theta\sin\varphi}$$

Directive Gain

$$G_D(\theta,\varphi) =$$

$$\frac{\sin^2\theta\sin^2[(kNd/2)\sin\theta\sin\varphi]\sin^2[(kd/2)\sin\theta\sin\varphi]}{2N/3 + 2\sum\limits_{m=1}^{N-1}(N-m)\{[(mkd)^2-1]\sin mkd\}/(mkd)^3 + (\cos mkd)/(mkd)^2)}$$

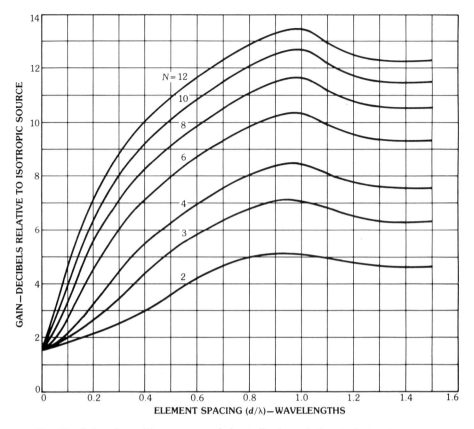

Fig. 10. Gain of a collinear array of short dipoles relative to isotropic source.

tributions are necessarily glossed over. For further information the reader is referred to Stratton [2], Silver [3], and Poggio and Miller [7].

The field equations in Chart 1 are a convenient point to begin the construction of the integral equations. The electric-field representations will be used since, for antennas, the driving source is most easily specified in terms of voltage or electric field. Later, the magnetic-field representation will be discussed in relation to large conducting surfaces.

The representations in Chart 1 are for the fields due to volumetric distributions of sources. For electric current sources constrained to a surface S (which may be considered to be the boundary of V in Chart 1) the integral representations are simply modified in that the volume densities become surface densities and the volume integral becomes a surface integral over S. The integral representations for the electric field due to electric sources over S are shown in Chart 13.

The general boundary-value problem of determining the current distribution on a perfect electric conducting surface is approached using an integral equation. The boundary condition on S is stated as

$$\hat{\mathbf{n}}(\mathbf{r}) \times \mathbf{E}_t(\mathbf{r}) = 0, \quad \mathbf{r} \in S \tag{62}$$

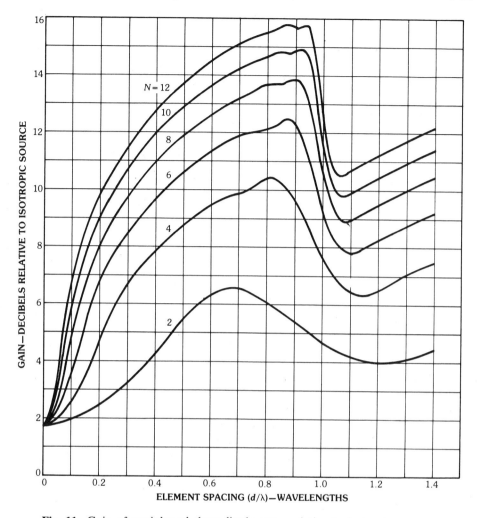

Fig. 11. Gain of equiphased short-dipole array relative to isotropic source.

with $\hat{\mathbf{n}}(\mathbf{r})$ the outwardly pointing normal to S and $\mathbf{E}_t(\mathbf{r})$, the total electric field at the surface. The total field is composed of an incident or driven portion $\mathbf{E}^i(\mathbf{r})$ and a portion generated by the induced surface sources $\mathbf{J}_s(\mathbf{r}')$ and $\mathbf{p}_s(\mathbf{r}')$ referred to as $\mathbf{E}(\mathbf{r})$. The boundary condition then requires

$$\hat{\mathbf{n}}(\mathbf{r}) \times \mathbf{E}(\mathbf{r}) = -\hat{\mathbf{n}}(\mathbf{r}) \times \mathbf{E}^i(\mathbf{r}), \qquad \mathbf{r} \in S \tag{63}$$

The field component $\mathbf{E}(\mathbf{r})$ is given by the integral representations in Chart 5 but with the observation point $\mathbf{r}$ on the surface. The procedure of taking the observation point to the surface must be performed delicately due to the singularity in $\varphi(\mathbf{r}, \mathbf{r}')$ when $|\mathbf{r} - \mathbf{r}'| \to 0$. These issues are dealt with in Stratton [2], Silver [3], Poggio and Miller [7], and Maue [28]. The integral equations most widely used and derived from the electric-field integral representations are presented in Chart

Broadside Array
$L = \lambda/2$
Polarization: Vertical

Theoretical Gain of Broadside ½λ Elements at Different Spacings *a*	
Spacing *a* (wavelengths)	Gain (dB above dipole)
5/8	4.8
3/4	4.6
1/2	4.0
3/8	2.4
1/4	1.0
1/8	0.3

Theoretical Gain of Broadside ½λ Elements for Different Numbers of Elements	
Number of Elements	Gain (dB above dipole)
2	4.0
3	5.5
4	7.0
5	8.0
6	9.0

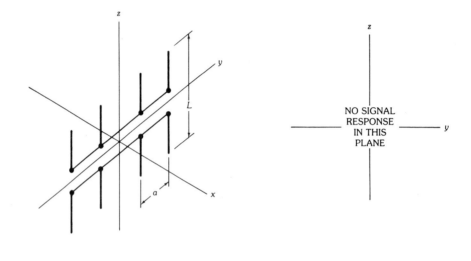

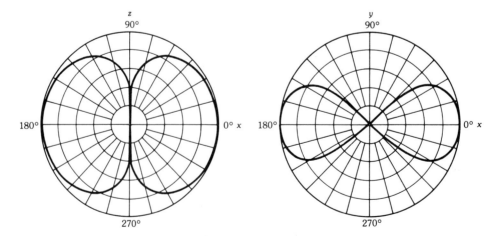

Fig. 12. Reference data for practical arrays. (*After Salati [21]*)

Techniques for Low-Frequency Problems

End-Fire Array
$L = \lambda/2$
Polarization: Vertical

Theoretical Gain of Two End-Fire $\tfrac{1}{2}\lambda$
Elements for Various Spacings a

a	Gain (dB above dipole)
5/8	1.7
1/2	2.2
3/8	3.0
1/4	3.8
1/20	4.1
1/8	4.3

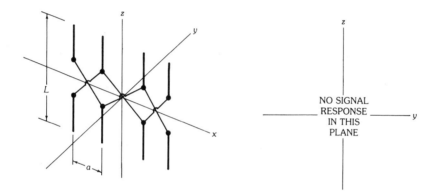

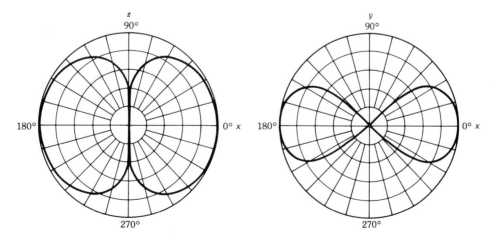

Fig. 12, *continued.*

Parasitic Array
$L = \lambda/2$
Polarization: Horizontal

Number of Elements	Gain (dB above dipole)	Front-to-Back Ratio (dB)
2	4 to 5	10 to 15
3	6 to 7	15 to 25
4	7 to 9	20 to 30
5	9	

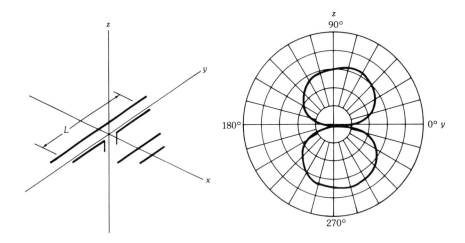

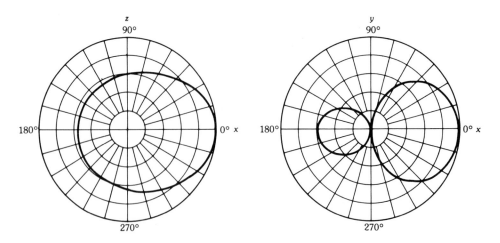

Fig. 12, *continued.*

Techniques for Low-Frequency Problems 3-43

Collinear Array
$L = \lambda/2$
$b = \lambda/4$

Spacing a Between Centers of Adjacent $\frac{1}{2}\lambda$ Elements	Number of $\frac{1}{2}\lambda$ Elements in Array Versus Gain in dB Above a Reference Dipole				
	2	3	4	5	6
$a = \frac{1}{2}\lambda$	1.8	3.3	4.5	5.3	6.2
$a = \frac{3}{4}\lambda$	3.2	4.8	6.0	7.0	7.8

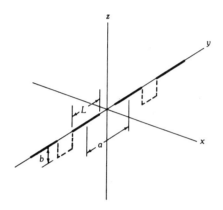
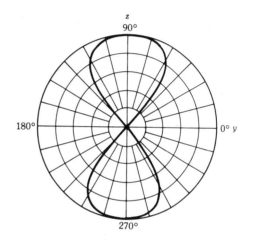
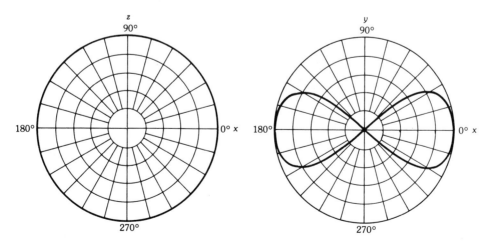

Fig. 12, *continued.*

Chart 13. Electric-Field Integral Representations for Electric Sources

$$E(r) = \frac{1}{4\pi j\omega\epsilon}(\nabla\nabla\cdot + k^2)\int_S J(r')\varphi(r,r')\,d^2r'$$

$$E(r) = -\frac{1}{4\pi}\int_S \left[j\omega\mu J(r')\varphi(r,r') + \frac{1}{j\omega\epsilon}\nabla'\cdot J(r')\nabla'\varphi(r,r')\right]d^2r'$$

$$E(r) = -\frac{1}{4\pi}\int_S j\omega\mu J(r')\varphi(r,r')\,d^2r' + \frac{1}{4\pi j\omega\epsilon}\nabla\int_S \nabla'\cdot J(r')\varphi(r,r')\,d^2r'$$

$$E(r) = \frac{1}{4\pi j\omega\epsilon}\int_S J(r')\cdot(\nabla\nabla + k^2\bar{\bar{I}})\varphi(r,r')\,d^2r'$$

14. The integrals should be interpreted in a principal value sense [7], i.e., the range of integration excludes an infinitesimal region around $r = r'$.

Thin-Wire Integral Equations

For conducting bodies composed of thin wires, i.e., structures composed of interconnected conducting cylinders whose radii a_i are small in terms of wavelengths ($a_i/\lambda \ll 1$), several approximations regarding the behavior of current and charge densities can be made. It can be assumed that:

1. Azimuthal or circumferential currents produce negligible effects when determining the net axially directed current on the wires.
2. The induced electric sources on the surface of the wires can be located on the axis of the wires, thus giving rise to a filamentary source representation.
3. The boundary condition on the electric field can be enforced on the surface of the conductor.

Using $\hat{s}(r)$ and $\hat{s}'(r)$ to denote unit vectors tangent to the conductor and parallel to its axis at the observation point and source point, respectively, the boundary condition and current density are

Chart 14. Electric-Field Integral Equations for Electric Sources on a Surface

$$\hat{n}\times E^i(r) = -\hat{n}\times\left\{\frac{1}{4\pi j\omega\epsilon}(\nabla\nabla\cdot + k^2)\int_S J(r')\varphi(r,r')\,d^2r'\right\}, \quad r\in S$$

$$\hat{n}\times E^i(r) = \hat{n}\times\left\{\frac{1}{4\pi}\int_S \left[j\omega\mu J(r')\varphi(r,r') + \frac{1}{j\omega\epsilon}\nabla'\cdot J(r')\nabla'\varphi(r,r')\right]d^2r'\right\}, \quad r\in S$$

$$\hat{n}\times E^i(r) = \hat{n}\times\left\{\frac{1}{4\pi}\int_S j\omega\mu J(r')\varphi(r,r')\,d^2r' - \frac{1}{4\pi j\omega\epsilon}\nabla\int_S \nabla'\cdot J(r')\varphi(r,r')\,d^2r'\right\}, \quad r\in S$$

$$\hat{n}\times E^i(r) = -\hat{n}\times\left\{\frac{1}{4\pi j\omega\epsilon}\int_S J(r')\cdot(\nabla\nabla + k^2\bar{\bar{I}})\varphi(r,r')\,d^2r'\right\}, \quad r\in S$$

Techniques for Low-Frequency Problems

$$\hat{s}(\mathbf{r}) \cdot \mathbf{E}_t(\mathbf{r}) = 0, \quad \mathbf{r} \in s \tag{64}$$

$$\mathbf{J}(\mathbf{r}') = \hat{s}'(\mathbf{r}')I(\mathbf{r}')/2\pi a \tag{65}$$

Because of the assumptions of the locations of the source and observation points, the distance variable $|\mathbf{r} - \mathbf{r}'|$ in $\varphi(\mathbf{r}, \mathbf{r}')$ is approximated by a distance R, which can never be zero since it is the distance from a point on the axis of the wire to a point on the surface and is thus never less than a. For example, for a z-aligned straight wire, $R = [(z - z')^2 + a^2]^{1/2}$. Such an approximation leads to the widely used thin wire or reduced kernel.

With s and s' denoting the axial coordinates at the observation and source points and $C(\mathbf{r})$ denoting the range of integration over the wires, the integral equations in Chart 14 can be written in their thin-wire forms as shown in Chart 15. Some of the integral equations have been widely used for thin wires. The "mixed-potential integral equation" has been used in the analysis of many structures [29]. Another equation which has found wide usage was derived from the first of the equations in Charts 14 and 15 by solving the differential equation for a

Chart 15. Thin-Wire Integral Equations

$$\hat{s} \cdot \mathbf{E}^i(s) = -\frac{1}{4\pi j\omega\epsilon} \left(\frac{\partial}{\partial s} \nabla \cdot + k^2 \hat{s} \cdot \right) \int_{C(\mathbf{r})} \hat{s}' I(s') \frac{e^{-jkR}}{R} ds', \quad s \in C(\mathbf{r})$$

$$\hat{s} \cdot \mathbf{E}^i(s) = \frac{1}{4\pi} \int_{C(\mathbf{r})} \left[\hat{s} \cdot \hat{s}' j\omega\mu I(s') \frac{e^{-jkR}}{R} - \frac{1}{j\omega\epsilon} \frac{\partial I(s')}{\partial s'} \hat{s} \cdot \nabla \frac{e^{-jkR}}{R} \right] ds', \quad s \in C(\mathbf{r})$$

$$\hat{s} \cdot \mathbf{E}^i(s) = \frac{1}{4\pi} \int_{C(\mathbf{r})} \hat{s} \cdot \hat{s}' j\omega\mu I(s') \frac{e^{-jkR}}{R} ds' - \frac{1}{4\pi j\omega\epsilon} \frac{\partial}{\partial s} \int_{C(\mathbf{r})} \frac{\partial I(s')}{\partial s'} \frac{e^{-jkR}}{R} ds', \quad s \in C(\mathbf{r})$$

Pocklington's Integral Equation

$$\hat{s} \cdot \mathbf{E}^i(s) = -\frac{1}{4\pi j\omega\epsilon} \int_{C(\mathbf{r})} I(s') \left[k^2 \hat{s} \cdot \hat{s}' - \frac{\partial}{\partial s} \frac{\partial}{\partial s'} \right] \frac{e^{-jkR}}{R} ds', \quad s \in C(\mathbf{r})$$

Magnetic Vector Potential Integral Equation for Arbitrarily Curved Wires

$$\int_{C(\mathbf{r})} I(s') \left\{ \frac{e^{-jkR}}{R} \hat{s} \cdot \hat{s}' + \frac{1}{2} \int_{C(\mathbf{r})} d\xi \, \theta(\xi - s) \left[\frac{\partial}{\partial s'} \frac{e^{-jkR}}{R} (\hat{\xi} \cdot \hat{s}') + \frac{\partial}{\partial \xi} \left[(\hat{\xi} \cdot \hat{s}') \frac{e^{-jkR}}{R} \right] \right] e^{-jk|s - \xi|} \right\} ds'$$

$$= Ae^{-jks} + Be^{jks} + \frac{1}{2\sqrt{\mu/\epsilon}} \int_{C(\mathbf{r})} \hat{s}' \cdot \mathbf{E}^i(s') e^{-jk|s - s'|} ds', \quad s \in C(\mathbf{r}), \quad \begin{array}{l} \theta(u) = 1, \ u \geq 0 \\ \theta(u) = 0, \ u < 0 \end{array}$$

Magnetic Vector Potential Integral Equation for Straight Wires
(Hallen's Integral Equation)

$$\int_{C(\mathbf{r})} I(s') \frac{e^{-jkR}}{R} ds' = Ae^{-jks} + Be^{jks} + \frac{1}{2\sqrt{\mu/\epsilon}} \int_{C(\mathbf{r})} \hat{s}' \cdot \mathbf{E}^i(s') e^{-jk|s - s'|} ds', \quad s \in C(\mathbf{r})$$

straight wire by Hallen [30] and generalized to curved wires by Mei [31]. Also, "Pocklington's integral equation" has found widespread usage [7, 13, 23].

Integral Equations for Solid Bodies

The integral equations in Chart 14 in terms of electric field have been reduced to forms appropriate for thin-wire analysis as shown in Chart 15. Of course, the equations in Chart 14 can be used for the analysis of the radiating characteristics of nonwire bodies. To this end the magnetic-field integral equation has been developed and discussed in detail [7, 28] and, with the various forms of the electric-field integral equation, leads to useful representations for describing the interaction of electromagnetic waves with conducting bodies.

The two integral equations for use with perfect electric conductors are given by

$$\hat{\mathbf{n}}(\mathbf{r}) \times \mathbf{E}^i(\mathbf{r}) = \frac{1}{4\pi} \hat{\mathbf{n}} \times \int_S \left[j\omega\mu \mathbf{J}_s(\mathbf{r}')\varphi(\mathbf{r},\mathbf{r}') + \frac{\nabla' \cdot \mathbf{J}(\mathbf{r}')}{j\omega\epsilon} \nabla'\varphi \right] d^2r' \quad (66)$$

and

$$\mathbf{J}_s(\mathbf{r}) = 2\hat{\mathbf{n}}(\mathbf{r}) \times \mathbf{H}^i(\mathbf{r}) + \frac{1}{2\pi} \hat{\mathbf{n}} \times \int_S \mathbf{J}_s(\mathbf{r}) \times \nabla\varphi \, d^2r' \quad (67)$$

where $\int$ represents the principal value integral. These equations are referred to as the *electric-field integral equation* and *magnetic-field integral equation*, respectively.

In a shorthand mathematical notation the electric-field integral equation is written as

$$\hat{\mathbf{n}} \times \mathbf{E}^i(\mathbf{r}) = -\hat{\mathbf{n}} \times \left\{ \frac{1}{4\pi j\omega\epsilon} \int_S \mathbf{J}_s(\mathbf{r}) \cdot \bar{\bar{\mathbf{G}}}(\mathbf{r},\mathbf{r}') d^2r' \right\} \quad (68)$$

and the magnetic-field integral equation as

$$\mathbf{J}_s(\mathbf{r}) = 2\hat{\mathbf{n}} \times \mathbf{H}^i(\mathbf{r}) + \frac{1}{2\pi} \hat{\mathbf{n}} \times \int \mathbf{J}_s(\mathbf{r}) \cdot \bar{\bar{\mathbf{\Gamma}}}(\mathbf{r},\mathbf{r}') d^2r' \quad (69)$$

where

$$\bar{\bar{\mathbf{G}}}(\mathbf{r},\mathbf{r}') = (\nabla\nabla + k^2 \bar{\bar{\mathbf{I}}})\varphi$$

$$\bar{\bar{\mathbf{\Gamma}}}(\mathbf{r},\mathbf{r}') = \bar{\bar{\mathbf{I}}} \times \nabla'\varphi(\mathbf{r},\mathbf{r}')$$

$$\varphi(\mathbf{r},\mathbf{r}') = \frac{e^{-jk|\mathbf{r}-\mathbf{r}'|}}{|\mathbf{r}-\mathbf{r}'|}$$

and where $\bar{\bar{\mathbf{G}}}(\mathbf{r},\mathbf{r}')$ and $\bar{\bar{\mathbf{\Gamma}}}(\mathbf{r},\mathbf{r}')$ are referred to as Green's dyads for electric and magnetic fields due to electric current sources.

The electric-field integral equation, as seen previously, is widely used for wire antenna analysis in part because of the driving-source definition in terms of electric

Techniques for Low-Frequency Problems

field. On the other hand, the magnetic-field integral equation has been used extensively for the analysis of nonwirelike structures such as closed surfaces. In part this is promoted by the fact that it is a Fredholm integral equation of the second kind with the unknown outside and within the integral. For flat surfaces of infinite extent it has a trivial solution:

$$\mathbf{J}_s(\mathbf{r}) = 2\hat{\mathbf{n}} \times \mathbf{H}^i(\mathbf{r})$$

so that one can infer that the integral is a correction term accounting for body size limitations or curvature. Examples of the application of the magnetic-field integral equation are provided in [7] while the inclusion of this equation in a widely used computer program is well documented by Burke and Poggio [23]. The electric-field integral equation, a Fredholm integral equation of the first kind, has also been applied to arbitrary bodies [45].

The Imperfectly Conducting Ground

The integral equations presented above have dealt with surface integrals over the surface of the antenna radiating structure. Of course, the presence of a conducting ground plane can, in principle, be treated in a manner similar to any conducting body in the system of interest (i.e., by solving for the sources induced on that plane). However, a simplification can be introduced into the solution of the problem by the use of appropriately modified kernels in the integral equation which take into account the effect of the ground plane. For a perfectly conducting ground plane the principle of images as illustrated previously for simple antenna models can be used to construct a rigorous modified kernel. For a ground plane of arbitrary electrical parameters, a rigorous treatment requires a more involved approach, although an approximate treatment can be achieved by a modified image theory via reflection coefficients (the reflection coefficient approximation). In the following we outline the various approaches, progressing from the implementation of image theory for a perfect ground to the reflection coefficient approximation and finally to the Sommerfeld integral approach.

Perfect Ground—Implementation of images for perfectly conducting grounds in the integral equation approach is straightforward [23]. The Green's function for a perfectly conducting ground is the sum of the free-space Green's function of the source current element and the negative of the free-space Green's function of the image of the source reflected in the ground plane. For the electric field the Green's dyad for a perfect ground in the $z = 0$ plane is

$$\bar{\bar{\mathbf{G}}}_{pg}(\mathbf{r},\mathbf{r}') = \bar{\bar{\mathbf{G}}}(\mathbf{r},\mathbf{r}') + \bar{\bar{\mathbf{G}}}_I(\mathbf{r},\mathbf{r}') \tag{70}$$

where

$$\bar{\bar{\mathbf{G}}}_I(\mathbf{r},\mathbf{r}') = -\bar{\bar{\mathbf{I}}}_R \cdot \bar{\bar{\mathbf{G}}}(\mathbf{r}, \bar{\bar{\mathbf{I}}}_R \cdot \mathbf{r}')$$
$$\bar{\bar{\mathbf{I}}}_R = \hat{\mathbf{x}}\hat{\mathbf{x}} + \hat{\mathbf{y}}\hat{\mathbf{y}} - \hat{\mathbf{z}}\hat{\mathbf{z}}$$

and $\bar{\bar{I}}_R$ is a dyad that produces a reflection in the $z = 0$ plane when used in a dot product. For the magnetic field with free-space Green's dyad $\bar{\bar{\Gamma}}(\mathbf{r}, \mathbf{r}')$ given in (69), the Green's dyad over a perfect ground is

$$\bar{\bar{\Gamma}}_{pg} = \bar{\bar{\Gamma}}(\mathbf{r}, \mathbf{r}') + \bar{\bar{\Gamma}}_I(\mathbf{r}, \mathbf{r}') \tag{71}$$

where

$$\bar{\bar{\Gamma}}_I(\mathbf{r}, \mathbf{r}') = -\bar{\bar{I}}_R \cdot \bar{\bar{\Gamma}}(\mathbf{r}, \bar{\bar{I}}_R \cdot \mathbf{r}') \tag{72}$$

The introduction of these dyads into (68) and (69) yields the integral equations for perfectly conducting bodies over perfectly conducting ground planes.

Imperfect Ground: Modified Image Theory or Reflection Coefficient Approximation—The Green's dyads for electric and magnetic fields over an imperfectly conducting ground resulting from the reflection coefficient approximation are, in keeping with the development in a previous section [23],

$$\bar{\bar{G}}_g(\mathbf{r}, \mathbf{r}') = \bar{\bar{G}}(\mathbf{r}, \mathbf{r}') + R_v \bar{\bar{G}}_I(\mathbf{r}, \mathbf{r}') + (R_h - R_v)[\bar{\bar{G}}_I(\mathbf{r}, \mathbf{r}') \cdot \hat{\mathbf{p}}]\hat{\mathbf{p}} \tag{73}$$

and

$$\bar{\bar{\Gamma}}'_g(\mathbf{r}, \mathbf{r}') = \bar{\bar{\Gamma}}(\mathbf{r}, \mathbf{r}') + R_h \bar{\bar{\Gamma}}_I(\mathbf{r}, \mathbf{r}') + (R_v - R_h)[\bar{\bar{\Gamma}}_I(\mathbf{r}, \mathbf{r}') \cdot \hat{\mathbf{p}}]\hat{\mathbf{p}} \tag{74}$$

where

$$\hat{\mathbf{p}} = \frac{(\mathbf{r} - \mathbf{r}') \times \hat{\mathbf{z}}}{|(\mathbf{r} - \mathbf{r}') \times \hat{\mathbf{z}}|}$$

The reflection coefficient approximation for finitely conducting grounds uses image fields modified by Fresnel plane-wave reflection coefficients as described earlier for simple antenna models. These reflection coefficients are strictly correct only for an infinite plane-wave field but have been used in the integral equation approach and have been shown to provide useful results for structures that are not too near the ground [23, 24, 25].

In the integral equation formulation a ground plane changes the solution in three ways: (1) by modifying the current distribution through the fields reflected from the ground; (2) by changing the field illuminating the structure; and (3) by changing the reradiated field. Effects 2 and 3 are easily analyzed by plane-wave reflection as a direct ray and a ray reflected from the ground. The reradiated field is not a plane wave when it reflects from the ground, but, as can be seen from reciprocity, plane-wave reflection gives the correct far-zone field. Analysis of the near-field interaction effect is, however, much more difficult in general. In the following we consider the rigorous treatment based on the work of Sommerfeld [32] and Banos [33].

Techniques for Low-Frequency Problems

Imperfect Ground: Rigorous Sommerfeld Treatment—The integral equation for an arbitrarily oriented straight wire antenna over an imperfectly conducting ground of complex relative dielectric constant ϵ_E has been derived in a rigorous manner in Miller et al. [24, 25]. For a straight wire of radius a and length L, the integral equation for the current excited by a field $\mathbf{E}^A$ can be written as

$$\frac{j\omega\mu}{4\pi}\int_L I(s')ds' \left[\left(1 + \frac{1}{k^2}\frac{\partial}{\partial s^2}\right)g_0 + \left(\cos 2\beta' + \frac{1}{k^2}\frac{\partial}{\partial s \partial s^*}\right)g_i \right.$$

$$+ \left(\cos\beta' + \frac{1}{k^2}\frac{\partial^2}{\partial s \partial z}\right)\left(g_{sHz}\sin\beta' - g_{sVz}\cos\beta'\right) \qquad (75)$$

$$\left. + \sin\beta'\left(\sin\beta + \frac{1}{k^2}\frac{\partial^2}{\partial s \partial s'}\right)g_{sHs}\right] = -E^A(s), \quad s \in L$$

where

$$\varrho' = [(x - x')^2 + (y - y')^2 + a^2]^{1/2}$$

$$g_{sHs} = 2\int_0^\infty \frac{\lambda}{\mu + \mu_E} J_0(\lambda\varrho')e^{-\mu(z+z')}d\lambda$$

$$g_{sHz} = \frac{\cos(\phi' - \alpha')}{k^2}\int_0^\infty \frac{\mu - \mu_E}{\epsilon_E\mu + \mu_E} J'_0(\lambda\varrho')e^{-\mu(z+z')}\lambda^2 d\lambda$$

$$g_{sVz} = 2\int_0^\infty \frac{\mu_E}{\epsilon_E\mu + \mu_0} J_0(\lambda\varrho')e^{-\mu(z+z')}\frac{\lambda}{\mu}d\lambda$$

$$g_0 = \frac{e^{-jkR}}{R}, \quad R = [(x - x')^2 + (y - y')^2 + (z - z')^2 + a^2]^{1/2}$$

$$g_i = \frac{e^{-jkR^*}}{R^*}, \quad R^* = [(x - x')^2 + (y - y')^2 + (z + z')^2]^{1/2}$$

$$\phi' = \tan^{-1}[(y - y')/(x - x')]$$

$$\mu = (\lambda^2 - k^2)^{1/2}, \quad k = \omega(\mu_o\epsilon_0)^{1/2}$$

$$\mu_E = (\lambda^2 - k_E^2)^{1/2}, \quad k_E = (\epsilon_E)^{1/2}k$$

In the preceding, α and β are the direction angles as shown in Fig. 13 with a prime used when referring to source coordinates, s and s' are observation and source coordinates of the wire, $\hat{s}^*$ is a unit vector in the direction of the image current with s^* the corresponding coordinate for the image current. The zeroth order Bessel function and its derivative with respect to argument are denoted by J_0 and J'_0. The description of the origin of the terms g_{sHs}, g_{sHz}, and g_{sVz} is beyond the scope of this chapter but is provided in the literature [24, 25].

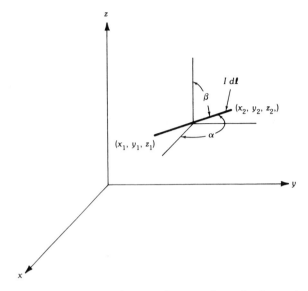

Fig. 13. Geometry for a current element above an imperfectly conducting ground.

For a wire antenna of arbitrary geometry and orientation, denote the wire contour $C(\mathbf{r})$, and $\alpha(\mathbf{r})$, $\beta(\mathbf{r})$, and $\hat{\mathbf{s}}(\mathbf{r})$ the direction angles with respect to the x and z axes and the tangent vector to the wire, respectively. Further, letting $\alpha = \alpha(\mathbf{r})$, $\alpha' = \alpha(\mathbf{r}')$, etc., we obtain the rigorous integral equation for the antenna current $I(s')$ as a function of position s' along the wire as

$$\frac{j\omega\mu}{4\pi} \int_{C(\mathbf{r})} I(s')ds' \left[\left(\hat{\mathbf{s}}\cdot\hat{\mathbf{s}} + \frac{1}{k^2}\frac{\partial^2}{\partial s \partial s'}\right) g_0 + \left(\hat{\mathbf{s}}\cdot\hat{\mathbf{s}}^* + \frac{1}{k^2}\frac{\partial^2}{\partial s \partial s^*}\right) g_i \right.$$

$$+ \left(\cos\beta + \frac{1}{k^2}\frac{\partial^2}{\partial s \partial z}\right)\left(g_{sHz}\sin\beta' - g_{sVz}\cos\beta'\right) \qquad (76)$$

$$\left. + \sin\beta' \left(\sin\beta \cos(\alpha - \alpha') + \frac{1}{k^2}\frac{\partial^2}{\partial s \partial s'}\right) g_{sHs} \right] = -E^i(s), \quad s \in C(\mathbf{r})$$

where $\int_{C(\mathbf{r})} ds'$ implies integration along the wire length over the contour $C(\mathbf{r})$.

The preceding discussion concerning integral equations for antennas over lossy grounds is condensed and necessarily sketchy since the related work has been widespread. The indicated references will, however, serve as convenient starting points which will ultimately lead the reader to more involved and detailed developments.

Integral Equations in the Time Domain

In the previous discussions a frequency-domain approach has been emphasized. All analyses were performed for time-harmonic fields so that the analyses were strictly valid only for the frequency for which they were performed.

Techniques for Low-Frequency Problems

It is sometimes desirable to consider the performance of systems such as antennas under transient conditions, i.e., when excited by either a transient local source or when receiving a nonsinusoidal electromagnetic wave. Analyses of the transient problem can proceed along two different avenues. The first is by performing the time-harmonic analyses over a broad spectrum of frequencies so that the spectral characterization can be used to construct the temporal characterization as one would when using the Fourier or Laplace transform to convert frequency-domain information into the time domain. The other is to perform the analyses directly in the time domain. While a detailed discussion of the issues associated with a direct solution in the time domain is beyond the scope of this text, a brief discussion is appropriate.

The theoretical development of time-domain integral equations is covered in Poggio and Miller [7], Mittra [34], and Sengupta and Tai [34]. Further, the application of the integral equation approach to radiation and scattering is presented in the first two references above, whereas Sengupta and Tai delve more deeply and with broader scope into the transient radiation and reception properties of linear antennas.

The time-domain equivalents to the integral equations presented earlier for electromagnetic wave interactions with perfect electric conductors are [7]

Electric-Field Integral Equation

$$\hat{\mathbf{n}} \times \mathbf{E}^i(\mathbf{r}, t) = \frac{1}{4\pi} \hat{\mathbf{n}} \times \oint_S \left\{ \mu \frac{\partial}{\partial \tau} \mathbf{J}_s(\mathbf{r}', \tau) \frac{1}{|\mathbf{r} - \mathbf{r}'|} \right.$$
$$\left. - \left[\frac{1}{|\mathbf{r} - \mathbf{r}'|} + \frac{1}{c} \frac{\partial}{\partial \tau} \right] \frac{\varrho_s}{\epsilon} \frac{\mathbf{r} - \mathbf{r}'}{|\mathbf{r} - \mathbf{r}'|^2} \right\} d^2 r' \qquad (77)$$

Magnetic-Field Integral Equation

$$\mathbf{J}_s(\mathbf{r}, t) = 2\hat{\mathbf{n}} \times \mathbf{H}^i(\mathbf{r}, t) = \frac{1}{2\pi} \hat{\mathbf{n}} \times \oint_S \left[\frac{1}{|\mathbf{r} - \mathbf{r}'|} \right.$$
$$\left. + \frac{1}{c} \frac{\partial}{\partial \tau} \right] \mathbf{J}_s(\mathbf{r}', \tau) \times \frac{(\mathbf{r} - \mathbf{r}')}{|\mathbf{r} - \mathbf{r}'|^2} d^2 r' \qquad (78)$$

where $\mathbf{r} \in S$ and $\tau = t - |\mathbf{r} - \mathbf{r}'|/c$.

Integral equations based on the magnetic vector potential for curved and straight wires are described in [7]. These later equations have not found extensive application as have (77) and (78).

The specialization of the electric-field integral equation to thin wires, which has found extensive applications, is detailed in [36] and [37]. In these works the thin-wire time-domain integral equation is derived for a geometry as described for the frequency domain earlier and is given as

$$\hat{\mathbf{s}}\cdot\mathbf{E}^i(\mathbf{r},t) = \frac{\mu_0}{4\pi}\int_{c(\mathbf{r})}\left[\frac{\hat{\mathbf{s}}\cdot\hat{\mathbf{s}}'}{R}\frac{\partial}{\partial\tau}I(s',\tau)\right.$$
$$\left.+ c\frac{\hat{\mathbf{s}}\cdot\mathbf{R}}{R^2}\frac{\partial}{\partial s'}I(s',\tau) - c^2\frac{\hat{\mathbf{s}}\cdot\mathbf{R}}{R^3}q(s',\tau)\right]ds' \quad (79)$$

with $\mathbf{r}\in C(\mathbf{r})$, $\mathbf{R}=\mathbf{r}-\mathbf{r}'$, $R=|\mathbf{R}|$, and $\tau = t - R/c$.

As in the frequency domain, the thin-wire approximation is used so that the distance R is always greater than zero with the integration path $C(\mathbf{r})$ along the wire contour displaced from the observation point path by the wire radius.

The numerical solution of (77), (78), and (79) will be discussed in a later section. To date, time-domain analysis for antenna systems has not received the attention allocated to frequency-domain analysis. But, increasing attention is being given because of the expanding interest in pulsed applications. Fast pulse radar and inverse scattering applications have led to interest in and numerical solution of (78), whereas pulse signal transmission and reception lead to interest in (79).

Part 2

Numerical Issues Involved in Integral Equations for Antenna Analysis

Now we turn to the numerical treatment of the integral equations that we have developed in Part 1.

1. Introduction

The advent of the digital computer has opened up new vistas in antenna analysis. In this section we present a summary of computational procedures for solving antenna problems of a more general nature than can be handled by some of the analytically based approaches described in the previous part.

We must caution the reader that in cataloging the various specific computer codes that might be considered, there can exist great differences among them with respect to documentation, validation, availability, support, and the like. A qualitative assessment of these aspects of such computer tools is given in the summary tables included at the end of this section. Before considering such specifics, however, we make a brief tour through the analytical and numerical issues which precede the actual development and use of a computer model.

2. Preliminary Discussion

The process of developing a numerical procedure for solving electromagnetic-field problems may be conveniently divided into the steps outlined in Fig. 14 and discussed below. Since approximations are an intrinsic part of all the steps, we summarize some of those most widely used in Chart 16.

Techniques for Low-Frequency Problems

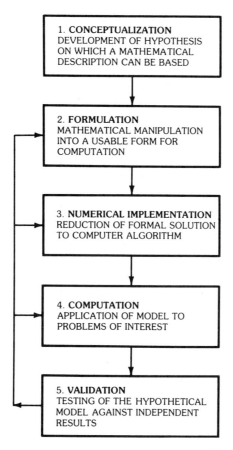

Fig. 14. The basic steps involved in developing a computer model of an electromagnetic field.

Step 1. Conceptualization

It is at this step where physical principles, experimental observations, and so on, are used to form hypotheses from which mathematical descriptions of the relevant phenomena can be devised.

Step 2. Formulation

This involves the evolution of the physical idea or mathematical description from its elementary form into one suitable for numerical or analytical evaluation. Various approximations may be utilized to make the subsequent analysis and/or computation easier.

Step 3. Numerical Implementation

At this stage the formulation is reduced to a form suitable for computation, leading to a computer code or algorithm. For the problems considered here, this

Chart 16. Examples of Approximations Used in the Various Steps of Developing a Computer Solution

Step	Approximation	Limitations/Impact
Conceptualization	Physical optics	Best for backscatter and main lobe region of reflector antennas, from resonance region ($ka > 1$, where a is the object radius) and up in frequency. Fields propagated via Green's function with amplitude established from tangential **H** on surface
	Geometrical theory of diffraction	Generally applicable for $ka > 2$ to 5, where a is the object radius. Fields propagated via a divergence factor with amplitude obtained from diffraction coefficient. Can involve complicated ray tracing
	Geometrical optics	Ray tracing without diffraction. Improves with increasing frequency
	Compensation theorem	Solution obtained in terms of perturbation from a reference, known solution
	Born-Rytov	Approach for low-contrast objects
	Rayleigh	Treats fields at surface of object as only having outward propagating components
Formulation	Surface impedance	Reduces number of field quantities by assuming an impedance relation between tangential **E** and **H**. May be used in combination with physical optics
	Thin-wire	Ignores circumferential sources, circumferential variations of longitudinal sources, and treats current as a filament on wire axis. Generally limited to $ka < 1$, with a the wire radius
Numerical implementation	$\dfrac{\partial f}{\partial x} \to \dfrac{\Delta f}{\Delta x}$	Differentiation and integration of continuous functions represented in terms of analytic operations on sampled approximations, for which polynomial or trigonometric functions are often used. Inherently a discretizing operation, for which typically $\Delta x < \lambda/2\pi$ for acceptable accuracy
	$\int f\,dx \to \sum f_i \Delta x_i$	
	$I(s') \to \sum_{n=1}^{N} a_n f_n(s')$	Representation of unknown is also approximated in terms of basis functions, the number and form of which affect solution accuracy and efficiency
	$L(s,s')f(s') = g(s) \to$ $\sum_{n=1}^{N} a_n \langle w_m(s), L(s,s')f_n(s') \rangle$ $= \langle w_m(s), g(s) \rangle$	Inner product operation with weight functions determines the manner in which the original equation is satisfied by the numerical model
Computation	Deviation of numerical model from physical reality	The greatest source of uncertainty for most problems. Model details affect solution in ways that are difficult to quantify
	Nonconverged solution	Discretized solutions usually converge using a global measure in proportion to $\exp(-AN)$ with A determined by the problem. At least two solutions using different values of N are needed to estimate A

Techniques for Low-Frequency Problems

step almost always involves developing a linear system of equations which are solved using matrix techniques.

Step 4. Computation/Application

In the computation/application the limitations and "bugs" are uncorked and accuracy measures are established. Computation also involves approximation, but in a more ambiguous way than the previous steps, because a model is employed by the user to represent the reality of interest, and there is rarely a simple or obvious way to model most real problems. Furthermore, the numerical model itself may not be solved exactly (see Section 4, under "Modeling Errors," below).

Step 5. Validation

This step is probably the most crucial as it establishes the degree to which the code can eventually be relied upon. It is an open-ended process since a code is usually applied to an expanding variety of problems. Several kinds of validation can be used, which include internal checks for self-consistency, and external checks, which can include independent analytical and numerical results and experimental measurement.

It can be seen that code developers are most intensively concerned with steps 1–3, while code users are more involved with steps 4 and 5.

3. Numerical Implementation

Maxwell's equations can be written in either integral or differential form and either as a function of frequency or time. Each of these can be solved numerically using the method of moments (MOM). This involves approximating the unknowns in a set of basis or expansion functions and satisfying the governing equations using a set of testing or weight functions. Solution of the problem is thus reduced to finding the coefficients of the expansion from a set of linear equations. For various reasons as discussed below, integral equations are more widely useful for antenna modeling, and after some preliminary discussion subsequent detailed attention will be limited to them.

The General Idea

A differential equation (or a set, e.g., Maxwell's equations) relates field and source quantities in a local, pointwise sense (see Chart 17). Field propagation through a medium is represented as a continuum of local interactions which are approximated in a discretized sequence (grid) by the numerical model. Boundary conditions are enforced by specifying the values of the relevant quantities at the appropriate places in the solution continuum in the analytical case or in a solution grid in the numerical case. Analytical solutions are obtainable for only a few separable geometries. Since numerical values must be found throughout the solution volume, the number of unknowns $N \propto A(D/\Delta D)^d$, where D is a characteristic dimension of the solution volume, ΔD is the spatial resolution that is sought, d is the spatial dimensionality (1, 2, or 3), and A is a problem-specific constant determined by the number of field quantities being found at each grid point.

Chart 17. Demonstration of the Difference between an Integral and Differential Formulation of an Electromagnetic Field in the Frequency Domain

Infinite Cylinder in z Direction (Normal to Paper)

(Geometry is two-dimensional for simplicity with TM polarization (E_z, H_x, and H_y components)

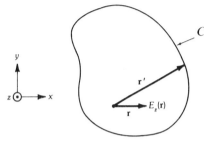

$C(\mathbf{r})$, boundary between electrically dissimilar media. For simplicity the outer medium is assumed to be perfectly conducting

Integral Description

$$E_z(\mathbf{r}) = \int_{C(\mathbf{r})} I_z(\mathbf{r}')K(\mathbf{r},\mathbf{r}')dr'$$

$$\cong \sum_{i=1}^{I} I_z(\mathbf{r}_i)K(\mathbf{r},\mathbf{r}_i)\Delta r_i$$

Field at $\mathbf{r}$ determined from knowledge of equivalent sources on boundary [at most $(\hat{\mathbf{n}} \times \mathbf{E}, \hat{\mathbf{n}} \times \mathbf{H})$], in this case $I_z\hat{\mathbf{z}} = \hat{\mathbf{n}} \times \mathbf{H}$

Differential Description

$$E_z(\mathbf{r}) = -\frac{j}{\omega}\left[\frac{\partial}{\partial x}H_y(\mathbf{r}) - \frac{\partial}{\partial y}H_x(\mathbf{r})\right]$$

$$\cong -\frac{j}{\omega}\left\{\left[\frac{H_y(\mathbf{r}+\hat{\mathbf{x}}\Delta x/2) - H_y(\mathbf{r}-\hat{\mathbf{x}}\Delta x/2)}{\Delta x}\right] - \left[\frac{H_x(\mathbf{r}+\hat{\mathbf{y}}\Delta y/2) - H_x(\mathbf{r}-\hat{\mathbf{y}}\Delta y/2)}{\Delta y}\right]\right\}$$

E_z field at $\mathbf{r}$ requires displaced $\mathbf{H}$ fields at $\mathbf{r} \pm \hat{\mathbf{x}}\Delta x/2$ and $\mathbf{r} \pm \hat{\mathbf{y}}\Delta y/2$, which requires in turn the displaced $\mathbf{E}$ fields

An integral equation, on the other hand, relates field and source quantities in a global sense (see Chart 17). Field propagation through a medium is described by a source-field relationship, which is commonly called a Green's function. Whereas analytical solutions can be found for a few separable boundaries (e.g., planes, cylinders, spheres, etc.) using special Green's functions, the integral representations and equations presented previously use an infinite, homogeneous-medium Green's function for the most part (an exception is the interface problem discussed previously), whatever the geometry. The evaluation of fields is then achieved by integrating the contribution of all sources within the volume of interest (known) and the contributions of all equivalent sources over the boundary (see Charts 1, 2, and 17). While the latter contributions are difficult to specify due to the unknown equivalent sources, the coincidence of this boundary with a surface whose boundary conditions are specified permits an integral equation to be written.

Obviously, the fields in space are then evaluated simply by summing the contributions from bounding surfaces. Since numerical values are needed only on these surfaces, the number of unknowns $N \propto A(D/\Delta D)^{d-1}$. (This relationship applies to objects having extended surfaces. Wire objects represent a special case with $N \propto (D/\Delta D)$, with D the total wire length and ΔD the desired resolution.)

An integral equation model for a given problem thus has a factor of approximately $\Delta D/D$ fewer unknowns than its differential equation counterpart. This can represent a significant difference in computer storage and time requirements, especially for wire objects. On the other hand, a differential-equation approach can be applied to anisotropic, inhomogeneous, nonlinear, and time-varying media, problems for which a Green's function is not easily derivable, if at all. In spite of these advantages, differential equation–based models have been less widely used, primarily because most problems of practical interest do not involve such media. Except for occasional comments the following presentation is therefore limited to an integral-equation formulation. A brief summary of differential and integral equation characteristics is given in Table 1, where N is the number of unknown field values (samples), $d = 1, 2$, or 3 is the problem dimensionality, D and ΔD the problem spatial size and resolution, and N_t the number of steps in a time-domain solution.

It was brought out earlier that various kinds of integral equations can be formulated even for the same problem. When the problems vary significantly, ranging from perfectly conducting to penetrable dielectric bodies for example, the variety can become even greater [7]. The integral equations for all such problems share certain characteristics, however.

First, they involve integrals over prescribed regions of known kernel (or Green's functions) functions operating on unknown field quantities. Second, they include the known source field outside the integral, and sometimes the unknown as well. If the latter is the case, they are called Fredholm integral equations of the second kind; otherwise, they are first-kind integral equations. Third, the required boundary conditions are satisfied implicitly by virtue of their enforcement during the construction of the integral equation. Finally, the unknown often, but not always, is subject to differential operators as a result of the kernel function.

Frequency-Domain Method of Moments

The method of moments (MOM) is an intuitively logical approach for solving operator equations numerically (see Table 2) [7, 38, 39]. Following commonly used notation, let a generic form of the integral equation of interest be written as

$$L(s,s')f(s') = g(s) \qquad (80)$$

where

L = the integral operator (in the frequency domain)
f = the unknown (response, e.g., a current) at source coordinate s'
g = the known (source or forcing function, e.g., a tangential electric field) at observation coordinate s

Table 1. Comparison of Some of the Characteristics of Differential- and Integral-Based Computer Models

Characteristic	Differential Form	Integral Form
Field propagator	Defining differential equation	Green's function, e.g., $e^{j\mathbf{k}\cdot\mathbf{R}}/4\pi R$
Boundary treatment:		
Radiation condition	Local "lookback" (approximate) via outward propagating Green's function Global "lookback" (rigorous) via match to model expansion or integral equation solution	Green's function
Boundary condition	Appropriate field values specified on grid boundaries to obtain staircase or piecewise linear approximation to boundary	Appropriate field values specified on object contour which can in principle be a general curvilinear surface, although this possibility seems to be seldom used
Medium properties that can be handled:		
Linear	Yes	Yes
Anisotropic	Yes	Yes
Inhomogeneous	Yes	Feasible at expense of increasing dimensionality of unknown (e.g., from surface to volume)
Nonlinear	Yes	Generally inapplicable
Time-varying	Yes	Generally inapplicable
Combination of above	Yes	Generally inapplicable (Except linear anisotropic)
Solution characteristics:		
Number of unknowns	$N \propto (D/\Delta D)^d$	$N \propto (D/\Delta D)^{d-1}$
Linear system	Sparse, but larger	Dense, but smaller
Approximate solution time:		
Time domain	$\propto NN_t \propto (D/\Delta D)^{d+1}$	$\propto N^2 N_t \propto (D/\Delta D)^{p(d-1)}$
Frequency domain	$\propto N^2 \propto (D/\Delta D)^{2d}$	$\propto N^p, \quad 2 \leq p \leq 3$

Note that the equations in Charts 14 and 15 can, with suitable interpretations, be written in this form.

Now expand f using a set of basis functions $\{f_n(s')\}$, $n = 1, \ldots, N$:

$$f(s') = \sum_{n=1}^{N} a_n f_n(s') \tag{81}$$

Techniques for Low-Frequency Problems

Table 2. Basic Steps in Developing a Moment-Method Solution (*After Harrington [38] and Miller [39]*)

Operation	Differential Equation	Integral Equation
Sampling of unknowns via basis-function expansion	Subdomain basis functions, usually of low order, are used for differential equations. When the basis functions are pulses, the differential-equation approach is referred to as a finite-difference procedure. When the basis functions are linear, the approach is referred to as a finite-element procedure.	Integral equations can use either entire-domain or subdomain basis functions. Use of the former is generally confined to bodies of rotation for sampling orthogonal to the axis of rotation. Subdomain sampling usually is of low order, with piecewise linear or sinusoidal being the maximum variation employed
Matching of equations via integration with weight functions	Pointwise matching is commonly employed, using a delta-function weight. Pulse and linear matching can also be used	Pointwise matching is commonly employed, using a delta-function weight. For wires, pulse, linear, and sinusoidal weight functions are frequently used as well
Numerical approximation of derivatives and integrals. Determined in part at least by the basis and/or weight functions	$\frac{d}{ds}f(s) \rightarrow [f(s_i + \Delta s_i/2)$ $- f(s_i - \Delta s_i/2)]/\Delta s_i$	$\int f(s)ds \rightarrow \sum f(s_i)\Delta s_i$
Solving for the unknown coefficients using matrix manipulation	ZA = B	ZA = B

which may be either a complete domain (each f_n is defined over the entire object or domain of interest as, for example, in a Fourier series) or subdomain expansion (each f_n defined over only a part of the object). From these equations there follows

$$L(s,s') \sum_{n=1}^{N} a_n f_n(s') = \sum_{n=1}^{N} a_n L(s,s') f_n(s') = g(s) \qquad (82)$$

On then forming an inner product* of (82) with a set of weight functions

$$\{w_m(s)\}, \quad m = 1, \ldots, M$$

there results

---

*An inner product of two functions $p(\mathbf{r}), q(\mathbf{r})$ over a surface S is defined as $\langle p(\mathbf{r}), q(\mathbf{r}) \rangle = \int_S p(\mathbf{r})q(\mathbf{r}) d^2r$.

$$\sum_{n=1}^{N} a_n \langle w_m(s), L(s,s') f_n(s') \rangle = \langle w_m(s), g(s) \rangle, \quad m = 1, \ldots, M \quad (83)$$

As for the basis function expansion, each weight function w_m can have either a complete domain or subdomain support. Some of the more commonly used basis-weight combinations are given in Table 3. The approaches most often used for wire problems involve a subdomain procedure of the form of subsectional collocation or Galerkin's method.

The form of (83) allows the representation

$$\sum_{n=1}^{N} Z_{mn} a_n = b_m, \quad m = 1, \ldots, M \quad (84)$$

with

$$Z_{mn} = \langle w_m(s), L(s,s') f_n(s') \rangle \quad \text{and} \quad b_m = \langle w_m(s), g(s) \rangle$$

Note that this equation represents a set of M simultaneous equations in N unknowns and can be written in matrix equation form as

$$\mathbf{ZA} = \mathbf{B} \quad (85)$$

where

$\mathbf{Z} = [Z_{mn}]$, an $M \times N$ matrix

$\mathbf{B} = [b_m]$, an $M \times 1$ vector

$\mathbf{A} = [a_n]$, an $N \times 1$ vector

Table 3. Representative Pairs of General Basis and Weight Functions Commonly Used in the Moment Method (*After Poggio and Miller [7],* © *1973 Pergamon Press*)

Method	nth Term of Basis Function	mth Term of Weight Function
Galerkin	$a_n f_n(x)$	$f_m(x)$
Least square	$a_n f_n(x)$	$Q(x) \dfrac{\delta \epsilon(x)}{\delta a_m}$
General collocation	$a_n f_n(x)$	$\delta(x - x_m)$
Point matching	$a_n \delta(x - x_n)$	$\delta(x - x_m)$
Subsectional collocation*	$U(x_n) \sum_{p=1}^{P} a_{np} f_p(x)$	$\delta(x - x_m)$
Subsectional Galerkin*	$U(x_n) \sum_{p=1}^{P} a_{np} f_p(x)$	$U(x_m) \sum_{p=1}^{P} f_{mp}(x)$

*Here $U(x_n) = 1$ for x in the subsection (segment) Δx_n, and $U(x_n) = 0$ otherwise.

Techniques for Low-Frequency Problems

A formal solution of (85) can be written as

$$\mathbf{A} = \mathbf{YB} \tag{86}$$

or

$$a_n = \sum_{m=1}^{M} Y_{nm} b_m, \quad n = 1, \ldots, N \tag{87}$$

where $\mathbf{Y} = [Y_{nm}] = [Z_{mn}]^{-1}$ for $M = N$. For $M > N$, a solution can be achieved by using the generalized inverse

$$\mathbf{A} = (\mathbf{Z}^T\mathbf{Z})^{-1}\mathbf{Z}^T\mathbf{B} \tag{88}$$

In the above the superscript T implies conjugate transpose.

Since the matrix $\mathbf{Z}$ relates a current to its electric field when this representation is used with the integral equations in Charts 14 and 15, it is called an *impedance matrix*. Similarly, the matrix $\mathbf{Y}$ is known as an *admittance matrix* and possesses some especially useful features. First, it provides a solution for arbitrary excitation. Second, it may be stored for subsequent reuse. Third, and most important, $\mathbf{Y}$ represents a complete electromagnetic characterization of a structure, within the limits of the approximations involved in its derivation. In this sense it has some characteristics that are similar to a hologram. The general expression for the field E_r radiated by the object obtained using reciprocity highlights this property:

$$E_r = \sum_{n=1}^{N} E_n^o \sum_{m=1}^{N} Y_{nm} E_m^i$$

with $E_n^o = \hat{\mathbf{s}}_n \cdot \mathbf{E}^o(s_n)$ and $E_m^i = \hat{\mathbf{s}}_m \cdot \mathbf{E}^i(s_m)$. Here, E^i is the field of the exciting source, and $\mathbf{E}^o$ is the field produced at the object by a point source located at the observation point. Clearly, it is possible to construct the spatial field produced anywhere in space by the object.

The computer time T_f involved in obtaining a frequency-domain solution of an integral equation at N_f frequencies for a general object (nonsymmetric) in free space can be approximated as*

$$T_f \cong (T_{f_1} + T_{f_2}N)N^2 N_f \tag{89a}$$

while the variable storage for a single frequency is approximately

$$S_f \cong (S_{f_1} + 2N)N \tag{89b}$$

where

---

*These approximate formulas include only the dominant terms and ignore symmetry effects.

$$N \geq \begin{cases} 2\pi L/\lambda & \text{for wires of length } L/\lambda \\ 2A(2\pi/\lambda)^2 & \text{for surfaces of area } A/\lambda^2 \end{cases}$$

The T_{f_1} term accounts for computation of the impedance matrix and the T_{f_2} term its solution via factorization or inversion where T_{f_1} and T_{f_2} are algorithm- and computer-dependent coefficients. It can be deduced that the computer time can vary with frequency over the range f^2 (wire object with impedance matrix computation time dominating) to f^6 (surface object with matrix solution time dominating). The constant S_{f_1} in (89b) accounts for geometry and other structure-dependent storage.

Time-Domain Method of Moments

A time-domain counterpart to the generic frequency-domain integral equation (80) can be written as

$$\tilde{L}(s,t; s',t')\tilde{f}(s',t') = \tilde{g}(s,t) \tag{90}$$

with $\tilde{g}(s,t)$ the time- and space-dependent forcing function, $\tilde{f}(s',t')$ the resulting response, and $\tilde{L}(s,t; s't')$ the integral operator which can be defined using (77), (78), and (79). It is important to observe that while $\tilde{L}$ is apparently a function of the space-time coordinates of the observation point (s,t) and source point (s',t'), the spatial integration over the source coordinate eliminates both source-point variables. The reason is that the spatial integration over s' for a given observation point (s,t) in space-time has an additional constraint imposed by $t' = t - R(s,s')/c$, where $R(s,s')$ represents the distance between the spatial source and observation points. Thus, as the integration over s' is performed the dependency on t' also vanishes. In essence, $\tilde{L}$ performs a spatial integration over the space-time cone with due regard paid to the retarded time restriction [7].

Because there are two independent variables, we approximate the unknown in a separable form:

$$\tilde{f}(s',t') = \sum_{n=1}^{N} \sum_{m=1}^{N_t} a_{nm}\tilde{f}_n(s')\tilde{q}_m(t') \tag{91}$$

which, with (90), leads to

$$\sum_{n=1}^{N} \sum_{m=1}^{N_t} a_{nm}\tilde{L}(s,t; s',t')\tilde{f}_n(s')\tilde{q}_m(t') = g(s,t) \tag{92}$$

On employing the weight function expansion

$$\{w_{pq}(s,t)\} = \{w_q(t)v_p(s)\}, \quad p = 1, \ldots, N, \quad q = 1, \ldots, N_t \tag{93}$$

there follows

Techniques for Low-Frequency Problems 3-63

$$\sum_{n=1}^{N} \sum_{m=1}^{N_t} a_{nm} \langle w_q(t), \langle v_p(s), \tilde{L}(s,t;s',t')\tilde{f}_n(s')\tilde{q}_m(t')\rangle\rangle$$
$$= \langle w_q(t), \langle v_p(s), g(s,t)\rangle\rangle \tag{94}$$

Most numerical solutions to wire problems in the time domain employ subsectional collocation [see a description of TWTD (thin-wire time domain) in a following section]. Then

$$\tilde{f}(s',t') = \sum_{n=1}^{N} \sum_{m=1}^{N_t} a_{nm} p_{nm}(s',t')\tilde{f}_n(s')\tilde{q}_m(t') \tag{95}$$

and

$$w_{pq}(s,t) = \delta(s-s_p)\delta(t-t_q) \tag{96}$$

where

$$p_{nm}(s',t') = \begin{bmatrix} 1 & \text{if } s' \in \Delta s'_n \text{ and } t' \in \Delta t'_m \\ 0 & \text{otherwise} \end{bmatrix}$$

The general form of the current [the most commonly used unknown as in (77), (78), and (79)] can be shown to be

$$I_{nm} = I(s_n, t_m) = Y_{nn} E^t_{nm} = Y_{nn}(E^i_{nm} + E^s_{nm})$$
$$= Y_{nn}(E^i_{nm} + \sum_{n'=1}^{N} A_{nn'} I_{n',m-f(n,n')}) \tag{97}$$

where E^t_{nm} = total E_{nm}, E^s_{nm} = scattered E_{nm}, and $f(n,n') = |n-n'|$ for a straight wire, but in general is a more complicated function of object geometry, and $[A_{nn'}]$ is an object-dependent interaction matrix. This form is similar to a multi-input, multi-output linear predictor. In the general case, the general solution is developed by time-stepping beginning with a set of initial (known) conditions. Many of the details of applying the technique to solid bodies and wires using the magnetic-field and electric-field integral equations in space-time domain are found in References 7, 34, 36, and 37.

The computer time and storage involved in obtaining a time-domain solution having N_t time steps are given approximately by

$$T_t \cong (T_{t_3}N + T_{t_4}N_a)NN_tN_i \tag{98a}$$

and

$$S_t \cong (N + N_t)N \tag{98b}$$

with N_i the number of incident fields or excitations, and N_a the number of angles at which the far field is evaluated. In (98a) the T_{t_3} term accounts for current-charge computation, and T_{t_4} for far-field evaluation.

The N-Port Analogy

Solutions based on the MOM in either the frequency domain or time domain can be viewed as leading to an N-port equivalent network because in either case an $N \times N$ matrix can be developed as an approximation to the actual boundary-value problem of interest. Thus, finding the response of the structure to an arbitrary incident field variation in either the time domain or frequency domain requires determining the currents induced at each of the N ports for as many specific exciting source distributions as are necessary to adequately describe the incident field variation. This viewpoint can provide useful physical insight [7]; an idea of the amount of computational effort required is shown in Table 4. In this table, N is the number of unknowns, N_f the number of frequency samples, and N_i the number of independent spatial source distributions (incident fields). Note that N_f and N_i can be related to N for specific problems based on a sampling theorem argument. A complex interaction term is considered to be composed of two real interaction terms.

Table 4. Comparison of Frequency-Domain and Time-Domain Models in Terms of the Number of Real Interaction Terms Involved (*After Poggio and Miller [7]*, © 1973 *Pergamon Press*)

Computation*	Equivalent Real Interaction Terms for N-Port Structure		
	Frequency Domain (FD) N_f Frequencies	Time Domain (TD) N_t Time Steps	FD/TD > 1 favors TD < 1 favors FD
Monostatic (N_i angles of incidence)			
Single frequency ($N_f = 1$)	$2N^2$	NN_tN_i	$2N/N_tN_i$
Time response** ($N_f = N_t/2$)	$2N^2N_f$	NN_tN_i	N/N_i
Bistatic (one angle of incidence)			
Single frequency ($N_f = 1$)	$2N^2$	NN_t	$2N/N_t$
Time response** ($N_f = N_t/2$)	$2N^2N_f$	NN_t	N

*A monostatic computation involves finding the far field for enough incident sources (N_i) to define the angle-dependent back-scattering pattern, while a bistatic computation involves finding the scattered field for one incident source. The latter is the situation most relevent to antenna analysis where only a few antenna feed points would be of interest.

**For equivalent information, $N_f = N_t$ in transforming between the frequency domain and time domain. But because the admittance samples for plus and minus frequencies are complex conjugate, only the plus-frequency values are needed so that $N_f = N_t/2$.

Techniques for Low-Frequency Problems

Comparison of Frequency- and Time-Domain Approaches

It is useful to compare the steps involved in performing frequency-domain and time-domain computations, as is done in Table 5. In this table the various constants for storage and time are computer and algorithm dependent, and n is a symmetry measure given by 2^d for objects having $d = 1$, 2, or 3 planes of mirror symmetry, and by the number of sectors for objects having discrete rotational symmetry. (Symmetry effects are discussed more thoroughly in Table 7.)

The most important points to note in Table 5 are that a frequency-domain approach yields a solution valid at a single frequency but for an arbitrary spatial source distribution. The time-domain approach, on the other hand, yields a solution valid over a wide band of frequencies but for only a single spatial source distribution. In addition, it is relevant to mention that the time-domain approach can handle nonlinear effects, while dispersive and/or dissipative media can be handled more readily in the frequency domain using integral equations.

Comparison of the computer-time and -storage expressions in (89) and (98) reveals further interesting differences. A summary of computer-time dependencies for several kinds of common problems is given in Table 6, which was developed assuming the dependence on L and C of the various sampling densities as indicated in the footnote. This table includes only the dominant terms of the computer-time dependencies; i.e., as frequency increases without limit, the solution time should eventually exhibit the indicated behavior. The multiplying factor needed to estimate the actual time, which is computer, algorithm, and problem dependent, is not included. It is enlightening, however, to see the trends that characterize these different problems, permitting the potentially most efficient approaches to be identified.

Benefits of Symmetry

Symmetry can substantially reduce both storage and computation time, essentially by eliminating redundant operations. Consider, for example, the case of the center-fed, straight dipole. Its current is symmetric about the feed point, so that finding the current on just one half of the antenna defines the current on the whole antenna. In addition, the impedance matrix for equal-sized segments exhibits a symmetry about the antenna's midpoint, a fact that can reduce the matrix fill time by one half. Actually, a further symmetry holds for this structure in that only the first row is needed to define the entire impedance matrix. This is because the structure is translation invariant, i.e., the interaction between any two points a given distance apart on the wire is independent of their actual positions.

There are three different kinds of symmetries that can be exploited: plane, rotational, and translation symmetry. Each is discussed briefly below and summarized in Table 7, which was constructed considering only multiply and divide operations.

Plane Symmetry—An object can exhibit up to three orthogonal symmetry planes. The dipole example discussed above possessed one. Each plane halves both the storage needed for the impedance matrix and the computer time needed to fill it. Solution time is reduced because the admittance matrix possesses the same kind of symmetry as the original impedance matrix. Finally, if the exciting field is either

Table 5. Comparison of the Steps Involved in Developing and Using Frequency-Domain and Time-Domain Computer Models (After Miller [40])

Frequency Domain			Development of Solution	Time Domain		
Time	Storage	Operation		Operation	Storage	Time
		$f(s, t) = e^{j\omega t} f(s)$	Maxwell's equations	Time dependent		
		$L(s, s') f(s') = g(s)$	Plus BC, etc.	$\tilde{L}(s, s') \tilde{f}(s', t') = g(s, t)$ $t' = t - R/c$		
$T_{f_1} N^2/n$	$2N^2/n + S_{f_1} N/n$	$\sum_{n=1}^{N} Z_{mn} f_n = g_m$	Apply MOM to get Nth-order system ($m = 1, \ldots, N$)	$\sum_{n=1}^{N} \tilde{Z}_{mn} \tilde{f}_{np} = \tilde{g}_{mp}$ $p = 1, \ldots, N_t$	$S_{t_1} N^2$	$T_{t_1} N^2$
$T_{f_2} N^3/n^2$		$(Z_{mn})^{-1} \to Y_{mn}$	Solve matrix	$[\tilde{Z}_{mn}]^{-1} \to \tilde{Y}_{mn}$		$T_{f_2} N^3/n^2$
$T_{f_3} N^2/n$	$2S_{f_3} N$	$f_m = \sum_{n=1}^{N} Y_{mn} g_n$	Obtain induced current	$\tilde{f}_{mp} = \sum_{n=1}^{N} \tilde{Y}_{mn} \tilde{g}_{np}$	NN_t	$T_{t_3} N^2 N_t$
$T_{f_4} NN_a$	$2N_a$	$E_0 = \sum_{m=1}^{N} E_{0m} f_m$	Compute far field for $0 = 1, \ldots, N_a$ observation angles	$\tilde{E}_{0p} = \sum_{m=1}^{N} E_{0m} \tilde{f}_{mp}$	$N_a N_t$	$T_{t_4} NN_t N_a$
$\times N_t$			Repeat for additional spatial sources (total N_t)	$\times N_t$		
$\times N_f$			Repeat for additional frequencies $q = 1, \ldots, N_f$ Observe that the two solutions are related by a Fourier transform $f_{mq} \overset{FT}{\longleftrightarrow} \tilde{f}_{mp}$			

Techniques for Low-Frequency Problems

Table 6. Dependence of Computation Time on Object Size for Wires of Length L and Surfaces with Enclosing Spheres of Circumference C Enclosing Them (*After Poggio and Miller [7]*)

Calculation*	Frequency Domain	Time Domain
	Transient	
Monostatic:		
Surface	$(C/\lambda)^7$	$(C/\lambda)^6$
Wire	$(L/\lambda)^4$	$(L/\lambda)^4$
Bistatic:		
Surface	$(C/\lambda)^7$	$(C/\lambda)^5$
Wire	$(L/\lambda)^4$	$(L/\lambda)^3$
	Monochromatic	
Monostatic:		
Surface	$(C/\lambda)^6$	$(C/\lambda)^6$
Wire	$(L/\lambda)^3$	$(L/\lambda)^4$
Bistatic:		
Surface	$(C/\lambda)^6$	$(C/\lambda)^5$
Wire	$(L/\lambda)^3$	$(L/\lambda)^3$

*These dependencies on object size are arrived at by determining the highest-order terms in (89a) and (98a), assuming that the various sampling densities depend on object size in a systematic way. For wires we use $N \propto L/\lambda$, the number of incidence angles $N_i \propto N$, the number of time steps $N_t \propto N$, and the number of frequencies $N_f \propto N$. Similar values for surfaces are $N \propto (C/\lambda)^2$, $N_i \propto C/\lambda$, $N_t \propto C/\lambda$, and $N_f \propto C/\lambda$, respectively.

Table 7. Effect of Various Kinds of Symmetry on Impedance Matrix Storage and Fill Time and Admittance Matrix Solution Time Relative to Same Number of Unknowns and Without Symmetry

Type of Symmetry	Impedance Matrix Storage and Fill Time	Admittance Matrix Solution Time	
		General Source	One Symmetric Source
None	1	1	1
Orthogonal plane (d planes)	$1/2^d$	$1/2^{2d}$	$1/2^{3d}$
Rotational (n sectors)	$1/n$	$1/n^2$	$1/n^3$
Translational	$1/N$	$1/N$	$1/N$

even or odd about the symmetry plane(s), only that part of the admittance matrix which gives an even or odd current is needed.

Rotational Symmetry—An object can be either continuously rotational about an axis (a circular loop, for example) or discretely rotational (an n-sided polygon). Since a wire code normally models curved objects using piecewise linear (or straight) segments, discrete rotational symmetry is the one usually encountered in wire modeling. Storage and impedance matrix fill time are reduced in proportion to the number of identical sectors (n) comprising the model. Because the admittance matrix is circulant in the same way as the impedance matrix, its solution time is decreased due to needing only a part of the factored or inverted matrix. Furthermore, if the exciting source has symmetries in common with the object, additional savings are possible. There is a close correspondence between continuously and discretely rotational objects, in that the continuous Fourier transform applicable to the former is analogous to the discrete Fourier transform applicable to the latter.

Translational Symmetry—This is the symmetry exhibited by an object that is space invariant along some line. It can be continuous or discrete, as demonstrated by a continuous shell of arbitrary contour that is modeled using a wire grid. Impedance matrix storage and fill time are reduced in this case by the fact that interactions along the structure are invariant with respect to absolute coordinates. The admittance matrix is a product of two matrices having the same structure as the impedance matrix, yielding a further, very substantial time savings. This kind of symmetry produces what is termed a *Toeplitz matrix*.

As a concluding comment it should be observed that the above symmetries can occur in combination. When this is the case, even more reductions of computer resources are possible.

4. Computation

Computation is, of course, the focal point of modeling, and is the purpose to which all the other efforts are ultimately directed. It is where a user first encounters a given code and may experience difficulties. Because these are often due to deficiencies in documentation, maintenance, and assistance for the EM codes, we discuss each of these areas below.

1. *Documentation*: At a minimum the documentation should include a written description of the code and a reasonable amount of comment statements within it and directions on how to use it. Also highly desirable are applications guidelines, a discussion of the code's limitations, and examples of its use to provide check cases for the new user. As an option (highly desirable) the documentation could also provide the theoretical basis for the code and a discussion of the numerical treatment.
2. *Maintenance and Updating*: Most codes that receive any significant amount of use benefit greatly from periodic updating. One of these benefits is to correct "bugs" and identify unanticipated or unknown limitations in the code. The other is to inform the users of new capabilities that have been added to it.
3. *User Assistance*: Perhaps the most crucial factor in making the user-code encounter a successful one is that of user assistance since the best engineered and documented codes can leave unanswered questions and produce inexplicable results. Access to an experienced user, or the code developer, can save a user much time and anxiety. User assistance is also beneficial as a feedback mechanism for the developer.

Modeling Errors

Although a variety of errors can arise in computer modeling, they can be conveniently put into just two categories defined according to their cause as physical modeling errors and numerical modeling errors [41]. The former, which we denote by ϵ_p, provides a measure of the mismatch which may occur between the numerical model and the physical reality it is intended to represent. The latter, which we denote by ϵ_N, provides a measure of the mismatch between the solution obtained from the computer model and an exact one for the numerical model being used. In Table 8 are listed examples of some commonly encountered physical and numerical modeling errors.

As a general rule, ϵ_N can be made acceptably small by simply increasing the number of unknowns (number of segments N in subsectional models) used for the numerical model. This is because various measures of ϵ_N (see Table 9, for example) demonstrate that it tends to be an exponential function of the number of samples per wavelength. For geometrically complex structures, unfortunately, the value of N needed before ϵ_N exhibits this behavior may be unaffordably large so that ϵ_N is not reducible to an acceptable level. Some results of increasing N are shown in Figs. 15 and 16.

In Fig. 15 the results are based on the backscatter radar cross section averaged over 4π steradians of incidence angles using the value obtained from a model having the maximum feasible number of segments used as a reference for convergence purposes. All objects tend to yield the same slope past some minimum sampling density per wavelength, N_λ, with $\epsilon_N \propto \exp(-AN_\lambda)$, where A is a value in the range from 0.15 to 0.30.

In Fig. 16a the rms current $I_{\text{rms}}(N)$ for N segments is shown, while Fig. 16b shows the rms current error relative to a reference: $I_{\text{rms}}(N)/I_{\text{rms}}(\text{ref})$. Models using sinusoidal bases are seen to be superior where a traveling-wave current can exist.

There are other ways of estimating ϵ_N besides increasing N, as discussed in the next section.

Physical modeling errors are intrinsically the more difficult to establish because they almost invariably require experimental measurements for their direct assessment. But because experimentation is subject to its own errors and uncertainties, differences between a measured physical model and a computed numerical one cannot necessarily be attributed to the computation alone. One way to handle this problem is to use two or more experimental models, with one intended to model the real physical problem as closely as possible and another which matches the numerical model. Assuming systematic errors affect both measured results in the same way, a comparison of the two measurements should provide an accurate indication of ϵ_p. Two (or more) numerical models might be used in a similar fashion to estimate ϵ_p. The important point to keep in mind is that comparing either two (or more) measurements or two (or more) calculations can provide a way to estimate ϵ_p. The value of such comparisons is not limited to this application, however. Other uses are discussed below.

Limitations

Every code without exception is subject to limitations of various kinds. These limitations may be known or unknown, intrinsic or imposed, and can arise from the formulation, numerical treatment, or other choices made by the developer. Whatever the cause, it is important that the user be informed about known and anticipated limitations which affect the code's applications.

Information on limitations should be included in the documentation and can be compiled by listing the modeling capabilities incorporated in the code or, less preferably, by listing the things it cannot or was not designed to do.

Table 8. Generic Examples of Physical and Numerical Modeling Errors and Specific Examples for Representative Problems

Physical Modeling Errors	
Cause	Effect
Smoothly curved object represented numerically by a piecewise linear contour (e.g., a circular loop modeled as a polygon)	Introduces a nonphysical discretization that can obscure the needed relationship between the physical and numerical models. One criterion is to use equal lengths for wire models and areas for surface models that equal those quantities for the object being modeled.
Surface (closed or open) represented numerically by a wire grid or mesh	Provides a nonphysical mechanism for fields to "leak through" the surface. A modeling criterion sometimes used here is to choose the grid parameters such that its surface area equals that of the actual object.
Neglect of end cap in thin-wire numerical model	Produces a negligible effect on the solution for the most part, except near the end of a wire
Omission of fine object detail in developing the numerical model (sometimes referred to as *model order reduction*)	Reduces geometry-dependent correlation of computed results. Object complexity can impose more severe sampling requirements than can size, which is why this error source can be so important.
Numerical Modeling Errors	
Thin-wire approximation	Limits minimum segment length to approximately the wire diameter, or sharply oscillatory solution will result. Also ignores possible circumferential current variation near bends and junctions.
Under sampling of the unknown and boundary conditions	Unpredictable, but can generally be expected to most affect near-field quantities
Basis and weight function incompatibility with formulation	Can lead to anomalous-appearing non-physical near-field behavior, due to discontinuities in the basis functions that are not smoothed by the weight functions

Table 8, *continued.*

Specific Examples of Modeling Errors		
Object	Physical Modeling Error	Numerical Modeling Error
Straight wire	—	Thin-wire approximation and neglect of end caps
Circular loop	Piecewise linear model is discrete representation of curved object	Thin-wire approximation and neglect of possible circumferential variation at the bends
Wire-grid model of continuous surface	"Lumped" filamentary model of continuous surface	Thin-wire approximation and neglect of possible circumferential variation at the junctions
Wire attached to wire-grid model of continuous surface	Lumped filamentary model of continuous surface plus constrained flow onto surface at connection point	Thin-wire approximation and neglect of possible circumferential variation at the junctions
Surface-patch model of continuous surface	None necessarily. Usually however, the surface is modeled with piecewise linear patches which neglect surface curvature.	Variation of source within surface patch is not always included
Wire attached to surface-patch model of continuous surface	None necessarily. Usually however, the surface is modeled with piecewise linear patches which neglect surface curvature.	Variation of source within surface patch is not always included; thin-wire approximation and neglect of possible circumferential variation at the junctions
Box or other object	As above for either wire-grid or surface-patch model	Variation of source within surface patch is not always included; for example, edge singularities in the induced sources are usually not explicitly modeled.
Wire attached to box near edge or corner	As above for either wire-grid or surface-patch model	Variation of source within surface patch is not always included; for example, edge singularities in the induced sources are usually not explicitly modeled.

Table 9. Examples of Convergence Measures for Evaluating Numerical Modeling Errors

ϵ_N Measure	Form	Demonstrates	Properties
Local convergence:			
Current	$\lim\limits_{N \to N\max} I(s)$	Convergence of input impedance, current, fields, etc.	Can yield non-monotonic measure of convergence
Field	$\lim\limits_{N \to N\max} E(\mathbf{r})$		
Global convergence:			
Current	$\int_{C(\mathbf{r})} I(s')I^*(s')ds'$ $\cong \sum I_n I_n^* \Delta s'_n$	Convergence over entire object $[C(\mathbf{r})]$	A more complete measure of convergence
Field	$\int_{R^n} \mathbf{E}(\mathbf{r}) \cdot \mathbf{E}^*(\mathbf{r}) dx^n$	Convergence of field over $n = 1$, 2, or 3 dimensions	Can be used for near or far field. If a near-field quantity, can be expensive computation
Random convergence (local or global)	$\frac{1}{R} \sum\limits_{n=1}^{R} F(\mathbf{r}_n)$ With $F(\mathbf{r}_n)$ a field quantity which is a function of a random variable $\mathbf{r}_n$	Convergence of any field quantity measured by a random observation variable	Permits estimation of convergence and uncertainty of convergence estimates

Information concerning limitations should also be incorporated in the code itself. This can be as simple as providing internal checks that prohibit unacceptable parameter values from being used or inappropriate problems from being run, at least without warning the user. Further checks might be performed at subsequent steps in the computation to determine whether other limitations have been violated.

5. Validation

Unless a code can be used with some acceptable degree of confidence that it produces reliable results, it will have little value to prospective users. The process of validation is a vital and ongoing one. Here we discuss several ways to validate a code.

Experimental Validation

In spite of the difficulties which attend experimental validation already alluded to in the section on modeling errors, measurements still remain the most satisfying

Techniques for Low-Frequency Problems

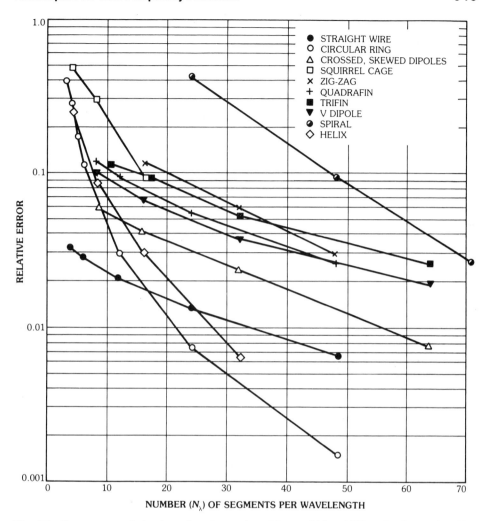

Fig. 15. Convergence behavior of various wire objects. (*After Miller, Burke, and Selden [42], © 1975 IEEE*)

way to demonstrate a code's validity. Experimental validation is most suitable where the measurement and computation can be made wholly congruent to reduce or eliminate any associated ambiguity that affects their comparison. If this is the case, it is then possible to make meaningful, absolute comparisons as opposed to the relative ones previously discussed.

The quality of the experimental validation can depend on the kinds of quantities being compared, specifically the relative degree of difficulty their measurement represents. Progressing in order of increasing difficulty, common quantities for validating a code are far fields (patterns and frequency dependence), radiation resistance (computed from integrating the far field), current and charge distribution, impedance resonances, and input reactance. An example of a resonance check is given in Fig. 17. Shown here is the resonance in the input impedance of a center-fed antenna consisting of a variable number of long zigzags

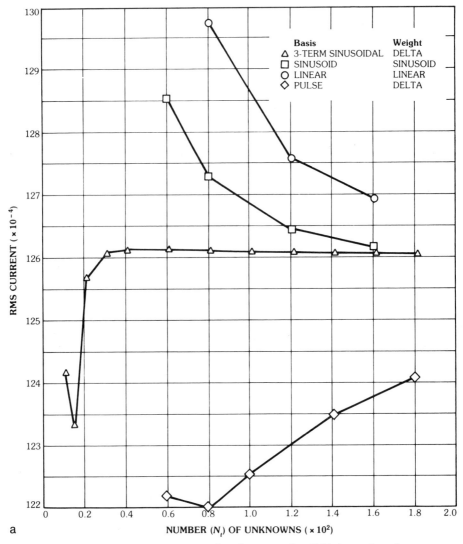

Fig. 16. Convergence results for scattering from a wire 5.8λ long, based on an rms-current measure. (*a*) Showing the rms current. (*b*) Showing the rms-current relative error. (*After Poggio, Bevensee, and Miller [43]*, © *1974 IEEE*)

as a function of the bend angle α, which is 180° when the antenna is a straight wire. The tip-to-tip length of the antenna is kept fixed at 29.6 in (175.18 cm). Good agreement is observed between the measured and computed results.

Because input reactance is so dependent on feed-region details, both in the experimental and numerical models, it is probably the most sensitive quantity to use for validation. But it is extremely common to observe shifts in frequency between measured and computed impedances. Shifts in far-field pattern details are also frequently observed both as a function of angle and frequency, as depicted graphically in Fig. 18. It is for this reason that relative comparisons, i.e., comparing a measurement with a measurement, or a computation with a computation, are

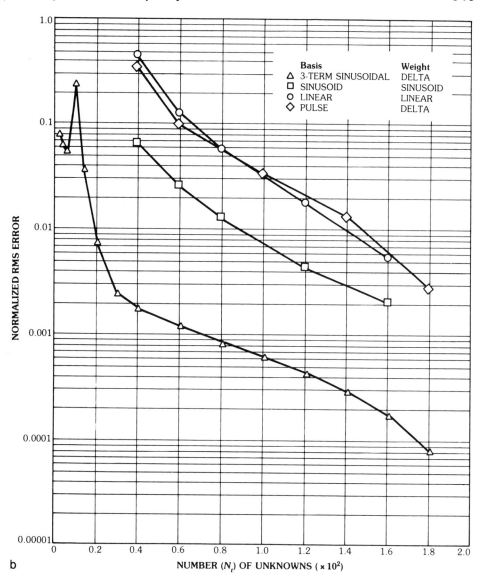

Fig. 16, *continued.*

often more appropriate. Besides their use in validating a code, experimental results can be invaluable in developing an appropriate model for a given problem. It may seem to be reasonable to conclude that a computer model can be developed merely from a description of the problem, but it often happens that the physical modeling error depends unpredictably on the details of the computer model. Without the insight provided by the experimental data, there is little to guide the modeler in selecting the model parameters best suited to the particular problem. The outcome may be that valid numerical results are obtained, but for a significantly different physical problem than the one of interest. Only by using the experimental data can the most appropriate model be chosen.

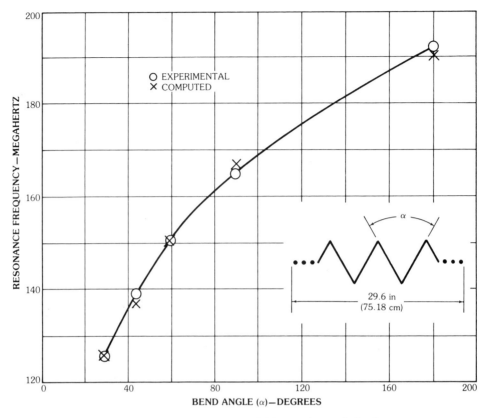

Fig. 17. Example of a near-field experimental check.

Analytical Validation

There are relatively few problems having analytical solutions that can be used as checks on computer models. In the case of wire objects there are accurate closed-form but approximate solutions for straight wire and circular loop antennas, both of which have been employed to validate computed results. Perhaps the only truly analytical solution available for validating wire codes is that for the two-wire transmission line. But since the transmission-line equations do not account for radiation, this check is also an approximate one. For surface objects, the sphere, spheroid, and ellipsoid provide useful check cases.

Unfortunately, all these problems tend to provide relatively undemanding test cases, as straight wires and spheres seem to be the simplest of problems to solve numerically. As checks on computer models they may be viewed more as necessary rather than sufficient conditions for establishing code validity. Analytical concepts, however, do provide the basis for other checks as described below under internal validation.

Numerical Validation

There are essentially two kinds of computations that can be used for numerical validation of a code. These are external checks, in which one code is used to vali-

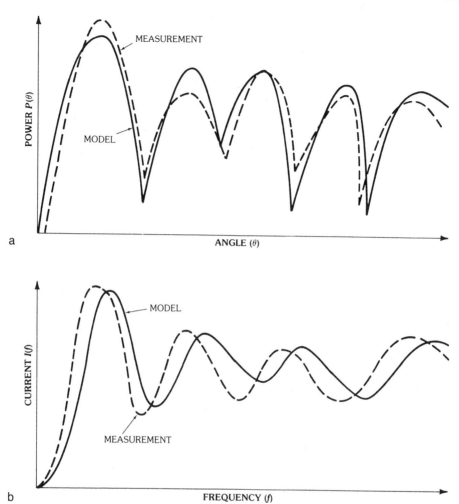

Fig. 18. Generic example of shifts frequently found between computed and measured results. (*a*) As a function of angle. (*b*) As a function of frequency.

date another, and internal checks, wherein various conditions required of a valid solution, but perhaps not routinely examined, are computed. Each kind of check is discussed further below.

External Validation—One of the most reassuring ways of validating a code's numerical performance is to find that, using some standard error measure, it agrees with a different code within some acceptable error bound, ϵ_E, when they are both applied to the same problem. To the extent that these codes are independent in their formulation and numerical treatment, and ϵ_E has a meaningfully small value, agreement between them will imply their mutual validity. However, the value of ϵ_E may signify nothing quantitative about ϵ_N, because neither code can be necessarily assumed to have converged. In addition, we emphasize that such comparisons should preferably utilize an integral or global error measure, as a pointwise or local comparison is less likely to provide a meaningful test.

Internal Validation—Internal validation is achieved by verifying that the code is self-consistent, i.e., satisfies Maxwell's equations in various necessary ways. These include power conservation, boundary conditions, reciprocity relations, and so on. Table 10 summarizes some of these tests, which are briefly discussed below. Most of these are necessary but not sufficient requirements for solution validity. Boundary-condition checks are probably the most convincing, but they are also the most expensive in computer time.

Table 10. Some Checks That Could Be Used for Internal Validation of a Computer Model

Check	Requires	Comments
Power balance	Power supplied by incident field to equal sum of dissipated plus radiated power [(99)–(100)]	Provides a good check on the antenna source model for the radiation resistance and validity of resistive loading. A necessary but not sufficient condition
Reciprocity	Interchanging observation and source locations to yield identical results [(101)]	A useful check for antenna patterns. The receiving pattern is likely to be more dependable than the transmitting pattern because it is an integral (global) quantity. A necessary but not sufficient condition
Boundary condition	The specified conditions on the boundary to be met [(103)–(104)]	The most fundamental check on the solution. Consistency requires that the same weight function be used for this check as are employed in the numerical model. Can be computationally expensive. A necessary and sufficient condition
Convergence	Computed result to approach a limiting value as N is increased	Exhibits behavior of a given quantity as number of unknowns is increased. Can be applied on a local or global basis to any observable. Near-field results provide a more sensitive measure especially on a local or pointwise basis
"Nonphysical behavior"	Computed result to exhibit a physically plausible behavior	Can be a subjective check. One example is provided by spatial oscillation of current over distance much less than the wavelength. Other examples are negative antenna resistance and anomalous near fields

Techniques for Low-Frequency Problems

Power Conservation—For lossless media, all the power supplied to an object must equal the sum of its resistive losses and integrated far-field power flow. This condition can be stated mathematically as [see (26)]

$$\text{Re}\left\{\int_{C(\mathbf{r})} \mathbf{I}(s')\cdot\mathbf{E}^{*i}(s')ds'\right\} = \text{Re}\left\{\int_{C(\mathbf{r})} \mathbf{I}(s')\cdot\mathbf{I}^*(s')R(s')ds'\right\} \\ + \text{Re}\left\{\lim_{R\to\infty}\oint_S \mathbf{E}(\mathbf{r})\times\mathbf{H}^*(\mathbf{r})\cdot d^2\mathbf{r}'\right\} \quad (99)$$

where the term on the left is the supplied power, and the right-side terms account for the dissipated and far-field power, respectively. Note that the supplied-power term arises from integrating the inward Poynting's vector normal to the object's surface due to the exciting field. Equation 99 applies to both antenna and scattering problems, but for the special case of an antenna excited at a single point (segment), the supplied power simplifies to

$$\text{Re}\left\{\int_{C(\mathbf{r})} \mathbf{I}^*(s')\cdot\mathbf{E}^i(s')d^2\mathbf{r}'\right\} = \text{Re}\{I_f^* V_{\text{in}}\} = |I_f|^2 R_{\text{in}} \quad (100)$$

Thus (100) provides an explicit test on the computed input resistance when the object is excited as an antenna.

Reciprocity Relations—Another general requirement of a valid solution to Maxwell's equations is that of reciprocity, for which a general mathematical statement is

$$\int_{C_2(\mathbf{r})} \mathbf{E}_1(s')\cdot\mathbf{I}_2(s')ds' = \int_{C_1(\mathbf{r})} \mathbf{E}_2(s')\cdot\mathbf{I}_1(s')ds' \quad (101)$$

A numerical model can be tested for reciprocity in various ways. Both far-field and near-field tests are useful. An example of the former is

$$\mathbf{E}(\theta_0, \theta_i) = \mathbf{E}(\theta_i, \theta_0) \quad (102)$$

i.e., the field scattered in direction θ_0 due to a plane wave incident from direction θ_i is the same as when the two angles are interchanged.

A near-field test is provided by subjecting the admittance matrix to a test defined by

$$\Delta_j Y_{ij} \cong \Delta_i Y_{ji}$$

where the source segment length is $\Delta_{(\cdot)}$. Because of reciprocity the impedance matrix is symmetric when a Galerkin's technique is used, but with point matching (delta-function weights) this is not the case unless the two segments in question are of equal length.

Boundary Conditions—Although the integral equation is derived with boundary conditions explicitly satisfied everywhere on the object, the numerical solution can only approximate that condition. It is therefore useful to compute the surface fields over the object using the currents (and charges) that have been computed. However, this check cannot be applied using the same testing-function locations as were used to generate the impedance matrix since the boundary conditions at these locations should be satisfied to computer accuracy. In between these locations, though, it can provide a demanding test of solution accuracy.

Such a check is best done in an integral sense, as a few isolated field values are not likely to provide a meaningful test. Furthermore, the errors have a quantitative significance only relative to the exciting field value. One logical measure of boundary-condition mismatch is given by

$$\epsilon_B = \frac{\int_{C(r)} [\mathbf{E}(s') - \mathbf{E}^i(s')]\cdot\hat{\mathbf{s}}' \ [\mathbf{E}^*(s') - \mathbf{E}^{*i}(s')]\cdot\hat{\mathbf{s}}' \ ds'}{\int_{C(r)} [\mathbf{E}^i(s')\cdot\hat{\mathbf{s}}'] \ [\mathbf{E}^{*i}(s')\cdot\hat{\mathbf{s}}'] \ ds'} \qquad (103)$$

This error functional does not include explicitly the possible role of the current in establishing the importance of boundary-condition mismatch. It can happen, for example, that large errors in the field are less important in places where the current is small than where it is large. This is because the field error when multiplied by the current is equivalent to a nonzero power flow normal to the object. Thus an alternative to (103) is

$$\epsilon'_B = \frac{\int_{C(r)} \{\mathbf{I}(s')\cdot[\mathbf{E}(s') - \mathbf{E}^i(s')] \ \mathbf{I}^*(s')\cdot[\mathbf{E}(s') - \mathbf{E}^i(s')]^*\} \ ds'}{\int_{C(r)} [\mathbf{I}(s')\cdot\mathbf{E}^i(s')] \ [\mathbf{I}(s')\cdot\mathbf{E}^i(s')]^* ds'} \qquad (104)$$

Any validation procedure which requires evaluation of near fields can be computationally expensive to implement. Thus, even though boundary-condition checking may be a quantitatively meaningful way to estimate solution validity, it is not viable for routine use. Such checks are generally most valuable when employed selectively, for example when a new kind of problem is being modeled.

Check Cases—It is frequently useful to have results available for standard check cases to use in establishing that a code is continuing to produce valid results. This is especially true when the code has been adapted to a new machine, an updated operating system is introduced, the user has made some changes, and the like. At the very least, some problems which exercise various of the code's features should be described and their output included in the documentation.

6. A Guided Tour of Some Codes and Their Features

In this section we will discuss some of the issues and trade-offs to be considered in developing and selecting a modeling code, and we will present a detailed catalog of representative examples.

Techniques for Low-Frequency Problems

The General Versus the Specific

Compromises must be made in developing any modeling code. Perhaps most immediate is that of deciding on the degree of generality (or complexity) versus specificity (or simplicity) the code is designed for in terms of problems to be treated and features it is to contain. By and large a code intended to handle a wide variety of problems will not be as efficient or accurate for a given problem as a code designed for that particular problem alone. Furthermore, other measures of code "usability," such as storage requirements, user ease, modification ease, and other user resource requirements, may favor the specific code over the general one, even at the expense of the limited scope [44]. Nevertheless, the general code can be an essential and important tool for a wide spectrum of users. The point is that there is a need for both kinds of codes, and the user should be aware of their relative advantages and disadvantages.

Generic Characteristics

Formulation—Although Maxwell's equations provide the starting point for any formulation, various significant differences in capability can arise in the process of developing a suitable mathematical description. Rather than reviewing this issue in detail, instead a few of the major options and their consequences that might be considered at this step are summarized in Chart 18.

Numerical Treatment—Once the formulation has been developed, the needs of the numerical treatment are fairly clear. Furthermore, whether a differential equation, integral equation, hybrid, or some other approach has been selected, these needs are quite similar, and involve the basic steps given in Table 2.

An issue of importance comparable to that of the formulation is the choice of basis and testing functions. We will discuss each briefly here, and summarize in Table 11 examples of some of the more commonly used combinations. Sinusoidal bases are preferable for objects having traveling-wave currents.

Basis Functions—The goal of the basis function is that it provide a match to the unknown to the degree of accuracy being sought using an acceptable number of unknown coefficients. There is quite a degree of latitude provided by this general goal, as several possibly conflicting factors may affect the choice, in a specific situation.

Perhaps the most obvious factor is that of the physical variation the basis function is intended to represent. Other things being equal, that basis function which can provide the closest match to the physical quantity being modeled while using the smallest number of unknowns would be preferable. The situation is hardly ever this simple, however, since the trade-off between impedance-matrix fill time and the subsequent solution time must also be considered. While the latter is smaller the fewer unknowns there are, the former may more than offset this savings by increasing the fill time because of the increasing basis-function complexity. Thus the spatial variability provided by the basis function and evaluation of its fields are its important characteristics.

Another factor is associated with the nature of the integral or differential

Chart 18. A Ranking of Time-Domain and Frequency-Domain Approaches Based on Integral Equation and Differential Equation Formulations with Respect to Various Kinds of Solution Capabilities and Issues

Time Domain		Issue	Frequency Domain	
Differential Equation	Integral Equation		Differential Equation	Integral Equation
		Medium		
√	√	Linear	√	√
−	−	Dispersive	√	√
√	−	Lossy	√	√
√	∼	Anisotropic	√	√
√	−	Inhomogeneous	√	−
√	−	Nonlinear	−	−
√	−	Time-varying	−	−
		Object		
∼	√	Line (Wire)	∼	√
√	√	Closed surface	√	√
√	√	Volume	√	√
∼	√	Open surface	∼	√
		Boundary Conditions		
√	√	Interior problem	√	√
∼	√	Exterior problem	∼	√
√	√	Linear	√	√
√	√	Nonlinear	−	−
√	√	Time-varying	−	−
−	−	Half-space	∼	√
		Other Aspects		
∼	∼	Symmetry exploitation	√	√
∼	√	Far-field evaluation	∼	√
−	∼	Number of unknowns	∼	√
√	∼	Length of code	∼	−
		Suitability for Combining with Other		
−	√	Numerical procedures	√	√
−	−	Analytical techniques	−	√
−	−	Geometrical theory of diffraction	−	√
		Legend		

√ = Highly suited or most advantageous
∼ = Moderately suited or neutral
− = Unsuited or least advantageous

Table 11. Specific Examples of Basis and Weight Function Combinations Used for Wire Models (*After Miller and Deadrick [41]*)

nth Basis	mth Weight	Comments
Subdomain		
Pulse $A_n P_n(s')$	Delta function $\delta(s - s_m)$	Produces point-matched fields. Current and charge are discontinuous at domain junctions
Linear $A_n \ell_n^{(+)} + B_n \ell^{(-)}$	Linear $\ell_m^{(+)} + \ell_m^{(-)}$	A Galerkin's procedure. Provides continuous current
Sinusoidal $A_n s_n^{(+)} + B_n s^{(-)}$	Sinusoidal $s_m^{(+)} + s_m^{(-)}$	A Galerkin's procedure. Provides continuous current. Especially good for longer $(L > 2\lambda)$ wires
Parabolic $A_n + B_n \ell_n^{(+)} + C_n \ell_n^{(+)2}$	Delta function $\delta(s - s_m)$	Produces point-matched fields. Can provide continuous current and charge
Three-term sinusoidal $A_n + B_n s_n + C_n c_n$	Delta function $\delta(s - s_m)$	Produces point-matched fields. Can provide continuous current and charge. Especially good for longer $(L > 2\lambda)$ wires
Entire Domain		
Fourier series $I(s') = \sum A_n e^{jns'}$	Delta function $\delta(s - s_m)$	Used for axisymmetric objects
Legend		
Δs_n = length of nth segment	$\ell_n^{(+)} = s' - s_n + \Delta s_n,$ $s_n = \sin[k(s' - s_n)]$	$s_n^{(+)} = \sin k\ell_n^{(+)},$
s_n = center coordinate of Δs_n along $C(\mathbf{r})$	$\ell_n^{(-)} = s' - s_n - \Delta s_n,$ $c_n = \cos[k(s' - s_n)]$	$s_n^{(-)} = \sin k\ell_n^{(-)},$

operator which generates the field from the sources. Care must be taken, for example, if the order of a differential operator exceeds the differentiability of the basis function. When this happens a nonphysical field behavior can occur, unless an extended interpretation of the differentiation is employed, as is done when using a pulse current basis and a pulse charge basis. If the differentiation is the more subtle kind that occurs from the multiplication of a source by its differentiated Green's function, it is no less a problem. Consequently, a basis function must possess the degree of continuity required by the formulation and numerical treatment.

Weight Functions—A third factor which affects the suitability of a given basis function is the weight function being considered. The weight functions may vary from a simple delta function (this is called *point matching*), to being the same as the basis function (this is called *Galerkin's Method*), to providing even more variability than the basis function. A weight function (other than a delta function) can be viewed, in general, as smoothing out the fields produced by the combination of the basis function and operator. For a given operator, it is the total variability (this might be quantified by the number of degrees of freedom) of the basis-weight function combination that is important. Thus, using a lower-order (fewer degrees of freedom) testing function might be compensated for by using a higher-order basis [45].

As a specific example of this possibility we observe that the combination of piecewise sinusoidal basis and testing functions seems closely comparable to the use of a three-term basis with a delta-function weight in solving the thin-wire electric-field integral equation. Curiously enough, it can also be shown that interchanging the basis and weight functions does not change the impedance matrix elements [45]. This emphasizes the need to evaluate the fields using a testing function that is compatible with the basis function being employed.

Features—While it is the formulation and numerical treatment that establish the features a code is fundamentally capable of providing, it is often the add-ons that make it truly useful. Summarized here, and in Table 12, are some of the more commonly included features to be found, divided into three categories as input, run time, and output features, respectively. Included are what could be expected as a minimum, as well as extended features for more fully developed codes.

Input Features—Aside from the task of developing the parameters of the numerical model to be used, the most time-consuming part of the modeling process for the user is that of actually putting the model data (which consists of both geometrical and electrical information) into the computer.

The problem to be modeled may be described to the computer via input data, stored modules, and computation. The simplest, but most laborious, method for inputting object geometry is to give the required parameters (for example, center coordinates, length, radius, and direction angles) of each individual wire segment (or surface patch) in the object. This segment-oriented and inefficient procedure can be significantly improved by adopting a wire-oriented approach, wherein only the geometry (straight or curved wire, etc.), end points, radius, and number of segments of each wire are given. The computer then computes the data required to specify each segment. An additional enhancement can be realized by storing geometry modules that may occur as parts of more complex objects.

For objects having planar or rotational symmetry the description can be further simplified. Reflections about one to three symmetry plans can expedite model development, as well as shorten the computation time. Rotation about an axis can be similarly used, either for axisymmetric objects or sectors of such objects.

Displaying the object geometry on a display terminal or plotting it out in hard copy is a check that is always worthwhile doing, even for simple objects. At the least a hard-copy plot provides a record that is valuable for interpreting the output data and keeping with it as a permanent record. More than that, these plots can reveal errors in the input data, or verify its validity. A tabular printout of the model data is also useful for quantitative values.

Besides object geometry, model data also includes (for most codes) information about electrical connectivity of the geometrical segments which make up the object. This information is needed to establish current reference directions, to relate the basis function on each segment to those segments connected to it, and to identify open-ended wires where the current goes to zero. Because segments that

Table 12. Examples of Generically Useful Code Features

Function	Minimum	Extended
Input		
Object definition	Specify object coordinates on a segment-by-segment basis	Provide a selection of "building blocks" from which a relatively complex object can be developed
Exciting source description	Specify on a segment basis	Develop exciting field over object from menu of user-selected options
Data check	None	Automatic checking of input data and parameters to identify user violation of recommended guidelines or inconsistent data
Run Time		
Admittance matrices	Verify that their product with the impedance matrix yields the identity matrix	Perform extended reciprocity check using admittance matrix
Current and charge evolution in the time domain	Produce error flag if the computed quantity becomes divergent	Estimate the exponential (SEM) parameters of the computed quantity to predict behavior and terminate computation as soon as possible
Output		
Induced current and charge distributions	Numerical printout	Monitor display or hard copy plotting. Could be interactive as user varies exciting field
Far fields	Numerical printout	Monitor display or hard copy plotting. Could be interactive as user varies exciting field
Near fields	Numerical printout	Monitor display or hard copy plotting. Could be interactive as user varies exciting field

are specified to be electrically connected must have their physical end points at common locations, electrical-connectivity data can be derived from the geometry. But, by providing both kinds of object data independently, a self-consistency check can then be made to validate the input data.

This object data is all that is needed to develop a model, as the N-port admittance matrix can be obtained from such data alone. Additional data is eventually required to specify the excitation, to select whatever run-time options the code provides, and to specify the output desired.

Run-Time Features—Until now, most codes have been designed for batch processing. Little or no control or monitoring of the actual computation is available to the user. The increasing interactivity and speed provided by present and future computers offers some interesting possibilities. Some of the more obvious are given below.

One run-time feature that would give the modeler more control of the computation is an interrupt and restart capability to enable the computation to be stored at an intermediate point. This could be done in order to examine its accuracy, for example, so that an otherwise expensive and valueless run could be terminated and appropriate changes made in the input parameters to redo the computation. Such a capability could be especially valuable when new problems are being modeled.

Another run-time feature of potential value is one which would enable the modeler to make changes in the model parameters during the computation. If an accuracy check revealed that some error measure exceeded a desired threshold value, then having the option of increasing the sampling in that area of the model could perhaps rectify the problem. In essence, the modeler becomes part of a feedback loop, making decisions about modeling that would be difficult to anticipate a priori. After gaining experience with that kind of problem, the feasibility of automating such decisions for similar problems could be established.

Output Features—Even the most elementary code can produce an overwhelming volume of output. Unless this output can be presented to the user in an understandable way, it is likely that not only will important results be missed, but also that wrong conclusions will be reached. At the least this makes the computation less useful than it should be, and at the worse, makes it counterproductive.

Output options provided by most models include a choice by the user of a variety of EM phenomena. For antennas these range from (most often) far-field patterns, input impedance, transfer or mutual impedances, current and charge distributions, and (least often) near fields. Auxiliary quantities derivable from the above include radiated power, input power, antenna loss, antenna efficiency, absolute and directive gain, polarization, field impedance, and insulator and impedance-load voltage drops. This is a broad range of quantities, many of which are not conveniently available from measurement.

The Importance of the Details

It is fair for the prospective user to ask why all the details of the formulation, numerical treatment, etc., should be of interest to him or her. After all, the average computer user does not need comparably detailed information about most of the software provided by a computer's operating system. The difference between these situations is that the typical EM modeling code is neither as simple nor as easy to use as the system routines. Therefore it is important for the user at least to be aware that there are trade-offs involved in designing the various codes that could be considered, and the implications thereof for his or her applications.

For example, consider the specific issue of the basis function used in a particular code. As has been discussed above, one criterion for the basis-function selection is that it provide a reasonable match to the expected current behavior so as to minimize the number of unknowns. In modeling long, traveling-wave

Techniques for Low-Frequency Problems

structures, such as the Beverage and sloping vee antennas, a basis which includes a sinusoid is preferable to one which uses polynomials [43].

A Code Catalog

Since even two codes designed for application to the same kinds of problems may differ greatly, it is worthwhile to identify a set of generic features by which they and other codes can be usefully compared. In this section we outline the properties of a representative collection of codes in these categories: (1) code description; (2) analytical foundation; (3) numerical treatment; and (4) user features. The codes summarized in Table 13 were chosen to illustrate the variability that may typically be encountered between such modeling codes. Equivalent information is not available in all cases so that one-to-one comparisons are not always possible. Nevertheless, information of the kind presented has been found to be useful and may provide a model for the interested reader in comparing codes being considered for his or her own use. Another source of code information, in this case for EMP applications, is given in [46].

A Closer Examination of Two Specific Codes

Two widely available computer codes are NEC (Numerical Electromagnetic Code) [23] and TWTD (Thin-Wire Time Domain) [47, 48, 49]. Their characteristics are summarized in Table 14. These are not the only computer tools that are applicable to wire modeling in the frequency and time domains, but they can be considered representative of this general kind of electromagnetic code. Also, NEC is one of the best documented and most widely used codes, and TWTD is almost the only one available for time-domain modeling of general wire objects.

For simplicity, and because of problem-caused variability, the computer and storage estimates listed in Table 14 do not include the effects of symmetry. Also, these aspects have been discussed previously.

In Table 15 are listed several representative problems to illustrate application of these two codes. Specific problems can always be found which reveal that one approach may have a computational advantage over another. It can be expected, however, that when each produces the equivalent information, without redundant or unneeded results, they generally will exhibit similar efficiency when developed using similar numerical and programming techniques.

Application Guidelines

It should be clear that the successful application of any EM modeling code depends on the user's adhering to the limitations that particular code may be subject to. These limitations may either be intrinsic (or generic) properties of the formulation, numerical implementation, and so on, or extrinsic constraints (specific) introduced by the developer. In either case it is vital that the user be made aware of guidelines appropriate for a given code to increase the probability of its producing valid results. While selecting a numerical model of a physical problem involves a high degree of art, choosing the parameters of the computer model that gets the best performance from a given code is hardly less artful. Table 16 summarizes some generic guidelines for using EM modeling codes.

Table 13. A Catalog of Some Representative Modeling Codes with Respect to Overall Description, Analytical Foundation, Numerical Treatment, and User Features

Issue	FDTD, A. Taflove, Northwestern U., Chicago, IL	Surface Patch Code, E. H. Newman, Ohio State U., Columbus, OH	Main Scattering Program, P. Barber, Clarkson Inst. of Tech.	EM Scattering, Wilton, Rao & Glisson, U. of Miss., University, MS	LASMN3, Hagman, Grandi & Durney, U. of Utah, Salt Lake City, UT	GEMACS, Balestri et al., BDM Corp., Albq., NM	NEC, Burke et al., LLNL, Livermore, CA	MFIE/Aircraft Real Earth Code, Sancer & Siegel, RDA, Santa Monica, CA	COBRA, Medgyesi-Mitchgang, McDonnell Douglas, St. Louis, MO	ROTSY, C. L. Bennett, Raytheon, Sudbury, MA	TWTD, J. A. Landt, Amtech, Los Alamos, NM
Machines	CDC STAR 100 CYBER 203 CYBER 205	IBM 370-165	UNIVAC 1108 CDC 7600 CRAY 1 Minicomp. LSI-11/2	DEC-10	PDP-10	HONEYWELL 6180	CDC 6600 7600 CRAY IBM 360 3033 VAX	CDC 6600 7600 PRIM 750	CDC 6600 7600	UNIVAC 1106 HONEYWELL 635	CDC 6600 CDC 7600 IBM 3601
Availability†	Author Research Code	Author	Author	Author	Author	Author	Author or LLNL	Author	Restricted	Author or LLNL	Author
Length (lines) Storage B = bytes W = words	350 $3(10^6)$ W	4000 26 KW $+ N(N+1)$	1500 50 KB	3000 85 KW $N < 140$	950 $\cong 9$ (no. cells)2 complex W	19000 93 KW	8000 83 KW (7600)	4000 52 KW (7600)	4600 Variable	1000 29 KW	1000–2000 30 KW
Time	$0.3\,\mu s$/cell per time step (CYBER 203)	$\lambda/4$ wire on $1-\lambda^2$ plate $\cong 20$ s (IBM 370-165)	23 min for prolate spheroid of $ka = 5.6$ (TERAK-8510/H)	15–20 min $N + 90$ (DEC-10)	$\cong 20$ hr for $N = 540$ (PDP-10)	3 hr for $N = 1000$ (HON. 6180)	$AN^2 + BN^3$ $A = 3(10^{-4})s$ $B = 2(10^{-6})s$ free space (CDC-7600)	$AN^2 + BN^3$ $A = 2.4(10^{-3})s$ $B = 2(10^{-5})s$ (CDC-7600)	$AN^2 + BN^3$ $A = 1.2(10^{-3})s$ $B = 2(10^{-5})s$ (CDC-7600)	Problem dependent (UNIVAC, HON)	$AN^2N_t + BNN_t$ $A = 2.5(10^{-5})s$ $B = 5(10^{-6})s$ (CDC 7600)
Documentation	Report	J. Article	J. Article	Report	Report	Report	Very thorough 3-part report	Report	Report	Report	Report
Scope, goals	Sinusoidal steady state penetration studies for 2D & 3D objects, using transient approach	Wires and plates (connected)	Scat for axisym dielectric	Conducting body with apertures OK	Absorption of EM by man	General wire modeling	General antennas and scatterers. Wires and surfaces, including ground.	EMP aircraft on real earth	Wire and slot radiators on a body of revolution	Time-domain scattering from closed, rotationally symmetric bodies	Wire objects in time domain

Code Description*

Analytical Formulation

Category	FDTD, A. Taflove, Northwestern U. Chicago, IL	Surface Patch Code, E. H. Newman, Ohio State U., Columbus, OH	Main Scattering Program, P. Barber, Clarkson Inst. of Tech.	EM Scattering, Wilton, Rao & Glisson, U. of Miss., University, MS	LASMN3, Hagman, Grandi & Durney, U. of Utah, Salt Lake City, UT	GEMACS, Balestri et al., BDM Corp., Albq., NM	NEC, Burke et al., LLNL, Livermore, CA	MFIE/Aircraft Real Earth Code, Sancer & Siegel, RDA, Santa Monica, CA	COBRA, Medgyesi-Mitchgang, McDonnell Douglas, St. Louis, MO	ROTSY, C. L. Bennett, Raytheon, Sudbury, MA	TWTD, J. A. Landt, Amtech, Los Alamos, NM
Domain (T = time, F = frequency)	T	F	F	F	F	F	F	F	F	T	T
Equations EFIE = electric integral eq. MFIE = magnetic field integral eq.	Maxwell 3D	Reaction Integral	EF-MF hybrid integral equation	EFIE mixed potential	EFIE	EFIE wires, MFIE for surfaces	EFIE wires, MFIE for surfaces	MFIE	EFIE	MFIE	EFIE
Media type:											
Lossless	√	√	√	√		√	√	√	√	√	√
Lossy	√		√	√	√		√	√	√		
Dispersive	√							√			
Inhomogeneous	√										
Nonlinear	√										
Object impedance	Volume and surface impedance	Lumped or distorted RLC on wires	Lossy	None	Lossy dielectric	Lumped or distributed RLC on wires	Lumped or distributed RLC on wires	None	Lumped or distributed RLC on wires	Surface impedance	Lumped or distributed RLC, lumped nonlinear
Geometry:											
Wire	Cables	√									
Surface‡	Closed	Open	Closed	Open		Closed	Closed	Closed	Closed	Closed	
Penetrable volume	√	√	√	√	√			√	√	√	√
Discontinuities:											
Wire bends	√	√				√	√		√	√	√
Wire junctions	√	√					√		√		√
Edges	√	√		√			√	√	√	√	
Apertures	√			√				√	√	√	
Approximations	Linear approx. of curved surfaces	≥ 24 basis functions/λ^2		Triangle patch edges ≤ $\lambda/8$	Cell size ≤ $\lambda/5$	Essentially same as NEC	Thin wire approx.; pulse basis functions on surface plates	$k \cdot$(body length) < 10–20	Norton fields for earth	Object size < 2 pulse-widths	Linear approx. of curved wire

Table 13, *continued.*

Issue	FDTD, A. Taflove, Northwestern U., Chicago, IL	Surface Patch Code, E. H. Newman, Ohio State U., Columbus, OH	Main Scattering Program, P. Barber, Clarkson Inst. of Tech.	EM Scattering, Wilton, Rao & Glisson, U. of Miss., University, MS	LASMN3, Hagman, Grandi & Durney, U. of Utah, Salt Lake City, UT	GEMACS, Balestr. et al., BDM Corp., Albq., NM	NEC, Burke et al., LLNL, Livermore, CA	MFIE/Aircraft Real Earth Code, Sancer & Siegel, Santa Monica, CA	COBRA, Medgyesi-Mitchgang, McDonnell Douglas, St. Louis, MO	ROTSY, C. L. Bennett, Raytheon, Sudbury, MA	TWTD, J. A. Landt, Amtech, Los Alamos, NM
Method	Finite difference	MOM	MOM, Volume	MOM	MOM, Volume	MOM	MOM	MOM	MOM	MOM Time Step	MOM Time Step
Basis functions	Pulse	Piecewise sine on wires and surfaces	Spherical expansion	Linear on pair of triangles	Pulse, interpolated	3-term trig. on wires, impulse on surfaces	3-term trig. on wires, impulse on surfaces	Pulse	Triangle and $e^{jn\phi}$	Pulse in space, polyn. in time	quadratic in space and time
Weight functions	Delta functions	Same	Same	Same	Impulse	Impulse	Impulse	Pulse	Same	Impulse	Impulse
Boundary conditions	Dirichlet, Neumann, impedance, radiation	Object surface	Extended bc	Object surface	Extended bc	E on wires, H on surfaces	Object surface	Object surface and earth interface	Object surface	Object surface	Wire surface
Symmetry exploitation	No	No	Plane, rotational	No	Plane	Plane rotational translation	Plane rotational	Plane	Rotational	Rotational discrete	Rotational discrete
Integration	N/A	Numerical	Numerical	Numerical	Numerical	Numerical	Analytical, adaptive numerical	Numerical self term	Numerical	Numerical self term	Analytical closed form
Antenna radiation	No	Yes	No	No	No	Yes	Yes	No	Yes	No	Yes
Scattering	Yes	Yes	Yes	Yes	Yes	Yes	Yes	Yes	No	Yes	Yes
Ground	No	No	No	No	No	Perfectly conducting	RCA, Sommerfeld	Sommerfeld	RCA	No	Perfect ground, dielectric plane
Other	—	Basis functions provide current cont. at edges	—	—	—	Banded matrix technique	—	—	—	—	—

User Features

Feature	FDTD, A. Taflove, Northwestern U., Chicago, IL	Surface Patch Code, E. H. Newman, Ohio State U., Columbus, OH	Main Scattering Program, P. Barber, Clarkson Inst. of Tech.	EM Scattering, Wilton, Rao & Glisson, U. of Miss., University, MS	LASMN3, Hagman, Grandi & Durney, U. of Utah, Salt Lake City, UT	GEMACS, Balestri et al., BDM Corp., Albq., NM	NEC, Burke et al., LLNL, Livermore, CA	MFIE/Aircraft Real Earth Code, Sancer & Siegel, Santa Monica, CA	COBRA, Medgyesi-Mitchgang, McDonnell Douglas, St. Louis, MO	ROTSY, C. L. Bennett, Raytheon, Sudbury, MA	TWTD, J. A. Landt, Amtech, Los Alamos, NM
Input data	Manual	Gross plate geometry	Cards (large computers); interactive (minicomp.)	Cards	Manual	Field-free language	Batch	Batch	Computer assisted	Manual	Manual batch
Output	Tabular, primitive graphics	Tab., orthographic plot of geometry	Tabular post-plotter	Tabular	Tabular	Line printer	Tabular	Menu sel. tabular	Menu selected	Tabular	Tabular plots
Input error checking	No	No	Yes	No	No	Yes	Input data checks	Minimal	On input data and during runs	Input checks, run-time checks	Input checks
Run-time checking	No	No	Yes	No	No	Yes	Integrated $E \times H$	No	No	Some	Some
Test problems	3-dimensional cylinders	Wires, wire on plates intersec. plates	Prolate spheroids	Sphere	Block man-model on ground plane	Wire grid sphere, reflectors	Yes	Sphere, cyl. aircraft model, cyl. slab	Monopoles on sphere; cone-sphere	Sphere	Dipole loop spiral
Debug help	No	No	No	Yes	?	Extensive, EMC/IAP ctr	LLNL	Yes	No	Yes	Yes
Documentation within code	Yes	?	?	No	?	?	Some	?	Yes	Yes	Some

*Information supplied here is based primarily on responses given to a questionnaire. It is intended more to be illustrative of what code characteristics might be of interest than to be definitive about the codes actually listed.

†Codes selected for inclusion here are thought to be available from the authors, or can be obtained from LLNL as indicated.

‡An open surface is a conducting plate or shell which can become closed.

Table 14. Closer Comparison of NEC and TWTD Wire Codes (*After Miller [40]*)

Feature	NEC	TWTD
Language	FORTRAN IV	FORTRAN IV
Machines	CDC-6800, 7600; UNIVAC-1110; IBM-360/65 VAX11 CRAY 1, IBM 3033, XEROX Sigma 9	CDC-6600, 7600; IBM-360
Number of statements (cards)	8000	1000–2000 (depending on features)
Storage required (decimal, words)		
Program	$\cong 83\,000$	$\cong 30\,000$
Variable	$\cong 30N + 2N^2$	$\cong (12)N^2 + 2NN_t$
Time required (CDC-7600, seconds)	$\cong (10^{-4}N^2 + 2 \times 10^{-6}N^3)N_f$	$\cong 2.5 \times 10^{-5}N^2 N_t N_i + 5 \times 10^{-6} NN_t N_i N_a$
Objects treated	Wires	Wires
	Surfaces (continuous or wire grid)	Wire-grid surface
	Wire-surface combination	
MOM technique used	Subsectional collocation	Subsectional collocation
Expansion (basis) functions	3-term trigonometric in space	9-term quadratic in space-time
Junction treatment, 2-wire	Amplitude and slope matching at segment juncture	Matches adjacent current samples
Junction treatment, multiwire	Charge density $\alpha \log(\alpha a)$ center	Matches equivalent segment at center
Testing (weight) function	Delta (point match)	Delta (point match)
Exciting (incident) field:		
Antenna	Tangential electric field, charge discontinuity, current source	Tangential electric field
Scatterer	Plane wave, local point source	Plane wave, local point source
Impedance loading:		
Lumped	Linear R, L, C	Nonlinear R, L, C
Distributed	Linear R, L, C; imperfectly conducting wire	Nonlinear R, L, C
Circuits	Nonradiating networks (including transmission lines)	Nonlinear networks

Symmetry included		
Planar	1 to 3 orthogonal planes	1 plane (storage and current computation)
Rotational	Up to 16 sectors	Up to n-fold (storage only)
Environments	Infinite medium (free-space or lossy). Lossy ground, wires on either side of or penetrating the interface	Free space. Perfect ground. Lossless, nondispersive, dielectric half-space
Input data preparation	Specify end points and connecting path (straight or circular arc), number of segments and radii. Provides for reflection or discrete rotation for object symmetry Excitation Options for output data	Specify end points and connecting (straight only), number of segments and radii. Also, special curved shapes (loop, spiral). Provides for reflection about ground plane. Time-variation of exciting field (functional or tabular). Options for output data
Output data options	Impedance, admittance, input power, current and charge distributions, efficiency, radiation patterns; directive gain, power gain, average gain, "worst case" coupling, scattering cross section, near **E** & **H** fields, receiving pattern NGF — file which can be used to add to the model and implement a partitioned matrix solution	Current and charge distributions near electric field, far fields in time domain and, via fast Fourier transform, impedance, admittance, far field, radiation pattern in frequency domain
Diagnostics	Data card error flags (geometry, dimension) Run-time error messages (integration, pivot element)	Checks for time-step consistency ($\Delta tc \leq \Delta_{max}$)
Documentation	Extensive, 3-volume set [23]: 1. Program description — theory 2. Program description — code 3. User's guide	Moderate: two somewhat redundant user's manuals [47, 48]

Note: N is number of space samples ($>2\pi L/\lambda$)
N_t is number of time steps ($\cong 2N$)
N_f is number of frequency steps ($\cong N$)
N_i is number of incident fields ($\cong 2\pi N$)
N_a is number of observation angles ($\cong 2\pi N$)

These are sampling estimates based on the need to resolve spatial, temporal, or spectral variations to resolutions approximately 2π samples per wavelength or hertz. For N_i and N_a, there are assumed to be two principle-plane pattern cuts of 2π radian coverage with a total number of lobes of approximately (4 to 8)L/λ. Thus, with approximately 2π samples per lobe, $N_i \cong N_a \cong 2\pi N$. Factors of 2 variation in any of these sampling rates would not be uncommon.

Table 15. A Set of Representative Problems and Computer Times Using NEC and TWTD (*After Miller [40]*)

Structure	Number N of Space Samples (NEC and TWTD)	Number N_t of Time Samples (TWTD)	Computer Time* CDC-7600 (s) TWTD ($N_i = 1$)	Computer Time CDC-7600 (s) NEC ($N_f = 1$)
Straight wire	48	480 (Well-converged solution)	$\cong 60$	$\cong 1$
Conical spiral	60	600 (Persistent ringing without resistive loading)	$\cong 100$	$\cong 1.5$
LP antenna	120	120	$\cong 45$	$\cong 8$
Truck (over perfect ground)	70	250	$\cong 60$	$\cong 4$
General			$\cong 2.5 \times 10^{-5} N^2 N_t N + 5 \times 10^{-6} N N_t N_a$	$\cong (5 \times 10^{-4} N^2 + 2 \times 10^{-6} N^3) N_f$

*Time is quite sensitive to i/o; a factor of 2 variation is not uncommon.

Table 16. Generic Modeling Guides

Parameter or Issue	Nominal value or Range	Reason
Wire length (L)	$L > 10d$	Neglect of end caps in thin-wire treatment
Wire diameter* (d)	$\lambda > \pi d$	Neglect of circumferential variation in current
Wire segment length* (Δ):		
As related to diameter	$\Delta > d$	Use to thin-wire kernel in integral equation
As related to wavelength	$\lambda > 2\pi\Delta$ ($\Delta < \lambda/2\pi$)	Necessity of sampling current densely enough in wavelengths
Step change in wire radius* (oa).	$\Delta > 10oa$	Neglect of treatment for sources on stepped surface (similar to end-cap problem)
Source location	Do not place on last segment of open-ended wire.	Avoids nonphysical situation of driving wire at open end
Angle of wire bend (α).	$a > 6a/k\Delta$	Keep adjacent wires from occupying too large a common volume
Axial separation of parallel wires.		Neglect of circumferential variation of current
With match points aligned	$r \gtrsim 5a$	Avoids placing one match point in error field of other wire
> otherwise	$r \gtrsim 10a$	
Surface patch area (Δ_s):		Need to sample surface currents densely enough in wavelengths
Frequency domain	$\lambda > 6\sqrt{\Delta_s}$	
Time domain	$\lambda > 4\sqrt{\Delta_s}$	
Piecewise linear model of curved wire or curved surface of radius of curvature R	$R > \Delta$ or $\sqrt{\Delta_s}$	Necessity of sampling a circular arc at least six times per 2π radians
Starting time in time-domain solution (t_{st})	$E^i_v(t)\big|_{max} \gtrsim 100\, E(t_{st})$	Achieves smooth buildup of exciting field
Stopping time† in time-domain solution (t_{sp})	$E(t)\big|_{max} \gtrsim 100|E(t) - E(t_{sp})|$ or $I(t)$ reaches steady state	Permits final current or field value to stabilize
Time step (ϱ) in time-domain solution (c = velocity of light)	$c\varrho \lesssim \Delta$	Satisfies Courant stability condition. Required for MFIE, but not for EFIE
Maximum frequency of transient source in time-domain solution (assume Gaussian source $e^{-g^2 t^2} \Rightarrow e^{-\omega^2/4g^2}$	$g \propto 2f_{max} = \omega_{max}/\pi$	Ensures source spectrum does not exceed upper frequency for model validity

*Wavelength of frequency-domain solution or approximate minimum wavelength for valid time-domain solution.
†If steady-state response is not needed, then to t_{sp} could be substantially shorter.

Acknowledgments

The authors are indebted to the following individuals for their gracious support in the preparation of this chapter:

Marian Holten
Rose O'Brien
Margaret Poggio
Diane Ray

7. References

[1] R. F. Harrington, *Time-Harmonic Electromagnetic Fields*, New York: McGraw-Hill Book Co., 1961.
[2] J. A. Stratton, *Electromagnetic Theory*, New York: McGraw-Hill Book Co., 1941.
[3] S. Silver, *Microwave Antenna Theory and Design*, New York: McGraw-Hill Book Co., 1949.
[4] S. A. Schelkunoff, *Advanced Antenna Theory*, New York: John Wiley & Sons, 1952.
[5] S. A. Schelkunoff and H. T. Friis, *Antennas: Theory and Practice*, New York: John Wiley & Sons, 1952.
[6] B. B. Baker and E. T. Copson, *The Mathematical Theory of Huygens' Principle*, 2nd ed., London: Oxford University Press, 1953.
[7] A. J. Poggio and E. K. Miller, "Integral equation solutions for three-dimensional scattering problems," in *Computer Techniques for Electromagnetics*, ed. by R. Mittra, New York: Pergamon Press, pp. 159–264, 1973.
[8] E. A. Wolff, *Antenna Analysis*, New York: John Wiley & Sons, 1966.
[9] W. L. Weeks, *Antenna Engineering*, New York: McGraw-Hill Book Co., 1968.
[10] H. Jasik, ed., *Antenna Engineering Handbook*, New York: McGraw-Hill Book Co., 1961.
[11] R. E. Collin and F. J. Zucker, eds., *Antenna Theory*, pt. 1, New York: McGraw-Hill Book Co., 1969.
[12] R. E. Collin and F. J. Zucker, eds., *Antenna Theory*, pt. 2, New York: McGraw-Hill Book Co., 1969.
[13] W. L. Stutzman and G. A. Thiele, *Antenna Theory and Design*, New York: John Wiley & Sons, 1981.
[14] J. D. Kraus, *Antennas*, New York: McGraw-Hill Book Co., 1950.
[15] S. Ramo and J. R. Whinnery, *Fields and Waves in Modern Radio*, New York: John Wiley & Sons, 1960.
[16] C. H. Walker, *Traveling-Wave Antennas*, New York: Dover Publications, 1965.
[17] R. M. Bevensee, *A Handbook of Conical Antennas and Scatterers*, New York: Gordon and Breach Science Publishers, 1973.
[18] C. H. Papas, "Input impedance of wide-angle conical antennas," *Cruft Lab Tech. Rep. 52*, Harvard Univ., 1948.
[19] P. D. P. Smith, "The conical dipole of wide angle," *J. Appl. Phys.*, vol. 19, 1948.
[20] C. T. Tai, "On the theory of biconical antennas," *J. Appl. Phys.*, vol. 20, 1949.
[21] O. M. Salati, "Antenna chart for system designers," *Electron. Engineer*, January 1968.
[22] A. B. Bailey, *TV and Other Receiving Antennas*, New York: J. F. Ryder Publisher, 1950.
[23] G. J. Burke and A. J. Poggio, "Numerical electromagnetic code (NEC)—method of moments, vol. 1, pt. I: program description—theory; vol. 1, pt. II: program description—code; vol. 2, pt. III: user guide," *Tech. Doc. 116*, San Diego, CA 92152: Naval Ocean Systems Center, 1977.
[24] E. K. Miller, A. J. Poggio, G. J. Burke, and E. S. Selden, "Analysis of wire antennas

in the presence of a conducting half space, pt. I: the vertical antenna in free space," *Can J. Phys.*, vol. 50, 1972.

[25] E. K. Miller, A. J. Poggio, G. J. Burke, and E. S. Selden, "Analysis of wire antennas in the presence of a conducting half space, pt. II: the horizontal antenna in free space," *Can. J. Phys.*, vol. 50, 1972.

[26] H. Bach and J. E. Hansen, "Uniformly spaced arrays," in *Antenna Theory*, pt. I, ed. by R. E. Collin and F. J. Zucker, New York: McGraw-Hill Book Co., 1969.

[27] M. T. Ma, *Theory and Application of Antenna Arrays*, New York: John Wiley & Sons, 1974.

[28] A. W. Maue, "The formulation of a general diffraction problem by an integral equation," *Z. für Phys.*, Bd. 126, pp. 601–618.

[29] H. H. Chao and B. J. Strait, "Radiation and scattering by configurations of bent wires with junctions," *IEEE Trans. Antennas Propag.*, vol. AP-19, no. 5, 1971.

[30] E. Hallen, "Theoretical investigation into transmitting and receiving antennae," *Nova Acta Regiae Societatis Scientiarum Upsaliensis* (Sweden), 1938.

[31] K. K. Mei, "On the integral equation of thin wire antennas," *IEEE Trans. Antennas Propag.*, vol. AP-13, 1965.

[32] A. Sommerfeld, *Partial Differential Equations*, New York: Academic Press, 1969.

[33] A. Banos, *Dipole Radiation in the Presence of a Conducting Half Space*, New York: Pergamon Press, 1966.

[34] R. Mittra, "Integral equation methods for transient scattering," in *Transient Electromagnetic Fields*, ed. by L. B. Felsen, New York: Springer-Verlag, 1973.

[35] D. L. Sengupta and C.-T. Tai, "Radiation and reception of transients by linear antennas," in *Transient Electromagnetic Fields*, New York: Springer-Verlag, 1973.

[36] E. K. Miller, A. J. Poggio, and G. J. Burke, "An integro-differential equation technique for the time-domain analysis of thin-wire structures, I: the numerical method," *J. Comput. Phys.*, vol. 12, no. 1, 1973.

[37] A. J. Poggio, E. K. Miller, and G. J. Burke, "An integro-differential equation technique for the time-domain analysis of thin-wire structures, II: numerical results," *J. Comput. Phys.*, vol. 12, no. 2, 1973.

[38] R. F. Harrington, *Field Computation by Moment Methods*, New York: Macmillan Co., 1968.

[39] E. K. Miller, "Some computational aspects of transient electromagnetics," *Rept. UCRL 51276*, Lawrence Livermore Lab., 1972.

[40] E. K. Miller, "Time-domain modeling of wires," in *Application of the Method Moments to Electromagnetic Fields*, ed. by B. J. Strait, Kissimmee, Florida: SCEEE Press, 1980.

[41] E. K. Miller and F. J. Deadrick, "Some computational aspects of thin-wire modeling," in *Numerical and Asymptotic Techniques in Electromagnetics*, ed. by R. Mittra, Berlin: Springer-Verlag, 1975.

[42] E. K. Miller, G. J. Burke, and E. S. Selden, "Accuracy modeling guidelines for integral-equation evaluation of thin wire structures," *IEEE Trans. Antennas Propag.*, vol. AP-19, pp. 534–536, 1975.

[43] A. J. Poggio, R. M. Bevensee, and E. K. Miller, "Evaluation of some thin-wire computer programs," *1974 Intl. IEEE/AP-S Symp. Dig.*, Georgia Tech, Atlanta, GA.

[44] E. K. Miller and A. J. Poggio, "Moment method techniques in electromagnetics from an applications viewpoint," in *Electromagnetic Scattering*, ed. by P. O. E. Uslenghi, New York: Academic Press, 1978.

[45] D. R. Wilton and C. M. Butler, "Effective methods for solving integral and integro-differential equations," *Electromagnetics*, vol. 1, pp. 289–308, 1981.

[46] R. M. Bevensee, J. N. Brittingham, F. J. Deadrick, T. H. Lehman, E. K. Miller, and A. J. Poggio, "Computer codes for EMP interaction and coupling," *IEEE Trans. Antennas Propag.*, vol. AP-26, pp. 156–165, 1978.

[47] M. Van Blaricum and E. K. Miller, "TWTD: a computer program for the time-domain analysis of thin-wire structures," *Rept. UCRL 5127*, Lawrence Livermore Laboratory, 1972.

[48] J. A. Landt, E. K. Miller, and M. L. Van Blaricum, "WT-MBA/LLL1B: a computer program for the time-domain response of thin-wire structures," *Rept. UCRL 51585*, Lawrence Livermore Laboratory, 1974.

[49] E. K. Miller and J. A. Landt, "Direct time-domain techniques for transient radiation and scattering from wires," *Proc. IEEE*, vol. 68, pp. 1396–1423, 1980.

Chapter 4
Techniques for High-Frequency Problems

P. H. Pathak
Ohio State University ElectroScience Laboratory

CONTENTS

1. Introduction — 4-3
2. Wavefronts, Rays, and the Geometrical Optics Field — 4-6
    The Ray Concept 4-6
    The Geometrical Optics Field 4-8
3. The Physical Optics Field — 4-18
    Three-Dimensional Case 4-19
    Two-Dimensional Case 4-20
4. The Geometrical Theory of Diffraction and Its Uniform Versions — 4-23
    General Form of the GTD Diffracted-Ray Fields E_k^d and H_k^d 4-23
    Uniform Version of the GTD 4-31
5. The Equivalent Current Method — 4-96
    ECM for Edge-Diffracted Ray Caustic Field Analysis 4-97
6. The Physical Theory of Diffraction and Its Modifications — 4-102
    PTD for Edged Bodies 4-104
7. References — 4-109

Prabhakar H. Pathak received the BSc degree in physics from the University of Bombay, India, in 1962, the BS degree in electrical engineering from Louisiana State University, Baton Rouge, in 1965, and the MS and PhD degrees in electrical engineering from Ohio State University, Columbus, in 1970 and 1973, respectively.

From 1965 to 1966 he was an instructor in the Department of Electrical Engineering at the University of Mississippi, Oxford. During the summer of 1966, he worked as an electronics engineer with the Boeing Company in Renton, Washington. Since 1968 he has been with the Ohio State University ElectroScience Laboratory, where his research interests have centered around mathematical methods, electromagnetic antenna and scattering problems, and uniform ray techniques. He is also an associate professor in the Department of Electrical Engineering at Ohio State University, where he teaches courses in electromagnetics, antennas, and linear systems.

Dr. Pathak has participated in invited lectures, and several short courses on the uniform geometrical theory of diffraction, both in the United States and abroad. He has also coauthored chapters on ray methods for four books. Dr. Pathak is a member of Commission B of the International Scientific Radio Union (URSI), and of Sigma Xi. He is a Fellow of the IEEE.

1. Introduction

Techniques based on the method of modal expansions, the Rayleigh-Stevenson expansion in inverse powers of the wavelength, and also the method of moments solution of integral equations are essentially restricted to the analysis of electromagnetic radiating structures which are small in terms of the wavelength. It therefore becomes necessary to employ approximations based on "high-frequency techniques" for performing an efficient analysis of electromagnetic radiating systems that are large in terms of the wavelength.

One of the most versatile and useful high-frequency techniques is the *geometrical theory of diffraction* (GTD), which was developed around 1951 by J. B. Keller [1, 2, 3]. A class of diffracted rays are introduced systematically in the GTD via a generalization of the concepts of classical geometrical optics (GO). According to the GTD these diffracted rays exist in addition to the usual incident, reflected, and transmitted rays of GO. The diffracted rays in the GTD originate from certain "localized" regions on the surface of a radiating structure, such as at discontinuities in the geometrical and electrical properties of a surface, and at points of grazing incidence on a smooth convex surface as illustrated in Fig. 1. In particular, the diffracted rays can enter into the GO shadow as well as the lit regions. Consequently, the diffracted rays entirely account for the fields in the shadow region where the GO rays cannot exist. Thus the GTD overcomes the failure of GO in the shadow region; it also improves the GO solution in the lit region. In this sense the GTD constitutes a significant improvement over GO. The initial amplitude of a diffracted ray is given in terms of a diffraction coefficient just as the initial values of GO reflected and transmitted rays are given in terms of reflection and transmission coefficients. Away from the points of diffraction the diffracted rays in the GTD propagate according to the laws of ordinary GO. The diffraction coefficients can be found from the asymptotic solutions to appropriate canonical problems. As a result, the GTD provides an efficient high-frequency solution to problems that cannot be solved rigorously. For example, an analysis of complex radiating systems such as antennas on aircraft, missiles, or ships can be performed efficiently by simulating those structures with approximate models which are "built up" from simpler shapes. These simpler shapes carefully simulate parts of the actual structure that dominate the reflection and diffraction effects, as is described later in Chapter 20 dealing with the engineering applications of GTD.

The ability of the GTD to accurately solve the problems of electromagnetic radiation and scattering from complex structures in a relatively simple, physically appealing, and efficient manner makes it a very powerful tool for antenna engineers. Moreover, the GTD can also provide information on ways to control the radiation and scattering. Although the GTD is a high-frequency technique it

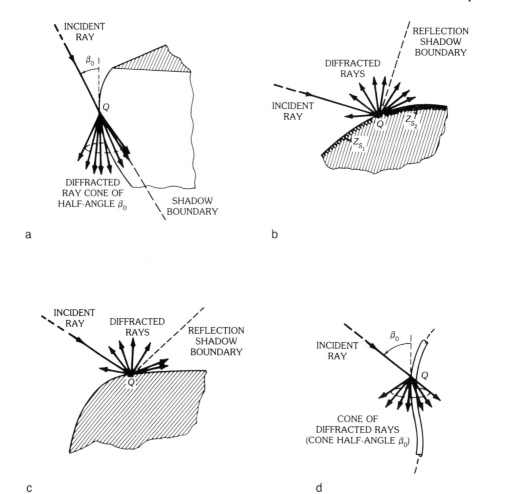
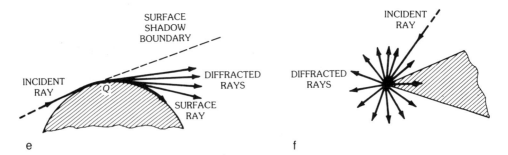

Fig. 1. Examples of diffracted rays. (*a*) Diffraction by a curved wedge. (*b*) Diffraction at a discontinuity in surface impedance (Z_{s_1} and Z_{s_2}). (*c*) Diffraction at a discontinuity in surface curvature. (*d*) Diffraction by a thin, curved wire. (*e*) Diffraction by a smooth, convex surface. (*f*) Diffraction by a vertex in a plane screen.

works surprisingly well in many situations, even for radiating objects almost as small as a wavelength in extent. Even though the GTD is not rigorous it yields the leading terms in the asymptotic expansions of many diffraction problems. However, because the GTD is a purely ray optical theory it fails within the transition regions adjacent to the GO incident and reflection shadow boundaries that divide the space surrounding an illuminated structures into the lit and shadow zones; it also at and near caustics* of diffracted rays just as GO fails at caustics of GO rays.

The failure of GTD within the incident and reflection shadow boundary transition regions can be overcome through the use of uniform ray techniques based on the *uniform geometrical theory of diffraction* (UTD) [4,5], and the *uniform asymptotic theory* (UAT) [6]. The UTD and UAT automatically reduce to the GTD outside the shadow boundary transition regions where the latter theory becomes valid. In engineering applications it is essential to use the uniform version of the GTD, namely the UTD or UAT, because the diffracted field generally assumes its strongest value at the shadow boundary where the GTD fails. The GTD, UTD, and UAT fail within the diffracted-ray caustic regions; in these regions the GTD, UTD, and UAT may be augmented by an *equivalent current method* (ECM) [7,8,9], which indirectly employs the GTD far from caustics to obtain equivalent currents that radiate fields valid at caustics. The ECM, which generally reduces to the GTD outside the caustic regions, is good only when the GO shadow boundary transition regions do not overlap with the diffracted-ray caustic regions. The GTD, UTD, UAT, and ECM all generally fail in regions where there is an overlap of the GO shadow boundary transition regions with the diffracted and/or GO ray caustic regions; in such regions the fields may be calculated via the *physical theory of diffraction* (PTD) [10,11] and its modifications. The PTD was developed by P. Ya. Ufimtsev in the Soviet Union at about the same time as Keller's GTD, and it constitutes a systematic extension of the method of physical optics (PO) just as GTD is an extension of GO. The PO field is calculated from a radiation integral [12] which employs a GO approximation for the surface currents on the structure. At the present time the PTD is developed only for edged bodies and is therefore not as general as the GTD and its uniform versions.

The format of this chapter is as follows. First the GO and PO high-frequency methods are reviewed in Sections 2 and 3. Then a discussion of the GTD, UTD, and UAT is presented in Section 4. Subsequently the ECM and PTD are discussed in Sections 5 and 6. The treatment in this chapter is restricted to perfectly conducting structures in an isotropic, homogeneous medium. Not included in this chapter are some recent generalizations of the ray methods, such as those dealing with Gaussian beams [13], and the hybrid combinations of ray methods with the moment method or modal techniques [14,15,16]. In this chapter the fields are assumed to be time harmonic with an $e^{j\omega t}$ time dependence, which is suppressed. Only here does ω refer to the angular frequency and t refer to time.

---

*Ray caustics or focii occur whenever a family of rays (i.e., ray congruences) merge or intersect to form a focal surface, a focal line, or a focal point.

2. Wavefronts, Rays, and the Geometrical Optics Field

In this section some background material is introduced briefly to indicate the connection between the wavefronts and rays at high frequencies, and to also provide examples of ray optical fields. In addition, the basic ideas of the geometrical optics (GO) ray technique are reviewed.

The Ray Concept

A wavefront is an equiphase surface. At high frequencies the electromagnetic energy flow in an isotropic medium is associated primarily with the propagation of its wavefront along curved paths which are everywhere normal to the wavefront. Such highly localized paths of wave propagation which are directed along the normals to the wavefronts are called *rays*; these ray paths are straight lines in a homogeneous medium. The family of rays associated with the propagation of a wavefront is usually referred to as a *normal congruence of rays*.

The above connection between wavefronts and rays may be established via an application of the method of stationary phase to the radiation integral over the equivalent (Huygens') sources which are associated with the field distribution on the wavefront; this procedure constitutes a straightforward generalization of the scalar treatment of rays and wavefronts given by Silver [12]. Let $\mathbf{E}(\mathbf{r}')$ and $\mathbf{H}(\mathbf{r}')$ refer to the electric and magnetic fields at any point $\mathbf{r}'$ on the wavefront surface S', and let dS' be an elemental area of S' at the point $\mathbf{r}'$ as shown in Fig. 2. Let $d\mathbf{E}(\mathbf{r})$ denote the field associated with a spherical wave at P which originates from dS'. Thus the total electric field $\mathbf{E}(\mathbf{r})$ at P is then obtained by superposing the spherical wave contributions from each elemental dS' over the entire surface S'; namely,

$$\mathbf{E}(\mathbf{r}) = \iint_{S'} d\mathbf{E}(\mathbf{r}) \tag{1a}$$

where

$$d\mathbf{E}(\mathbf{r}) = \frac{jkZ_0}{4\pi} [\hat{\mathbf{R}} \times \hat{\mathbf{R}} \times \mathbf{J}_s(\mathbf{r}') + Y_0 \hat{\mathbf{R}} \times \mathbf{M}_s(\mathbf{r}')] \frac{e^{-jkR}}{R} dS' \tag{1b}$$

in which the equivalent (or Huygens') electric and magnetic current sources on S' are given by $\mathbf{J}_s(\mathbf{r}') = \hat{\mathbf{n}}' \times \mathbf{H}(\mathbf{r}')$ and $\mathbf{M}_s(\mathbf{r}') = \mathbf{E}(\mathbf{r}') \times \hat{\mathbf{n}}'$. It is noted that $Z_0 = (Y_0)^{-1}$ denotes the wave impedance in the medium, k constitutes the wave number of the medium, and $\hat{\mathbf{n}}'$ denotes the unit outward normal vector to S' at $\mathbf{r}'$.

For convenience the point O is chosen in the present development to correspond to a point on S' which is nearest to a given observation point P such that $OP = \hat{\mathbf{z}}|OP|$, and also such that the unit outward normal vector $\hat{\mathbf{n}}'$ to the surface S' is given by $\hat{\mathbf{n}}' = \hat{\mathbf{z}}$ at O, as in Fig. 2. Generally, there exists at least one such point O. To start with, it is assumed that there exists only one such point O on S'; otherwise, the point P is said to be a *caustic* or *focal point*.

It can be shown via the principle of stationary phase [12] that e^{-jkR} in the integrand of (1b) oscillates rapidly for large k to produce a destructive interference or cancellation between the different spherical wave contributions to P in (1a) that

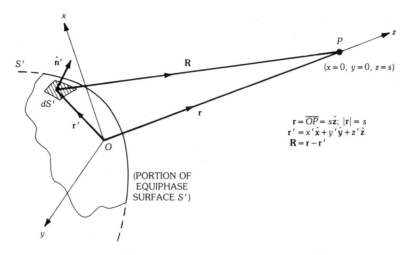

Fig. 2. Wavefront geometry.

arise from each of the equivalent sources on dS' at points on S' which do not lie in the neighborhood of O, whereas e^{-jkR} changes slowly to produce a constructive interference between the various spherical wave contributions to the radiation integral of (1a) from points on S' which lie in the neighborhood of O. The point O is thus referred to as the *stationary phase point*. Therefore, at high frequencies (or large k) the dominant contribution to the integral in (1a) comes from the *stationary point at O*.

In particular, an evaluation of (1a) via the method of stationary phase [12] yields

$$\mathbf{E}(P) \sim \mathbf{E}(O) \sqrt{\frac{\varrho_1 \varrho_2}{(\varrho_1 + s)(\varrho_2 + s)}} e^{-jks}, \qquad |OP| = s \tag{2}$$

where ϱ_1 and ϱ_2 are the principal radii of curvature of the wavefront surface S' at the stationary point O. It is important to note that

$$\sqrt{\frac{\varrho}{\varrho + s}} = \left|\sqrt{\frac{\varrho}{\varrho + s}}\right| e^{j\{{}_{\pi/2}^{0}\}} \qquad \text{if } \frac{\varrho}{\varrho + s} \begin{cases} > 0 \\ < 0 \end{cases} \tag{3}$$

in which ϱ can be either ϱ_1 or ϱ_2. The above result in (2) indicates that the electric field $\mathbf{E}(P)$ arrives at P along a "ray" from O to P, i.e., this field is a continuation of the electric field $\mathbf{E}(O)$ at the point O on the initial wavefront S' to the wavefront containing the observation point P. It is noted that the amplitude of $\mathbf{E}(P)$ varies as $\sqrt{\varrho_1\varrho_2/(\varrho_1 + s)(\varrho_2 + s)}$ along the ray, and e^{-jks} represents the phase path delay along that ray. Furthermore, $\mathbf{E}(O)$ is polarized transverse to the ray path OP, and $\mathbf{E}(P)$ also has the same polarization as $\mathbf{E}(O)$ because the ray path is straight in a homogeneous medium. Since energy in the high-frequency electromagnetic (EM) field is transported along rays, it is evident from geometrical considerations that energy must be conserved in a tube of rays (or a ray bundle). This may be verified by considering a tube or a narrow bundle of rays about the central (or axial) ray OP as shown in Fig. 3. First, the principal wavefront radii of curvature ϱ_1 and ϱ_2 at O, which are also shown in Fig. 3, are referred to as the *ray caustic distances*. Now, the energy flux in the field which crosses the area dA_O of the ray tube at O is given by $|E(O)|^2 dA_O$, and likewise, the energy flux crossing the area dA of the same ray tube at P is $|E(P)|^2 dA_P$; however, it is clear from Fig. 3 that $dA_O \cong |(\varrho_1 d\psi_1)(\varrho_2 d\psi_2)|$, and $dA_P \cong |[(\varrho_1 + s)d\psi_1][(\varrho_2 + s)d\psi_2]|$, so that conservation of energy flux within the ray tube requires one to satisfy $|E(P)|^2 dA_P = |E(O)|^2 dA_O$. It then follows immediately that

$$|E(P)| = |E(O)| \left| \sqrt{\frac{\varrho_1 \varrho_2}{(\varrho_1 + s)(\varrho_2 + s)}} \right| \qquad (4)$$

which is in agreement with (2).

The field in (2) associated with the ray OP of Fig. 3 is referred to as an *arbitrary ray optical field* since ϱ_1 and ϱ_2 can be arbitrary. Also, the ray tube in Fig. 3 is commonly referred to as an *astigmatic ray tube* or a *quadratic ray pencil* because of the quadratic wavefront surface approximation at O which is employed in the stationary phase method to obtain the result in (2). It can be easily verified that plane, cylindrical, conical, and spherical wave fields are special cases of an arbitrary ray optical field, i.e., each of those wave fields is also a ray optical field.

The Geometrical Optics Field

Geometrical optics (GO) is a high-frequency approximation which employs rays to describe the fields that are directly incident from the source, and to describe the fields which are reflected and refracted (or transmitted) at an interface between two different media. According to classical GO, the high-frequency electromagnetic field is assumed to propagate along ray paths which satisfy Fermat's principle, and the family of rays is everywhere orthogonal to the wavefronts in an isotropic medium. The ray paths are straight lines in a homogeneous medium, but they can

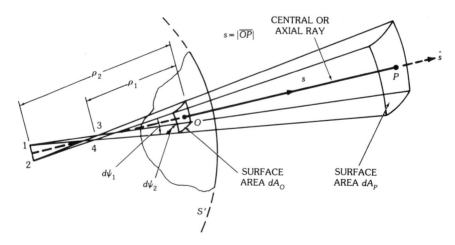

Fig. 3. Geometrical (ray tube) interpretation of the result in (2).

change directions at an interface between two different media according to the well-known (Snell's) laws of reflection and refraction, which can be deduced from Fermat's principle. The amplitude variation of the classical GO field is obtained by requiring conservation of energy flux in an astigmatic ray tube associated with a central (or axial) ray as indicated previously with the aid of Fig. 3. These considerations lead to an expression identical with that in (4), as might be expected. The information on the field polarization (as being transverse to the ray) and on the wave nature of the field is then introduced heuristically in the classical GO approximation of (4) to arrive at a more complete expression for the GO field. The latter, more complete expression is found to be identical with that obtained in (2) via stationary phase considerations. Indeed, the surface S' in Fig. 3 could be associated with any wavefront; thus, if S' represents the wavefront of an incident or reflected type ray congruence, then (2) would represent the GO incident or reflected electric field, respectively.

An alternative development for the GO field is based on an asymptotic high-frequency solution to Maxwell's equations in which the electromagnetic fields are expanded in inverse powers of the wave number k, as done by Luneberg and Kline [17, 18]. The leading term in their expansion corresponds to the GO field. According to Luneberg and Kline the electric field $\mathbf{E}$ in a source-free, homogeneous, isotropic medium can be expressed at high frequencies by

$$\mathbf{E}(\mathbf{r}, k) \sim e^{-jk\psi(\mathbf{r})} \sum_{n=0}^{\infty} \frac{\mathbf{E}_n(\mathbf{r})}{(-jk)^n} \tag{5}$$

in which $\mathbf{r}$ is the position vector of the field point. Substituting (5) into the vector Helmholtz equation satisfied by $\mathbf{E}$, namely $(\nabla^2 + k^2)\mathbf{E} = 0$ (in which ∇^2 is the Laplacian operator), leads to the usual eikonal and transport equations $|\nabla \psi|^2 = 1$ and $[\nabla^2 \psi + 2\nabla \psi \cdot \nabla]\mathbf{E}_n = -\nabla^2 \mathbf{E}_{n-1}$ (with $\mathbf{E}_{-1} \equiv 0$), respectively. The surfaces of constant ψ are defined as the wavefronts. Integrating the transport equation from some reference point $\mathbf{r}_O$ to $\mathbf{r}$ for the $n = 0$ case yields the leading term in the Luneberg-Kline expansion as

$$\mathbf{E}(\mathbf{r}) \sim \mathbf{E}(\mathbf{r}_O) \sqrt{\frac{\varrho_1 \varrho_2}{(\varrho_1 + s)(\varrho_2 + s)}} e^{-jks} \tag{6}$$

in which

$$\mathbf{E}(\mathbf{r}_O) \equiv \mathbf{E}_0(\mathbf{r}_O) e^{-jk\psi(\mathbf{r}_O)}$$

where $\mathbf{E}_0$ implies $\mathbf{E}_n$ for $n = 0$ in (5). The quantities ϱ_1, ϱ_2, and s in (6) have the same meaning as in Fig. 3, with $\mathbf{r}$ being the position vector of P and $\mathbf{r}_O$ being the position vector of O. Furthermore, from the requirement $\nabla \cdot \mathbf{E}(\mathbf{r}) = 0$ for source-free regions, one obtains $\hat{\mathbf{s}} \cdot \mathbf{E}(\mathbf{r}) = 0$, which implies that $\mathbf{E}(\mathbf{r})$ in (6) is polarized transverse to the ray direction $\hat{\mathbf{s}}$ (see Fig. 3). It is thus evident that (6) is the same as (2). A Luneberg-Kline expansion for $\mathbf{H}$ yields

$$\mathbf{H}(\mathbf{r}) \sim \hat{\mathbf{s}} \times Y_0 \mathbf{E}(\mathbf{r}) \tag{7}$$

In (6) the distance s is measured as positive in the direction of ray propagation. The caustic distances ϱ_2 and ϱ_1 are positive if the associated caustics at 1–2 and 3–4 in Fig. 3 occur before the reference point O as one propagates along the ray; otherwise, they are negative. The positive branch of the square roots is chosen in (6). Therefore, if $\varrho_{1,2} < 0$ and $s > -|\varrho_2|$, or $s > -|\varrho_1|$, then a caustic is crossed at 1–2 or 3–4 (see Fig. 3), respectively, and $(\varrho_2 + s)$ or $(\varrho_1 + s)$ changes sign within the square root of (6) so that a phase jump of $\pi/2$ results naturally in each case; namely,*

$$\sqrt{\frac{\varrho_i}{\varrho_i + s}} = \left|\frac{\varrho_i}{\varrho_i + s}\right| e^{j\pi/2}, \quad \text{if } s > -|\varrho_i| \quad \text{for } i = 1, 2$$

The above result is consistent with (3), which is implied in (2). The GO field in (6) is therefore completely consistent with the expression in (2) which was obtained via the method of stationary phase. At a ray caustic, $s = -|\varrho_i|$ for $i = 1, 2$, and the GO field in (6) and (2) becomes infinite. Consequently, the GO field representation fails at and near the GO ray caustics.

For a perfectly conducting surface illuminated by a source (antenna), only the fields directly incident from the source and the fields reflected by that surface in just the specular direction can be described by GO. These GO incident and reflected fields are discussed next for the two-dimensional (2-D) and three-dimensional (3-D) cases.

The GO Incident Field (3-D Case)—The GO incident field is the field directly radiated by the source (antenna) to the observation point; this incident ray optical (or GO) field exists "in the presence of" the surface which it illuminates. Part of the GO incident ray system (or congruence) which strikes an impenetrable surface is blocked by that surface; as a result the surface creates a shadow region behind it where the incident rays cannot exist, and consequently GO predicts a zero field in the shadow zone! A shadow boundary is created naturally by the presence of the shadow region; this boundary divides the space surrounding the surface into the lit and shadow zones corresponding to regions where the source is directly visible and where it is not, respectively. In particular, there is an incident shadow boundary (ISB) for a perfectly conducting structure with an edge, such as a wedge as in Fig. 4a, and likewise there is a surface shadow boundary (SSB) associated with a smooth, perfectly conducting, convex surface as shown in Fig. 4b. Thus the GO incident field is truncated to zero at the incident shadow boundary (ISB or SSB) so that it vanishes within the shadow region of an obstacle illuminated by that field.

If the incident wavefront is characterized by two different radii of curvature ϱ_1^i and ϱ_2^i, then the GO incident field is given by (6) as

$$\mathbf{E}^i(\mathbf{r}) = \mathbf{E}^i(\mathbf{r}_O) \sqrt{\left(\frac{\varrho_1^i}{\varrho_1^i + s^i}\right)\left(\frac{\varrho_2^i}{\varrho_2^i + s^i}\right)} e^{-jks^i} \qquad (8)$$

---

*Actually, if both caustics (at 1–2 and 3–4) are crossed, then a total phase jump of $\pi/2 + \pi/2 = \pi$ results.

Techniques for High-Frequency Problems

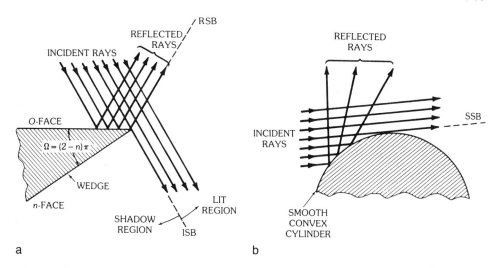

Fig. 4. Shadow boundaries ISB and SSB. (*a*) Incident shadow boundary. (*b*) Surface shadow boundary.

where the superscript i is employed to denote quantities associated with the incident ray as in Fig. 5. In the case of plane-wave illumination, ϱ_1^i and ϱ_2^i are infinite, so that (8) reduces to

$$\mathbf{E}^i(\mathbf{r}) = \mathbf{E}^i(\mathbf{r}_O) e^{-jks^i} \qquad (9)$$

for a *plane-wave* incidence. Likewise, in the case of spherical-wave illumination for which $\varrho_1^i = \varrho_2^i \equiv \varrho^i$, and for which the reference point P_O in Fig. 5 is moved to the point caustic (so that $\varrho^i \to 0$), it is seen that for a *spherical-wave* or *point-source* type illumination, (8) reduces to

$$\mathbf{E}^i(\mathbf{r}) = \mathbf{C}^i \frac{e^{-jks^i}}{s^i} \qquad (10)$$

with

$$\mathbf{C}^i \equiv \lim_{\varrho^i \to 0} \varrho^i \mathbf{E}^i(\mathbf{r}_O)$$

in which $\mathbf{C}^i$ is related to the strength of the point source. The plane-wave and spherical-wave type of GO incident fields are illustrated in Fig. 6.

It is noted that the incident magnetic field $\mathbf{H}^i(\mathbf{r})$ associated with $\mathbf{E}^i(\mathbf{r})$ of (8) through (10) is simply given by

$$\mathbf{H}^i(\mathbf{r}) \sim Y_0 \hat{\mathbf{s}}^i \times \mathbf{E}^i(\mathbf{r}) \qquad (11)$$

The GO Incident Field (2-D Case)—In the case of cylindrical wave or 2-D line

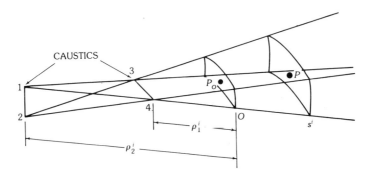

Fig. 5. Incident-ray tube.

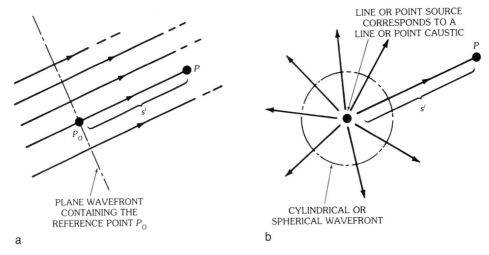

Fig. 6. Plane-wave and cylindrical- or spherical-wave type GO incident fields. (*a*) Plane-wave GO incident field. (*b*) Cylindrical- or spherical-wave GO incident field.

source illumination, one may let $\varrho_1^i \to \infty$ in (8), and allow the reference point P_O (at $\mathbf{r}_O$) in Fig. 5 be moved to the caustic 1–2 for convenience so that $\varrho_2^i \to 0$, together with the condition that

$$\lim_{\varrho_2^i \to 0} \sqrt{\varrho_2^i}\, \mathbf{E}^i(\mathbf{r}_O) \equiv \mathbf{C}^i$$

where $\mathbf{C}^i$ is finite and related to the strength of the line source. Incorporating this limit in (8) yields

$$\mathbf{E}^i(\mathbf{r}) = \mathbf{C}^i \frac{e^{-jks^i}}{\sqrt{s^i}} \tag{12}$$

Techniques for High-Frequency Problems

for a *cylindrical wave* or *line source* type illumination.

The magnetic field $\mathbf{H}^i(\mathbf{r})$ can be obtained from $\mathbf{E}^i(\mathbf{r})$ via (11).

The GO Reflected Field (3-D Case)—When a family of rays incident from a source strike a perfectly conducting surface, they are transformed at the surface into a family of reflected rays as shown in Figs. 4a and 4b. Such a reflected ray congruence is discontinuous across the reflection shadow boundary (RSB) for the edge type structure shown in Fig. 4a; on the other hand, the incident and reflection shadow boundaries merge into the surface shadow boundary (SSB) for the convex surface as shown in Fig. 4b. Furthermore, the GO reflected field vanishes at the SSB and also within the shadow region for the problem in Fig. 4b. The GO field associated with a reflected ray has the form indicated in (6). In particular, the electric field $\mathbf{E}^r(\mathbf{r})$ associated with a GO reflected ray is given by

$$\mathbf{E}^r(P) = \mathbf{E}^r(Q_R)\sqrt{\frac{\varrho_1^r}{(\varrho_1^r + s^r)}\frac{\varrho_2^r}{(\varrho_2^r + s^r)}}\, e^{-jks^r} \tag{13}$$

where the reference position Q_R on the ray is chosen here to be at the point of reflection on the surface as shown in Fig. 7. The distances ϱ_1^r, ϱ_2^r, and s^r are associated with the reflected ray from Q_R to P. The reflected field $\mathbf{E}^r(Q_R)$ is related to the incident field $\mathbf{E}^i(Q_R)$ at the point of reflection Q_R by the boundary condition

$$\hat{\mathbf{n}} \times [\mathbf{E}^i(Q_R) + \mathbf{E}^r(Q_R)] = 0 \tag{14}$$

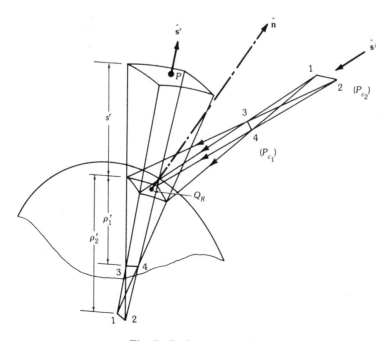

Fig. 7. Reflected-ray tube.

which requires the total tangential electric field to vanish on the perfectly conducting surface. In (14), $\hat{\mathbf{n}}$ is the unit normal vector to the surface at Q_R. As a consequence of (14), the following relationship holds:

$$\mathbf{E}^r(Q_R) = \bar{\bar{\mathbf{R}}} \cdot \mathbf{E}^i(Q_R) \tag{15}$$

where $\bar{\bar{\mathbf{R}}}$ denotes the dyadic surface reflection coefficient at Q_R.
Incorporating (15) into (13) yields

$$\mathbf{E}^r(P) = \mathbf{E}^i(Q_R) \cdot \bar{\bar{\mathbf{R}}} \sqrt{\frac{\varrho_1^r}{\varrho_1^r + s^r} \frac{\varrho_2^r}{\varrho_2^r + s^r}} \, e^{-jks^r} \tag{16}$$

Likewise, the associated reflected magnetic field $\mathbf{H}^r(P)$ is simply given by

$$\mathbf{H}^r(P) \sim Y_0 \hat{s}^r \times \mathbf{E}^r(P) \tag{17}$$

The expression for $\bar{\bar{\mathbf{R}}}$ simplifies if one expresses the fields in terms of an appropriate set of unit vectors which are fixed in the incident and reflected rays. Thus, let $\hat{\mathbf{e}}_\parallel^i$ be a unit vector transverse to $\hat{s}^i$ such that it lies in the plane of incidence defined by $\hat{\mathbf{n}}$ and $\hat{s}^i$ as in Fig. 8. It is noted that the reflected ray along $\hat{s}^r$ also lies in the plane of incidence. Likewise, let $\hat{\mathbf{e}}_\parallel^r$ be a unit vector in the plane of incidence but transverse to the reflected ray direction $\hat{s}^r$ as in Fig. 8, and let $\hat{\mathbf{e}}_\perp$ be a unit vector perpendicular to the plane of incidence. The $\hat{\mathbf{e}}_\perp$ is then automatically transverse to the directions of incidence and reflection, respectively. In this ray coordinate system,

$$\bar{\bar{R}} = \hat{\mathbf{e}}_\parallel^i \hat{\mathbf{e}}_\parallel^r R_h + \hat{\mathbf{e}}_\perp \hat{\mathbf{e}}_\perp R_s \tag{18}$$

with $R_{\hat{s}} = -1$ and $R_{\hat{h}} = +1$. In matrix notation, $\bar{\bar{\mathbf{R}}}$ in (18) is equivalent to

$$\begin{bmatrix} R_h & 0 \\ 0 & R_s \end{bmatrix} = \begin{bmatrix} 1 & 0 \\ 0 & -1 \end{bmatrix}$$

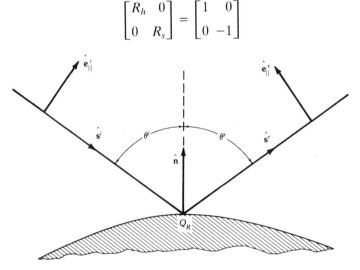

Fig. 8. Ray unit vectors associated with the reflection problem.

or (16) is equivalent via (18) to

$$\begin{bmatrix} E_\parallel^r(P) \\ E_\perp^r(P) \end{bmatrix} = \begin{bmatrix} 1 & 0 \\ 0 & -1 \end{bmatrix} \begin{bmatrix} E_\parallel^i(Q_R) \\ E_\perp^i(Q_R) \end{bmatrix} \sqrt{\frac{\varrho_1^r \varrho_2^r}{(\varrho_1^r + s')(\varrho_2^r + s')}} \, e^{-jks'}$$

where $\mathbf{E}^i(Q_R) = E_\parallel^i(Q_R)\hat{\mathbf{e}}_\parallel^i + E_\perp^i(Q_R)\hat{\mathbf{e}}_\perp$ and $\mathbf{E}^r(P) = E_\parallel^r(P)\hat{\mathbf{e}}_\parallel^r + E_\perp^r(P)\hat{\mathbf{e}}_\perp$ in the ray coordinates of Fig. 8.

The principal radii of curvature $(\varrho_1^i, \varrho_2^i)$ and $(\varrho_1^r, \varrho_2^r)$ of the incident and reflected wavefronts and their associated principal directions $(\hat{\mathbf{X}}_1^i, \hat{\mathbf{X}}_2^i)$ and $(\hat{\mathbf{X}}_1^r, \hat{\mathbf{X}}_2^r)$ are given in [19], and $(\varrho_1^r, \varrho_2^r)$ are expressed more compactly here as*

$$\frac{1}{\varrho_{1,2}^r} = \frac{1}{\varrho_m^i} + \frac{f}{\varrho_g(Q_R)\cos\theta^i} \left(1 \pm \left\{\frac{\varrho_g^2(Q_R)\cos^2\theta^i}{4f^2} \cdot \left(\frac{1}{\varrho_1^i} - \frac{1}{\varrho_2^i}\right)^2\right.\right.$$

$$+ \frac{\varrho_g^2(Q_R)\cos\theta^i}{f^2} \left(\frac{1}{\varrho_1^i} - \frac{1}{\varrho_2^i}\right) \cdot \left[\frac{g\cos 2\alpha_0}{\varrho_g(Q_R)} - \sin 2\alpha_0 \sin 2\omega_0 \cos\theta^i \right.$$

$$\left.\left.\times \left(\frac{1}{R_1} - \frac{1}{R_2}\right)\right] + 1 - \frac{4\varrho_g^2(Q_R)\cos^2\theta^i}{f^2 R_1 R_2}\right\}^{1/2}\right) \tag{19a}$$

where $R_{1,2} > 0$ for a convex surface and $R_{1,2} < 0$ for a concave surface, and

$$\frac{1}{\varrho_m^i} = \frac{1}{2}\left(\frac{1}{\varrho_1^i} + \frac{1}{\varrho_2^i}\right) \tag{19b}$$

$$\begin{Bmatrix} f \\ g \end{Bmatrix} = \left\{1 \pm \frac{\varrho_g(Q_R)}{\varrho_t(Q_R)} \cos^2\theta^i\right\} \tag{19c}$$

The quantities R_1 and R_2 constitute the principal radii of curvatures of the surface at Q_R, and $\hat{\mathbf{U}}_1$ and $\hat{\mathbf{U}}_2$ denote the corresponding principal surface directions at Q_R. The quantity ϱ_g denotes the radius of curvature of the surface at Q_R in the plane of incidence which contains $\hat{\mathbf{s}}^i$, $\hat{\mathbf{n}}$, and $\hat{\mathbf{t}}$, where $\hat{\mathbf{t}}$ is tangent to the surface. Also, ϱ_t is the radius of curvature of the surface at Q_R in the plane containing $\hat{\mathbf{n}}$ and the binormal vector $\hat{\mathbf{b}}$. The unit vectors $\hat{\mathbf{t}}$, $\hat{\mathbf{n}}$, $\hat{\mathbf{b}}$, $\hat{\mathbf{U}}_1$, and $\hat{\mathbf{U}}_2$ are shown in Fig. 9a together with the angle ω_0 between $\hat{\mathbf{t}}$ and $\hat{\mathbf{U}}_2$.

Note that ω_0 corresponds to the angle between the plane of incidence and one of the principal surface directions $(\hat{\mathbf{U}}_2)$ at the point of reflection.

The principal directions $(\hat{\mathbf{X}}_1^i, \hat{\mathbf{X}}_2^i)$ of the incident wavefront and α_0 are shown in Fig. 9b. The angle of incidence θ^i is defined by $\hat{\mathbf{n}} \cdot \hat{\mathbf{s}}^i = -\cos\theta^i = -\hat{\mathbf{n}} \cdot \hat{\mathbf{s}}^r$.

A matrix formulation to calculate $(\varrho_1^r, \varrho_2^r)$ is given in [20, 21, 22]; in particular, [20, 22] contain results for both reflected and transmitted ray caustics in the case of penetrable media.

---

*The principal radii of curvature $\varrho_\ell^{i,r} = \pm|\varrho_\ell^{i,r}|$ for a diverging (converging) pair of adjacent rays. Here, $\ell = 1, 2$.

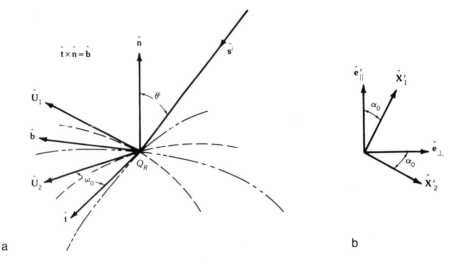

Fig. 9. Geometry for description of wavefront reflected from a curved surface. (*a*) Showing unit vectors. (*b*) Showing principal directions. (*After Sitka, Burnside, Chu, and Peters [27]*, © *1983 IEEE*)

The incident spherical wavefront is commonly encountered; for this case it can be shown that

$$\frac{1}{\varrho'_{1,2}} = \frac{1}{s^i} + \frac{1}{\cos\theta^i}\left(\frac{\sin^2\theta_2}{R_1} + \frac{\sin^2\theta_1}{R_2}\right) \pm \sqrt{\frac{1}{\cos^2\theta^i}\left(\frac{\sin^2\theta_2}{R_1} + \frac{\sin^2\theta_1}{R_2}\right)^2 - \frac{4}{R_1 R_2}}$$

in which s^i is the radius of curvature of the incident spherical wavefront at Q_R, θ_1 is the angle between $\hat{s}^i$ and $\hat{U}_1$, and θ_2 is the angle between $\hat{s}^i$ and $\hat{U}_2$.

It is easily verified that

$$\sin^2\theta_1 = \cos^2\omega_0 + \sin^2\omega_0\cos^2\theta^i$$
$$\sin^2\theta_2 = \sin^2\omega_0 + \cos^2\omega_0\cos^2\theta^i$$

Thus a further simplification is possible in the case of spherical wave incidence if $\omega_0 = 0$ so that the preceding expressions for $(\varrho'_1)^{-1}$ and $(\varrho'_2)^{-1}$ become

$$\frac{1}{\varrho'_1} = \frac{1}{s^i} + \frac{2\cos\theta^i}{R_1}, \quad \frac{1}{\varrho'_2} = \frac{1}{s^i} + \frac{2}{R_2\cos\theta^i}, \quad \text{if } \omega_0 = 0$$

It is noted that $\omega_0 = 0$ implies that the plane of incidence is aligned with one of the principal directions ($\hat{U}_2$) on the surface at Q_R.

Explicit results are given in [23] for the GO field reflected from an arbitrary surface due to an incident spherical wave.

It is clear that the GO representation of (16) fails at caustics which are the intersection of the paraxial rays (associated with the ray tube or pencil) at the lines 1–2 and 3–4 as shown in Figs. 5 and 7. On crossing a caustic in the direction of propagation, $(\varrho^{i,r} + s^{i,r})$ changes sign under the radical in (16), and a phase jump of $+\pi/2$ results. Furthermore, the reflected field $\mathbf{E}^r$ of (16) fails in the transition region adjacent to SSB of Fig. 4b. It is important to note that near the SSB (i.e., as $\theta^i \to \pi/2$), ϱ_1^r and ϱ_2^r of (19a) approach the following limiting values:

$$\varrho_2^r \to \left(\frac{\varrho_g(Q_R) \cos \theta^i}{2}\right) \to 0 \qquad (20a)$$

$$\varrho_1^r \to \varrho_b^i \qquad \text{for } \theta^i \to \pi/2 \qquad (20b)$$

where

$$1/\varrho_b^i = (\sin^2\alpha_0/\varrho_1^i) + (\cos^2\alpha_0/\varrho_2^i) \qquad (21)$$

and ϱ_b^i is the radius of curvature of the incident wavefront in the $(\hat{\mathbf{t}}\hat{\mathbf{b}})$ plane (i.e., in the plane tangent to the surface) at Q_R for $\theta^i \to \pi/2$. Furthermore, the principal directions $\hat{\mathbf{X}}_1^r$ and $\hat{\mathbf{X}}_2^r$ of the reflected wavefront approach the following values for grazing incidence:

$$\hat{\mathbf{X}}_1^r \to \hat{\mathbf{b}} \quad (\text{at } Q_R) \qquad (22a)$$

$$\hat{\mathbf{X}}_2^r = (-\hat{\mathbf{s}}^r \times \hat{\mathbf{X}}_1^r) \to \hat{\mathbf{n}} \ (\text{at } Q_R) \qquad \text{for } \theta^i \to \pi/2 \qquad (22b)$$

The limiting values in (22a) and (22b) are independent of α_0. The total GO electric field $\mathbf{E}^{GO}$ at P_L in the lit region is the sum of the incident and reflected ray optical fields; hence

$$\mathbf{E}^{GO}(P_L) \sim \mathbf{E}^i(P_L) + \mathbf{E}^i(Q_R) \cdot \bar{\bar{\mathbf{R}}} \sqrt{\frac{\varrho_1^r \varrho_2^r}{(\varrho_1^r + s^r)(\varrho_2^r + s^r)}} e^{-jks^r} \qquad (23)$$

In summary, it is noted that the GO incident and reflected fields are *discontinuous* across their associated shadow boundaries, such as the ISB, RSB, and SSB in Figs. 4a and 4b. These GO field discontinuities cannot be removed by including additional terms in the Luneberg-Kline expansion, which loses validity when calculating the fields at and near such boundaries, even though these additional terms may be useful for improving the GO approximation elsewhere in the deep lit zone. The failure of GO to account for a proper nonzero field within the shadow region behind an impenetrable obstacle can be overcome through the GTD and its uniform versions, the UTD and UAT, which are discussed in Section 4. Nevertheless, GO generally yields the dominant contribution to the total high-frequency fields, and, as will be seen in Section 4, it constitutes the leading term in the GTD solution.

The GO Reflected Field (2-D Case)—The GO reflected field $\mathbf{E}^r(P_L)$ for the 2-D case can be deduced directly from the 3-D case by allowing ϱ_1^r to approach infinity. Thus one may let $\varrho_2^r \equiv \varrho^r$ and $\varrho_1^r \to \infty$ in (16) to arrive at the 2-D reflected GO field $\mathbf{E}^r(P_L)$ as

$$\mathbf{E}^r(P_L) = \mathbf{E}^i(Q_R) \cdot \bar{\bar{\mathbf{R}}} \sqrt{\frac{\varrho^r}{\varrho^r + s^r}} e^{-jks^r} \qquad (24)$$

in which the incident ray optical field $\mathbf{E}^i(Q_R)$ is a cylindrical wave at Q_R in the 2-D case, and the caustic distance ϱ^r in (24) for the 2-D case is given by

$$\frac{1}{\varrho^r} = \frac{1}{s^i} + \frac{2}{\varrho_g(Q_R)\cos\theta^i} \qquad (25)$$

where θ^i has the same meaning as before, and s^i is the radius of curvature of the incident cylindrical wavefront at Q_R. If the cylindrical wave is produced by a 2-D line source, then s^i in (25) can be chosen to be the distance from that line source to the point Q_R of reflection on the 2-D boundary. The quantity $\varrho_g(Q_R)$ in (25) denotes the radius of curvature of the 2-D boundary at the point Q_R of reflection.

3. The Physical Optics Field

The total electromagnetic field of a source (antenna) which radiates in the presence of a perfectly conducting surface as in Fig. 10 may be expressed as a superposition of the incident fields $(\mathbf{E}^i, \mathbf{H}^i)$ and the fields $(\mathbf{E}^s, \mathbf{H}^s)$ which are scattered by the surface. The incident fields referred to above are chosen here to denote the electric and magnetic fields of the source which exist everywhere, i.e., they exist as if the scatterer was "absent"; this is unlike the GO incident field, which exists in the presence of the surface of the scatterer (and which hence becomes discontinuous at the ISB or SSB, as in Figs. 4a and 4b). The scattered

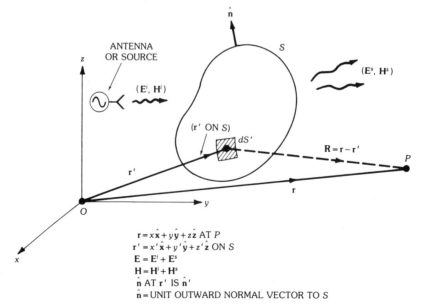

Fig. 10. Geometry pertaining to a 3-D obstacle (surface S) illuminated by an antenna.

Techniques for High-Frequency Problems

fields in this case can be expressed in terms of the radiation integrals [12] over the actual currents induced on the surface S of the scatterer. These currents also radiate the scattered fields in the absence of the scatterer, i.e., these currents are now viewed as equivalent sources which are impressed over the mathematical boundary of S but with the perfectly conducting scatterer removed. The medium internal and external to the mathematical surface S in this equivalent problem is the same as that which exists external to S in the original problem. Such an equivalent configuration produces the same scattered fields as in the original problem outside the mathematical surface S, whereas it produces scattered fields which exactly cancel the incident fields inside S (to yield a zero total field inside S).

In the physical optics (PO) method, that radiation integral for the scattered field is calculated by employing a GO approximation for the currents induced on S, which is assumed to be electrically large; hence the PO method is also a high-frequency method. The PO approximation for the 3-D and 2-D cases is discussed below.*

Three-Dimensional Case

The total electric field $\mathbf{E}(\mathbf{r})$ at a point P exterior to S which may be evaluated in the near zone of S but not extremely close to S (in the problem of Fig. 10) is given by

$$\mathbf{E}(\mathbf{r}) = \mathbf{E}^i(\mathbf{r}) + \mathbf{E}^s(\mathbf{r}) \tag{26a}$$

where

$$\mathbf{E}^s(\mathbf{r}) \cong \frac{jkZ_0}{4\pi} \iint_S \hat{\mathbf{R}} \times \hat{\mathbf{R}} \times d\mathbf{p}_e(\mathbf{r}') \frac{e^{-jkR}}{R} \tag{26b}$$

as in [12]. Likewise, the total magnetic field $\mathbf{H}(\mathbf{r})$ at P is

$$\mathbf{H}(\mathbf{r}) = \mathbf{H}^i(\mathbf{r}) + \mathbf{H}^s(\mathbf{r}) \tag{27a}$$

where

$$\mathbf{H}^s(\mathbf{r}) \cong \frac{-jk}{4\pi} \iint_S \hat{\mathbf{R}} \times d\mathbf{p}_e(\mathbf{r}') \frac{e^{-jkR}}{R} \tag{27b}$$

In (26) and (27) the source distribution $d\mathbf{p}_e(\mathbf{r}')$ at each point $(\mathbf{r}')$ on S is given by

$$d\mathbf{p}_e(\mathbf{r}') = \mathbf{J}_s(\mathbf{r}')dS' = \hat{\mathbf{n}} \times \mathbf{H}(\mathbf{r}')dS' \tag{28}$$

The $\mathbf{J}_s(\mathbf{r}')$ in (28) is the induced electric current density at $\mathbf{r}'$ on S, and dS' is an elemental area of S at $\mathbf{r}'$. If P (at $\mathbf{r}$) is extremely close to S, then it is necessary to include additional terms which depend on R^{-2} and R^{-3} in the integrands of (26b)

---

*Additional important comments on the PO method are given after the 3-D and 2-D cases are discussed.

and (27b) as indicated in [12]; these additional terms are neglected in the present development. If $\mathbf{J}_s(\mathbf{r}')$ in (28) is approximated by its value $\mathbf{J}_s^{GO}(\mathbf{r}')$ as predicted by geometrical optics (GO), then

$$\mathbf{J}_s(\mathbf{r}') \cong \mathbf{J}_s^{GO}(\mathbf{r}') = \begin{cases} 2\hat{\mathbf{n}}' \times \mathbf{H}^i(\mathbf{r}') & \text{on the lit portion of } S \\ 0 & \text{on the shadowed portion of } S \end{cases} \quad (29)$$

Here, $\mathbf{E}^i$ and $\mathbf{H}^i$ refer to the "incident" electric and magnetic fields which are produced by the source (antenna) in the "absence" of the surface S. If the above GO approximation of (29) is employed in (28), then the resulting $\mathbf{E}$ and $\mathbf{H}$ so obtained from (26b) and (27b) are defined as the physical optics (PO) fields $\mathbf{E}^{PO}$ and $\mathbf{H}^{PO}$, respectively. Thus

$$\mathbf{E}^{PO}(\mathbf{r}) = \mathbf{E}^i(\mathbf{r}) + \frac{jkZ_0}{4\pi} \iint_{S_{lit}} \hat{\mathbf{R}} \times \hat{\mathbf{R}} \times [2\hat{\mathbf{n}}' \times \mathbf{H}^i] \frac{e^{-jkR}}{R} dS' \quad (30)$$

and

$$\mathbf{H}^{PO}(\mathbf{r}) = \mathbf{H}^i(\mathbf{r}) + \frac{-jk}{4\pi} \iint_{S_{lit}} \hat{\mathbf{R}} \times [2\hat{\mathbf{n}}' \times \mathbf{H}^i] \frac{e^{-jkR}}{R} dS' \quad (31)$$

where S_{lit} in the integrals of (30) and (31) refers to the part of S which is directly illuminated by the source, i.e., to the area of only the lit portion of S in accordance with (29). Let $\mathbf{E}_{PO}^s$ and $\mathbf{H}_{PO}^s$ denote the integrals in (30) and (31); thus they denote the scattered fields based on the PO approximation. In the event that the observation point P at $(\mathbf{r})$ is in the far zone of S, then $\hat{\mathbf{R}}$ can be replaced by $\hat{\mathbf{r}}$, which can be removed outside the integrals of (30) and (31); furthermore,

$$\frac{e^{-jkR}}{R} \cong \frac{e^{-jkr}}{r} e^{+jk\hat{\mathbf{r}} \cdot \mathbf{r}'}$$

in the far zone as usual, and the e^{-jkr}/r can also be factored outside those integrals. In the far zone the scattered fields in (26), (27), (30), and (31) are related by

$$\begin{aligned} \mathbf{E}^s(\mathbf{r}) &\sim -\hat{\mathbf{r}} \times Z_0 \mathbf{H}^s(\mathbf{r}) \\ \mathbf{H}^s(\mathbf{r}) &\sim Y_0 \hat{\mathbf{r}} \times \mathbf{E}^s(\mathbf{r}) \end{aligned} \quad (32)$$

and

$$\begin{aligned} \mathbf{E}_{PO}^s(\mathbf{r}) &\sim -\hat{\mathbf{r}} \times Z_0 \mathbf{H}_{PO}^s(\mathbf{r}) \\ \mathbf{H}_{PO}^s(\mathbf{r}) &\sim Y_0 \hat{\mathbf{r}} \times \mathbf{E}_{PO}^s(\mathbf{r}) \end{aligned} \quad (33)$$

Two-Dimensional Case

Next, consider the two-dimensional (2-D) geometry illustrated in Fig. 11, where a perfectly conducting scatterer bounded by a contour C is illuminated by a

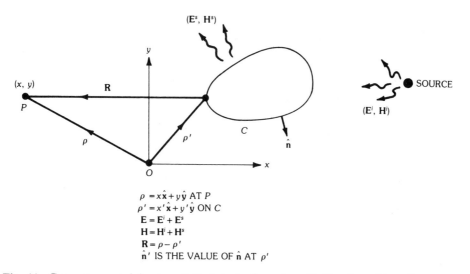

$\rho = x\hat{x} + y\hat{y}$ AT P
$\rho' = x'\hat{x} + y'\hat{y}$ ON C
$E = E^i + E^s$
$H = H^i + H^s$
$R = \rho - \rho'$
$\hat{n}'$ IS THE VALUE OF $\hat{n}$ AT ρ'

Fig. 11. Geometry pertaining to a 2-D obstacle (boundary C) illuminated by a line source.

2-D source. The total fields $(\mathbf{E}, \mathbf{H})$ exterior to the boundary C are given in a manner analogous to (26b) and (27b) as follows:

$$\mathbf{E}(\varrho) \cong \mathbf{E}^i(\varrho) + \frac{kZ_0}{4} \int_C [\hat{\mathbf{R}} \times \hat{\mathbf{R}} \times \mathbf{J}(\ell')] H_0^{(2)}(kR) \, d\ell' \tag{34}$$

and

$$\mathbf{H}(\varrho) \cong \mathbf{H}^i(\varrho) - \frac{k}{4} \int_C [\hat{\mathbf{R}} \times \mathbf{J}(\ell')] H_0^{(2)}(kR) \, d\ell' \tag{35}$$

where $\mathbf{R} = \varrho - \varrho'$ in the 2-D case, and $H_0^{(2)}(kR)$ corresponds to the cylindrical Hankel function of the second kind and of order zero. The argument of this Hankel function is $kR = k|\varrho - \varrho'|$. Following (26) and (27) for the 3-D case, the expressions in (34) and (35) for the 2-D case are likewise valid if the observation point P (at ϱ) is not too close to the boundary C. In the PO approximation the electric current density $\mathbf{J}(\ell') = \mathbf{J}(\varrho')$ in (34) and (35) must be replaced by its GO approximation,

$$\mathbf{J}(\ell') \cong \mathbf{J}^{GO}(\ell') = \begin{cases} 2\hat{\mathbf{n}}' \times \mathbf{H}^i(\varrho') & \text{in the lit zone of } C \\ 0 & \text{in shadowed zone of } C \end{cases}$$

as indicated previously in (29). With this GO approximation for $\mathbf{J}(\ell')$ in (34) and (35), one obtains the total fields $(\mathbf{E}^{PO}, \mathbf{H}^{PO})$ based on PO for the 2-D problem of Fig. 11 as

$$\mathbf{E}^{PO}(\varrho) = \mathbf{E}^i(\varrho) + \frac{kZ_0}{4} \int_{C_{lit}} (\hat{\mathbf{R}} \times \hat{\mathbf{R}} \times [2\hat{\mathbf{n}}' \times \mathbf{H}^i(\varrho')]) H_0^{(2)}(kR)\, d\ell' \quad (36)$$

and

$$\mathbf{H}^{PO}(\varrho) = \mathbf{H}^i(\varrho) - \frac{k}{4} \int_{C_{lit}} \{\hat{\mathbf{R}} \times [2\hat{\mathbf{n}}' \times \mathbf{H}^i(\varrho')]\} H_0^{(2)}(kR)\, d\ell' \quad (37)$$

in which C_{lit} refers to the lit portion of C, i.e., C_{lit} is that portion of C which is directly illuminated by the source. If the observation point is in the far zone of the scatterer, $\hat{\mathbf{R}}$ may be replaced by $\hat{\varrho}$, which can be removed outside the integrals in (36) and (37); furthermore, the Hankel function $H_0^{(2)}(kR)$ in those integrals may be approximated by

$$H_0^{(2)}(kR) \cong \sqrt{\frac{2j}{\pi k\varrho}}\, e^{-jk\varrho} e^{jk\hat{\varrho}\cdot\varrho'}$$

in the far-zone case. The quantity $\sqrt{2j/\pi k\varrho}\, e^{-jk\varrho}$ in the preceding far-zone approximation can also be removed outside the integrals of (36) and (37).

In general, the PO integrals in (30) and (31) [and also in (36) and (37)] must be evaluated numerically. If one can approximate the PO integral asymptotically via the method of stationary phase, then one obtains the GO reflected field which is associated with the stationary phase point on S_{lit} (or C_{lit}). In addition, there is a contribution to the integral arising from the boundary of S_{lit} (or C_{lit}), and also from regions where $\mathbf{J}_s^{GO}$ (or $\mathbf{J}^{GO}$) is discontinuous; the latter contribution can provide an approximate description for the fields diffracted by edged bodies which are otherwise smooth. The edge-diffracted fields given by PO are accurate only near the GO shadow boundary (or ISB and RSB directions in Fig. 4). Furthermore, the fields in the PO approximation do not satisfy reciprocity except in the direction of specular reflection [24]. The above limitations are to be anticipated from the GO current approximation inherent in the PO procedure. Thus the PO approximation is expected to be inaccurate in regions where the GO current does not produce the dominant scattering, e.g., for grazing angles of incidence, and for the fields in the deep shadow region behind S or C. It is noted that the truncation of the GO currents at the shadow boundary on S (i.e., at the boundary of S_{lit}) is one of the main sources of error in the PO approximation. On the other hand, PO does yield useful estimates in directions where the GO currents produce the dominant contribution to the scattered fields, e.g., as in the calculation of the main beam and the first few side lobes of a parabolic reflector antenna. Also, unlike GO, the PO integral provides field values even in directions away from specular reflection. Furthermore, the GO field amplitude is independent of the wave number k, whereas the PO integral is dependent on k and it may therefore provide an improvement over GO. The PO approximation can be improved by including a diffraction correction to the GO currents as in the PTD procedure, which is described later in Section 6. It is important to recall once again that in the PO calculation described here, the incident fields ($\mathbf{E}^i$, $\mathbf{H}^i$) exist everywhere as if the scatterer were absent; this is in contrast to the GO calculation, in which the incident

Techniques for High-Frequency Problems

fields are truncated at the incident shadow boundaries (ISB/SSB of Fig. 4) so that they vanish in the geometrical shadow region of the scatterer.

4. The Geometrical Theory of Diffraction and Its Uniform Versions

As pointed out in Section 1, the geometrical theory of diffraction (GTD) is a systematic extension of the ideas of GO in which diffracted rays are introduced via a generalization of Fermat's principle. While the GO rays exist only in the lit zone the diffracted rays, in general, enter into both the shadow as well as the lit zones.

The total GTD field consists of a superposition of the GO (incident and reflected) field and the field of all the diffracted rays which can reach the observation point. In particular, the total electric and magnetic fields as predicted by GTD are denoted here by $\mathbf{E}^{GTD}$ and $\mathbf{H}^{GTD}$, respectively, and

$$\mathbf{E}^{GTD} = \mathbf{E}^{GO} + \mathbf{E}_k^d \tag{38}$$

$$\mathbf{H}^{GTD} = \mathbf{H}^{GO} + \mathbf{H}_k^d \tag{39}$$

In (38) and (39) the $\mathbf{E}^{GO}$ and $\mathbf{H}^{GO}$ denote the GO component of the electric and magnetic fields; likewise, $\mathbf{E}_k^d$ and $\mathbf{H}_k^d$ refer to the corresponding diffracted-ray field components as given by Keller's GTD. The general form of $\mathbf{E}_k^d$ and $\mathbf{H}_k^d$ is discussed below.

General Form of the GTD Diffracted-Ray Fields E_k^d and H_k^d

It is appropriate to begin a discussion of diffracted-ray fields after summarizing the postulates on which the GTD is based, namely:

(a) Diffraction, like reflection, is a local phenomenon at high frequencies.
(b) The diffracted rays satisfy a generalized Fermat's principle.
(c) Away from the point of diffraction the diffracted rays behave according to the laws of conventional geometrical optics (GO).

It was mentioned in Section 1 that the initial value of the diffracted-ray field is given in terms of a diffraction coefficient in the GTD, and as a result of postulate *a*, the relevant diffraction coefficients can be found from the asymptotic solutions to simpler canonical problems which model the geometrical and electrical properties of the original configuration only in the local vicinity of the points of diffraction.

As a result of postulate *b*, the rays diffracted from a line of discontinuity (e.g., an edge), as in Figs. 1 and 12, lie on a cone about the tangent to the line of discontinuity with a cone half-angle β_0, which is identical with the angle that the incident ray makes with that same tangent at the point of diffraction. Likewise, postulate *b* requires that an incident ray which grazes a smooth convex surface must launch rays which follow geodesic paths on that surface. As an additional consequence of postulate *b*, rays are shed along the forward tangents to each geodesic surface ray path, thereby giving rise to surface-diffracted rays as in Fig. 13. Thus the surface-ray field attenuates as it propagates due to the continual shedding of energy from the surface-ray field. The example in Fig. 13 illustrates the GO

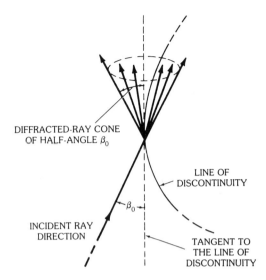

Fig. 12. Diffracted-ray cone from a line of discontinuity.

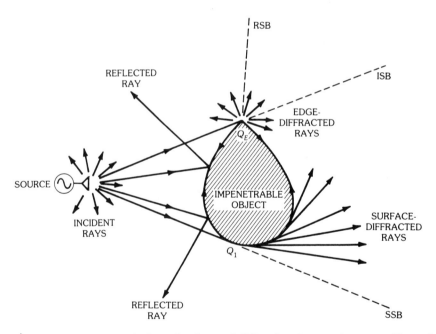

Fig. 13. Rays associated with the reflection and diffraction from an impenetrable surface.

incident and reflected rays as well as the edge- and surface-diffracted rays that can contribute to the total field at any exterior point when a source (antenna) radiates in the presence of an impenetrable structure consisting of an edge in an otherwise smooth convex body. It is shown in Fig. 13 that the incident ray at the edge can also

Techniques for High-Frequency Problems

excite surface rays and, conversely, surface rays can diffract from the edge. Through postulate *b*, the diffracted-ray optical path length is an extremum—usually a minimum. As mentioned previously the total field is the sum of the fields of all the individual rays that reach an observation point. The contribution from multiply interacting rays is usually much weaker and may thus be ignorable in such cases. In some cases it may be sufficient to include second-order interactions if they are important in comparison with the remaining higher-order multiply interacting rays, whose effects may be negligible. More is said about these interactions in Chapter 20. Next, the general form of the GTD diffracted fields ($\mathbf{E}_k^d, \mathbf{H}_k^d$) associated with the diffraction by edges, convex surfaces, and vertices will be obtained. However, the explicit forms of the various diffraction functions present in the expressions for these GTD diffracted fields will be indicated later in the second part of Section 4, which deals with the uniform versions of the GTD.

GTD for Edges—Consider a perfectly conducting wedge which is illuminated by a source at O as shown in Fig. 14. The total electric field $\mathbf{E}^{\text{GTD}}(P)$ at any point P exterior to the wedge is given by

$$\mathbf{E}^{\text{GTD}}(P) = \mathbf{E}^{\text{GO}}(P) + \mathbf{E}_k^d(P) \tag{40}$$

where the GO field component $\mathbf{E}^{\text{GO}}(P)$ is given as

$$\mathbf{E}^{\text{GO}}(P) = \mathbf{E}^i(P) U_i + \mathbf{E}^r(P) U_r \tag{41}$$

in which the GO incident and reflected fields $\mathbf{E}^i$ and $\mathbf{E}^r$ exist only in the lit regions. The domains of existence of these incident and reflected GO fields are indicated by the step functions U_i and U_r, respectively. In the case of the wedge, U_i and U_r are defined by:

$$U_i = \begin{cases} 1 & \text{if } 0 < \phi < \pi + \phi' \\ 0 & \text{if } \pi + \phi' < \phi < n\pi \end{cases}$$

and

$$U_r = \begin{cases} 1 & \text{if } 0 < \phi < \pi - \phi' \\ 0 & \text{if } \pi - \phi' < \phi < n\pi \end{cases}$$

The GO incident and reflected fields have been discussed previously in Section 2.

The diffracted field $\mathbf{E}_k^d(P)$ (and likewise $\mathbf{H}_k^d$) exists everywhere exterior to the wedge, i.e., for $0 < \phi < n\pi$. The diffracted rays lie on cones about the edge as in Figs. 12 and 14. Away from the edge, $\mathbf{E}_k^d(P)$ behaves as a ray optical (or GO) type field via Keller's postulate (*c*) above; the validity of this postulate can be observed directly via the discussion in Section 2, under "The Ray Concept," by considering S' (of Fig. 3) to be the wavefront associated with the edge-diffracted field (since S' could represent any wavefront). Thus, from (2) one may write

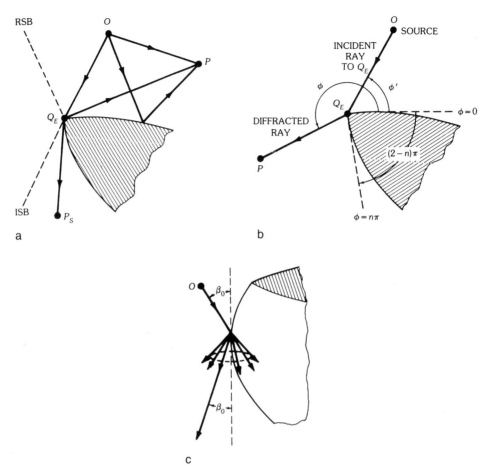

Fig. 14. Rays associated with the problem of edge diffraction. (*a*) Incident, reflected, and diffracted rays. (*b*) Side view, for edge-diffracted ray. (*c*) Perspective view, for edge-diffracted ray.

$$\mathbf{E}_k^d(P) \sim \mathbf{E}_k^d(P_0) \sqrt{\frac{\varrho_1^d \varrho_2^d}{(\varrho_1^d + s_0^d)(\varrho_2^d + s_0^d)}} \, e^{-jks_0^d} \qquad (42a)$$

The diffracted-ray tube corresponding to (42) is shown in Fig. 15, where the superscript d on ϱ_1^d, ϱ_2^d, s_0^d, and s^d denotes that these quantities are associated with the diffracted-ray field component. In order to relate $\mathbf{E}_k^d(P)$ to the incident field at the point Q_E of edge diffraction, one moves the reference P_0 in Fig. 15 to the point Q_E of diffraction on the edge (i.e., by letting $\varrho_1^d \to 0$) so that

$$\mathbf{E}_k^d(P) = \lim_{\varrho_1^d \to 0} [\sqrt{\varrho_1^d} \, \mathbf{E}_k^d(P_0)] \sqrt{\frac{\varrho_2^d}{(\varrho_1^d + s_0^d)(\varrho_2^d + s_0^d)}} \, e^{-jks_0^d} \qquad (42b)$$

Since $\mathbf{E}_k^d(P)$ is independent of the reference point P_0, the above limit exists and one defines it as

Techniques for High-Frequency Problems

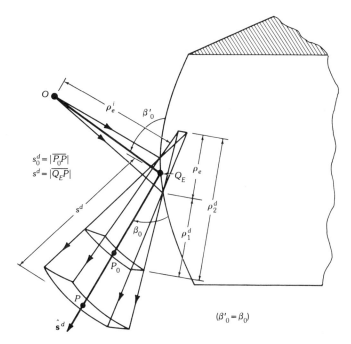

Fig. 15. Edge-diffracted ray tube.

$$\lim_{\varrho_1^d \to 0} \sqrt{\varrho_1^d} \, \mathbf{E}_k^d(P_0) \equiv \mathbf{E}^i(Q_E) \cdot \bar{\bar{\mathbf{D}}}_e^k \quad (43)$$

where $\bar{\bar{\mathbf{D}}}_e^k = \bar{\bar{\mathbf{D}}}_e^k(\phi, \phi', \beta_0; k)$ is the *dyadic edge-diffraction coefficient*, which indicates how the energy is distributed in the diffracted field as a function of the angles ϕ, ϕ', and β_0; the dyadic $\bar{\bar{\mathbf{D}}}_e^k$ also depends on n and the wave number k. From (42) and (43) one obtains

$$\mathbf{E}_k^d(P) \sim \mathbf{E}^i(Q_E) \cdot \bar{\bar{\mathbf{D}}}_e^k(\phi, \phi', \beta_0; k) \sqrt{\frac{\varrho_e}{s^d(\varrho_e + s^d)}} \, e^{-jks^d} \quad (44)$$

where

$$\lim_{\varrho_1^d \to 0} \varrho_2^d \equiv \varrho_e$$

is the edge-diffracted ray caustic distance, and likewise

$$\lim_{\varrho_1^d \to 0} s_0^d \equiv s^d$$

as shown in Fig. 15. The field $\mathbf{E}_k^d(P)$ is polarized transverse to the diffracted-ray direction $\hat{\mathbf{s}}^d$ since $\mathbf{E}_k^d(P)$ is ray optical; thus the associated magnetic field can be expressed as

$$\mathbf{H}_k^d(P) \sim Y_0 \hat{\mathbf{s}}^d \times \mathbf{E}_k^d(P) \quad (45)$$

In Fig. 15 Q_E is one of the caustics of the edge-diffracted field.

The $\bar{\bar{\mathbf{D}}}_e^k$ for the curved edge in a curved surface of Fig. 15 may be found from the asymptotic solutions to some canonical problems of EM diffraction by a wedge [25, 26]. The exact form of $\bar{\bar{\mathbf{D}}}_e^k$ will be indicated subsequently in the part of Section 4 which deals with the more useful, uniform version of the GTD.

GTD for Convex Surfaces—The total GTD electric field $\mathbf{E}^{\mathrm{GTD}}$ exterior to a perfectly conducting convex surface in Fig. 16 which is excited by an incident wave is given (as in the case of edge diffraction) by

$$\mathbf{E}^{\mathrm{GTD}}(P) = \begin{cases} \mathbf{E}^{\mathrm{GO}}(P_L)(= \mathbf{E}^i(P_L)U_i \\ \qquad\qquad + \mathbf{E}^r(P_L)U_r) + \mathbf{E}_k^d(P_L) & \text{if } P = P_L \text{ in the lit zone} \\ \mathbf{E}_k^d(P_S) & \text{if } P = P_S \text{ in the shadow zone} \end{cases}$$

However, in the case of the convex surface in Fig. 16, $\mathbf{E}_k^d(P_L)$ as predicted by Keller's GTD is zero in the lit zone. It is noted that U_i and U_r have the same meaning as in (41). For the convex surface the U_i and U_r can be defined by:

$$U_i = U_r = \begin{cases} 1 & \text{in lit zone above SSB} \\ 0 & \text{in shadow zone below SSB} \end{cases}$$

since the ISB and RSB merge into the SSB. Therefore, for the convex surface in Fig. 16 the above expression for $\mathbf{E}^{\mathrm{GTD}}$ becomes

$$\mathbf{E}^{\mathrm{GTD}}(P) = \begin{cases} \mathbf{E}^i(P_L) + \mathbf{E}^r(P_L) & \text{if } P = P_L \text{ in the lit zone} \\ \mathbf{E}_k^d(P_S) & \text{if } P = P_S \text{ in the shadow zone} \end{cases} \quad (46)$$

The ray paths associated with $\mathbf{E}^i$, $\mathbf{E}^r$, and $\mathbf{E}_k^d$ have been discussed previously in Sections 1 (see Fig. 1) and 2. The surface shadow boundary (SSB) is an extension of the incident ray which grazes the surface at Q_1; it divides the exterior space into the lit and shadow zones. The explicit forms of $\mathbf{E}^i(P_L)$ and $\mathbf{E}^r(P_L)$ have been

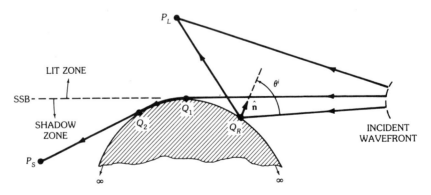

Fig. 16. Rays associated with the scattering and diffraction by a convex surface.

introduced earlier in Section 2, "The Geometrical Optics Field"; thus it remains only to consider the form of the diffracted-ray field. The field $\mathbf{E}_k^d(P_S)$ can be expressed ray optically via Keller's postulate c (or by making use of the arguments indicated in Section 2, under "The Ray Concept," which allow one to consider S' of Fig. 3 to be the wavefront of the surface-diffracted ray from Q_2 to P_S); therefore $\mathbf{E}_k^d(P_S)$ becomes

$$\mathbf{E}_k^d(P_S) \sim \mathbf{E}_k^d(P_0) \sqrt{\frac{\varrho_1^d \varrho_2^d}{(\varrho_1^d + s_0^d)(\varrho_2^d + s_0^d)}} e^{-jks_0^d} \qquad (47)$$

The usual distances ϱ_1^d, ϱ_2^d, s_0^d, and s^d associated with a surface-diffracted ray tube corresponding to (47) are illustrated in Fig. 17. In order to relate $\mathbf{E}_k^d(P_0)$ to the incident field at the point of grazing Q_1, one moves the reference point P_0 to the point Q_2 from where the surface-diffracted ray is shed tangentially to arrive at the point P_S. Since $\mathbf{E}_k^d(P_S)$ is independent of the reference point P_0, one may write

$$\mathbf{E}_k^d(P_S) = \lim_{\varrho_1^d \to 0} [\sqrt{\varrho_1^d}\, \mathbf{E}_k^d(P_0)] \sqrt{\frac{\varrho_2^d}{(\varrho_1^d + s_0^d)(\varrho_2^d + s_0^d)}} e^{-jks_0^d} \qquad (48)$$

together with

$$\lim_{\varrho_1^d \to 0} \sqrt{\varrho_1^d}\, \mathbf{E}_k^d(P_0) \equiv \mathbf{E}^i(Q_1)\cdot\bar{\bar{\mathbf{T}}}^k(Q_1, Q_2)\, e^{-jkt}\sqrt{\frac{d\eta(Q_1)}{d\eta(Q_2)}} \qquad (49)$$

so that (47) becomes

$$\mathbf{E}_k^d(P_S) \sim \mathbf{E}^i(Q_1)\cdot\bar{\bar{\mathbf{T}}}^k(Q_1, Q_2)\, e^{-jkt}\sqrt{\frac{d\eta(Q_1)}{d\eta(Q_2)}} \sqrt{\frac{\varrho_s}{s^d(\varrho_s + s^d)}} e^{-jks^d} \qquad (50)$$

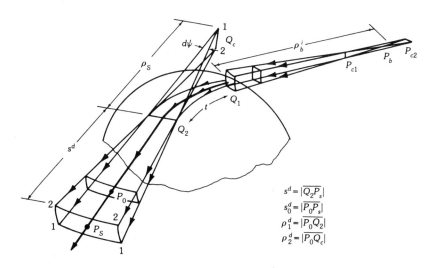

Fig. 17. Surface-diffracted ray tube.

In arriving at (50), use is also made of the definitions

$$\lim_{\varrho_1^d \to 0} \varrho_2^d \equiv \varrho_s \quad \text{and} \quad \lim_{\varrho_1^d \to 0} s_0^d = s^d$$

In (49) and (50) $d\eta(Q_1)$ and $d\eta(Q_2)$ refer to the widths of the surface-ray strip at Q_1 and Q_2, respectively, as illustrated in Fig. 17. Furthermore, the factor $\sqrt{d\eta(Q_1)/d\eta(Q_2)}\,e^{-jkt}$ indicates a conservation of the energy flux in the surface-ray strip from Q_1 to Q_2, in which e^{-jkt} denotes the dominant phase delay along that surface ray path. However, since rays are shed along the forward tangents to the geodesic surface-ray paths to give rise to the surface-diffracted rays, energy is lost from the surface rays and the surface-ray field attenuates. Thus the dyadic transfer function $\bar{\bar{T}}^k(Q_1, Q_2)$ in (48) and (49) is introduced to indicate the launching of the surface-ray field at Q_1, the attenuation of the surface-ray field between Q_1 and Q_2, and also the amount of diffraction of the surface-ray field from Q_2. The explicit form of $\bar{\bar{T}}^k(Q_1, Q_2)$ will be indicated subsequently in Section 4, under "Uniform Version of the GTD." The shadowed part of the surface in Fig. 16 is a caustic of the surface-diffracted rays. The diffracted-ray optical field $\mathbf{E}_k^d(P_S)$ has an associated magnetic field $\mathbf{H}_k^d(P_S)$ given as usual by

$$\mathbf{H}_k^d(P_S) \sim Y_0 \hat{s}^d \times \mathbf{E}_k^d(P_S) \tag{51}$$

GTD for Vertices—A vertex is formed, for example, at the tip of a cone, the tip of a pyramid, or by the corner of a plane angular sector. From Keller's postulate a, an incident ray which strikes a perfectly conducting vertex produces a continuum of diffracted rays which emanate in all directions from the vertex as shown in Fig. 1f. One such vertex-diffracted ray to P is shown in Fig. 18. Using postulate c, as before, allows one to express the electric field $\mathbf{E}_k^d(P)$ diffracted by the corner (or vertex) as in (2) or (42a); thus,

$$\mathbf{E}_k^d(P) \sim \mathbf{E}_k^d(P_0) \sqrt{\frac{\varrho_1^d \varrho_2^d}{(\varrho_1^d + s_0^d)(\varrho_2^d + s_0^d)}}\, e^{-jks_0^d}$$

$$\sim \mathbf{E}_k^d(P_0) \frac{\varrho_v^d}{\varrho_v^d + s_0^d}\, e^{-jks_0^d} \tag{52}$$

since $(\varrho_1^d = \varrho_2^d) \equiv \varrho_v^d$ for a vertex as in Fig. 18. In order to relate $\mathbf{E}_k^d(P)$ of (52) to the incident field $\mathbf{E}^i(Q_v)$ at the corner or tip Q_v, one defines

$$\lim_{\varrho_v^d \to 0} \varrho_v^d \mathbf{E}_k^d(P_0) = \mathbf{E}^i(Q_v) \cdot \bar{\bar{\mathbf{D}}}_v \tag{53}$$

where $\bar{\bar{\mathbf{D}}}_v = \bar{\bar{\mathbf{D}}}_v(\hat{s}^d, \hat{s}^i; k)$ is the dyadic diffraction coefficient for a vertex or corner; it indicates the manner in which the energy in the field incident at Q_v is distributed in the rays diffracted in all directions from the point Q_v. This $\bar{\bar{\mathbf{D}}}_v$ depends on the angles of incidence and diffraction, $\hat{s}^i$ and $\hat{s}^d$, respectively, and also on the wave number k besides depending on the local geometry of the corner at Q_v. The limit in

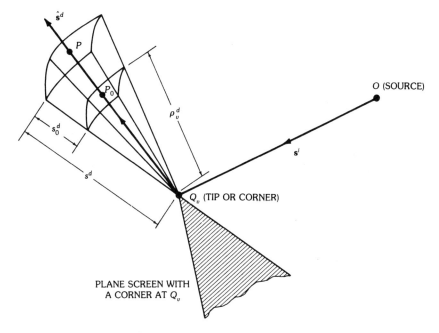

Fig. 18. Corner- or vertex-diffracted ray tube.

(53) exists because $\mathbf{E}_k^d(P)$ in (52) is independent of the position of the reference point at P_0. From Fig. 18 it is evident that

$$\lim_{\varrho_v^d \to 0} s_0^d = s^d$$

Using this information and (53) allows one to write (52) as

$$\mathbf{E}_k^d(P) \sim \mathbf{E}^i(Q_v) \cdot \bar{\bar{\mathbf{D}}}_v(\hat{\mathbf{s}}^i, \hat{\mathbf{s}}^d, k) \frac{e^{-jks^d}}{s^d} \tag{54}$$

when $\varrho_v^d \to 0$. The vertex- or corner-diffracted ray optical field in (54) is polarized transverse to the ray direction $\hat{\mathbf{s}}^d$ and, as before, the associated magnetic field $\mathbf{H}_k^d(P)$ is given by

$$\mathbf{H}_k^d(P) \sim Y_0 \hat{\mathbf{s}}^d \times \mathbf{E}_k^d(P) \tag{55}$$

The point Q_v is a caustic of the vertex-diffracted rays. Also, the total GTD field surrounding the vertex or corner is the sum of the GO field and the corner-diffracted field together with other diffracted fields (e.g. edge and/or surface diffracted components) depending on the nature of the surface containing the corner, as in (38). The explicit form of $\bar{\bar{\mathbf{D}}}_v$ will be indicated later in the following section dealing with the uniform version of the GTD.

Uniform Version of the GTD

A uniform version of the GTD which is referred to as the *uniform geometrical theory of diffraction*, and which is abbreviated as *UTD* [4, 5], is described below for analyzing the problems of electromagnetic diffraction by perfectly conducting edges, vertices, and convex surfaces. An alternative uniform theory of diffraction referred to as the *UAT* [6], which has been developed primarily for analyzing the problem of electromagnetic diffraction by an edge, is also described below.

Both the *UTD* and *UAT* are "uniform" in the sense that, firstly, they overcome the failure of the GTD within the GO shadow boundary transition regions away from the point of diffraction and, secondly, they reduce automatically to the GTD exterior to these transition regions where the GTD becomes valid and accurate.

UTD for Edges—Curved wedges occur as part of many practical antenna and scattering configurations, e.g., the edge of a reflector antenna, the bases of cylindrical and conical structures, and the trailing edges of wings and stabilizers. The UTD diffraction coefficients which are useful in the analysis of the radiation by antennas in the presence of such structures with edges are described below. Specifically, UTD diffraction coefficients are presented for both 3-D and 2-D curved wedge configurations which serve to locally model 3-D and 2-D structures with edges. Only field points exterior to the wedge are considered here.

3-D Case—The total UTD electric field $\mathbf{E}(P)$ at a point exterior to a perfectly conducting wedge, which is illuminated by a source at O as in Fig. 14, is given according to UTD by

$$\mathbf{E}(P) = \mathbf{E}^{GO}(P) + \mathbf{E}^d(P) \tag{56}$$

The field $\mathbf{E}^{GO}(P)$ is as defined earlier in (41). The UTD field $\mathbf{E}(P)$ in (56) remains valid even within the transition regions adjacent to the incident and reflection shadow boundaries (ISB and RSB of Fig. 4a), respectively, where the GTD fails. Furthermore, exterior to these transition regions, $\mathbf{E}(P)$ reduces automatically to $\mathbf{E}^{GTD}(P)$ of (40), where the GTD becomes valid; i.e., the uniform edge-diffracted field $\mathbf{E}^d(P)$ in (56) reduces properly to Keller's edge-diffracted field $\mathbf{E}_k^d(P)$ [in (40)] exterior to the transition regions. Since $\mathbf{E}^{GO}(P)$ has been described earlier in Section 2, under "The Geometrical Optics Field," it only remains to give explicit expressions for the diffracted field $\mathbf{E}^d(P)$ in (56). Firstly, the edge-diffracted field $\mathbf{E}^d(P)$ is associated with the edge-diffracted rays of the GTD. If an incident ray strikes the edge obliquely, making an angle β_0' with the unit vector $\hat{e}$ which is tangent to the edge at the point of diffraction, then the diffracted rays lie on a Keller cone whose half-angle is $\beta_0 = \beta_0'$ (law of edge diffraction) as shown in Figs. 12, 14, and 15. Secondly, the UTD expression for $\mathbf{E}^d(P)$ associated with an edge-diffracted ray can also be expressed in the format of the GTD as in (44) for $\mathbf{E}_k^d(P)$; in particular,

$$\mathbf{E}^d(P) = \mathbf{E}^i(Q_E) \cdot \bar{\bar{\mathbf{D}}}_e(\phi, \phi', \beta_0; k) \sqrt{\frac{\varrho_e}{s^d(\varrho_e + s^d)}} e^{-jks^d} \tag{57}$$

Thus the GTD dyadic edge-diffraction coefficient $\bar{\bar{\mathbf{D}}}_e^k$ in (44) is replaced with the uniform dyadic edge-diffraction coefficient $\bar{\bar{\mathbf{D}}}_e$ in the UTD expression of (57) for the edge-diffracted field. In addition, the UTD dyadic edge-diffraction coefficient $\bar{\bar{\mathbf{D}}}_e$ can be expressed conveniently in terms of unit vectors fixed in the incident and diffracted rays as follows.* Note that the Keller cone is defined by the law of edge diffraction, $\hat{s}^i \cdot \hat{e} = \hat{s}^d \cdot \hat{e}$, which results from Keller's generalization of Fermat's principle (Section 4, under "General Form of the GTD Diffracted-Ray Fields $\mathbf{E}_k^d$ and $\mathbf{H}_k^d$"). Let the incident-ray direction $\hat{s}^i$ and $\hat{e}$ define an edge-fixed plane of incidence; likewise, let the direction of the diffracted ray $\hat{s}^d$ and $\hat{e}$ define an edge-fixed plane of diffraction. The unit vectors $\hat{\boldsymbol{\beta}}_0'$ and $\hat{\boldsymbol{\beta}}_0$ are parallel to the edge-fixed plane of incidence and diffraction, respectively, and

$$\hat{\boldsymbol{\beta}}_0' = \hat{s}^i \times \hat{\boldsymbol{\phi}}' \tag{58a}$$

$$\hat{\boldsymbol{\beta}} = \hat{s}^d \times \hat{\boldsymbol{\phi}} \tag{58b}$$

as shown in Fig. 19a. Here, $\hat{\boldsymbol{\phi}}'$ and $\hat{\boldsymbol{\phi}}$ point in the direction of increasing angles ϕ' and ϕ, respectively, which indicate the azimuthal (or circumferential) angular coordinates of the incident and diffracted rays about the edge tangent $\hat{e}$ as illustrated in Fig. 19b. The field $\mathbf{E}^d(P)$ can now be expressed invariantly in terms of these triads of unit vectors $(\hat{s}^i, \hat{\boldsymbol{\beta}}_0', \hat{\boldsymbol{\phi}}')$ and $(\hat{s}^d, \hat{\boldsymbol{\beta}}_0, \hat{\boldsymbol{\phi}})$, respectively, which are fixed in the incident and diffracted rays. In this ray coordinate system [25, 26] one may write

$$\mathbf{E}^d(P) = \hat{\boldsymbol{\beta}}_0 E_{\beta_0}^d + \hat{\boldsymbol{\phi}} E_\phi^d \tag{59a}$$

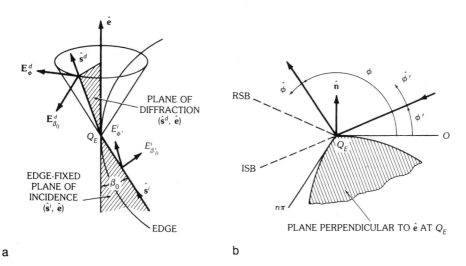

Fig. 19. Edge-fixed planes of incidence and diffraction. (*a*) Planes of incidence and diffraction. (*b*) Angular coordinates about $\hat{e}$. (*After Kouyoumjian and Pathak [26], © 1974 IEEE*)

---

*The main difference between $\bar{\bar{\mathbf{D}}}_e$ and $\bar{\bar{\mathbf{D}}}_e^k$ is that the former is range dependent, whereas the latter is not. As a result, the expression in (57) is not strictly ray optical within the shadow boundary transition regions; outside these regions, (57) reduces to (44), as shown later.

and

$$\mathbf{E}^i(Q_E) = \hat{\boldsymbol{\beta}}'_0 E^i_{\beta'_0} + \hat{\boldsymbol{\phi}}' E^i_\phi \qquad (59b)$$

with

$$\bar{\bar{\mathbf{D}}}_e = -\hat{\boldsymbol{\beta}}'_0\hat{\boldsymbol{\beta}}_0 D_{es} - \hat{\boldsymbol{\phi}}'\hat{\boldsymbol{\phi}} D_{eh} \qquad (59c)$$

A matrix representation for (57) using (59a)–(59c) is given by

$$\begin{bmatrix} E^d_{\beta_0} \\ E^d_\phi \end{bmatrix} = \begin{bmatrix} -D_{es} & 0 \\ 0 & -D_{eh} \end{bmatrix} \begin{bmatrix} E^i_{\beta'_0} \\ E^i_{\phi'} \end{bmatrix} \sqrt{\frac{\varrho_e}{s^d(\varrho_e + s^d)}} e^{-jks^d} \qquad (60)$$

in which [5, 26]

$$\begin{aligned} D_{es,eh}(\phi,\phi';\beta_0) = \frac{-e^{-j\pi/4}}{2n\sqrt{2\pi k}\sin\beta_0} &\left[\cot\left(\frac{\pi + (\phi - \phi')}{2n}\right) F[kL^i a^+(\phi - \phi')] \right. \\ &+ \cot\left(\frac{\pi - (\phi - \phi')}{2n}\right) F[kL^i a^-(\phi - \phi')] \\ &\mp \left\{\cot\left(\frac{\pi + (\phi + \phi')}{2n}\right) F[kL^{rn} a^+(\phi + \phi')] \right. \\ &+ \left.\left.\cot\left(\frac{\pi - (\phi + \phi')}{2n}\right) F[kL^{ro} a^-(\phi + \phi')]\right\}\right] \end{aligned} \qquad (61)$$

where the asymptotic large parameter kL is required to be sufficiently large (generally greater than 3) and

$$a^\pm(\beta) = 2\cos^2\left(\frac{2n\pi N^\pm - \beta}{2}\right) \qquad (62a)$$

in which $N^\pm$ are the integers which most nearly satisfy the equation

$$2n\pi N^\pm - \beta = \pm\pi \qquad (62b)$$

with

$$\beta = \phi \pm \phi' \qquad (62c)$$

In (61) [and (62a, 62b)] n defines the exterior wedge angle as in Fig. 19b; hence $n = 2$ for a half-plane, and $n = 3/2$ for an exterior right-angle wedge, etc. The two faces of the wedge are thus referred to as the "o" face at $\phi = 0$ and the "n" face at $\phi = n\pi$, respectively, as in Fig. 19b.

For exterior edge diffraction, $N^+ = 0$ or 1, and $N^- = -1, 0,$ or 1. The values of $N^\pm$ at the shadow and reflection boundaries as well as their associated transition regions are given in Table 1 for exterior wedge angles ($1 < n < 2$).

Table 1. Values of N for Exterior Wedge Angles

	The Cotangent Is Singular When	Value of N at the Boundary
$\cot\left[\dfrac{\pi + (\phi - \phi')}{2n}\right]$	$\phi = \phi' - \pi$, an SB Surface $\phi = 0$ is shadowed	$N^+ = 0$
$\cot\left[\dfrac{\pi - (\phi - \phi')}{2n}\right]$	$\phi = \phi' + \pi$, an SB Surface $\phi = n\pi$ is shadowed	$N^- = 0$
$\cot\left[\dfrac{\pi + (\phi + \phi')}{2n}\right]$	$\phi = (2n - 1)\pi - \phi'$, an RB Reflection from surface $\phi = n\pi$	$N^+ = 1$
$\cot\left[\dfrac{\pi - (\phi + \phi')}{2n}\right]$	$\phi = \pi - \phi'$, an RB Reflection from surface $\phi = 0$	$N^- = 1$

For a point-source (or spherical-wave) type illumination the distance parameter L^i is

$$L^i = \frac{s^i s^d}{s^i + s^d} \sin^2\beta_0 \tag{63}$$

in which s^i and s^d are the distances from the point of edge diffraction at Q_E to the source and observation points, respectively. Only for a straight wedge with planar faces that is illuminated by a point source does

$$L^{ro} = L^{rn} = L^i = \frac{s^i s^d}{s^i + s^d} \sin^2\beta_0$$

as in (63). In the case of an arbitrary ray-optical type of illumination which is characterized by two distinct principal wavefront radii of curvature, ϱ_1^i and ϱ_2^i, the above L^i must be modified as shown below in the general expressions for L^{ro} and L^{rn} pertaining to a curved wedge; thus

$$L^i = \left[\frac{s^d(\varrho_e^i + s^d)\varrho_1^i \varrho_2^i \sin^2\beta_0}{\varrho_e^i(\varrho_1^i + s^d)(\varrho_2^i + s^d)}\right]_{\text{ISB}} \tag{64a}$$

$$L^r = \left[\frac{s^d(\varrho_e^r + s^d)\varrho_1^r \varrho_2^r \sin^2\beta_0}{\varrho_e^r(\varrho_1^r + s^d)(\varrho_2^r + s^d)}\right]_{\text{RSB}} \tag{64b}$$

Here, L^{ro} and L^{rn} are the values of L^r associated with the o and n faces of the wedge, respectively. Furthermore, ϱ_e^r is given by

$$\frac{1}{\varrho_e^r} = \frac{1}{\varrho_e^i} - \frac{2(\hat{n}\cdot\hat{n}_e)(\hat{s}^i\cdot\hat{n})}{a \sin^2\beta_0} \tag{65a}$$

Also ϱ_e in (57) is given by

$$\frac{1}{\varrho_e} = \frac{1}{\varrho_e^i} - \frac{\hat{n}_e \cdot (\hat{s}^i - \hat{s}^d)}{a \sin^2 \beta_0} \tag{65b}$$

The unit vector $\hat{n}$ is defined in Fig. 19b, whereas $\hat{n}_e$ is a unit vector normal to the edge which is directed away from the center of edge curvature at Q_E. The radius of edge curvature is denoted by a in (65). The term ϱ_e^i is the radius of curvature of the incident wavefront at Q_E which lies in the edge-fixed plane of incidence. In the far zone when $s^d \gg \varrho_{1,2}^i$, $s^d \gg \varrho_{1,2}^r$, and $s^d \gg \varrho_e$, then the L^i and L^r in (64a) and (64b) simplify to $L \cong (\varrho_1 \varrho_2 \sin^2 \beta_0)/\varrho_e$ in which the appropriate superscripts on L, ϱ_1, and ϱ_2 are omitted for convenience. It is noted that L^i and L^r in (64a) and (64b) are calculated on the appropriate shadow boundaries. The transition function F which appears in (61) contains a Fresnel integral; it is defined by

$$F(x) = 2j\sqrt{x}\, e^{jx} \int_{\sqrt{x}}^{\infty} e^{-j\tau^2} d\tau \tag{66}$$

A plot of the above $F(x)$ is illustrated in Fig. 30. In (66), $\sqrt{x} = |\sqrt{x}|$ if $x > 0$ and $\sqrt{x} = -j|\sqrt{x}|$ if $x < 0$. If $x < 0$, then $F(x)|_{x<0} = F^*(|x|)$ where * denotes the complex conjugate. Exterior to the ISB and RSB transition regions, x becomes large and $F(x) \to 1$ so that the uniform $D_{es,eh}$ in (61) then reduces to Keller's form [1] as it should; namely,

$\mathbf{E}^d(P) \to \mathbf{E}_k^d(P)$, outside the transition region, or

$\bar{\bar{\mathbf{D}}}_e \to \bar{\bar{\mathbf{D}}}_e^k$, outside the transition region, where

$$\bar{\bar{\mathbf{D}}}_e^k = -\hat{\boldsymbol{\beta}}_0' \hat{\boldsymbol{\beta}}_0 D_{es}^k - \hat{\boldsymbol{\phi}}' \hat{\boldsymbol{\phi}} D_{eh}^k \tag{67}$$

$$D_{es,eh}^k (\phi, \phi'; \beta_0) = \frac{e^{-j\pi/4} \sin(\pi/n)}{n\sqrt{2\pi k}\, \sin \beta_0} \left\{ \frac{1}{\cos(\pi/n) - \cos[(\phi - \phi')/n]} \right.$$
$$\left. \mp \frac{1}{\cos(\pi/n) - \cos[(\phi + \phi')/n]} \right\} \tag{68}$$

Note that $D_{es,eh}^k$ in (68) becomes singular at the shadow boundaries ISB ($\phi - \phi' = \pi$) and RSB ($\phi + \phi' = \pi$) for the o face. On the other hand, $D_{es,eh}$ in (61) is well behaved at these shadow boundaries. Indeed, at these (ISB and RSB) boundaries the small-argument approximation for $F(x)$ may be employed (since $x = 0$ on the ISB and RSB), namely, one may insert

$$F(x) \underset{x \to 0}{\to} \sqrt{\pi x}\, e^{j(\pi/4 + x)} \tag{69}$$

into (61) to arrive at the following result:

$$\mathbf{E}^d|_{\text{ISB;RSB}} = \mp \frac{1}{2} \mathbf{E}^{i;r} + \text{continuous higher-order terms in } k,$$

$$\text{if } \begin{cases} \text{on just the lit side of ISB;RSB} \\ \text{on just the shadow side of ISB;RSB} \end{cases} \tag{70}$$

Techniques for High-Frequency Problems

The above result in (70) ensures the continuity of the total high-frequency field in (56) at the ISB and RSB. The field contributions arising from the edge-excited "surface-diffracted rays" are not included in (56); these may be important for observation points close to the surface shadow boundaries (SSB) associated with the tangents to the o and n faces of a curved wedge at Q_E if the o and n faces are convex boundaries. The result in (56) and (57) together with (61) is valid away from any diffracted-ray caustics and away from the edge caustic at Q_E.

In the case of grazing angles of incidence on a wedge with planar faces, $D_{es} = 0$, and D_{eh} must be replaced by $(1/2) D_{eh}$. The reason for the factor $1/2$ in the latter case may be argued as follows. The incident and reflected GO fields tend to combine into a single "total incident field" as one approaches grazing angles of incidence; consequently, only half of this "total field" illuminating the edge at grazing constitutes the incident GO field, while the other half constitutes the reflected GO field. The case of grazing angles of incidence at an edge on a curved surface cannot be handled as easily as the case of a wedge with planar faces. In fact, work needs to be done to extend the UTD to grazing angles of incidence on a curved surface forming the wedge; at present one can only treat angles of incidence that are greater than $[2/k\varrho_g(Q_E)]^{1/3}$, where $\varrho_g(Q_E)$ is the radius of curvature of the surface in the direction of the incident ray at the point Q_E of edge diffraction.

Under the above restrictions the general result in (61) for $D_{es,eh}$ simplifies in the case of a plane or curved screen ($n = 2$ case) to

$$D_{es,eh}(\phi, \phi'; \beta_0) = \frac{-e^{-j\pi/4}}{2\sqrt{2\pi k} \sin\beta_0} \left\{ \sec\left(\frac{\phi - \phi'}{2}\right) F[kL^i a(\phi - \phi')] \right.$$
$$\left. \mp \sec\left(\frac{\phi + \phi'}{2}\right) F[kL^r a(\phi + \phi')] \right\}$$

where $a(\beta) \equiv 2\cos^2(\beta/2)$ and $L^{i,r}$ are as in (64a, 64b) with the understanding that L^r is evaluated at the RSB corresponding to the face which is illuminated; hence the superscripts o and n in L^r are dropped for this $n = 2$ case.

2-D Case—As in the 3-D case the total electric field $\mathbf{E}(P)$ at a point exterior to a 2-D perfectly conducting wedge illuminated by a 2-D line source is described by (56) in which the GO field $\mathbf{E}^{GO}(P_L)$ is given by (41), and $\mathbf{E}^i$ and $\mathbf{E}^r$ are found via (12) and (24). The 2-D edge-diffracted field $\mathbf{E}^d(P)$ can be obtained from (57) by allowing ϱ_e to approach infinity and by requiring $\beta_0 = \pi/2$; thus, for the 2-D case,

$$\mathbf{E}^d(P) = \mathbf{E}^i(Q_E) \cdot \overline{\overline{\mathbf{D}}}_e(\phi, \phi', \pi/2; k) \frac{e^{-jks^d}}{\sqrt{s^d}} \quad (71)$$

The $\overline{\overline{\mathbf{D}}}_e$ in (71) for the 2-D case is available from (59c) and (61), with $\beta_0 = \pi/2$ (or $\sin\beta_0 = 1$). Also, L^i for the 2-D case is as in (63), with $\beta_0 = \pi/2$; in particular,

$$L^i = \frac{s^i s^d}{s^i + s^d} \quad (72)$$

Likewise, L^r is obtained from (64b), with $\beta_0 = \pi/2, \varrho_1^r \to \infty, \varrho_2^r \equiv \varrho^r$ [as in (24)], and $\varrho_e^r \to \infty$; therefore, in the 2-D case,

$$L^r = \frac{\varrho^r s^d}{\varrho^r + s^d} \tag{73}$$

Note that ϱ^r in (73) is the same as the one in (24); however, ϱ^r is in general different for the o and n faces of the wedge, with L^{ro} and L^{rn} denoting the values of L^r for these two different faces. While the expression for L^r in (64b) is fixed to its value on the RSB for convenience [26], the one in (73) can be evaluated as a function of the observation point with almost the same ease as if one had approximated the value of L^r by its value at the RSB. The values of L^i and L^r in (64a) and (64b) involve several quantities for the general 3-D case, and since these parameters L^i and L^r are slowly varying within the ISB and RSB transition regions, it is therefore convenient in the 3-D case to approximately fix their values to those at the ISB and RSB, respectively, as done in (64a) and (64b). Outside their respective transition regions, $kL^i a^i$ and $KL^{rn} a^{rn}$ and $kL^{ro} a^{ro}$ become large anyway so that $F(kL^i a^i) \to 1$ and $F(kL^{rn} a^{rn}) \to 1$ and $F(kL^{ro} a^{ro}) \to 1$, unaffected by the aforementioned approximations.

The comments under (70) pertaining to grazing angles of incidence on a wedge also apply to the 2-D case.

Slope Diffraction by Edges Based on UTD—The UTD results for the edge-diffracted field $\mathbf{E}^d(P)$ in (57) and (71) for the 3-D and 2-D cases are accurate if the incident field $\mathbf{E}^i(Q_E)$ in those expressions exhibits a slow spatial variation (in its amplitude but not necessarily in its phase along the incident ray) at Q_E. If the field $\mathbf{E}^i(Q_E)$ is not slowly varying, then a higher-order term, referred to as the *slope diffraction term*, must be added to (57) and (71), respectively, as indicated in [4, 5]. This slope diffraction contribution pertaining to 3-D and 2-D cases is given below.

3-D Case—The UTD edge-diffracted field $\mathbf{E}^{dt}$ produced by a rapidly varying field $\mathbf{E}^i(Q_E)$ which is incident on a wedge may be expressed as

$$\mathbf{E}^{dt}(P) = \mathbf{E}^d(P) + \mathbf{E}^{sd}(P) \tag{74}$$

in which $\mathbf{E}^d(P)$ is as in (57) for the 3-D case, and $\mathbf{E}^{sd}(P)$ is the additional higher-order slope diffraction contribution. The total UTD field exterior to the wedge is as usual given by

$$\mathbf{E}(P) = \mathbf{E}^i(P)U_i + \mathbf{E}^r(P)U_r + \mathbf{E}^{dt}(P) \tag{75}$$

The field $\mathbf{E}^{sd}(P)$ in (74) may be expressed as in (59a) by

$$\mathbf{E}^{sd}(P) = \hat{\boldsymbol{\beta}}_0 E^{sd}_{\beta_0} + \hat{\boldsymbol{\phi}} E^{sd}_{\phi} \tag{76}$$

where

$$\begin{bmatrix} E^{sd}_{\beta_0} \\ E^{sd}_{\phi} \end{bmatrix} = \left\{ \begin{bmatrix} -d^i & 0 \\ 0 & -d^i \end{bmatrix} \begin{bmatrix} \dfrac{\partial E^i_{\beta'_0}}{\partial n^i} \\ \dfrac{\partial E^i_{\phi'}}{\partial n^i} \end{bmatrix} + \begin{bmatrix} -d^r_s & 0 \\ 0 & -d^r_h \end{bmatrix} \begin{bmatrix} \dfrac{\partial E^r_{\beta'_r}}{\partial n^r} \\ \dfrac{\partial E^r_{\phi'_r}}{\partial n^r} \end{bmatrix} \right\} \sqrt{\dfrac{\varrho_e}{s^d(\varrho_e + s^d)}} e^{-jks^d} \tag{77}$$

following the matrix notation of (60). In (77)

$$d^i = \frac{1}{jk\sin\beta_0}\left\{A_0\left[\csc^2\left(\frac{\pi+\beta^-}{2n}\right)F_s[kL^i a^+(\beta^-)]\right.\right.$$
$$\left.\left. - \csc^2\left(\frac{\pi-\beta^-}{2n}\right)F_s[kL^i a^-(\beta^-)]\right]\right\} \tag{78}$$

and

$$d^r_{s,h} \cong \frac{(\pm 1)}{jk\sin\beta_0}\left\{A_0\left[\csc^2\left(\frac{\pi+\beta^+}{2n}\right)F_s[kL^{rn}a^+(\beta^+)]\right.\right.$$
$$\left.\left. - \csc^2\left(\frac{\pi-\beta^+}{2n}\right)F_s[kL^{ro}a^-(\beta^+)]\right]\right\} \tag{79}$$

with $\beta^\mp = \phi \mp \phi'$, and $a^\pm(\beta)$ as in (62a). Also,

$$A_0 \equiv \frac{-e^{-j\pi/4}}{4n^2\sqrt{2\pi k}\sin\beta_0} \tag{80}$$

and

$$F_s(x) \equiv 2jx[1 - F(x)] \tag{81}$$

with $F(x)$ as in (66).

The terms within the curly braces in (78) and (79) are the ϕ' derivatives of those present in $D_{es,eh}$ of (61). The partial derivative $\partial/\partial n^i$ in (77) is evaluated at Q_E and $\hat{\mathbf{n}}^i = \varrho^i_{\phi'}\hat{\boldsymbol{\phi}}'$, where $\varrho^i_{\phi'}$ as the radius of curvature of the incident wavefront in the $\hat{\boldsymbol{\phi}}'$ direction. For a point source, $\varrho^i_{\phi'} = s^i(Q_E)$, where $s_i(Q_E)$ is simply the distance from the point Q_E of edge diffraction to the source point. The terms involving d^i in the first matrix on the right-hand side of (77) ensure that the slope of the total field is continuous across the ISB, in addition to the total field being continuous there. The continuity of the total field is already ensured by the term $\mathbf{E}^d$ in (74), as discussed previously. If the field $\mathbf{E}^i$ exhibited a null in the direction of incidence at Q_E, then $\mathbf{E}^d$ of (57) would vanish [since $\mathbf{E}^d$ is proportional to $\mathbf{E}^i(Q_E)$]; in this special case the higher-order slope diffraction term $\mathbf{E}^{sd}$ becomes the main contributor to the edge-diffracted field. The terms involving d^i are also important near the ISB if the incident field has a pattern which varies rapidly near Q_E. Likewise, the terms involving $d^r_{s,h}$ in the second matrix on the right-hand side of (77) are important at and near the RSB, where they ensure that the slope of the total field is continuous there. In order to evaluate these terms involving $d^r_{s,h}$ in (77), the reflected field $\mathbf{E}^r$ may be expressed for convenience as in (59b) by

$$\mathbf{E}^r = \hat{\boldsymbol{\beta}}'_r E_{\hat{\beta}'_r} + \hat{\boldsymbol{\phi}}'_r \mathbf{E}^r_{\phi'_r} \tag{82}$$

so as to view the reflected field $\mathbf{E}^r$ as being "incident" from virtual or image space. The direction of incidence from image space is of course $\hat{\mathbf{s}}^r$. Then $\hat{\boldsymbol{\beta}}'_r$ and $\hat{\boldsymbol{\phi}}'_r$ may be given the same meaning with respect to $\hat{s}^r$ (the direction of incidence from image space) as $\hat{\beta}'_0$ and $\hat{\phi}'$ have with respect to $\hat{s}^i$. Thus $\hat{\beta}'_r = \hat{s}^r \times \hat{\phi}'_r$, and $\mathbf{n}^r = \varrho^r_{\phi'_r} \hat{\phi}'_r$ (for evaluating $\partial/\partial n^r$) in which $\varrho^r_{\phi'_r}$ is the radius of curvature of the reflected wavefront in the direction $\hat{\phi}'_r$, etc. In the case of a very blunt wedge for which the RSB corresponding to the two faces may be close, the results in (77)–(79) must be appropriately modified; that modification is not given here.

In the special case of a *half-plane* illuminated at the edge by a rapidly varying spherical wave, (77) simplifies to

$$\begin{bmatrix} E^{sd}_{\beta_0} \\ E^{sd}_{\phi} \end{bmatrix} = \begin{bmatrix} -d_s & 0 \\ 0 & -d_h \end{bmatrix} \begin{bmatrix} \dfrac{\partial}{\partial n'} E^i_{\beta'_0} \\ \dfrac{\partial}{\partial n'} E^i_{\phi'} \end{bmatrix} \sqrt{\dfrac{\varrho_e}{s^d(\varrho_e + s^d)}} e^{-jks^d} \tag{83}$$

where

$$d_{s,h} = \dfrac{e^{-j\pi/4}}{4\sqrt{2\pi k} \sin\beta_0} \left\{ \dfrac{\sin(\beta^-/2)}{\cos^2(\beta^-/2)} F_s[2kL^i \cos^2(\beta^-/2)] \right.$$

$$\left. \pm \dfrac{\sin(\beta^+/2)}{\cos^2(\beta^+/2)} F_s[2kL^r \cos^2(\beta^+/2)] \right\} \tag{84}$$

in which $L^i = L^r$ as given previously in (63).

The UTD result in (75) is valid only if $\mathbf{E}^i$ is a ray optical field. If $\mathbf{E}^i$ is not a ray optical field, then one may have to decompose it into its ray optical components so that (74) can be employed to each of these incident-ray optical components, and these results may then be superposed. For example, if one is dealing with a large discrete antenna array near Q_E, then the near field of this array may not be ray optical at Q_E; however, each element of that array could still yield a ray optical field at Q_E. Likewise, a large-aperture antenna near Q_E can be quantized so that each quantized source in the aperture could yield a ray optical field at Q_E.

2-D Case—The results in (77) for the 3-D case can be reduced directly to treat the corresponding 2-D situation by requiring $\beta_0 = \pi/2$ and $\varrho_e \to \infty$, etc., exactly as done earlier to obtain (71) from (57). Thus, for the 2-D case the factor $\sqrt{\varrho_e/s^d(\varrho_e + s^d)}$ in (77) is replaced by $1/\sqrt{s^d}$. Also, the result in (84) for the special case of a *half-plane* illuminated at the edge by a rapidly varying spherical wave can be employed to deal with rapidly varying cylindrical- and plane-wave illumination if $\beta_0 = \pi/2$ and $\varrho_e \to \infty$ in (84), and if $s^i \to \infty$ for the case of an incident plane wave.

UAT for Edges—From the preceding discussions one may recall that the GTD solution for edge diffraction is expressed as $\mathbf{E}^{GTD} = \mathbf{E}^{GO} + \mathbf{E}^d_k$ [see (40)], while the corresponding UTD solution [3, 4] is expressed as $\mathbf{E} = \mathbf{E}^{GO} + \mathbf{E}^d$ [see (56)]. It is therefore evident that $\mathbf{E}^d_k$ in the GTD solution is replaced by $\mathbf{E}^d$ in the UTD, whereas the $\mathbf{E}^{GO}$ present in the GTD solution is left unchanged in the UTD. Also, as noted earlier, $\mathbf{E}^d_k$ in the GTD solution for edge diffraction becomes singular at the GO shadow boundaries (ISB and RSB) and it is therefore not valid within the transition regions adjacent to these boundaries; on the other hand, the $\mathbf{E}^d$ in the corresponding UTD solution is bounded and it properly compensates for the

Techniques for High-Frequency Problems

discontinuities of $\mathbf{E}^{GO}$ (at ISB and RSB) to yield a total field which is continuous everywhere away from the edge and other diffracted-ray caustics. Consequently the UTD overcomes the failure of the GTD within the GO shadow boundary transition regions; furthermore, outside these transition regions, $\mathbf{E}^d$ of the UTD reduces automatically to $\mathbf{E}_k^d$ of the GTD so that the total UTD solution in (56) also reduces uniformly to the GTD solution in (40), where the latter is valid and accurate.

In an alternative solution based on the uniform asymptotic theory (UAT) for edge diffraction [6], the total field $\mathbf{E}$ exterior to a perfectly conducting wedge is expressed as

$$\mathbf{E} = \mathbf{E}^G + \mathbf{E}_k^d \quad \text{(in the UAT)} \tag{85}$$

in which $\mathbf{E}^{GO}$ of the GTD is modified to $\mathbf{E}^G$ in the UAT, while $\mathbf{E}_k^d$ is left unchanged. The field $\mathbf{E}^G$ in the UAT now contains the necessary Fresnel integrals to ensure the uniform nature of $\mathbf{E}$ in (85), just as $\mathbf{E}^d$ contains those Fresnel integrals in the F type transition functions of the UTD solution of (60) and (61). Thus, exterior to the GO shadow boundary transition regions, $\mathbf{E}^G$ reduces automatically to $\mathbf{E}^{GO}$, thereby allowing $\mathbf{E}$ of the UAT to also reduce properly to $\mathbf{E}^{GTD}$ outside those transition regions.

Since $\mathbf{E}_k^d$ is as defined previously in (44) [together with (67) and (68)], it is only necessary to define the remaining quantity $\mathbf{E}^G$ in the UAT expression of (85). In particular,

$$\mathbf{E}^G(P) = [f(\xi^i) - \tilde{f}(\xi^i)]\mathbf{E}^i(P) + [f(\xi^r) - \tilde{f}(\xi^r)]\mathbf{E}^r(P) \tag{86}$$

where $\mathbf{E}^i(P)$ represents the unperturbed "incident" field which exists in the absence of the wedge; in contrast, the GO incident field $\mathbf{E}^i U_i$ [see (41)] exists only in the lit zone. Likewise, $\mathbf{E}^r(P)$ in (86) is interpreted here as the unperturbed "reflected" field which is incident from "image space." The $\mathbf{E}^r$ plays the same role as $\mathbf{E}^i$ (as opposed to the truncated GO reflected field $\mathbf{E}^r U_r$). It is important to note that $\mathbf{E}^i$ and $\mathbf{E}^r$ in the UAT expression for $\mathbf{E}^G$ in (86) are not discontinuous at the ISB and RSB, respectively, whereas the $\mathbf{E}^i U_i$ and $\mathbf{E}^r U_r$ in $\mathbf{E}^{GO}$ [see (41)] are of course discontinuous there as explicitly indicated by the step functions U_i and U_r. It is quite easily seen that $\mathbf{E}^i$ in (86) is a continuous vector field since it exists as if the wedge were absent. On the other hand, $\mathbf{E}^r$ is the field reflected from the o or the n face of the wedge which is illuminated by $\mathbf{E}^i$; consequently it is necessary to construct a smooth extension of $\mathbf{E}^r$ beyond the edge of the illuminated surface of the wedge to ensure that $\mathbf{E}^r$ does not become discontinuous across the RSB. For the wedge in Fig. 19b, if the o face alone is illuminated, this face is extended smoothly past the edge into a surface o from which $\mathbf{E}^i$ continues to be reflected to yield an $\mathbf{E}^r$ past the edge. The same arguments apply if the n face alone is illuminated, in which case the surface n is constructed as a smooth extension of the n face past the edge. If both faces of the wedge are illuminated, then (86) should be replaced by

$$\mathbf{E}^G(P) = [f(\xi^i) - \tilde{f}(\xi^i)]\mathbf{E}^i(P) + [f(\xi^{ro}) - \tilde{f}(\xi^{ro})]\mathbf{E}^{ro}(P)$$
$$+ [f(\xi^{rn}) - \tilde{f}(\xi^{rn})]\mathbf{E}^{rn}(P) \tag{87}$$

where ξ^{ro} and ξ^{rn} are the values of ξ^r pertaining to the o and n faces of the wedge, respectively, and $\mathbf{E}^{ro}$ and $\mathbf{E}^{rn}$ likewise refer to the values of $\mathbf{E}^r$ for these two faces (and their smooth extensions).

The function $f(\xi^{i,r})$ in (86) is defined by

$$f(\xi^{i,r}) = \frac{e^{j\pi/4}}{\sqrt{\pi}} \int_{\xi^{i,r}}^{\infty} e^{-j\tau^2} d\tau \tag{88}$$

and $\tilde{f}(\xi^{i,r})$ represents the first term in the large-argument approximation of $f(\xi^{i,r})$, namely,

$$\tilde{f}(\xi^{i,r}) = \frac{e^{-j\pi/4}}{2\xi^{i,r}\sqrt{\pi}} e^{-j(\xi^{i,r})^2} \tag{89}$$

The $\xi^{i,r}$ are referred to as the *detour parameters* defined by

$$\xi^i(P) = \varepsilon^i(P)|\sqrt{k[s^d(P) + s^i(Q_E) - s^i(P)]}| \tag{90}$$

and

$$\xi^r(P) = \varepsilon^r(P)|\sqrt{k[s^d(P) + s^i(Q_E) - s^i(Q_R) - s^r(P)]}| \tag{91}$$

where the $s^d(P)$, $s^i(P)$, and $s^r(P)$ are the distances s^d, s^i, and s^r to the observation point P as defined previously. Also, $s^i(Q_R)$ and $s^i(Q_E)$ are the values of s^i at the point of reflection Q_R, and the point of edge diffraction Q_E as in Fig. 20.

$$\varepsilon^i(P) = \begin{cases} +1 & \text{if } P \text{ is on the shadow side of the ISB} \\ -1 & \text{if } P \text{ is on the lit side of the ISB} \end{cases} \tag{92}$$

and

$$\varepsilon^r(P) = \begin{cases} +1 & \text{if } P \text{ is on the shadow side of the RSB} \\ -1 & \text{if } P \text{ is on the lit side of the RSB} \end{cases} \tag{93}$$

In the UTD solution for the field in the presence of a wedge expressed as $\mathbf{E} = \mathbf{E}^{GO} + \mathbf{E}^d$ [see (56)], both $\mathbf{E}^{GO}$ and $\mathbf{E}^d$ are finite everywhere (except at caustics and the edge) including the ISB and RSB, and hence $\mathbf{E}$ is also finite there as indicated previously. In contrast, $\mathbf{E}_k^d$ in the UAT solution of (85) becomes singular at the ISB and RSB; however, $\mathbf{E}^G$ in (85) also becomes singular at these shadow boundaries such that it exactly cancels the singularity in $\mathbf{E}_k^d$, thereby keeping the total UAT solution in (85) finite and continuous at these boundaries. Like the UTD the UAT edge-diffraction solution is also valid away from the edge and ray caustics. Care must be exercised in evaluating the UAT field of (85) in the vicinity of the image sources [i.e., the virtual sources or caustics of $\mathbf{E}^r$ in (86)]; otherwise a spurious singularity may result at those points. Note that the general form in (85) is valid for both 3-D and 2-D cases. The $\mathbf{E}_k^d$ in (85) is given by (44) and (67) in the 3-D

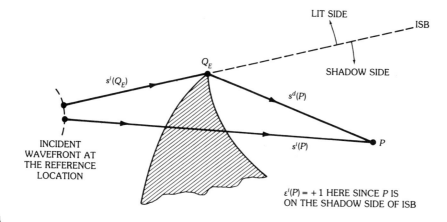

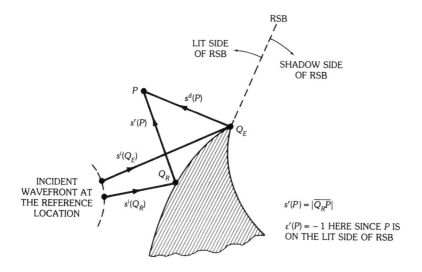

Fig. 20. Path lengths in the calculation of the detour parameters. (*a*) Calculation of $\varepsilon^i(P)$. (*b*) Calculation of $\varepsilon^r(P)$.

case; whereas $\sqrt{\varrho_e/s^d(\varrho_e + s^d)}$ in (44) is replaced by $1/\sqrt{s^d}$ for $\mathbf{E}_k^d$ in the 2-D case (since $\varrho_e \to \infty$ in the 2-D case), just as (71) is obtained from (57) for the 2-D situation in evaluating $\mathbf{E}^d$ of the UTD.

Explicit results are given in [23] for the UAT solution to the problem of the diffraction by an arbitrary smooth surface bounded by an arbitrary smooth edge.

UTD for Vertices—A vertex or a corner can be formed by the truncation of an edge. Figs. 21a and 21b illustrate a corner formed by the truncation of edges in planar as well as nonplanar geometries. An empirical UTD corner-diffraction solution proposed in [27] is described below. The general form of the corner-diffracted field was discussed under "GTD for Vertices" in Section 4. Even though the present corner-diffraction coefficient is still in its development stages, it has been shown to be very successful in predicting the fields diffracted by a corner for a number of plate structures.

In its present form the total UTD corner-diffracted field is the sum of the corner-diffracted fields associated with each of the edges which terminate at that corner. For example, in the case of a cube there are always three edges which terminate to form every vertex or corner of that cube; hence the field diffracted by each corner or vertex of a cube is a superposition of the corner-diffracted fields associated with every one of the three edges which terminate into that corner.

For simplicity, consider the geometry of Fig. 21a, which consists of a right-angled corner in a perfectly conducting planar screen. The corner-diffracted electric field $\mathbf{E}^{dc}$ associated with only *one* of the two edges of the geometry in Fig. 21a, when the illumination is due to a spherical wave (or point source), may be expressed as

$$\mathbf{E}^{dc} = E^c_{\beta_0} \hat{\boldsymbol{\beta}}_0 + E^c_\phi \hat{\boldsymbol{\phi}} \tag{94}$$

in which

$$\begin{bmatrix} E^c_{\beta_0} \\ E^c_\phi \end{bmatrix} = \begin{bmatrix} -D^c_s & 0 \\ 0 & -D^c_h \end{bmatrix} \begin{bmatrix} E^i_{\beta'_0} \\ E^i_{\phi'} \end{bmatrix} \sqrt{\frac{s'}{s''(s'+s'')}} \sqrt{\frac{s(s+s_c)}{s_c}} \frac{e^{-jks}}{s} \tag{95}$$

as in (60) for edge diffraction,* except for the additional factor $\sqrt{s(s+s_c)/s_c}$, and the corner-diffraction coefficient $D^c_{s,h}$ which is given by

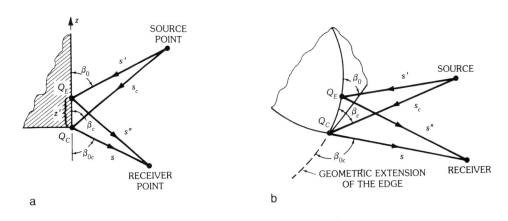

Fig. 21. Geometry for corner-diffraction problem. (*a*) Corner in a planar surface. (*b*) Corner in a nonplanar surface. (*After Sitka, Burnside, Chu, and Peters [27], © 1983 IEEE*)

---

*The incident field in (95) is evaluated at Q_c. Moreover, the corner-diffracted field in (94) and (95) is in terms of $\hat{\beta}_0$ and $\hat{\phi}$ associated with the ray reaching the receiver from Q_c (see Fig. 21).

$$D_{s,h}^c = \frac{e^{-j\pi/4}}{\sqrt{2\pi k}} C_{s,h}(Q_E) \frac{\sqrt{\sin\beta_c \sin\beta_{0c}}}{\cos\beta_{0c} - \cos\beta_c} F[kL_c a(\pi + \beta_{0c} - \beta_c)] \qquad (96)$$

For the case of a corner in a planar geometry as in Fig. 21a, $C_{s,h}(Q_E)$ simplifies to

$$C_{s,h}(Q_E) = \frac{-e^{-j\pi/4}}{2\sqrt{2\pi k}\sin\beta_0} \left\{ \frac{F[kLa(\beta^-)]}{\cos(\beta^-/2)} \left| F\left[\frac{kLa(\beta^-)/2\pi}{kL_c a(\pi + \beta_{0c} - \beta_c)}\right] \right| \right.$$
$$\left. \mp \frac{F[kLa(\beta^+)]}{\cos(\beta^+/2)} \left| F\left[\frac{kLa(\beta^+)/2\pi}{kL_c a(\pi + \beta_{0c} - \beta_c)}\right] \right| \right\} \qquad (97)$$

where

$$a(\psi) \equiv 2\cos^2(\psi/2) \qquad (98)$$

Note that F in (97) is the same as in (66) and $\beta^{\mp} = \phi \mp \phi'$ is as in (62c). In addition,

$$L = \frac{s's''}{(s' + s'')} \sin^2\beta_0 \qquad (99a)$$

and

$$L_c = \frac{s_c s}{s_c + s} \qquad (99b)$$

for spherical-wave incidence. Since the corner in Fig. 21a is formed by the intersection of *two* edges, a corner-diffracted field contribution of the type in (95) and (96), which is associated with the *other* edge, must also be included; it is found in a similar fashion. The total UTD corner-diffracted field ensures the continuity of the total high-frequency field across the edge-diffraction shadow boundary (Fig. 22), where the edge-diffracted field becomes discontinuous past the corner.

Note that the corner-diffracted field should be added to any edge-diffracted and geometrical optics field components which can exist at a given observation point. In Fig. 22 only one edge-diffracted field component due to diffraction from $Q_E^{(b)}$ on edge (b) contributes, because the diffraction from $Q_E^{(a)}$ on the geometric extension of edge (a) past the corner does not fall within the physical limits of that edge. Another situation is shown in Fig. 23, where both $Q_E^{(a)}$ and $Q_E^{(b)}$ lie on their edge extensions; therefore no edge-diffracted field component is present. Fig. 24 shows the case where edge- and corner-diffracted fields from both edges are incident on the observation point or the receiver.

In order to treat corners formed by the truncation of nonplanar edged structures as in Fig. 21b, it is necessary only to generalize the $C_{s,h}(Q_E)$ and L_c for a planar corner in (97) and (99b). Such a generalization of $C_{s,h}(Q_E)$ and L_c for any *one* of the edges terminating into a vertex in a *nonplanar* geometry with spherical-wave illumination is given by

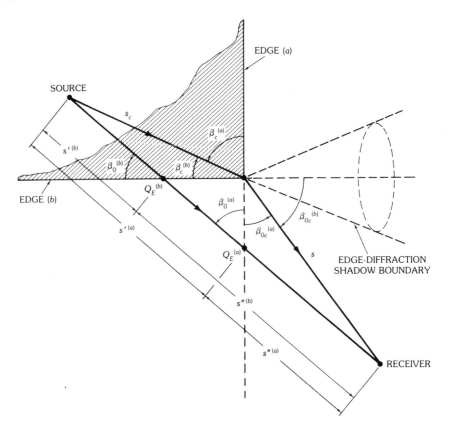

Fig. 22. Case where only one edge-diffracted field and both corner fields are received. (*After Sitka, Burnside, Chu, and Peters [27], © 1983 IEEE*)

$$C_{s,h} = \frac{-e^{-j\pi/4}}{2n\sqrt{2\pi k}\sin\beta_0}\left(\left\{\cot\left[\frac{\pi+(\phi-\phi')}{2n}\right]F[kL^i a^+(\phi-\phi')]\right.\right.$$

$$\left.\times\left|F\left[\frac{L^i a^+(\beta^-)/\lambda}{kL_c a(\pi+\beta_{0c}-\beta_c)}\right]\right| + \cot\left[\frac{\pi-(\phi-\phi')}{2n}\right]F[kL^i a^-(\phi-\phi')]\right.$$

$$\left.\times\left|F\left[\frac{L^i a^-(\beta^-)/\lambda}{kL_c a(\pi+\beta_{0c}-\beta_c)}\right]\right|\right\} \mp \left\{\cot\left[\frac{\pi+(\phi+\phi')}{2n}\right]F[kL^{rn} a^+(\phi+\phi')]\right.$$

$$\left.\times\left|F\left[\frac{L^{rn} a^+(\beta^+)/\lambda}{kL_c a(\pi+\beta_{0c}-\beta_c)}\right]\right| + \cot\left[\frac{\pi-(\phi+\phi')}{2n}\right]F[kL^{ro} a^-(\phi+\phi')]\right.$$

$$\left.\left.\times\left|F\left[\frac{L^{ro} a^-(\beta^+)/\lambda}{kL_c a(\pi+\beta_{0c}-\beta_c)}\right]\right|\right\}\right) \quad (100)$$

where

$$L_c = \frac{\varrho_c s}{\varrho_c + s} \quad (101a)$$

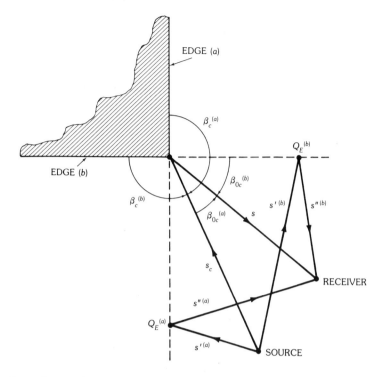

Fig. 23. Case where only corner fields are received. (*After Sitka, Burnside, Chu, and Peters [27],* © *1983 IEEE*)

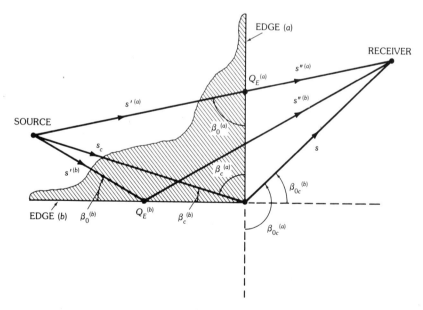

Fig. 24. Case where all edge- and corner-diffracted fields are received. (*After Sitka, Burnside, Chu, and Peters [27],* © *1983 IEEE*)

$$\frac{1}{\varrho_c} = \frac{1}{\varrho_e^i(Q_c)} - \frac{\hat{\mathbf{n}}_e \cdot (\hat{\mathbf{s}}_c - \hat{\mathbf{s}})}{a_c \sin\beta_c \sin\beta_{0c}} \qquad (101b)$$

and $a^{\pm}(\beta)$ and $a(\psi)$ are as defined in (62a) and (98), respectively.

The usefulness of the present UTD corner-diffraction coefficient is illustrated in Figs. 25a and 25b.

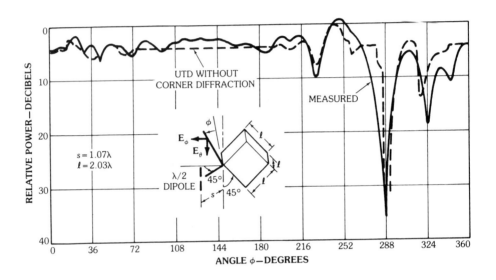

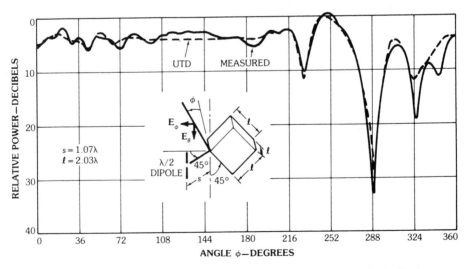

Fig. 25. Comparison of measured and calculated E_ϕ radiation pattern for a dipole near a box in the indicated plane.

UTD for Analyzing Problems of the Radiation by Antennas Either On or Off a Smooth, Perfectly Conducting, Convex Surface

The UTD solutions to the problems associated with the prediction of high-frequency radiation from sources which are located either near or directly on smooth convex surfaces are useful in analyzing a variety of antenna problems. For example, these solutions are useful for the following problems: (*a*) in predicting the scattering from the smooth convex portions of an aircraft fuselage, a ship mast, or a satellite shape, which is illuminated by a nearby antenna (that is located on some other parts of these structures); and (*b*) in predicting the radiation patterns of conformal antennas which are mounted directly on the smooth convex portions of an aircraft, a missile, or a spacecraft, etc.

The situation that arises in dealing with problems of type *a* above is depicted in Fig. 26. The problem indicated in Fig. 26 will henceforth be referred to as the "scattering problem"; in this problem, the source and observer are both located off the convex surface. The observer in Fig. 26 may be located either at a point P_L in the lit region or at a point P_S in the shadow region.

The situation that typically arises in dealing with problems of type *b* indicated above is illustrated in Fig. 27, and it defines what will henceforth be referred to as the "radiation problem." The source is positioned directly on the convex surface in this case, whereas the observation point is always located off the surface, either at P_L in the lit zone or at P_S in the shadow zone as indicated in Fig. 27. Note that the radiation problem of Fig. 27 is directly related to finding the electric current density, or the charge density, that is induced at the point Q' on a perfectly conducting convex surface by an appropriate source at P_L or P_S via the reciprocity theorem for electromagnetic fields. This reciprocal problem of determining the currents and charges induced on a perfectly conducting convex surface by an external source is of importance, for example, in electromagnetic pulse applications.

Another situation of interest which arises in dealing with problems of type *b* above is shown in Fig. 28; this problem in Fig. 28 will henceforth be referred to as the "mutual coupling problem." The source and the observation points are both positioned directly on the surface in this situation. It is clear that the problem in Fig. 28 is related directly to the calculation of the mutual coupling between a pair of

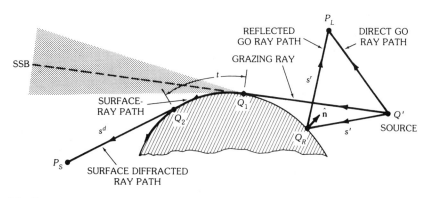

Fig. 26. Geometry associated with the scattering problem.

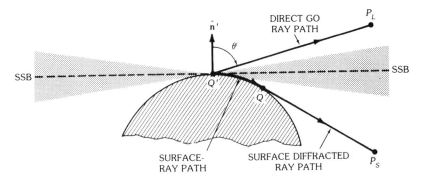

Fig. 27. Geometry associated with the radiation problem.

conformal antennas on a convex surface; such a calculation essentially reduces to finding the "surface fields" at Q due to a source at Q' on the same surface.

The scattering problem will be treated first, to be followed by similar treatments for the radiation and mutual coupling problems. It is of course implied that the principal surface radii of curvature at points of reflection and diffraction are sufficiently large in terms of the wavelength so that these high-frequency solutions remain valid. Some examples are provided which shed information as to the approximate bounds on how large in terms of the wavelength the surface has to be for obtaining accurate results based on these solutions. Note that the UTD solutions* to be described below for the scattering, radiation, and mutual coupling associated with a convex surface all employ the rays of the GTD.

UTD Analysis of the Scattering by a Smooth, Convex Surface (3-D Case)—A uniform GTD (or UTD) solution which is convenient and accurate for engineering applications is presented below for the scattering problem in Fig. 26 such that it overcomes the limitations of the GTD within the SSB transition region (shaded in Fig. 26); furthermore, this UTD solution automatically reduces to the GTD solution exterior to the SSB transition region, where the latter solution is indeed valid. The angular extent of the SSB transition region is of the order $[2/k\varrho_g(Q_1)]^{1/3}$

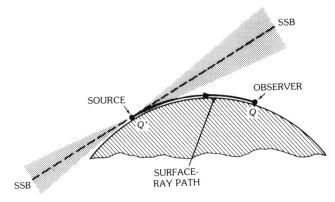

Fig. 28. Geometry associated with the mutual coupling problem.

---

*These UTD solutions employ Fock type transition functions which have been studied extensively by M. A. Logan in "General Research in Diffraction Theory", Vols. I and II, Lockheed Aircraft Corp., 1959.

Techniques for High-Frequency Problems

radians, where $\varrho_g(Q_1)$ is the surface radius of curvature along the incident-ray direction at Q_1. The incident field can be an arbitrary ray optical type field as in (8); it is noted that plane- and spherical-wave fields are special cases of this arbitrary ray optical field (see "The GO Incident Field" in Section 2). The UTD solution given below is applicable, provided that the field point and the caustics of the incident-ray system are not in the close vicinity of the surface. It is also assumed that the amplitude (or the pattern) of the incident field does not exhibit a rapid spatial variation in the vicinity of Q_1, otherwise additional slope diffraction contributions to the total field need to be included; expressions for the latter have not yet been developed for the arbitrary convex surface. Two additional minor restrictions on the solution are indicated later on, when the explicit field expressions are given.

The total electric field **E** for the problem in Fig. 26 is given as

$$\mathbf{E}(P) = \begin{cases} \mathbf{E}^i(P_L)U + \mathbf{E}^r(P_L)U + \mathbf{E}^d(P_L) & \text{if } P = P_L \text{ in the lit zone} \\ \mathbf{E}^d(P_S)[1 - U] & \text{if } P = P_S \text{ in the shadow zone} \end{cases} \quad (102)$$

The incident and reflected fields $\mathbf{E}^i$ and $\mathbf{E}^r$ are associated with the incident and reflected GO rays shown in Fig. 26. The step function U in (102) is defined as

$$U = \begin{cases} 1 & \text{in the lit region which lies above the SSB} \\ 0 & \text{in the shadow region which lies below the SSB} \end{cases} \quad (103)$$

where the surface shadow boundary (SSB) is as illustrated in Fig. 26. The field $\mathbf{E}^d(P_S)$ in (102) within the shadow region follows the surface-diffracted ray path also shown in Fig. 26, whereas the field $\mathbf{E}^d(P_L)$ in (102) which is diffracted into the lit region follows the reflected-ray path (of $\mathbf{E}^r$) in this solution. Therefore it is convenient in this problem to combine the GO reflected field $\mathbf{E}^r(P_L)U$ and the diffracted field $\mathbf{E}^d(P_L)$ into a "generalized reflected field" $\mathbf{E}^{gr}(P_L)U$ in the lit region so that (102) becomes

$$\mathbf{E}(P) = \begin{cases} \mathbf{E}^i(P_L)U + \mathbf{E}^{gr}(P_L)U & \text{if } P = P_L \text{ in the lit zone} \\ \mathbf{E}^d(P_S)[1 - U] & \text{if } P = P_S \text{ in the shadow zone} \end{cases} \quad (104)$$

The surface-diffracted ray path (see Fig. 26) associated here with $\mathbf{E}^d(P_S)$ was described earlier under "GTD for Convex Surfaces" in Section 4. Also, the incident GO field $\mathbf{E}^i(P_L)$ was described previously under "The GO Incident Field" in Section 2. The fields $\mathbf{E}^{gr}(P_L)$ and $\mathbf{E}^d(P_S)$ are obtained in [28, 29] and are given by

$$\mathbf{E}^{gr}(P_L) \sim \mathbf{E}^i(Q_R) \cdot [\mathcal{R}_s \hat{\mathbf{e}}_\perp \hat{\mathbf{e}}_\perp + \mathcal{R}_h \hat{\mathbf{e}}_\parallel^i \hat{\mathbf{e}}_\parallel^r] \sqrt{\frac{\varrho_1^r \varrho_2^r}{(\varrho_1^r + s^r)(\varrho_2^r + s^r)}} e^{-jks^r} \quad (105)$$

$$\mathbf{E}^d(P_S) \sim \mathbf{E}^i(Q_1) \cdot [\mathcal{D}_s \hat{\mathbf{b}}_1 \hat{\mathbf{b}}_2 + \mathcal{D}_h \hat{\mathbf{n}}_1 \hat{\mathbf{n}}_2] \sqrt{\frac{\varrho_s}{s^d(\varrho_s + s^d)}} e^{-jks^d} \quad (106)$$

where the points Q_R and Q_1 and the distances s^r and s^d are indicated in Fig. 26. The quantities ϱ_1^r and ϱ_2^r have the same meaning as in (16) for the ordinary reflected GO field $\mathbf{E}^r(P_L)$, which is discussed under "The GO Reflected Field" in Section 2. Also, ϱ_s in (106) is the same as in (50) of Section 4. The quantities within brackets involving $\mathcal{R}_{s,h}$ in (105) and $\mathcal{D}_{s,h}$ in (106) may be viewed as generalized dyadic coefficients for surface reflection* and diffraction, respectively. It is noted that (105) and (106) are expressed invariantly in terms of the unit vectors fixed in the reflected and surface-diffracted ray coordinates. The unit vectors $\hat{\mathbf{e}}_\parallel^i$, $\hat{\mathbf{e}}_\parallel^r$, and $\hat{\mathbf{e}}_\perp$ in (105) have been defined in "The GO Reflected Field" in Section 2 (see Fig. 8). At Q_1 let $\hat{\mathbf{t}}_1$ be the unit vector in the direction of incidence, $\hat{\mathbf{n}}_1$ be the unit outward normal vector to the surface, and $\hat{\mathbf{b}}_1 = \hat{\mathbf{t}}_1 \times \hat{\mathbf{n}}_1$; likewise at Q_2 let a similar set of unit vectors $(\hat{\mathbf{t}}_2, \hat{\mathbf{n}}_2, \hat{\mathbf{b}}_2)$ be defined with $\hat{\mathbf{t}}_2$ in the direction of the diffracted ray as in Fig. 29. In the case of surface rays with zero torsion, $\hat{\mathbf{b}}_1 = \hat{\mathbf{b}}_2$. It is noted that ϱ_s in (106) is the wavefront radius of curvature of the surface-diffracted ray evaluated in the $\hat{\mathbf{b}}_2$ direction at Q_2. The $\mathcal{R}_{s,h}$ and $\mathcal{D}_{s,h}$ in (105) and (106) are given by [28, 29]:

$$\mathcal{R}_{s,h} = -\left(\sqrt{\frac{-4}{\xi^L}} e^{-j(\xi^L)^3/12} \left\{\frac{e^{-j\pi/4}}{2\sqrt{\pi}\xi^L}[1 - F(X^L)] + \tilde{P}_{s,h}(\xi^L)\right\}\right) \quad (107)$$

and for the lit region

$$\mathcal{D}_{s,h} = -\left[\sqrt{m(Q_1)m(Q_2)}\sqrt{\frac{2}{k}}\left\{\frac{e^{-j\pi/4}}{2\sqrt{x}\xi}[1 - F(X^d)] + \tilde{P}_{s,h}(\xi)\right\}\right] \cdot \sqrt{\frac{d\eta(Q_1)}{d\eta(Q_2)}} e^{-jkt} \quad (108)$$

for the shadow region. The function F appearing above has been introduced in the section dealing with edge diffraction [see (66)]. The Fock type surface-reflection functions $\tilde{P}_s$ and $\tilde{P}_h$ are related to the soft and hard Pekeris functions p^* and q^* by [28, 29] (note that $\delta = 0$ at SSB):

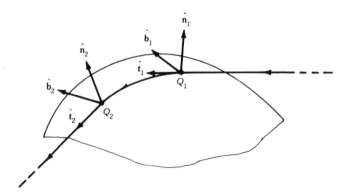

Fig. 29. Surface-ray unit vectors $\hat{\mathbf{t}}$, $\hat{\mathbf{n}}$, and $\hat{\mathbf{b}}$.

---

*Actually, cross terms exist in the above generalized dyadic reflection coefficient; but in general their effect is seen to be weak within the SSB transition region. Also these terms vanish in the deep lit region and on the SSB, hence they have been ignored in (105).

Techniques for High-Frequency Problems

$$\tilde{P}_{s,h}(\delta) = \begin{Bmatrix} p^*(\delta) \\ q^*(\delta) \end{Bmatrix} e^{-j\pi/4} - \frac{e^{-j\pi/4}}{2\sqrt{\pi}\delta} \quad \begin{array}{c}(109a)\\(109b)\end{array}$$

where p^* and q^* are finite and well behaved even when $\delta = 0$; these universal functions are plotted in Figs. 30, 31, and 32. Also,

$$\tilde{P}_{s,h}(\delta) = \frac{e^{-j\pi/4}}{\sqrt{\pi}} \int_{-\infty}^{\infty} \frac{\tilde{Q}V(\tau)}{\tilde{Q}W_2(\tau)} e^{-j\delta\tau} d\tau$$

where

$$\tilde{Q} = \begin{cases} 1 & \text{soft case} \\ \partial/\partial\tau & \text{hard case} \end{cases} \quad \begin{array}{c}(110a)\\(110b)\end{array}$$

in which the Fock type Airy functions $V(\tau)$ and $W_2(\tau)$ are [28, 29]

$$2jV(\tau) = W_1(\tau) - W_2(\tau) \tag{111a}$$

$$W_1(\tau) = \frac{1}{\sqrt{\pi}} \int_{\infty\exp(-j2\pi/3)}^{\infty} e^{\tau t - t^3/3} dt \tag{111b}$$

$$W_2(\tau) = \frac{1}{\sqrt{\pi}} \int_{\infty\exp(+j2\pi/3)}^{\infty} e^{\tau t - t^3/3} dt \tag{111c}$$

The remaining quantities occurring in (107) and (108) are*

$$\xi^L = -2m(Q_R)\cos\theta^i \tag{112}$$

$$\xi = \int_{Q_1}^{Q_2} \frac{m(t')}{\varrho_g(t')} dt' \tag{113}$$

$$m(\cdot) = \left[\frac{k\varrho_g(\cdot)}{2}\right]^{1/3} \tag{114}$$

$$t = \int_{Q_1}^{Q_2} dt' \tag{115}$$

$$X^L = 2kL\cos^2\theta^i \tag{116}$$

and

---

*In [29] ξ^L is more precisely given by $\xi^L = 2m(Q_R)f^{-1/3}\cos\theta^i$, where $f^{-1/3}$ depends on the principal surface radii of curvatures at Q_R and θ^i in a complicated manner. However, it appears that replacing $f^{-1/3}$ by unity as is done in (112) for all θ^i does not impair the accuracy of the solution. It is noted that $f = 1$ when $\theta^i = \pi/2$ (i.e., at SSB), and it differs from unity as $\theta^i \to 0$; however, as $\theta^i \to 0$, $\xi^L \ll 0$ and $\tilde{P}_{s,h} \to \pm\sqrt{-\xi^L/4}\,e^{j(\xi^L)^{3/2}}$ so that $\mathcal{R}_{s,h} \to \mp 1$ as it should with either definition for ξ^L.

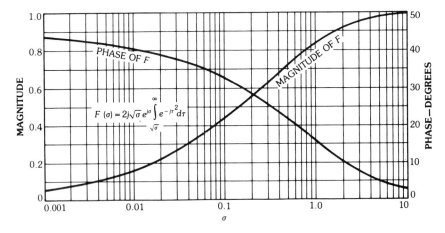

Fig. 30. A plot of $F(\sigma)$ versus σ. (*After Pathak [28], © 1979 American Geophysical Union*)

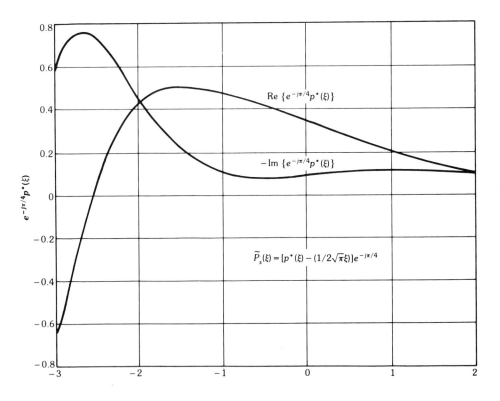

Fig. 31. A plot of $e^{-j\pi/4}p^*(\xi)$ versus ξ. (*After Pathak [28], © 1979 American Geophysical Union*)

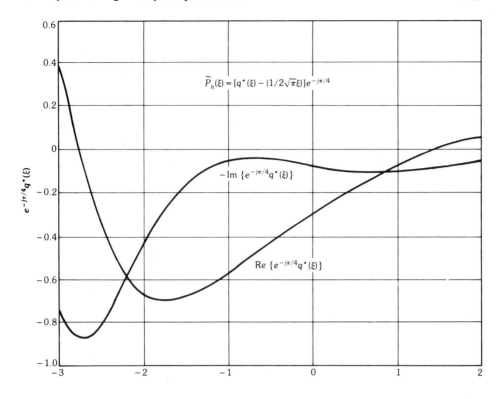

Fig. 32. A plot of $e^{-j\pi/4}q^*(\xi)$ versus ξ. (*After Pathak [28],* © *1979 American Geophysical Union*)

$$X^d = \frac{kL(\xi)^2}{2m(Q_1)m(Q_2)} \qquad (117)$$

The $\varrho_g(Q_R)$ in $m(Q_R)$ is the surface radius of curvature at Q_R in the plane of incidence, whereas $\varrho_g(Q_i)$ for $i = 1$ or 2 is the surface radius of curvature at Q_i in the $\hat{t}_i$ direction. Here, dt' is an incremental arc length along the surface-ray path. The angle of incidence θ^i is shown in Fig. 8. The $d\eta(Q_1)$ and $d\eta(Q_2)$ denote the widths of the surface-ray tube at Q_1 and Q_2, respectively; the surface-ray tube is formed by considering a pair of rays adjacent to the central ray as in Fig. 17. It is noted that the geodesic surface-ray paths are easy to find on cylinders, spheres, and cones. For example, the geodesic paths on a convex cylinder are helical, whereas they are great circle paths on a sphere. The geodesic surface-ray paths on cylinders and cones are illustrated later in Figs. 45 and 46. For more general types of convex surfaces the geodesic surface-ray paths must be found numerically with the aid of differential geometry. The distance parameter L in (116) and (117) is given by

$$L = \frac{\varrho_1^i(Q_1)\,\varrho_2^i(Q_1)}{[\varrho_1^i(Q_1) + s][\varrho_2^i(Q_1) + s]} \cdot \frac{s[\varrho_b^i(Q_1) + s]}{\varrho_b^i(Q_1)} \qquad (118)$$

where

$$s \equiv s^r\bigg|_{SSB} = s^d\bigg|_{SSB} \quad (119)$$

$$\varrho_b^i(Q_1) = \begin{cases} \text{incident wavefront radius of curvature} \\ \text{in the } \hat{b}_1 \text{ direction at } Q_1 \end{cases} \quad (120)$$

For any arbitrary point within the lit or shadow part of the SSB transition region, the distance s required in the calculation of L in (118) may be obtained by projecting s^r or s^d on the SSB *if* that observation point does *not* move in a predetermined manner. If the observation point moves across the SSB in a predetermined fashion, then it is clear that s in (118) and (119) can be found unambiguously. Note that $\varrho_1^i(Q_1)$ and $\varrho_2^i(Q_1)$ in (118) above denote the principal radii of curvature of the incident wavefront at Q_1, and ϱ_b^i, which is defined in (120), has been introduced previously in (21). In the *special case of point-source or spherical-wave illumination*, the L in (118) simplifies to*

$$L = \frac{s's}{s' + s} \quad \text{for spherical-wave illumination} \quad (121)$$

where

$$s' \equiv [\varrho_1^i(Q_1) = \varrho_2^i(Q_1) = \varrho_b^i(Q_1)] = \begin{cases} \text{distance from the point} \\ \text{source to the point of} \\ \text{grazing incidence at } Q_1 \end{cases} \quad (122)$$

In the case of *plane-wave illumination*, $s' \to \infty$ and hence (121) above simplifies to

$$L = s \quad \text{for plane-wave illumination} \quad (123)$$

If the incident wavefront is of the converging ($\varrho_{1,2}^i < 0$) or converging-diverging ($\varrho_1^i > 0$ and $\varrho_2^i < 0$; or $\varrho_1^i < 0$ and $\varrho_2^i > 0$) type, then the parameter L in (118) can become negative. It has not been investigated in detail how the general solution can be completed when L becomes negative. However, if one of the principal directions of the incident wavefront coincides with one of the principal planes of the surface at grazing, then one can treat a converging or converging-diverging type wavefront for which $L < 0$ by replacing $F(X^{L,d})$ with $F^*(|X^{L,d}|)$. Note that the asterisk on F^* denotes the complex conjugate operator. The use of $F^*(|X^{L,d}|)$ when $L < 0$ leads to a continuous total field at the SB in this case.

The above UTD result remains accurate outside the paraxial (i.e., near axial) regions of quasi-cylindrical or elongated convex surfaces; a different solution is required in these regions and it has not yet been completed.

---

*In general the present solution appears to be accurate even for kL as small as 3; in some special cases kL can be made as small as 1.

Techniques for High-Frequency Problems

The surface-diffracted field of the type $\mathbf{E}^d(P_S)$ can also be present in the lit zone if the surface is closed; this may be visualized by noting that the field of the type $\mathbf{E}^d$ can propagate around the closed surface. Also, additional contributions to $\mathbf{E}^d(P_S)$ can be present in the shadow zone for a closed surface because surface-diffracted rays can be initiated at all points of grazing incidence on that closed surface; furthermore, these surface rays can undergo multiple encirclements around the closed body. These additional surface-diffracted ray contributions, however, are generally quite weak in comparison with the $\mathbf{E}^{gr}$ contribution within the lit zone for surfaces which are quite large in terms of the wavelength; hence their contributions may be neglected in such cases. Fig. 33 indicates the rays associated with these additional contributions for a closed surface.

It is clear that the parameters ξ^L, ξ, X^L, and X^d become small as one approaches the surface shadow boundary, SSB, from both the lit and shadow regions. As one approaches the SSB the small-argument limiting form of the transition function $F(X)$ which has been introduced earlier in (66) becomes helpful for verifying the continuity of the total high-frequency field at the SSB. On the other hand, the above parameters become large as one moves outside the SSB transition region; in this case $F(X) \to 1$ for large X and

$$\tilde{P}_{s,h}(\delta)|_{\delta \ll 0} \sim \pm \sqrt{\frac{-\delta}{4}}\, e^{j\delta^3/12} \tag{124}$$

$$\tilde{P}_{s,h}(\delta)|_{d \gg 0} = \begin{cases} -\dfrac{e^{-j\pi/4}}{\sqrt{\pi}} \displaystyle\sum_{n=1}^{N} \dfrac{e^{j\pi/6}\, e^{\delta q_n \exp(-j5\pi/6)}}{2[Ai'(-q_n)]^2} \\[2ex] -\dfrac{e^{-j\pi/4}}{\sqrt{\pi}} \displaystyle\sum_{n=1}^{N} \dfrac{e^{j\pi/6}\, e^{\delta \bar{q}_n \exp(-j5\pi/6)}}{2\bar{q}_n[Ai(-\bar{q}_n)]^2} \end{cases} \tag{125}$$

where $N = 2$ is generally sufficient to compute $\tilde{P}_{s,h}(\delta)$ accurately for $\delta \gg 0$ via (125). In (125) the Miller type Airy function $Ai(\tau) = V(\tau)/\sqrt{\pi}$, and $Ai'(\tau) =$

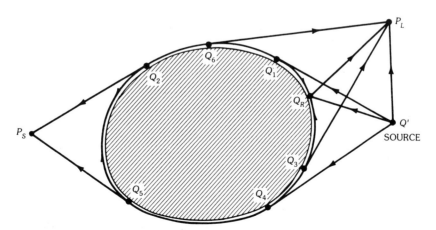

Fig. 33. Multiply encircling surface rays and contribution to P_L from surface-diffracted rays.

$dAi(\tau)/d\tau$. The parameters q_n and $\bar{q}_n$ in (125) are defined by $Ai(-q_n) = 0$ and $Ai'(-\bar{q}_n) = 0$, respectively. Explicit values of q_n and $\bar{q}_n$ are available in [30]; they are also given for $n = 1$ and $n = 2$ in Chart 1. Thus on incorporating the limiting values of (124) and (125), which are valid outside the SSB transition region, into (107) and (108) and replacing $F(X)$ by its asymptotic value of unity, it is clear that (105) and (106) properly reduce to $\mathbf{E}^{\text{GTD}}$ of (46). Indeed, $\mathscr{R}_{s,h}$ in (107) reduces to $\mathscr{R}_{s,h} = \mp 1$ outside the SSB transition region so that $\mathbf{E}^{gr}(P_L) \to \mathbf{E}^r(P_L)$, and likewise $\mathbf{E}^d(P_S) \to \mathbf{E}_k^d(P_S)$ therein, respectively, in which $\mathbf{E}^r(P_L)$ and $\mathbf{E}_k^d(P_S)$ of (46) are as given by (16) and (50). The $\bar{\bar{\mathbf{T}}}^k(Q_1, Q_2)$ appearing in (50) for $\mathbf{E}_k^d$ is seen to be

$$\bar{\bar{\mathbf{T}}}^k(Q_1, Q_2) = \hat{\mathbf{b}}_1 \hat{\mathbf{b}}_2 T_s + \hat{\mathbf{n}}_1 \hat{\mathbf{n}}_2 T_h \tag{126a}$$

where

$$T_{s,h} = \sum_{n=1}^{N} D_n^{s,h}(Q_1) \exp\left[-\int_{Q_1}^{Q_2} \alpha_n^{s,h}(t')\,dt'\right] D_n^{s,h}(Q_2) \tag{126b}$$

The $D_n^{s,h}$ and $\alpha_n^{s,h}$ are the Keller's GTD diffraction coefficients and attenuation constants for the nth soft (s) or hard (h) surface-ray mode. Thus, in the GTD the surface-ray field consists of surface-ray modes which propagate independently of one another. Also, this surface-ray field is not the true field on the surface; it is a boundary layer field which gives rise to the interpretation of $\bar{\bar{\mathbf{T}}}^k$ in (50) and (126a) as a dyadic transfer function to indicate the launching of the surface-ray field at Q_1 [via $D_n^{s,h}(Q_1)$], the attenuation of the surface-ray field between Q_1 and Q_2 {via $\exp[-\int_{Q_1}^{Q_2} \alpha_n^{s,h}(t)dt]$}, and the diffraction of the surface-ray field at Q_2 [via $D_n^{s,h}(Q_2)$] to arrive at P_S via the surface-diffracted ray path $\overline{Q_2 P_s}$ as in Fig. 26. From (125) it can be seen that $D_n^{s,h}$ and $\alpha_n^{s,h}$ are given by

$$[D_n^s(Q)]^2 = \sqrt{\frac{1}{2\pi k}} m(Q) \frac{e^{-j(\pi/12)}}{[Ai'(-q_n)]^2} \tag{127a}$$

$$[D_n^h(Q)]^2 = \sqrt{\frac{1}{2\pi k}} m(Q) \frac{e^{-j(\pi/12)}}{\bar{q}_n[Ai(-\bar{q}_n)]^2} \tag{127b}$$

and

$$\alpha_n^s(Q) = \left(\frac{q_n}{\varrho_g(Q)}\right) m(Q) \exp(j\pi/6) \tag{128a}$$

$$\alpha_n^h = \left(\frac{\bar{q}_n}{\varrho_g(Q)}\right) m(Q) \exp(j\pi/6) \tag{128b}$$

in which Q is any point on the geodesic surface-ray path. The expressions in (127a), (127b), (128a), and (128b) are the same as those obtained in Keller's GTD [3].

The GTD result of (46) in terms of (127) and (126) is not valid within the SSB transition region; it is valid only outside that transition region, as pointed out under

Chart 1. Zeroes of the Airy Function and Its Derivative

Zeroes of the Airy Function	Zeroes of the Derivative of the Airy Function
$Ai(-q_p) = 0$	$Ai'(-\bar{q}_p) = 0$
$q_1 \quad = 2.338\,11$	$\bar{q}_1 \quad = 1.018\,79$
$q_2 \quad = 4.087\,95$	$\bar{q}_2 \quad = 3.248\,20$
$Ai'(-q_1) = 0.701\,21$	$Ai(-\bar{q}_1) = 0.535\,66$
$Ai'(-q_2) = -0.803\,11$	$Ai(-\bar{q}_2) = -0.419\,02$

"GTD for Convex Surfaces" in Section 4. Therefore one must use the more general UTD result of (104) together with (105) and (106), which not only remains valid within the SSB transition region but which also reduces automatically and uniformly to the usual GTD result of (46) exterior to that transition region. Some examples indicating the usefulness of the UTD result for the scattering problem are shown in Figs. 34, 35, and 36. In Fig. 34 the UTD solution is compared with an independent moment method (MM)–based solution for the same configuration.

Note that the magnetic field $\mathbf{H}(P)$ can be obtained as usual from the corresponding $\mathbf{E}(P)$ using the local plane-wave approximation to each component of $\mathbf{E}$ as in (11), (17), and (51).

UTD Analysis of the Scattering by a Smooth, Convex Surface (2-D Case)—The above UTD result for the 3-D scattering configuration can be simply modified to recover the corresponding UTD result for the 2-D case by allowing the caustic distances ϱ'_1 and ϱ_s in (105) and (106) to recede to infinity. Then, let

$$\varrho^r_2 \equiv \varrho^r, \qquad \text{if } \varrho'_1 \to \infty \text{ and } \varrho_s \to \infty \qquad (129)$$

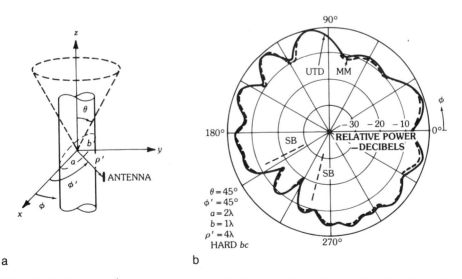

Fig. 34. Radiation pattern of a magnetic dipole located parallel to the axis of an elliptic cylinder. (a) Cylinder and antenna. (b) Radiation pattern. (*After Pathak, Burnside, and Marhefka [29], © 1980 IEEE*)

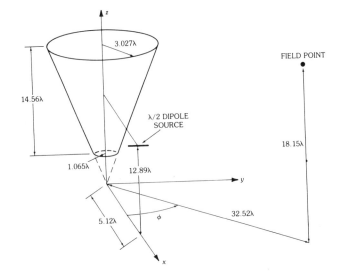
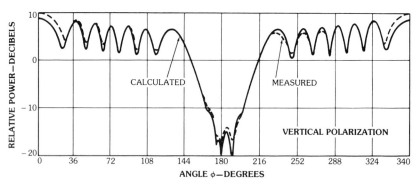
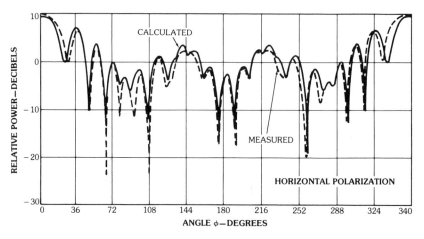

Fig. 35. Radiation patterns of an electric dipole mounted near a circular cone (see Fig. 37b). (*a*) Geometry for circular cone. (*b*) Comparison of measured and calculated radiation patterns. (*After Pathak, Burnside, and Marhefka [29], © 1980 IEEE*)

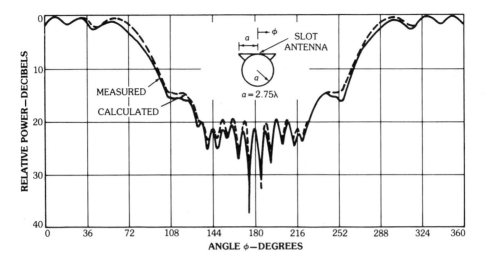

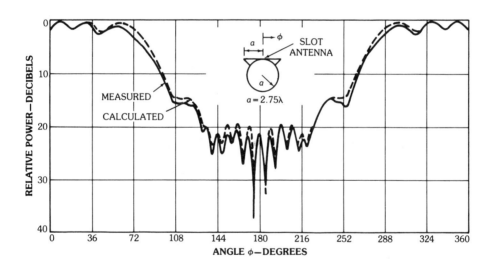

Fig. 36. Comparison of measured and calculated radiation patterns for a slot mounted in a plate-cylinder configuration with the slot parallel to the cylinder axis. (*After Pathak, Burnside, and Marhefka [29], © 1980 IEEE*)

so that $\mathbf{E}(P)$ is still given by (104), with (105) and (106) now changed to the following for the 2-D case:

$$\mathbf{E}^{gr}(P_L) \sim \mathbf{E}^i(Q_R) \cdot [\mathcal{R}_s \hat{\mathbf{e}}_\perp \hat{\mathbf{e}}_\perp + \mathcal{R}_h \hat{\mathbf{e}}_\parallel^i \hat{\mathbf{e}}_\parallel^r] \sqrt{\frac{\varrho^r}{\varrho^r + s^r}} \, e^{-jks^r} \qquad (130)$$

in which ϱ^r is discussed under "The GO Reflected Field" in Section 2, and

$$\mathbf{E}^d(P_S) \sim \mathbf{E}^i(Q_1) \cdot [\mathscr{D}_s \hat{\mathbf{b}} \hat{\mathbf{b}} + \mathscr{D}_h \hat{\mathbf{n}}_1 \hat{\mathbf{n}}_2] \frac{e^{-jks^d}}{\sqrt{s^d}} \tag{131}$$

since $\hat{\mathbf{b}}_1 = \hat{\mathbf{b}}_2 \equiv \hat{\mathbf{b}}$ for the 2-D case (note: $\hat{\mathbf{b}} = \hat{\mathbf{e}}_\perp$). The $\mathscr{R}_{s,h}$ and $\mathscr{D}_{s,h}$ in (130) and (131) are as defined in (107) and (108), respectively, except that the L appearing in (107) and (108) is given by

$$L = \frac{s' s^d}{s' + s^d}, \qquad \text{for the 2-D case} \tag{132}$$

where s' is the distance from the 2-D line source to the point of grazing incidence at Q_1 and $s \equiv s^d|_{\text{SSB}}$ as before. A comparison of the UTD and GTD solutions for a 2-D circular cylinder is illustrated in Figs. 37 and 38; those UTD solutions are then

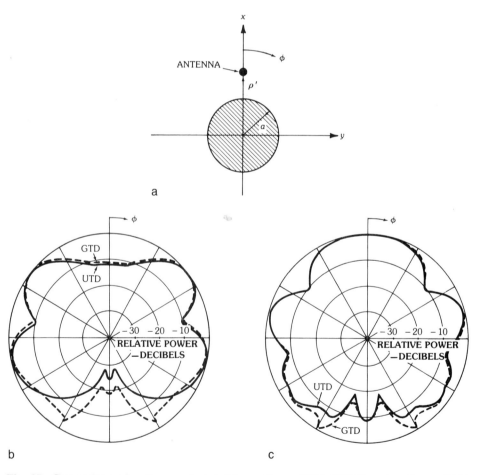

Fig. 37. Comparison of patterns calculated by uniform GTD (UTD) and ordinary GTD solutions for radiation by dipoles parallel to the axis of a perfectly conducting cylinder with $a = 1\lambda$ and $\varrho' = 2\lambda$. (a) Geometry. (b) Electric line source case. (c) Magnetic line source case. (*After Pathak, Burnside, and Marhefka [29], © 1980 IEEE*)

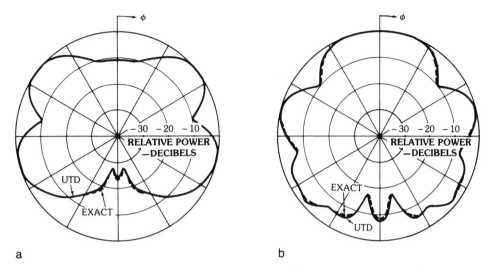

Fig. 38. Comparison of UTD solution of Fig. 37a with exact model series solution. (*a*) Radiation by electric line source. (*b*) Radiation by magnetic line source. (*After Pathak, Burnside, and Marhefka [29], © 1980 IEEE*)

compared with the corresponding exact eigenfunction solutions, indicating the better accuracy of the UTD result over the ordinary GTD result, especially in the SSB transition region.

The magnetic field $\mathbf{H}(P)$ can be found from $\mathbf{E}(P)$ as in (11), (17), and (51) via the local plane-wave approximation.

UTD Analysis of the Radiation by Antennas on a Smooth, Convex Surface (3-D Case)—A compact uniform GTD (or UTD) solution is described below for the problem in Fig. 27 such that it remains uniformly valid across the SSB transition region. This UTD solution for the radiation problem employs the ray coordinates of the GTD. In the shadow region the field radiated by a source at Q' propagates along Keller's surface-diffracted ray path to the point P_S (Fig. 39a), whereas in the lit region the field propagates along the GO ray path directly from the source at Q' to the field point P_L (Fig. 39b). These ray fields reduce to the GO field in the deep lit region and remain uniformly valid across the SSB transition region into the deep shadow region. Surface-ray torsion, which affects the radiated field in both the shadow and the SSB transition regions, appears "explicitly" in the solution as a torsion factor. The solution to the scattering problem described earlier did not explicitly contain such a torsion factor to first order; thus, explicit effects of surface-ray torsion appear to be localized only to regions in the neighborhood of sources on a convex surface. Since the field in the deep lit region is essentially that obtained from GO, and the field in the deep shadow region is relatively weak, the practical importance of the UTD solution described here is its ability to predict complex, surface-dependent field and polarization effects in the transition region adjacent to the shadow boundary (SSB) of a convex surface (Fig. 27).

The UTD solution for the radiated field is given below for the lit and shadow regions. These solutions for the two regions join smoothly at the shadow boundary

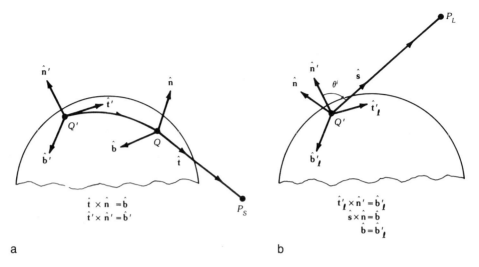

Fig. 39. Ray paths in the shadow and lit regions. (*a*) Field point in shadow region. (*b*) Field point in lit region. (*After Pathak, Wang, Burnside, and Kouyoumjian [31], © 1981 IEEE*)

(SSB). The UTD solution is presented here for the electromagnetic field radiated by an aperture in or by a short, thin monopole on a smooth, perfectly conducting, convex surface surrounded by a homogeneous isotropic medium. For an aperture in a convex surface it is convenient to define an infinitesimal magnetic current moment $d\mathbf{p}_m(Q')$ at any point Q' in the aperture as

$$d\mathbf{p}_m(Q') = \mathbf{E}^a(Q') \times \hat{\mathbf{n}}' \, dA' \tag{133}$$

where $\mathbf{E}^a(Q')$ is the electric field, $\hat{\mathbf{n}}'$ is the outward unit normal vector to the surface, and dA' is an area element at Q'. The tangential electric field in the aperture is assumed to be known. The $d\mathbf{p}_m(Q')$ radiating in the presence of the perfectly conducting convex surface, which now covers the aperture as well, constitutes the equivalent source of the electric field $d\mathbf{E}_m(P|Q')$ produced at any point P exterior to the surface. The total radiated electric field $\mathbf{E}_m(P)$ is then found by integrating the incremental field $d\mathbf{E}_m(P|Q')$ over the total area A of the aperture. Thus

$$\mathbf{E}_m(P) = \iint_A d\mathbf{E}_m(P|Q') \tag{134}$$

In the present context one notes that an aperture antenna frequently occurs in the form of a slot which is cut in a conducting surface. Slot radiators of rectangular, circular, or annular shapes are commonly employed in practice. Following the above development for the equivalent sources in the aperture radiation problem, one may similarly define an infinitesimal electric current moment $d\mathbf{p}_e(\ell')$ in dealing with the radiation by a monopole on a convex surface as

$$d\mathbf{p}_e(Q') = I(\ell')d\ell' \, \hat{\mathbf{n}}' \tag{135}$$

Techniques for High-Frequency Problems

where $I(\ell')$ denotes the electric current distribution on the monopole, and ℓ' is the distance along its length measured from the base at Q'. It is assumed here that the monopole is a short, thin wire whose total length h is such that $h \ll R_{1,2}(Q')$, where $R_1(Q')$ and $R_2(Q')$ denote the principal surface radii of curvatures at Q'. It is also assumed that the current distribution $I(\ell')$ is known. The current moment $d\mathbf{p}_e(\ell')$ radiating in the presence of the perfectly conducting surface then constitutes the equivalent source of the electric field $d\mathbf{E}_e(P|\ell')$ produced at P exterior to the surface. The total electric field $\mathbf{E}_e(P)$ radiated by the short monopole can be approximately calculated from a knowledge of only the field $d\mathbf{E}_e(P|Q')$ upon simply replacing the source strength $d\mathbf{p}_e(Q')$, which occurs in the latter solution, by $\int_0^h d\mathbf{p}_e(\ell')\cos(k\ell'\cos\theta^i)$ if P is in the lit region, or by $\int_0^h d\mathbf{p}_e(\ell')$ if P is in the shadow region.* The above integrals serve to properly incorporate the effects of the wire length and the current distribution of the short monopole into the radiation calculation based only on $d\mathbf{E}_e(P|Q')$. It is noted that the term $\cos\theta^i$ inside the integral is defined by $\cos\theta^i = \hat{\mathbf{n}}' \cdot \hat{\mathbf{s}}$, where $\hat{\mathbf{s}}$ is the radiation direction in the lit zone from any point ℓ' on the monopole, with $0 \leq \ell' \leq h$. For calculating the field radiated by monopoles which are longer than the one being considered here, one must employ the integral $\int_0^h d\mathbf{E}_e(P|\ell')$ over the length of the monopole. The latter integral may also be employed to calculate the field radiated by arbitrary curved wire monopoles; however, in this case the $d\mathbf{E}_e(P|\ell')$ is produced by an arbitrarily oriented $d\mathbf{p}_e(\ell')$ above the convex surface. The cases of long, straight, or curved wire monopoles will not be considered here. From the preceding discussion note that it suffices to present a uniform GTD (or UTD) formulation for only the incremental fields $d\mathbf{E}_m(P|Q')$ and $d\mathbf{E}_e(P|Q')$ radiated by the sources $d\mathbf{p}_m(Q')$ and $d\mathbf{p}_e(Q')$, respectively. The present formulation or ansatz of the uniform GTD solution leads to separate representations for the radiated field $d\mathbf{E}_{m,e}(P|Q')$ in the shadow and lit regions, respectively. As mentioned earlier, however, these different representations match exactly in polarization, amplitude, and phase, at the shadow boundary. The shadow region field representation for $d\mathbf{E}_{m,e}(P|Q')$ is presented first; a corresponding lit region field representation for $d\mathbf{E}_{m,e}(P|Q')$ is given next.

The field $d\mathbf{E}_{m,e}(P_S|Q')$ for the field at P_S in the shadow region arrives from Q' along the surface-diffracted ray path shown in Fig. 39a; it is given by [31]

$$d\mathbf{E}_{m,e}(P_S|Q') = d\mathbf{p}_{m,e}(Q') \cdot [\bar{\bar{\mathbf{T}}}_{m,e}(Q'|Q) + \bar{\bar{\boldsymbol{\Delta}}}_{m,e}] \sqrt{\frac{\varrho^d}{s^d(\varrho^d + s^d)}} e^{-jks^d} \quad (136)$$

with

$$\bar{\bar{\mathbf{T}}}_m(Q'|Q) = \frac{-jk}{4\pi} [\hat{\mathbf{b}}'\hat{\mathbf{n}}T_1(Q')H + \hat{\mathbf{t}}'\hat{\mathbf{b}}T_2(Q')S + \hat{\mathbf{b}}'\hat{\mathbf{b}}T_3(Q')S$$
$$+ \hat{\mathbf{t}}'\hat{\mathbf{n}}T_4(Q')H] e^{-jkt} \sqrt{\frac{d\psi_0}{d\eta(Q)}} \left[\frac{\varrho_g(Q)}{\varrho_g(Q')}\right]^{1/6} \quad (137)$$

---

*Here, $d\mathbf{E}_e(P|Q')$ is the value of $d\mathbf{E}_e(P|\ell')$ at $\ell' = 0$.

for the slot or $d\mathbf{p}_m$ case, and

$$\bar{\bar{\mathbf{T}}}_e(Q'|Q) = \frac{-jkZ_0}{4\pi}[\hat{\mathbf{n}}'\hat{\mathbf{n}}T_5(Q')H + \hat{\mathbf{n}}'\hat{\mathbf{b}}T_6(Q')S]\, e^{-jkt}\sqrt{\frac{d\psi_0}{d\eta(Q)}}\left[\frac{\varrho_g(Q)}{\varrho_g(Q')}\right]^{1/6} \quad (138)$$

for the monopole or $d\mathbf{p}_e$ case. The $\bar{\bar{\boldsymbol{\Delta}}}_{m,e}$ constitute the higher-order terms which are needed in special cases where the contribution from the leading terms in $\bar{\bar{\mathbf{T}}}_{m,e}$ is weak. The $\bar{\bar{\boldsymbol{\Delta}}}_{m,e}$ will be given below after the terms in $\bar{\bar{\mathbf{T}}}_{m,e}$ are defined in Table 2.

Let $\hat{\boldsymbol{\tau}}'_1$ and $\hat{\boldsymbol{\tau}}'_2$ denote the principal directions on the surface at Q' as shown in Fig. 40 along with the angle α' between $\hat{\mathbf{t}}'$ and $\boldsymbol{\tau}'_1$. Then R_1 and R_2 in Table 2 denote the principal surface radii of curvatures in the $\hat{\boldsymbol{\tau}}'_1$ and $\hat{\boldsymbol{\tau}}'_2$ directions at Q'. For the sake of definiteness one chooses $R_1 \geq R_2$ in this development. One notes that the expression for torsion $T(Q')$ given in Table 2 becomes negative if $\pi/2 < \alpha' < \pi$ or if $3\pi/2 < \alpha' < 2\pi$. The ϱ^d is expressed in terms of the quantities E and G in Table 2 which denote two of the three coefficients E, F, and G that occur in the "first fundamental form" representing the differential arc length along a curve on a surface (see, for example, D. J. Struik, *Differential Geometry*, 2nd ed., Addison-Wesley Publishing Co., Reading, Massachusetts, 1961). The functions H and S in $\bar{\bar{\mathbf{T}}}_{m,e}$ of (137) and (138) are

$$H = g(\xi) \quad (139)$$

$$S = \frac{-j}{m(Q')}\tilde{g}(\xi) \quad (140)$$

Here $g(\xi)$ and $\tilde{g}(\xi)$ denote the acoustic hard and soft Fock functions (or Fock integrals); they are defined as

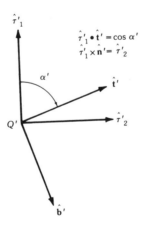

Fig. 40. Principal surface directions at the source. (*After Pathak, Wang, Burnside, and Kouyoumjian [31], © 1981 IEEE*)

Table 2. Terms for the Shadow Region

Type of Convex Surface	Slot or $d\mathbf{p}_m$ Case				Monopole or $d\mathbf{p}_m$ Case			Surface-Ray Torsion	Surface Radius of Curvature in $\hat{\mathbf{t}}'$ Direction	Surface-Diffracted Ray Caustic Distance
	$T_1(Q')$	$T_2(Q')$	$T_3(Q')$	$T_4(Q')$	$T_5(Q')$	$T_6(Q')$	$T(Q')$	$\varrho_g(Q')$	ϱ_c
Sphere	1	1	0	0	1	0	0	a	$a\tan(t/a)$
Circular cylinder	1	1	$\dfrac{\sin 2\alpha'}{2a}\dfrac{a}{\sin^2\alpha'}$	0	1	$\dfrac{\sin 2\alpha'}{2a}\dfrac{a}{\sin^2\alpha'}$	$\dfrac{\sin 2\alpha'}{2a}$	$\dfrac{a}{\sin^2\alpha'}$	t
Arbitrary convex surface	1	1	$T(Q')\varrho_g(Q')$	0	1	$T(Q')\varrho_g(Q')$	$\dfrac{\sin 2\alpha'}{2}\left[\dfrac{1}{R_2(Q')} - \dfrac{1}{R_1(Q')}\right]$ with $R_1(Q') \geq R_2(Q')$	$\left[\dfrac{\cos^2\alpha'}{R_1(Q')} + \dfrac{\sin^2\alpha'}{R_2(Q')}\right]^{-1}$	$\dfrac{2\sqrt{E}\,G}{\partial G/\partial t}$

4-67

$$g(\xi) = \frac{1}{\sqrt{\pi}} \int_{\infty \exp(-j2\pi/3)}^{\infty} e^{-j\tau\xi} [W_2'(\tau)]^{-1} d\tau \qquad (141)$$

and

$$\tilde{g}(\xi) = \frac{1}{\sqrt{\pi}} \int_{\infty \exp(-j2\pi/3)}^{\infty} e^{-j\tau\xi} [W_2(\tau)]^{-1} d\tau \qquad (142)$$

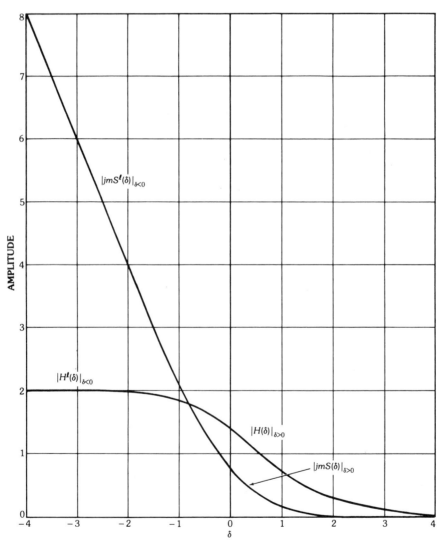

a

Fig. 41. Plots of the Fock radiation functions. (*a*) Amplitude. (*b*) Phase.

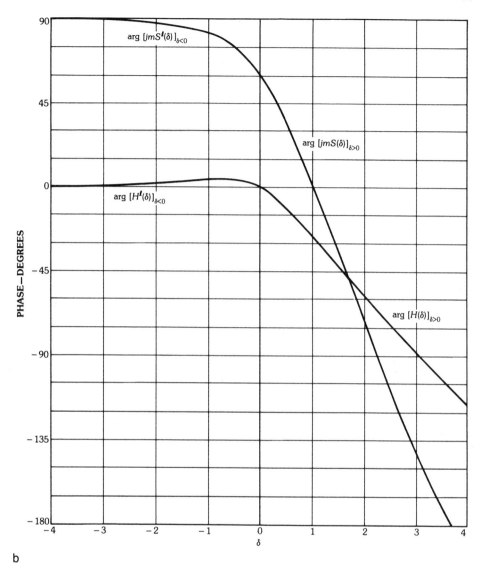

Fig. 41, *continued*

in which the Fock type Airy functions W_2 and W_2' are given by (111c) and $dW_2/d\tau$, respectively. The Fock parameter ξ for the "shadow zone" is defined as in (113) together with (114), but with Q_1 and Q_2 in (113) replaced by Q' and Q, respectively. Noted that $\xi > 0$ in the shadow zone. Also the distance t in e^{-jkt} of (137) and (138) is the length of the surface-ray geodesic path from Q' to Q. The functions $H(\xi)$ and $S(\xi)$ are plotted in Fig. 41 for $\xi > 0$. It is interesting to note that in the deep shadow zone, where $\xi \gg 0$, H and S can alternatively be expressed as

$$H(\xi)|_{\xi \gg 0} \sim \sum_{n=1}^{N} \left\{ L_n^h(Q') \exp\left[-\int_{Q'}^{Q} \alpha_n^h(t')dt'\right] D_n^h(Q) \right\} \left[\frac{\varrho_g(Q)}{\varrho_g(Q')}\right]^{-1/6} \quad (143)$$

and

$$S(\xi)|_{\xi \gg 0} \sim \sum_{n=1}^{N} \left\{ L_n^s(Q') \exp\left[-\int_{Q'}^{Q} \alpha_n^s(t')dt'\right] D_n^s(Q) \right\} \left[\frac{\varrho_g(Q)}{\varrho_g(Q')}\right]^{-1/6} \quad (144)$$

in which $D_n^{s,h}$ and $\alpha_n^{s,h}$ are as defined previously in (127) and (128), whereas

$$L_n^s(Q') = \sqrt{2\pi k}\, \frac{e^{-j\pi/12}}{m^2(Q')}\, Ai'(-q_n) D_n^s(Q')$$

and

$$L_n^h(Q') = \sqrt{2\pi k}\, \frac{e^{j\pi/12}}{m(Q')}\, Ai'(-\bar{q}_n) D_n^h(Q')$$

The $L_n^{s,h}$ are defined as the surface-ray launching coefficients [31]; they allow one to extend the ordinary GTD to the problem of the radiation by sources on a convex surface. Thus, outside the (SSB) transition region which lies in the deep shadow, UTD → GTD. As in (126b) $N = 2$ in (143) and (144) generally yields accurate results.

The quantity $d\eta(Q) = \varrho^d d\psi$, and the angles $d\psi_0$ at Q' and $d\psi$ at Q, respectively, which are associated with the surface-diffracted ray path from Q' to P_S, are shown in Fig. 42. Note that

$$\sqrt{\frac{d\psi_0}{d\psi}} = \begin{cases} \sqrt{1/\cos(t/a)} & \text{for a sphere} \\ 1 & \text{for a cylinder} \end{cases}$$

For an arbitrary convex surface, $\sqrt{d\psi_0/d\psi}$ may be determined numerically [31]. If $d\psi_0/\varrho^d d\psi < 0$, then

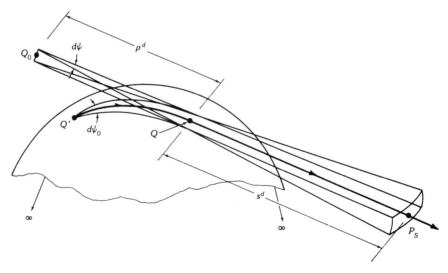

Fig. 42. Surface-diffracted ray tube associated with the surface-diffracted ray from Q' to P_s.

Techniques for High-Frequency Problems

$$\sqrt{d\psi_0/\varrho^d d\psi} = \sqrt{d\psi_0/d\eta(Q)} = +j|\sqrt{d\psi_0/\varrho^d d\psi}|$$

Finally, the higher-order terms $\bar{\bar{\Delta}}_{m,e}$ in (136) are given by [31]

$$\bar{\bar{\Delta}}_m = \left(\frac{-jk}{4\pi}\right)\left\{(\hat{\mathbf{b}}'\hat{\mathbf{n}}T_0 + \hat{\mathbf{t}}'\hat{\mathbf{n}})\left[T_0\frac{jH'}{2m^2(Q')}\sqrt{\frac{d\psi_0}{d\psi}}\right] + \hat{\mathbf{t}}'\hat{\mathbf{b}}\frac{jH}{2m^3(Q')}\Lambda_s\left(\frac{d\psi_0}{d\psi}\right)\right\}$$
$$\times \left[\frac{\varrho_g(Q)}{\varrho_g(Q')}\right]^{1/6}\sqrt{\frac{d\psi_0}{d\eta(Q)}}\, e^{-jkt} \tag{145}$$

and

$$\bar{\bar{\Delta}}_e = \left(\frac{-jkZ_0}{4\pi}\right)\hat{\mathbf{n}}'\hat{\mathbf{n}}\Lambda_c(1 + T_0^2)\frac{jH'}{2m^2(Q')}\sqrt{\frac{d\psi_0}{d\psi}}\left[\frac{\varrho_g(Q)}{\varrho_g(Q')}\right]^{1/6}\sqrt{\frac{d\psi_0}{d\eta(Q)}}\, e^{-jkt} \tag{146}$$

where $T_0 = T(Q')\varrho_g(Q')$, in which $T(Q')$ and $\varrho_g(Q')$ are defined in Table 2. Also, the function H' is defined as $dH/d\xi$, in which H has been defined previously in (139). A reasonable choice for Λ_s and Λ_c is given by [31]

$$\Lambda_s = R_2(Q')/R_1(Q') \tag{147a}$$

and

$$\Lambda_c = 1 - \Lambda_s \tag{147b}$$

with $R_1(Q') \geqq R_2(Q')$, in order to interpolate smoothly between the higher-order terms in the canonical circular cylinder and sphere solutions.

Next, the field $\mathbf{E}_{m,e}(P_L|Q')$ at a point P_L in the lit region arrives from the point Q' along the direct-ray path as shown in Fig. 39b; it is given by [31]

$$d\mathbf{E}_{m,e}(P_L|Q') = d\mathbf{p}_{m,e}(Q') \cdot (\bar{\bar{\mathbf{T}}}^\ell_{m,e} + \bar{\bar{\Delta}}^\ell_{m,e})\, e^{-jks} \tag{148}$$

with

$$\bar{\bar{\mathbf{T}}}^\ell_m = \frac{-jk}{4\pi}(\hat{\mathbf{b}}'_\ell \hat{\mathbf{n}} A + \hat{\mathbf{t}}'_\ell \hat{\mathbf{b}} B + \hat{\mathbf{b}}'_\ell \hat{\mathbf{b}} C + \hat{\mathbf{t}}'_\ell \hat{\mathbf{n}} D) \tag{149}$$

for the slot or $d\mathbf{p}_m$ case, and

$$\bar{\bar{\mathbf{T}}}^\ell_e = \frac{-jkZ_0}{4\pi}(\hat{\mathbf{n}}'\hat{\mathbf{n}} M + \hat{\mathbf{n}}'\hat{\mathbf{b}} N) \tag{150}$$

for the monopole or $d\mathbf{p}_e$ case. Again the $\bar{\bar{\Delta}}^\ell_{m,e}$ in (148) constitute the higher-order terms in the lit zone; whereas the $\bar{\bar{\mathbf{T}}}^\ell_{m,e}$ are the leading terms, which are generally more important than $\bar{\bar{\Delta}}^\ell_{m,e}$. The terms in $\bar{\bar{\mathbf{T}}}^\ell_{m,e}$ are defined below in Table 3. It is noted that $T(Q')$ and $\varrho_g(Q')$ in Table 3 have the same meaning as in Table 2 for the shadow zone. However, the angle α' in the definition of $T(Q')$ for the *lit zone* is now the angle between $\hat{\mathbf{t}}'_\ell$ and $\hat{\mathbf{t}}'_1$. The functions H^ℓ and S^ℓ in Table 3 are

Table 3. Terms for the Lit Region

Slot or $d\mathbf{p}_m$ Case			
A	B	C	D
$H^\ell + T_0^2 \Upsilon \cos\theta^i$	$S^\ell - T_0^2 \Upsilon \cos^2\theta^i$	$T_0 \Upsilon$	$T_0 \Upsilon \cos\theta^i$

Monopole or $d\mathbf{p}_{\hat{e}}$ Case			
M	N	T_0	T
$\sin\theta^i(H^\ell + T_0^2 \Upsilon \cos\theta^i)$	$T_0 \Upsilon \sin\theta^i$	$T(Q')\varrho_g(Q')$	$\dfrac{S^\ell - H^\ell \cos\theta^i}{1 + T_0^2 \cos\theta^i}$

$$H^\ell = g(\xi_\ell)e^{-j(\xi_\ell^3/3)} \tag{151}$$

$$S^\ell = \frac{-j}{m_\ell(Q')}\tilde{g}(\xi_\ell)e^{-j(\xi_\ell^3/3)} \tag{152}$$

in which the hard and soft Fock functions g and $\tilde{g}$ have been defined previously in (141) and (142). The Fock parameter ξ_ℓ for the lit region is given by

$$\xi_\ell = -m_\ell(Q')\cos\theta^i \tag{153}$$

where

$$m_\ell(Q') = \frac{m(Q')}{(1 + T_0^2\cos^2\theta^i)^{1/3}} \tag{154}$$

The angle θ^i is defined by $\hat{\mathbf{n}}' \cdot \hat{\mathbf{s}} = \cos\theta^i$; this angle is also shown in Fig. 43. It is noted that the Fock parameter $\xi_\ell < 0$ in the lit region. In the deep lit region it can be shown from the asymptotic properties of g and $\tilde{g}$ for $\xi_\ell \ll 0$ that $\Upsilon \to 0$. The factor Υ has been defined in Table 3. One may view $T_0\Upsilon$ as the "surface depolarization factor" since it plays an important role within the lit portion of the shadow boundary transition region, where it serves as a measure of the extent to which the polarization of the radiated field is affected by the surface geometry near Q'. Clearly, in the deep lit region where the geometrical optics approximation becomes valid, the polarization of the radiated field is dictated by the source rather than the surface, and $T_0\Upsilon \to 0$ as one would expect.

The functions $H^\ell(\xi_\ell)$ and $S^\ell(\xi_\ell)$ are plotted in Fig. 41 for $\xi_\ell < 0$. It is easily verified that the $d\mathbf{E}_{m,e}$ of (148) reduces to the geometrical optics field solution in the deep lit region, where $T_0\Upsilon \to 0$, $H^\ell \to 2$, and $S^\ell \to 2\cos\theta^i$ for $\xi_\ell \ll 0$, i.e., the leading terms yield

$$d\mathbf{E}_m(P_L|Q') \sim \frac{-jk}{4\pi}d\mathbf{p}_m(Q') \cdot (\hat{\mathbf{b}}'_\ell\hat{\mathbf{n}}\,2 + \hat{\mathbf{t}}'_\ell\hat{\mathbf{b}}\,2\cos\theta^i)\frac{e^{-jks}}{s}, \quad \text{for } \xi_\ell \ll 0 \tag{155}$$

and

$$d\mathbf{E}_e(P_L|Q') \sim \frac{-jkZ_0}{4\pi}d\mathbf{p}_e(Q') \cdot (\hat{\mathbf{n}}'\hat{\mathbf{n}}\,2\sin\theta^i)\frac{e^{-jks}}{s}, \quad \text{for } \xi_\ell \ll 0 \tag{156}$$

Finally, the higher-order terms $\bar{\bar{\Delta}}^\ell_{m,e}$ in (148) are given by [31]

$$\bar{\bar{\Delta}}^\ell_m = \frac{-jk}{4\pi}\left\{(\hat{\mathbf{b}}'_\ell\hat{\mathbf{n}}\,T_0 + \hat{\mathbf{t}}'_\ell\hat{\mathbf{n}})\left[T_0\frac{\sin^2\theta^i}{(1+T_0^2\cos^2\theta^i)^2} - \frac{jH^{\ell'}}{2m_\ell^2(Q')}\right]\right.$$
$$\left. + \hat{\mathbf{t}}'_\ell\hat{\mathbf{b}}\left[\frac{jH^\ell}{2m_\ell^3(Q')}\Lambda_s - \frac{T_0^2\cos\theta^i\sin^2\theta^i}{(1+T_0^2\cos^2\theta^i)^2}\frac{jH^{\ell'}}{2m_\ell^2(Q')}\right]\right\} \tag{157}$$

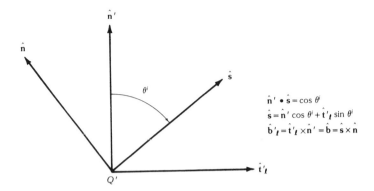

Fig. 43. Unit vectors for the lit region. (*After Pathak, Wang, Burnside, and Kouyoumjian [31], © 1981 IEEE*)

and

$$\bar{\bar{\Delta}}_e^\ell = \left(\frac{-jkZ_0}{4\pi}\right)\left[\hat{n}'\hat{n}\,\frac{\sin\theta^i\Lambda_c}{1+T_0^2\cos^2\theta^i}\,\frac{1+T_0^2}{1+T_0^2\cos^2\theta^i}\,\frac{jH^{\ell'}}{2m_\ell^2} \right.$$
$$\left. + \hat{n}'\hat{b}\,\frac{(-\sin\theta^i\cos\theta^i T_0)\Lambda_c}{1+T_0^2\cos^2\theta^i}\,\frac{1+T_0^2}{1+T_0^2\cos^2\theta^i}\,\frac{jH^{\ell'}}{2m_\ell^2}\right] \quad (158)$$

where T_0 is specified in Table 3. Also $H^{\ell'} = dH^\ell/d\xi_\ell$ in which H^ℓ is as defined in (151). The remaining quantities in (157) and (158) are identical with those defined earlier.

At the shadow boundary (SSB) it can be readily shown that $\bar{\bar{\Delta}}_{m,e}^\ell = \bar{\bar{\Delta}}_{m,e}$, so that the higher-order terms, like the leading terms, are continuous at $\theta^i = \pi/2$.

The solution presented above is employed here to compute the radiation from slots in cylinders, spheres, cones, and spheroids, and also from a monopole on a spheroid. In these computations the pertinent rays which pass through the field point in the lit and shadow regions are depicted in Fig. 44. The ray which is launched at Q' and then propagates in a clockwise direction around the convex surface before it sheds to the field point P_L in the lit region is not shown in Fig. 44a. The field of this ray, as well as that of rays shed from multiple encirclements, can be neglected for electrically large surfaces. The fields of rays which traverse long distances on surfaces with radii of curvature large in terms of the wavelength become highly attenuated and hence their effect on the resulting pattern is negligible. One notes that in the case of a prolate spheroidal surface there are two additional surface-diffracted ray contributions besides the ones shown in Fig. 44a; however, these contribute as significantly as the two rays in Fig. 44b only in the deep shadow region. Elsewhere, their contribution is negligible.

The various functions contained in the present solution are rather easy to compute; on the other hand, the associated geodesic surface-ray paths can be easily determined only in the case of some simple shapes, such as planes, cylinders, spheres, and cones; otherwise, the surface geodesics must be determined

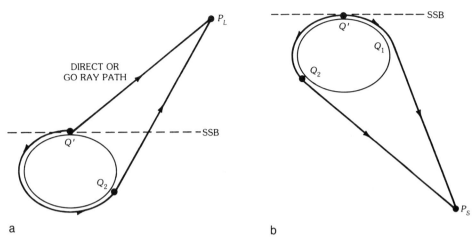

Fig. 44. Ray paths in the shadow and lit regions for a closed convex body. (*a*) Lit region. (*b*) Shadow region. (*After Pathak, Wang, Burnside, and Kouyoumjian [31], © 1981 IEEE*)

numerically from their governing differential equations (or from an alternative integral expression). Fig. 45 indicates the dominant helical geodesic surface-ray paths for a convex cylinder. In Fig. 45a angles α_1' and α_2' stay the same with respect to the generator (or the $\hat{z}$ direction) of the cylinder. For a spherical surface the geodesic surface-ray paths are great circle paths. A discussion of the numerical computation of geodesic paths on circular and elliptic cylinders may be found in [32], whereas the geodesics on a cone may be computed as described in [33, 34]. A typical geodesic surface-ray path on a cone is illustrated in Fig. 46. Also, a procedure for finding the geodesic paths corresponding to a "given" radiation direction ($\hat{t}$ or $\hat{s}$) is discussed in [33] for the more general prolate spheroidal surface.

Radiation pattern calculations using only the leading terms of this asymptotic (UTD) solution are presented in Figs. 47b, 47c, and 47d for circumferential and axial slots in circular and elliptic cylinders. The rectangular slots in these cases are short and thin so that the dominant mode cosine distribution can be assumed in the slot. The cylinder geometry is shown in Fig. 47a. It is observed that the agreement between the asymptotic (UTD) calculations and the exact calculations in Figs. 47b, 47c, and 47d is very good even though the cylinders are not too large in terms of the wavelength. The exact results in Figs. 47c and 47d for the elliptic cylinder case have been obtained from the work of [35].

The next two radiation pattern calculations presented in Figs. 48 and 49 illustrate the importance of the higher-order terms in the UTD solution.* While the inclusion of the higher-order terms is in general not essential for accuracy, there are

---

*A caustic of the surface-diffracted rays is present at $\theta = 0°$ and $180°$ for the spherical geometry. The field of this caustic is weak in comparison with the direct or geometrical optics field of the source at $\theta = 0°$; however, it is the only field which exists at $\theta = 180°$. The present UTD solution is not directly valid at caustics, hence the pattern is not shown close to $\theta = 180°$ in Fig. 48; more will be said about modifications of the UTD solution at caustics later.

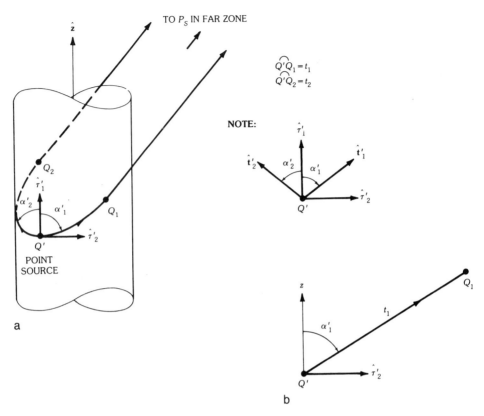

Fig. 45. Helical geodesic surface-ray paths from Q'. (*a*) Dominant path to points Q_1 and Q_2 of diffraction on a convex cylinder. (*b*) Helical geodesic surface ray path from Q' to point Q_1 of Fig. 45a on a developed cylinder.

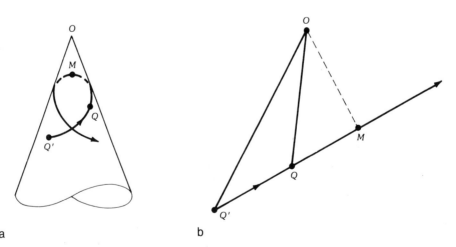

Fig. 46. Typical surface-ray path on a semi-infinite cone. (*a*) Geodesic surface-ray path on a cone. (*b*) Geodesic path on a developed cone.

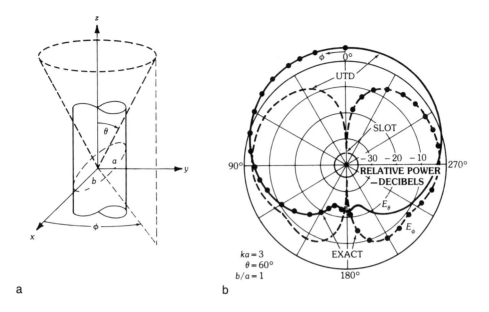

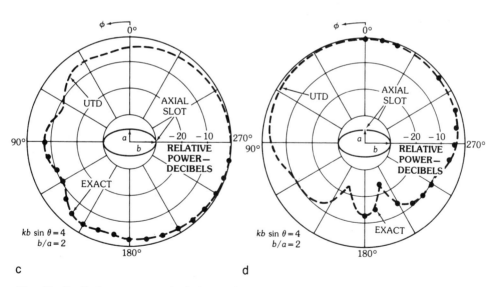

Fig. 47. Radiation pattern calculations using only the leading terms of UTD solution for circumferential and axial slots in circular and elliptic cylinders. (*a*) Elliptic cylinder geometry. (*b*) E_θ and E_ϕ radiation patterns of circumferential slot on circular cylinder. (*c*) $|E_\phi|$ radiation patterns of axial slot at major axis of elliptic cylinder. (*d*) $|E_\phi|$ radiation patterns of axial slot at minor axis of cylinder. (*After Pathak, Wang, Burnside, and Kouyoumjian [31], © 1981 IEEE*)

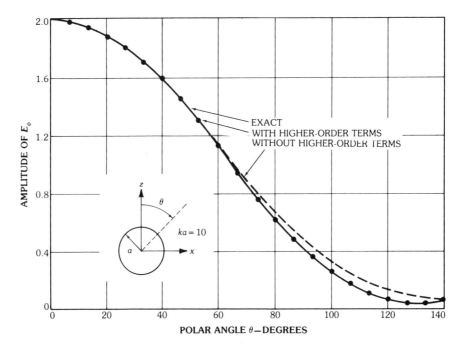

Fig. 48. The $|E_\phi|$ radiation pattern in the xz plane of a circumferential or $\hat{x}$-directed slot in a sphere. (*After Pathak, Wang, Burnside, and Kouyoumjian [31], © 1981 IEEE*)

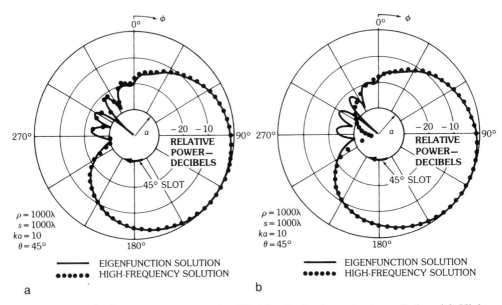

Fig. 49. The $|E_\phi|$ radiation pattern of a 45° (tilted) slot in a circular cylinder. (a) High-frequency solution with higher-order terms. (b) High-frequency solution without higher-order terms. (*After Pathak, Wang, Burnside, and Kouyoumjian [31], © 1981 IEEE*)

Techniques for High-Frequency Problems

special situations where these terms become as important as the leading terms and they must then be included to maintain accuracy.

The UTD radiation pattern calculations for short, thin, radial slots in a cone are illustrated in Figs. 50 and 51. These calculations are compared with the corresponding exact (eigenfunction) results and measurements which are given in [36]. In Fig. 50 the patterns corresponding to both the $\hat{\theta}$ and the $\hat{\phi}$ polarized components of the electric field are presented for $\theta = 80°$ and $0° \leq \phi \leq 360°$. In Fig. 51 the $\hat{\phi}$ component of the electric field is plotted as a function of θ in the $\phi = 40°$ and $\phi = 220°$ planes. It is noted for the cone configuration of Fig. 51 that the field point in the $\phi = 40°$ plane lies in the lit region, whereas for the $\phi = 220°$ case it lies primarily in the shadow region.

Figs. 52a–52d illustrate the UTD radiation pattern calculation for antennas on prolate spheroids together with measured patterns for comparison. The geometry of this problem is shown in Fig. 53. This close agreement between calculations

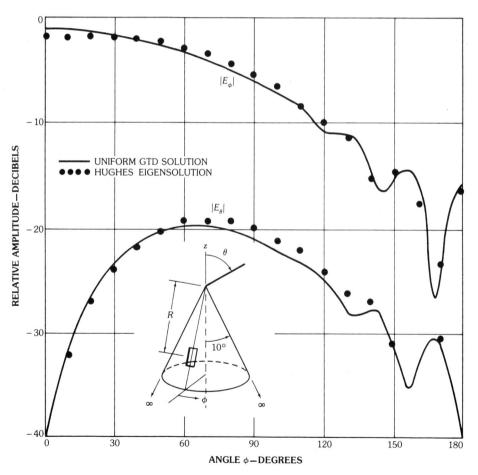

Fig. 50. The $|E_\phi|$ radiation pattern of a radial slot in a cone. (*After Pathak, Wang, Burnside, and Kouyoumjian [31], © 1981 IEEE*)

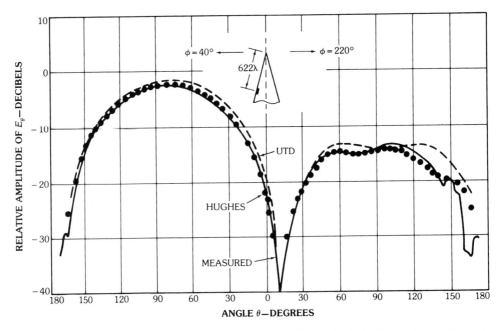

Fig. 51. Radiation patterns of a radial slot in a cone. (*After Pathak, Wang, Burnside, and Kouyoumjian [31], © 1981 IEEE*)

and measurements for the spheroidal shape is very gratifying because it confirms the accuracy of the UTD solution even when the surface is far more general than a cylinder, cone, or sphere.

The (UTD) ray solution presented above for calculating the radiation from sources on a large, perfectly conducting, smooth, convex surface of slowly varying curvature appears to work surprisingly well even for surfaces which are only moderately large in terms of the wavelength, as seen from some of the numerical results which have been presented. This solution is not expected to be accurate within the paraxial regions of almost cylindrical surfaces (i.e., along or near the axial direction of cylindrically shaped surfaces), and it must be modified for this special case. Such a modification has not been obtained at this time. Another special case where this solution cannot be directly employed is near a caustic of the surface-diffracted rays; however, if such a caustic region does exist, then it is generally possible to modify the UTD solution in the vicinity of the caustic via the equivalent-ring current method (ECM) [37], which is discussed in Section 5.

The magnetic field $d\mathbf{H}(P)$ can be obtained directly from the corresponding $d\mathbf{E}(P)$ via the local plane-wave relationship

$$d\mathbf{H}(P_{L,S}) = \begin{Bmatrix} \hat{\mathbf{s}} \\ \hat{\mathbf{t}} \end{Bmatrix} \times Y_0 \, d\mathbf{E}(P_{L,S})$$

UTD Analysis of the Radiation by Antennas on a Smooth, Convex Surface (2-D Case)—The UTD solution for analyzing the radiation from 2-D sources on a 2-D convex surface can be obtained from the 3-D solution by noting that $\hat{\mathbf{b}}' = \hat{\mathbf{b}}$

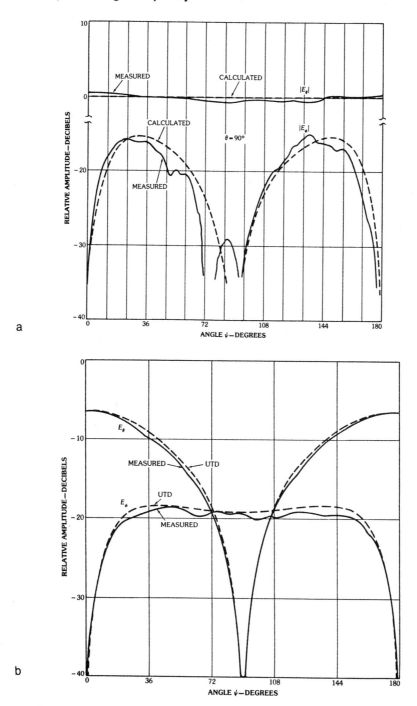

Fig. 52. Radiation patterns of antennas on a prolate spheroid (at the source location in Fig. 54). (*a*) Of $\hat{n}$-directed monopole. (*b*) Of $\hat{x}$-directed rectangular slot with $\theta = 90°$. (*c*) Of $\hat{x}$-directed rectangular slot with $\theta = 100°$. (*d*) Of $\hat{x}$-directed rectangular slot with various θ. (*After Pathak, Wang, Burnside, and Kouyoumjian [31], © 1981 IEEE*)

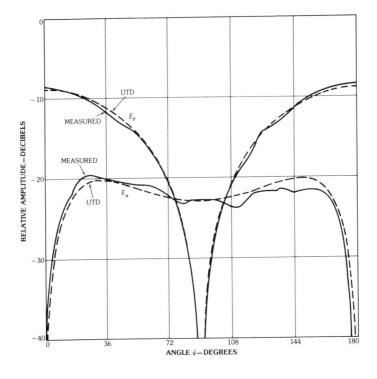

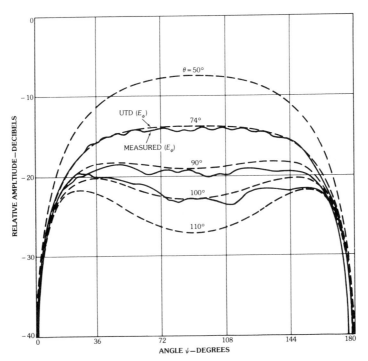

Fig. 52, *continued*

Techniques for High-Frequency Problems

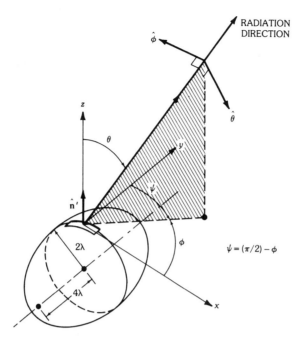

Fig. 53. Prolate spheroidal geometry. (*After Pathak, Wang, Burnside, and Kouyoumjian [31], © 1981 IEEE*)

and $\varrho^d \to \infty$ in the 2-D case, and by allowing the cross terms in the dyads $\bar{\bar{T}}_{e,m}$ and $\bar{\bar{T}}^{\ell}_{e,m}$ to vanish [31, 38]. Thus the electric field radiated by 2-D electric and magnetic line currents of strength $dI_m \hat{n}'$ and $d\mathbf{M}$ at Q', respectively, on a 2-D convex surface are given by

$$dE(P_L) = C_0 Z_0 \, dI_m(Q') \hat{n}' \cdot [\hat{n}' \hat{n}_\ell H^\ell(\xi_\ell)] \frac{e^{-jks}}{\sqrt{s}} \sin \theta^i \tag{159}$$

and

$$dE(P_S) = C_0 Z_0 \, dI_m(Q') \hat{n}' \cdot [\hat{n}' \hat{n} H(\xi)] \left[\frac{\varrho_g(Q)}{\varrho_g(Q')} \right]^{1/6} \frac{e^{-jkt-jks^d}}{\sqrt{s^d}} \tag{160}$$

for the electric line monopole source case, and by

$$dE(P_L) = C_0 \, d\mathbf{M}(Q') \cdot [\hat{b}' \hat{n}_\ell H^\ell(\xi_\ell) + \hat{t}' \hat{b} S^\ell(\xi_\ell)] \frac{e^{-jks}}{\sqrt{s}} \tag{161}$$

and

$$dE(P_S) = C_0 \, d\mathbf{M}(Q') \cdot [\hat{b}' \hat{n} H(\xi) + \hat{t}' \hat{b} S(\xi)] \left[\frac{\varrho_g(Q)}{\varrho_g(Q')} \right]^{1/6} \frac{e^{-jkt-jks^d}}{\sqrt{s^d}} \tag{162}$$

for the magnetic line source case. It is noted that $d\mathbf{M}$ in (161) and (162) can be expressed as $d\mathbf{M} = \hat{\mathbf{z}}dM_z + \hat{\mathbf{t}}dM_d$, where $\hat{\mathbf{z}}$ denotes the direction along the axis or generator of the 2-D convex cylinder and $\hat{\mathbf{t}}$ denotes the circumferential direction on that cylinder;* also $\hat{\mathbf{b}} = \hat{\mathbf{b}}$ in (161) and (162) for the 2-D situation. The quantity C_0 in (159)–(162) is given by

$$C_0 = -\sqrt{\frac{k}{8\pi}} \, e^{+j\pi/4} \tag{163}$$

Finally, the magnetic field $d\mathbf{H}(P)$ can be found as usual from $d\mathbf{E}(P)$ via

$$d\mathbf{H}(P_{L,S}) = \begin{Bmatrix} \hat{\mathbf{s}} \\ \hat{\mathbf{t}} \end{Bmatrix} \times Y_0 \, d\mathbf{E}(P_{L,S})$$

on employing the local plane-wave approximation.

UTD Analysis of the Mutual Coupling Between Antennas on a Smooth, Convex Surface (3-D Case)—A uniform GTD (or UTD) solution is presented below for analyzing the problem in Fig. 28, and it is uniform in the sense that it can also be used within the SSB transition region. This solution employs surface rays of the GTD, and it is assumed, once again, that the convex surface is electrically large. In particular, this UTD solution specifically describes in a simple fashion the surface field at Q which is excited by an infinitesimal magnetic current moment $d\mathbf{p}_m(Q')$, or an infinitesimal electric current moment $d\mathbf{p}_e(Q')$, respectively, at Q' on the same convex surface.

The electromagnetic surface fields excited by an aperture in or a short, thin monopole on a convex conducting surface may be obtained in terms of the surface fields due to infinitesimal magnetic or electric current moments on the same surface as follows. In the aperture case, let $\mathbf{E}^a(Q')$ denote the electric field at any point Q' in the aperture. One may then define an infinitesimal magnetic current moment $d\mathbf{p}_m(Q')$ as in (133) for the radiation problem. The $d\mathbf{p}_m(Q')$ radiating in the presence of the perfectly conducting surface which covers the aperture as well now constitutes an equivalent source in the incremental electric and magnetic fields $d\mathbf{E}_m(Q|Q')$ and $d\mathbf{E}_m(Q|Q')$, respectively, at Q on the surface. The surface fields $\mathbf{E}_m(Q)$ and $\mathbf{H}_m(Q)$ due to the aperture are then found via an integration of $d\mathbf{E}_m$ and $d\mathbf{H}_m$ over the aperture area S_a. Thus,

$$\mathbf{E}_m(Q) = \iint_{S_a} d\mathbf{E}_m(Q|Q') \tag{164a}$$

$$\mathbf{H}_m(Q) = \iint_{S_a} d\mathbf{H}_m(Q|Q') \tag{164b}$$

---

*Note that $d\mathbf{M}(Q') = \lim_{dt' \to 0} \mathbf{E}^a(Q') \times \hat{\mathbf{n}}' \, dt'$, where dt' is the infinitesimal circumferential width of an infinitesimal 2-D aperture and $\mathbf{E}^a(Q')$ is the electric field at Q' in the aperture.

Techniques for High-Frequency Problems

Let t be the arc length of the geodesic surface-ray path from Q' to Q as in Fig. 54. The expressions in (164a) and (164b) and those for $d\mathbf{E}_m$ and $d\mathbf{H}_m$ obtained here are valid even for $t \to 0$. One may likewise define an infinitesimal electric current moment $d\mathbf{p}_e(\ell')$ at Q' again, as done previously in (135) for the radiation problem. The $d\mathbf{p}_e(\ell')$ now constitutes an equivalent source of the incremental electric and magnetic fields $d\mathbf{E}_e(Q|\ell')$ and $d\mathbf{H}_e(Q|\ell')$, respectively, on the surface at Q. The total fields $\mathbf{E}_e(Q)$ and $\mathbf{H}_e(Q)$ are obtained by integrating the incremental fields over the total length h of the monopole as follows:

$$\mathbf{E}_e(Q) = \int_0^h d\mathbf{E}_e(Q|\ell') \tag{165a}$$

and

$$\mathbf{H}_e(Q) = \int_0^h d\mathbf{H}_e(Q|\ell') \tag{165b}$$

If the distance t of the observation point Q is far enough away from Q' such that the monopole is not directly visible at Q, then

$$\mathbf{E}_e(Q)|_{t>t_0} \cong \frac{d\mathbf{E}_e(Q|Q')}{d\mathbf{p}_e(Q')} \left[\int_0^h d\mathbf{p}_e(\ell') \right]; \tag{166a}$$

$$\mathbf{H}_e(Q)|_{t>t_0} \cong \frac{d\mathbf{H}_e(Q|Q')}{d\mathbf{p}_e(Q')} \left[\int_0^h d\mathbf{p}_e(\ell') \right] \tag{166b}$$

where $d\mathbf{p}_e(Q')$ is the value of $d\mathbf{p}_e(\ell')$ at the base of the monopole, i.e., at $\ell' = 0$ or at Q'. Since $d\mathbf{p}_e(\ell') = I(\ell')d\ell'$, it follows that $d\mathbf{p}_e(Q') = I(Q')d\ell'$, and $I(Q')$ is the current at the base of the monopole. The fields $d\mathbf{E}_e(Q|Q')$ and $d\mathbf{H}_e(Q|Q')$ are directly proportional to $d\mathbf{p}_e(Q')$, and the distance t_0 at which (166a) and (166b) may be used is given by $t_0 \cong \varrho_g(Q')\cos^{-1}\{\varrho_g(Q')/[\varrho_g(Q') + h]\}$. Clearly t_0 can be made small only if h is sufficiently small. Even though the $\mathbf{E}_e$ and $\mathbf{H}_e$ in

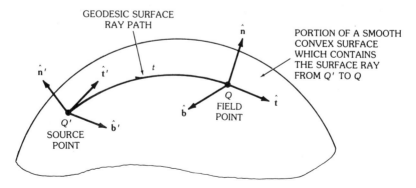

Fig. 54. Surface-ray coordinates. (*After Pathak and Wang [39], © 1981 IEEE*)

(166a) and (166b) are valid for $t > t_0$, the $d\mathbf{E}_e(Q|Q')$ and $d\mathbf{H}_e(Q|Q')$ due to $d\mathbf{p}_e(Q')$ on the surface are valid even for $t \to 0$ (i.e., $Q \to Q'$).

Explicit expressions for the surface fields $d\mathbf{H}_m$ and $d\mathbf{E}_m$ due to a source on an arbitrary convex surface are as follows [39]:

$$d\mathbf{H}_m(Q|Q') = \frac{-jk}{4\pi} d\mathbf{p}_m(Q') \bigg(2Y_0 \bigg\{ \hat{\mathbf{b}}'\hat{\mathbf{b}} \bigg[\bigg(1 - \frac{j}{kt}\bigg) \tilde{V}(\xi)$$
$$+ D^2 \bigg(\frac{j}{kt}\bigg)^2 (\Lambda_s \tilde{U}(\xi) + \Lambda_c \tilde{V}(\xi)) + \tilde{T}_0^2 \frac{j}{kt}(\tilde{U}(\xi) - \tilde{V}(\xi)) \bigg]$$
$$+ \hat{\mathbf{t}}'\hat{\mathbf{t}} \bigg[D^2 \frac{j}{kt} \tilde{V}(\xi) + \frac{j}{kt} \tilde{U}(\xi) - 2\bigg(\frac{j}{kt}\bigg)^2 (\Lambda_s \tilde{U}(\xi) + \Lambda_c \tilde{V}(\xi)) \bigg]$$
$$+ (\hat{\mathbf{t}}'\hat{\mathbf{b}} + \hat{\mathbf{b}}'\hat{\mathbf{t}}) \bigg[\frac{j}{kt} \tilde{T}_0(\tilde{U}(\xi) - \tilde{V}(\xi)) \bigg] \bigg\} \bigg) D\, G_0(kt) \quad (167)$$

and

$$d\mathbf{E}_m(Q|Q') = \frac{-jk}{4\pi} d\mathbf{p}_m(Q') \bigg(2\bigg\{ \hat{\mathbf{b}}'\hat{\mathbf{n}} \bigg[\bigg(1 - \frac{j}{kt}\bigg) \tilde{V}(\xi) + \tilde{T}_0^2 \frac{j}{kt}(\tilde{U}(\xi) - \tilde{V}(\xi)) \bigg]$$
$$+ \hat{\mathbf{t}}'\hat{\mathbf{n}} \bigg[\tilde{T}_0 \frac{j}{kt}(\tilde{U}(\xi) - \tilde{V}(\xi)) \bigg] \bigg\} \bigg) D\, G_0(kt) \quad (168)$$

Similarly, the surface fields $d\mathbf{H}_e$ and $d\mathbf{E}_e$ due to $d\mathbf{p}_e$ on an arbitrary convex surface are given by [39]:

$$d\mathbf{E}_e(Q|Q') = \frac{-jk}{4\pi} d\mathbf{p}_e(Q') \bigg\{ 2Z_0 \hat{\mathbf{n}}'\hat{\mathbf{n}} \bigg[\tilde{V}(\xi) - \frac{j}{kt} \tilde{V}(\xi)$$
$$+ \bigg(\frac{j}{kt}\bigg)^2 (\Lambda_s \tilde{V}(\xi) + \Lambda_c \tilde{U}(\xi)) + \tilde{T}_0^2 \frac{j}{kt}(\tilde{U}(\xi) - \tilde{V}(\xi)) \bigg] \bigg\} D\, G_0(kt) \quad (169)$$

and

$$d\mathbf{H}_e(Q|Q') = \frac{-jk}{4\pi} d\mathbf{p}_e(Q') \bigg(2\bigg\{ \hat{\mathbf{n}}'\hat{\mathbf{b}} \bigg[\bigg(1 - \frac{j}{kt}\bigg) \tilde{V}(\xi) + \tilde{T}_0^2 \frac{j}{kt}(\tilde{U}(\xi) - \tilde{V}(\xi)) \bigg]$$
$$+ \hat{\mathbf{n}}'\hat{\mathbf{t}} \bigg[\tilde{T}_0 \frac{j}{kt}(\tilde{U}(\xi) - \tilde{V}(\xi)) \bigg] \bigg\} \bigg) D\, G_0(kt) \quad (170)$$

Here,

$$G_0(kt) = \frac{e^{-jkt}}{t} \quad (171)$$

and ξ is as defined previously in (113) together with (114). Furthermore,

$$D = \sqrt{t \, d\psi_0/d\eta(Q)} \qquad (172a)$$

$$d\eta(Q) = \varrho^d \, d\psi \qquad (172b)$$

In Fig. 55 a pair of infinitesimally separated surface rays adjacent to the central ray from Q' at Q is shown; these adjacent rays constitute a surface-ray strip (or a surface-ray tube), and $d\psi_0$ in (172a) then refers to the angle between these adjacent surface rays at the source point Q'. On the other hand, $d\psi$ is the angle between the backward tangents to the same pair of adjacent surface rays at the field point Q. The distance ϱ^d between Q and Q_c is the geodesic circle at Q. Thus, $d\eta(Q)$ in (172b) denotes the width of the surface-ray strip at Q as discussed previously in the radiation configuration (see Table 3 and Fig. 42). In the case of a sphere of radius a, one obtains the following simplifications: $\xi = mt/a$, with $m = (ka/2)^{1/3}$, and $D = \sqrt{(t/a)/\sin(t/a)}$. Also, $\varrho^d = a\tan(t/a)$ for a sphere. For a circular cylinder of radius a, one obtains $\xi = mt/\varrho_g$ along a helical geodesic surface-ray path with $m = (k\varrho_g/2)^{1/3}$ and $\varrho_g = a/\sin^2\alpha'$, with α' constant along a given helical geodesic path on a convex cylinder.* Also, $D = 1$, and $T_0 = T\varrho_g$ in which $T = (\sin 2\alpha')/2a$ for a circular cylinder. The generalized torsion factor $\tilde{T}_0$ for the arbitrary convex surface is given by

$$\tilde{T}_0 = \mp |\sqrt{T_0(Q')T_0(Q)}| \qquad (173)$$

where the negative sign in (173) is chosen if either $T_0(Q') < 0$ or $T_0(Q) < 0$; otherwise, the positive sign is chosen. Note that for a general convex surface, one employs

$$T_0(Q') = T(Q')\varrho_g(Q') \qquad (174a)$$

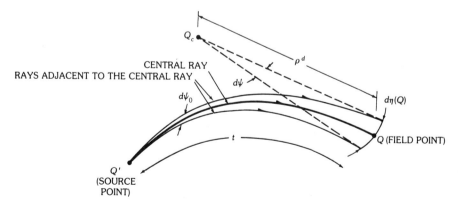

Fig. 55. Surface-ray strip (or tube). (*After Pathak and Wang [39], © 1981 IEEE*)

---

*Here α' could be either α_1' or α_2', as in Fig. 45a.

$$T(Q') = \frac{\sin 2\alpha'}{2} \left[\frac{1}{R_2(Q')} - \frac{1}{R_1(Q')} \right] \tag{174b}$$

and

$$\frac{1}{\varrho_g(Q')} = \frac{\cos^2\alpha'}{R_1(Q')} + \frac{\sin^2\alpha'}{R_2(Q')} \tag{174c}$$

as in Table 2, where α' is shown in Fig. 40, and $R_1(Q') \geq R_2(Q')$ as in the radiation problem. Note that $T_0(Q)$ is given by (174a)–(174c) on replacing Q' by Q therein. A reasonable choice for Λ_s and Λ_c is [39]

$$\Lambda_s = \left| \sqrt{\frac{R_2(Q')}{R_1(Q')} \frac{R_2(Q)}{R_1(Q)}} \right| \tag{175a}$$

$$\Lambda_c = 1 - \Lambda_s \tag{175b}$$

in order to interpolate smoothly between the higher-order terms in the canonical circular cylinder and sphere solutions. The above choice of Λ_s and Λ_c is also essential for recovering the field solution for the planar surface case when $R_{1,2} \to \infty$. The generalized Fock integrals $\tilde{U}(\xi)$ and $\tilde{V}(\xi)$ for the surface fields on an arbitrary convex surface which are present in (167)–(170) are given by

$$\tilde{U}(\xi) = \left[\frac{kt}{2m(Q')m(Q)\xi} \right]^{3/2} U(\xi) \tag{176a}$$

$$\tilde{V}(\xi) = \left[\frac{kt}{2m(Q')m(Q)\xi} \right]^{1/2} V(\xi) \tag{176b}$$

in which

$$U(\xi) \equiv \frac{\xi^{3/2} e^{j3\pi/4}}{\sqrt{\pi}} \int_{\infty \exp(-j2\pi/3)}^{\infty} \frac{W'_2(\tau)}{W_2(\tau)} e^{-j\xi\tau} d\tau \tag{177}$$

and

$$V(\xi) \equiv \frac{\xi^{1/2} e^{j\pi/4}}{2\sqrt{\pi}} \int_{\infty \exp(-j2\pi/3)}^{\infty} \frac{W_2(\tau)}{W'_2(\tau)} e^{-j\xi\tau} d\tau \tag{178}$$

The above functions $U(\xi)$ and $V(\xi)$ are tabulated in [40]; furthermore, they can be easily computed for large ξ in terms of a rapidly converging residue series as follows [40]:

$$V(\xi) = e^{-j\pi/4} \sqrt{\pi} \xi^{1/2} \sum_{n=1}^{\infty} (\tau'_n)^{-1} e^{-j\xi\tau'_n} \tag{179}$$

Techniques for High-Frequency Problems

$$U(\xi) = e^{j\pi/4} 2\sqrt{\pi}\, \xi^{3/2} \sum_{n=1}^{\infty} e^{-j\xi\tau_n} \tag{180}$$

$$V_1(\xi) = e^{j\pi/4} 2\sqrt{\pi}\, \xi^{3/2} \sum_{n=1}^{\infty} e^{-j\xi\tau'_n} \tag{181}$$

$$V'(\xi) = \frac{1}{2} e^{-j\pi/4} \sqrt{\pi}\, \xi^{-1/2} \sum_{n=1}^{\infty} (1 - j2\xi\tau'_n)(\tau'_n)^{-1} e^{-j\xi\tau'_n} \tag{182}$$

$$U'(\xi) = e^{j\pi/4} 3\sqrt{\pi}\, \xi^{1/2} \sum_{n=1}^{\infty} (1 - j2\xi\tau_n/3) e^{-j\xi\tau_n} \tag{183}$$

where τ_n and τ'_n are zeros of $W_2(\tau)$ and $W'_2(\tau)$, respectively; they are tabulated in Chart 2. Note that the sums in (179)–(183) can be truncated at $n = 10$ without losing accuracy. The additional functions $U'(\xi)$, $V'(\xi)$, and $V_1(\xi)$ in (183), (182), and (181) are defined by $U'(\xi) = dU/d\xi$, $V'(\xi) = dV(\xi)/d\xi$, and

$$V_1(\xi) = \frac{e^{j3\pi/4}\, \xi^{3/2}}{\sqrt{\pi}} \int_{\infty \exp(-j2\pi/3)}^{\infty} \frac{W_2(\tau)}{W'_2(\tau)} \tau e^{-j\xi\tau}\, d\tau$$

If ξ is small and positive, one may employ a small-argument asymptotic expansion for the Fock functions as follows [40]:

$$V(\xi) \sim 1 - \frac{\sqrt{\pi}}{4} e^{j\pi/4}\, \xi^{3/2} + \frac{7j}{60} \xi^3 + \frac{7\sqrt{\pi}}{512} e^{-j\pi/4}\, \xi^{9/2} - \cdots \tag{184}$$

$$U(\xi) \sim 1 - \frac{\sqrt{\pi}}{2} e^{j\pi/4}\, \xi^{3/2} + \frac{5j}{12} \xi^3 + \frac{5\sqrt{\pi}}{64} e^{-j\pi/4}\, \xi^{9/2} - \cdots \tag{185}$$

$$V_1(\xi) \sim 1 + \frac{\sqrt{\pi}}{2} e^{j\pi/4}\, \xi^{3/2} - \frac{7j}{12} \xi^3 - \frac{7\sqrt{\pi}}{64} e^{-j\pi/4}\, \xi^{9/2} + \cdots \tag{186}$$

Chart 2. Zeros of $W_2(\tau)$ and $W'_2(\tau)$

	$\tau_n = \|\tau_n\|\, e^{-j\pi/3}$	$\tau'_n = \|\tau'_n\|\, e^{-j\pi/3}$
n	$\|\tau_n\|$	$\|\tau'_n\|$
1	2.338 11	1.018 79
2	4.087 95	3.248 19
3	5.520 56	4.820 10
4	6.786 61	6.163 31
5	7.944 13	7.372 18
6	9.022 65	8.488 49
7	10.040 2	9.535 45
8	11.008 5	10.527 7
9	11.930 0	11.475 1
10	12.828 8	12.384 8

$$V'(\xi) \sim \frac{3\sqrt{\pi}}{8} e^{-j3\pi/4} \xi^{1/2} + \frac{7j}{20} \xi^2 + \frac{63\sqrt{\pi}}{1024} e^{-j\pi/4} \xi^{7/2} - \ldots \quad (187)$$

$$U'(\xi) \sim \frac{3}{4} \sqrt{\pi} e^{-j3\pi/4} \xi^{1/2} + \frac{5j}{4} \xi^2 + \frac{45\sqrt{\pi}}{128} e^{-j\pi/4} \xi^{7/2} - \ldots \quad (188)$$

For $\xi \geq \xi_0$ the residue series representation with the first ten terms in the summation may be used. For $\xi \leq \xi_0$ the small-argument asymptotic expression with the first three terms may be used. Here, ξ_0 is taken to be 0.6, as shown in [40].

The above results in (167)–(170), which are available from [39], can be modified for even greater accuracy in the paraxial (near axial) regions of quasi-cylindrical (or elongated) convex surfaces by adding higher-order terms in m^{-1}. In particular, these higher-order terms to be added to (167), (168), (169), and (170) are

$$\frac{-jk}{4\pi} d\mathbf{p}_m \cdot [2Y_0 \hat{\mathbf{b}}' \hat{\mathbf{b}} C] DG_0(kt)$$

$$\frac{-jk}{4\pi} d\mathbf{p}_m \cdot [2\hat{\mathbf{b}}' \hat{\mathbf{n}} C] DG_0(kt)$$

$$\frac{-jk}{4\pi} d\mathbf{p}_e \cdot [2Z_0 \hat{\mathbf{n}}' \hat{\mathbf{n}} C] DG_0(kt)$$

and

$$\frac{-jk}{4\pi} d\mathbf{p}_e \cdot [2Y_0 \hat{\mathbf{n}}' \hat{\mathbf{b}} C] DG_0(kt)$$

respectively, in which

$$C = j(4/3)(\sqrt{2} k \tilde{\varrho}_g)^{-2/3} (\tilde{\varrho}_g/\tilde{\varrho}_b) \tilde{\tau} V'(\xi)$$

with $V'(\xi) = dV/d\xi$, and with $\tilde{\tau}$ and $\tilde{\varrho}_g$ as defined later in (189b) and (189c).

The expressions given above for the surface fields of infinitesimal magnetic and electric current moments on an arbitrary convex surface are employed to calculate the mutual coupling between a pair of antennas on such a surface. Before proceeding to an illustration of the numerical results based on these calculations, it is useful to also give an alternative expression for $d\mathbf{H}_m(Q|Q')$ developed in [34] as compared with the one in (167) developed in [39], namely

$$d\mathbf{H}_m(Q|Q') = \frac{k^2 Y_0}{2\pi j} d\mathbf{p}_m(Q') \cdot \left[\hat{\mathbf{b}}' \hat{\mathbf{b}} \left\{ (1 - j/kt) \tilde{V}(\xi) - (1/kt)^2 \tilde{U}(\xi) \right. \right.$$
$$\left. + j(\sqrt{2} k \tilde{\varrho}_g)^{-2/3} \left[\tilde{\tau} V'(\xi) + \frac{\tilde{\varrho}_g}{\tilde{\varrho}_b} \tilde{\tau}^3 U'(\xi) \right] \right\} + \hat{\mathbf{t}}' \hat{\mathbf{t}} (j/kt)[\tilde{V}(\xi)$$
$$\left. + (1 - 2j/kt) \tilde{U}(\xi) + j(\sqrt{2} k \tilde{\varrho}_g)^{-2/3} \tilde{\tau}^3 U'(\xi)] \right] DG_0(t) \quad (189)$$

with

$$\tilde{\tau} = \left[\frac{kt}{2m(Q')m(Q)\xi}\right]^{1/2} \quad (189b)$$

$$\tilde{\varrho}_g = [\varrho_g(Q')\varrho_g(Q)]^{1/2} \quad (189c)$$

$$\tilde{\varrho}_b = [\varrho_b(Q')\varrho_b(Q)]^{1/2} \quad (189d)$$

Here ϱ_g and ϱ_b are the surface radii of curvature in the $\hat{\mathbf{t}}$ and $\hat{\mathbf{b}}$ directions as usual, which are associated with the surface-ray path. Alternative expressions for $d\mathbf{E}_m$, $d\mathbf{H}_e$, and $d\mathbf{E}_e$ are not available in [34]. It is noted that these alternative expressions for $d\mathbf{H}_m$ in (167) and (189) yield almost the same accuracy.

Consider a pair of rectangular slots in a smooth, perfectly conducting, convex surface. Let the electric field in the aperture of the first slot antenna be $\mathbf{E}_1^a$; likewise, let $\mathbf{E}_2^a$ denote the aperture distribution in the second slot antenna. If the slots are sufficiently short and thin, the dominant mode approximation may be employed for $\mathbf{E}_1^a$ and $\mathbf{E}_2^a$. Thus,

$$\mathbf{E}_1^a = V_{11}\mathbf{e}_1, \qquad \mathbf{E}_2^a = V_{22}\mathbf{e}_2 \quad (190)$$

where V_{11} and V_{22} are the modal voltages associated with the dominant electric vector mode functions $\mathbf{e}_1$ and $\mathbf{e}_2$ [41]. If the two slots have identical dimensions, then $\mathbf{e}_1 = \mathbf{e}_2$. The expression for the mutual admittance Y_{21} between the two slots with the first slot transmitting and the second slot short-circuited is [42]

$$Y_{21} = -\iint_{S_2}\iint_{S_1} d\mathbf{H}_m(Q|Q') \cdot d\mathbf{p}_m(Q)/V_{11}V_{22} \quad (191)$$

where Q' is any point in the aperture of the first slot, and Q is any point in the aperture of the second slot which is short-circuited. As before, $d\mathbf{H}_m(Q|Q')$ is the surface-magnetic field at Q due to the equivalent source $d\mathbf{p}_m(Q')$ at Q'; furthermore,

$$d\mathbf{p}_m(Q') = \mathbf{E}_1^a(Q') \times \hat{\mathbf{n}}' \, dS_1 \quad \text{and} \quad d\mathbf{p}_m(Q) = \mathbf{E}_2^a(Q) \times \hat{\mathbf{n}} \, dS_2$$

The double integrals in (191) are evaluated over the surface areas S_1 and S_2 of the two apertures. One notes that $Y_{21} = Y_{12}$. The result in (191) is also applicable to a pair of slots in an array environment provided that the field distribution in each of the slot array elements can be approximated by the dominant mode as in (190) and the array elements are not too closely spaced. Equation (191) is employed here to calculate the isolation S_{12} (which can be expressed in terms of Y_{12}) between a pair of rectangular slots in a perfectly conducting cone; the developed cone is shown in Fig. 56 and the results of the calculations are shown in Figs. 57a and 57b. The results for $d\mathbf{H}_m(Q|Q')$ based on (167) are employed in (191) to obtain the curves designated by OSU in Figs. 57a and 57b, whereas those based on (189) are

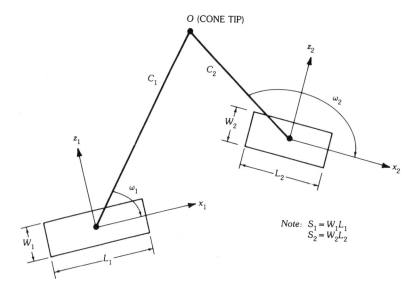

Fig. 56. Two rectangular slots on a developed cone.

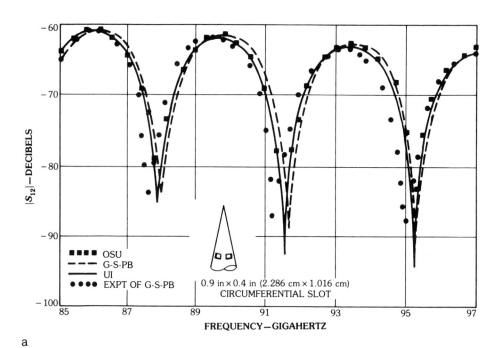

a

Fig. 57. Coupling coefficient S_{12} between two circumferential slots on a cone versus frequency. (*After Pathak and Wang [39]*) (a) Radial separation between slots is $|C_1 - C_2| = |45.53 - 45.53| = 0$ cm, angular separation is $\phi_0 = 60.8°$, and cone half-angle is $\theta_0 = 12.2°$. (b) Radial separation between slots is $|C_1 - C_2| = |27.03 - 25.88| = 1.15$ cm, angular separation is $\phi_0 = 80°$, and cone half-angle is $\theta_0 = 11°$. (*After Pathak and Wang [39]*, © 1981 IEEE)

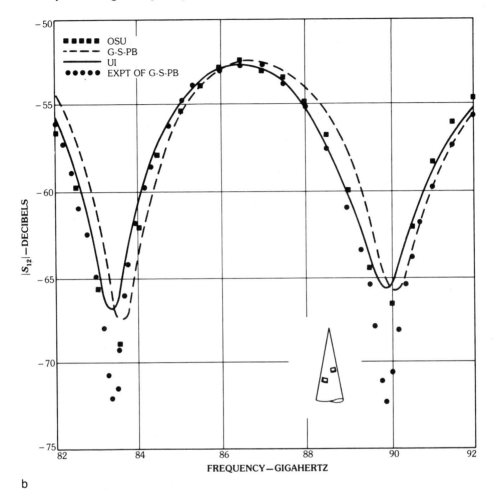

Fig. 57, *continued*

designated by UI in these figures. In addition, the OSU and UI solutions are compared in those figures with the measured results and also with calculations based on an equivalent cylinder model, both of which were obtained by Golden, Stewart, and Pridmore-Brown [43]. The latter measured results are designated by EXPT of G-S-PB, and the approximate cylinder model–based calculations are designated by G-S-PB in these figures. The diffraction by the cone tip has been included in the calculations designated as OSU and UI, respectively. This tip-diffraction contribution is essential for obtaining the interference pattern which is present in these figures, and it has been calculated via the interpolation formula given in [34] pertaining to the cone tip diffraction contribution available from [43]. In particular, $Y_{12}^{\text{total}} = Y_{12}$ [of (190)] $+ Y_{12}^{\text{tip}}$, for the tipped cone geometry, where [34, 43]

$$Y_{12}^{\text{tip}} = \tilde{Y} \sin \omega_1 \sin \omega_2 \tag{192}$$

and

$$\tilde{\Upsilon} = \sigma_0 \frac{(S_1 S_2)^{1/2}}{30\pi^4 C_1 C_2 \sin\theta_0} \left(\frac{\tan\theta_0}{2\pi}\right)^{1/2} \cdot \frac{\sin(kW_1/2)\sin(kW_2/2)}{(kW_1/2)(kW_2/2)} e^{j(\pi/4 - kC_1 - kC_2)} \quad (193)$$

in which S_1 and S_2 are the areas of the rectangular slots, and the other dimensions are illustrated in Fig. 56. The quantity σ_0 in (193) is given by [34, 43]

$$\sigma_0 = A e^{jB} \quad (194)$$

where

$$A = 1.3057 \theta_0^{-1} - 1.755 + 2.772\, \theta_0 - 1.459\, \theta_0^2 \quad (195)$$

and

$$B = 2.7195 + 1.4608\, \theta_0 - 1.1295\, \theta_0^2 + 0.6566\, \theta_0^3 \quad (196)$$

in which θ_0 = half-cone angle.

The mutual impedance Z_{21} between a pair of thin, short monopoles on a convex conducting surface may be calculated as follows. The transmitting monopoles are assumed to be sufficiently short and thin such that the currents $\mathbf{I}_1$ and $\mathbf{I}_2$ on these monopoles of length h_1 and h_2 may be approximated by the dominant (sinusoidal) mode currents ($\mathbf{i}_1$ and $\mathbf{i}_2$, respectively). Thus,

$$\mathbf{I}_1 = I_{11} \mathbf{i}_1 \quad (197a)$$

$$\mathbf{I}_2 = I_{22} \mathbf{i}_2 \quad (197b)$$

where I_{11} and I_{22} are the monopole base currents. Let the first monopole be $\hat{\mathbf{n}}'$-directed on the surface Q'; likewise, let the second monopole be $\hat{\mathbf{n}}$-directed on the surface at Q. Then Z_{21} for the case when the first monopole is transmitting and the second one is open circuited is given by

$$Z_{21} = -\int_0^{h_2}\int_0^{h_1} d\mathbf{E}_e(Q|Q') \cdot d\mathbf{p}_e(Q)/I_{11} I_{22} \quad (198)$$

where $d\mathbf{E}_e(Q|Q')$ is the field at Q due to the equivalent source $d\mathbf{p}_e(Q')$ at Q' and

$$d\mathbf{p}_e(Q') = I_{11} i_1(Q') d\ell_1(Q') \hat{\mathbf{n}}' \quad (199a)$$

$$d\mathbf{p}_e(Q) = I_{22} i_2(Q) d\ell_2(Q) \hat{\mathbf{n}} \quad (199b)$$

in which $d\ell_1(Q')$ and $d\ell_2(Q)$ are the incremental lengths of the two monopoles at points Q' and Q, respectively. It is assumed that the open-circuited monopole at Q scatters negligibly if it is short.

UTD Analysis of the Mutual Coupling Between Antennas on a Smooth, Convex Surface (2-D Case)—The tangential surface magnetic field **H** at a point Q [or (x, y)] on a perfectly conducting convex cylinder due to a magnetic line source of strength $d\mathbf{M} = \hat{\mathbf{z}} dM_z + \hat{\mathbf{t}} dM_d$ [see (161) and (162)] which is located at Q' [or (x', y')] on the same surface is given in terms of the surface-ray coordinates as [38]

$$d\mathbf{H}(Q) \sim C_0 Y_0 d\mathbf{M} \cdot [\hat{\mathbf{b}}'\hat{\mathbf{b}} F_s + \hat{\mathbf{t}}'\hat{\mathbf{t}} G_s] e^{-jkt} \qquad (200)$$

in which $\hat{\mathbf{b}}' = \hat{\mathbf{b}}$ for the 2-D case, and t denotes the surface-ray distance from Q' to Q. The functions G_s and F_s for the TM_z (or dM_d excitation) and TE_z (or dM_z excitation) cases, respectively, describe the behavior of the surface field. These functions are related to the mutual coupling Fock functions, U and V, respectively, as follows:

$$F_s = \left(\frac{jk}{2}\right)^{1/2} \left\{ m(Q') \left[\frac{\varrho_g(Q)}{\varrho_g(Q')}\right]^{1/6} \right\}^{-1} 2\xi^{-1/2} e^{-j\pi/4} V(\xi) \qquad (201)$$

and

$$G_s = -\left(\frac{jk}{2}\right)^{1/2} \left\{ m(Q') \left[\frac{\varrho_g(Q)}{\varrho_g(Q')}\right]^{1/6} \right\}^{-3} \xi^{-3/2} e^{-j3\pi/4} U(\xi) \qquad (202)$$

in which $U(\xi)$ and $V(\xi)$ are defined earlier in (177) and (178). Near the source, $Q \to Q'$ (i.e., $\xi \to 0$ and $t \to 0$); in this case one may use the small-argument approximation for $U(\xi)$ and $V(\xi)$ as in (185) and (184) to obtain

$$\hat{\mathbf{b}} \cdot d\mathbf{H}(Q) \bigg|_{\xi\,\text{small}} \cong -\frac{kY_0 dM_z}{2} \sqrt{\frac{2j}{\pi k t}} e^{-jkt} \left(1 - e^{j\pi/4} \frac{\sqrt{\pi}}{2} \xi^{3/2} \right.$$
$$\left. + j\frac{7}{60}\xi^3 + \frac{7\sqrt{\pi}}{512} e^{-j\pi/4} \xi^{9/2} \ldots \right) \qquad (203)$$

and

$$\hat{\mathbf{t}} \cdot d\mathbf{H}(Q) \bigg|_{\xi\,\text{small}} \cong \frac{kY_0 dM_d}{2} \frac{j}{kt} \sqrt{\frac{2j}{\pi k t}} e^{-jkt} \left(1 - e^{j\pi/4} \frac{\sqrt{\pi}}{2} \xi^{3/2} \right.$$
$$\left. + j\frac{5}{12}\xi^3 + \frac{5\sqrt{\pi}}{64} e^{-j\pi/4} \xi^{9/2} - \ldots \right) \qquad (204)$$

A more accurate representation for $\xi \to 0$ (and $t \to 0$) is possible if one replace the large-argument asymptotic form of the Hankel function identified as $\sqrt{2j/\pi kt}\, e^{-jkt}$ in (203) by $H_0^{(2)}(kt)$, and likewise if one replaces the large-argument asymptotic form $(j/kt)\sqrt{2j/\pi kt}\, e^{-jkt}$ in (204) by its corresponding original function $H_1^{(2)}(kt)/kt$.

Such a heuristic modification allows one to obtain the proper singular behavior for $\hat{\mathbf{b}} \cdot d\mathbf{H}(Q)$ and $\hat{\mathbf{t}} \cdot d\mathbf{H}(Q)$ in the immediate neighborhood of the source at Q'.

Far from the source, where $\xi > 0.6$, the rapidly convergent residue series or the surface-ray modal expansion representation for V and U in (184) and (185) may be used to obtain

$$F_s \bigg|_{\xi \gg 0} \cong \sum_{n=1}^{10} L_n^h(Q') \exp\left[-\int_{Q'}^{Q} \alpha_n^h(t') dt'\right] A_n^h(Q) \tag{205}$$

and

$$G_s \bigg|_{\xi \gg 0} \cong \sum_{n=1}^{10} L_n^s(Q') \exp\left[-\int_{Q'}^{Q} \alpha_n^s(t') dt'\right] A_n^s(Q) \tag{206}$$

in which $L_n^{s,h}$ and $\alpha_n^{s,h}$ have been employed earlier in this section [see (143) and (144)], whereas $A_n^{s,h}$ are referred to as the *attachment coefficients* [37]. It can be shown that [37]

$$L_n^h = A_n^h \tag{207a}$$

$$L_n^s = -A_n^s \tag{207b}$$

The attachment coefficient transforms the boundary layer GTD surface-ray field to the true field on the surface. The above relationships in (207a) and (207b) are consistent with reciprocity. Incorporating (205) and (206) into (200) yields a GTD representation for $d\mathbf{H}(Q)$ which is valid outside the SSB transition region (see Fig. 28).

If an electric line monopole current source of strength $\hat{\mathbf{n}}' dI_m$ as in (159) and (160) is placed at Q', then the $\hat{\mathbf{n}}$-directed electric field of this source which is observed at Q may be expressed as [38]

$$d\mathbf{E}(Q) \sim C_0 Z_0 (dI_m \hat{\mathbf{n}}') \cdot [\hat{\mathbf{n}}' \hat{\mathbf{n}} F_s] e^{-jkt} \tag{208}$$

5. The Equivalent Current Method

The equivalent current method (ECM) is primarily useful for evaluating the fields at and near the caustics of diffracted rays since the GTD and its uniform versions (UTD, UAT, etc., which remain valid within the GO shadow boundary transition regions) predict infinite fields at these caustics and are therefore not valid there. It is important, however, to note at the outset that the ECM which employs GTD to calculate its equivalent sources (or currents) is a valid procedure for evaluating the fields only if the caustic region is not close to the GO shadow boundary transition regions; otherwise, one may have to resort to the use of PO or its refined version, the PTD. Concepts based on the ECM are also useful for correcting the PO solution in a simple manner via a slight modification of the original PTD ansatz, as will be described in the next section, which deals with the

PTD. In addition, an approach based on the ECM is useful in analyzing many other special problems, such as in estimating the modal reflection coefficients of an open-ended waveguide, in the formulation of the UTD solution for edge-excited surface rays, and in formulating a UTD solution to the problem of the diffraction by a vertex which is given under "UTD for Vertices" in Section 4. A detailed description of the ECM may be found in [7, 8, 9].

Even though the GTD fails at and near the caustics of diffracted rays, it can be employed far from the caustics to calculate the necessary equivalent sources (or currents) of the ECM; these equivalent sources are then incorporated into the radiation integral [12] to estimate the fields at and near the caustics of diffracted rays. Outside the shadow boundary transition regions where this ECM is valid, the UTD and UAT reduce to the GTD. Thus the *ECM employs GTD* to find the equivalent sources which radiate the fields at the caustics of diffracted rays in a manner analogous to PO, which employs GO to find the source distribution of the scattered fields; in fact, the PO method is useful for calculating the fields at the caustics of GO rays. Also, the ECM requires an integration over the equivalent line sources (or currents) similar to the PO source integration. However, the ECM integration is over a line current, whereas the PO integral is over a surface current distribution for three-dimensional (3-D) problems. On the other hand, the ECM requires no integration for 2-D applications, whereas the PO surface integral reduces to a line integral in the 2-D case. It is important to note that away from the caustics an asymptotic evaluation of the integration in the ECM yields results which generally blend into the GTD solution. Since the ECM corrects the GTD only through the caustic regions of the diffracted rays, it is thus preferable to switch from the ECM solution to the GTD ray solution away from the diffracted-ray caustic regions, where the integration in the ECM becomes inefficient and unnecessary.

ECM for Edge-Diffracted Ray Caustic Field Analysis

Consider a curved edge of a general curved wedge structure which is illuminated by an incident electric field $\mathbf{E}^i$, as shown in Fig. 58a. Let $\mathbf{H}^i$ denote the incident magnetic field which is associated with $\mathbf{E}^i$. As mentioned previously, it is assumed in the GTD approximation that the curved wedge (containing the curved edge of Fig. 58a) can be modeled locally by a straight wedge whose faces are tangent to the surfaces forming the curved wedge at the point of diffraction, and whose edge is tangent to the curved edge at that point. Let $\hat{\mathbf{e}}$ denote the unit vector tangent to the edge at the point Q of diffraction as in Fig. 58a. The $\hat{\mathbf{e}}$-directed components of the GTD edge-diffracted electric and magnetic fields ($\mathbf{E}_k^d$ and $\mathbf{H}_k^d$) at P are given by (44), (45), (67), and (68) as

$$\hat{\mathbf{e}} \cdot \mathbf{E}_k^d(P) = [\hat{\mathbf{e}} \cdot \mathbf{E}^i(Q)] \, D_{es}^k(\phi, \phi'; \beta_0) \sqrt{\frac{\varrho_e}{s^d(\varrho_e + s^d)}} \, e^{-jks^d} \quad (209)$$

and

$$\hat{\mathbf{e}} \cdot \mathbf{H}_k^d(P) = [\hat{\mathbf{e}} \cdot \mathbf{H}^i(Q)] \, D_{eh}^k(\phi, \phi'; \beta_0) \sqrt{\frac{\varrho_e}{s^d(\varrho_e + s^d)}} \, e^{-jks^d} \quad (210)$$

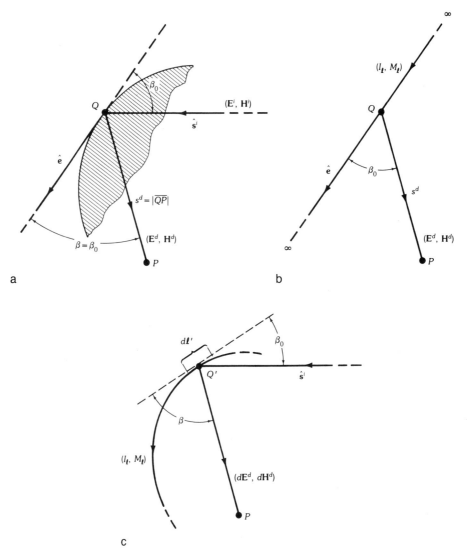

Fig. 58. Illustration of the equivalent edge current concept. (*a*) Diffraction by a curved wedge. (*b*) Equivalent line currents I_ℓ and M_ℓ at Q which generate edge-diffracted ray fields $(\mathbf{E}^d, \mathbf{H}^d)$ as in (*a*). (*c*) Contribution to P from an element $d\ell'$ at any point is Q' on the equivalent edge current.

In the case of a uniform wedge it can be proved readily that a knowledge of these two independent solutions $(\hat{\mathbf{e}} \cdot \mathbf{E}^d)$ and $(\mathbf{e} \cdot \mathbf{H}^d)$ is sufficient for obtaining all the remaining field components (which are transverse to $\hat{\mathbf{e}}$) [25]; this concept can be generalized to hold true asymptotically for the curved edge (in a curved wedge) via the principle of locality of high-frequency diffraction. If $s^d \ll \varrho_e$, even though s^d may be large in terms of the wavelength, then the two independent edge $(\hat{\mathbf{e}})$ directed field components in (209) and (210) can be approximated as

$$\hat{\mathbf{e}} \cdot \mathbf{E}_k^d(P) \cong [\hat{\mathbf{e}} \cdot \mathbf{E}^i(Q)] \, D_{es}^k \, \frac{e^{-jks^d}}{\sqrt{s^d}} \quad (211)$$

and

$$\hat{\mathbf{e}} \cdot \mathbf{H}_k^d(P) \cong [\hat{\mathbf{e}} \cdot \mathbf{H}^i(Q)] \, D_{eh}^k \, \frac{e^{-jks^d}}{\sqrt{s^d}} \quad (212)$$

Next, consider a traveling-wave electric or magnetic line current of strength I_ℓ or M_ℓ, respectively, which is located along the edge tangent ($\hat{\mathbf{e}}$) at Q as illustrated in Fig. 58b. Also, let I_ℓ and M_ℓ be of constant magnitude and let them possess a constant traveling-wave propagation factor equal to $k \cos \beta_0$ (in the $\hat{\mathbf{e}}$ direction), i.e., let $I_\ell = I_0 e^{-jk\ell\cos\beta_0}$ and $M_\ell = M_0 e^{-jk\ell\cos\beta_0}$, where the path ℓ is measured in the $\hat{\mathbf{e}}$ direction. Each of these traveling-wave currents radiates a conical wave of cone half-angle β_0. Furthermore, the field at P can be shown asymptotically to be associated with a ray from the point Q on the line source to the point P as shown in Fig. 58b. In particular, the line current strengths $I_\ell(Q)$ and $M_\ell(Q)$, at Q, generate the following $\hat{\mathbf{e}}$-directed components of the conical-wave electric and magnetic fields $\mathbf{E}^I(P)$ and $\mathbf{H}^M(P)$ at P, respectively:

$$\hat{\mathbf{e}} \cdot \mathbf{E}^I(P) \sim \left[\frac{-kZ_0 I_\ell(Q) e^{j\pi/4} \sin \beta_0}{\sqrt{8\pi k}} \right] \frac{e^{-jks^d}}{\sqrt{s^d}} \quad (213)$$

$$\hat{\mathbf{e}} \cdot \mathbf{H}^M(P) \sim \left[\frac{-kY_0 M_\ell(Q) e^{j\pi/4} \sin \beta_0}{\sqrt{8\pi k}} \right] \frac{e^{-jks^d}}{\sqrt{s^d}} \quad (214)$$

On comparing (211) and (212) with (213) and (214) outside the shadow boundary transition regions,* respectively, one obtains readily the strengths of the electric and magnetic line currents $I_\ell(Q)$ and $M_\ell(Q)$ which will generate the edge-diffracted fields $\mathbf{E}_k^d(P)$ and $\mathbf{H}_k^d(P)$ of (211) and (212); in particular,

$$I_\ell(Q) = \left(-\frac{1}{Z_0} \sqrt{\frac{8\pi}{k}} \, e^{-j\pi/4} \right) \frac{[\hat{\mathbf{e}} \cdot \mathbf{E}^i(Q)]}{\sin \beta_0} \, D_{es}^k(\phi, \phi'; \beta_0) \quad (215)$$

and

$$M_\ell(Q) = \left(-\frac{1}{Y_0} \sqrt{\frac{8\pi}{k}} \, e^{-j\pi/4} \right) \frac{[\hat{\mathbf{e}} \cdot \mathbf{H}^i(Q)]}{\sin \beta_0} \, D_{eh}^k(\phi, \phi'; \beta_0) \quad (216)$$

---

*The D_{es}^k and D_{eh}^k are not valid within the ISB and RSB (see Fig. 4a) transition regions; they must be replaced by the D_{es} and D_{eh} of the UTD. These D_{es} and D_{eh} are range (s^d) dependent because they contain Fresnel integrals which involve s^d; consequently, the edge-diffracted field within the shadow boundary transition regions is not a conical wave as in (211) and (212), and a comparison with (213) and (214) is therefore not possible within such regions.

It is evident from the above expressions that the currents I_ℓ and M_ℓ not only depend on the position (Q) on the edge, but they also depend on the angles of incidence (ϕ') and diffraction (ϕ) at the point of diffraction (Q). On the other hand, true currents do not depend on the angle of observation; consequently, the above aspect-dependent I_ℓ and M_ℓ are referred to as the strengths of "equivalent" currents which generate the edge-diffracted fields. It was mentioned earlier that a knowledge of $\hat{\mathbf{e}} \cdot \mathbf{E}_k^d$ and $\hat{\mathbf{e}} \cdot \mathbf{H}_k^d(P)$ is sufficient for generating all the remaining field components (which are then transverse to $\hat{\mathbf{e}}$) via Maxwell's equations. Similarly, when both the equivalent currents $I_\ell(Q)$ and $M_\ell(Q)$, of (215) and (216) above, are incorporated into the radiation integrals [12] for obtaining the fields radiated by these currents, then they also yield a complete description of the high-frequency edge-diffracted fields. It follows that one can employ both the equivalent edge currents I_ℓ and M_ℓ to radiate from every point Q on a curved edge as illustrated in Fig. 58c. Unlike the edge-diffracted fields as given by GTD or its uniform versions (UTD and UAT), the fields radiated by the "equivalent edge currents" [$I_\ell(Q)$ and $M_\ell(Q)$] remain valid even within the caustic regions of the edge-diffracted rays; such a procedure constitutes the ECM for edge-diffraction.

While the equivalent edge currents I_ℓ and M_ℓ radiate the edge-diffracted fields everywhere exterior to the curved wedge (via the radiation integral [12]), it is evident from (215) and (216) that these currents are defined only for observation points lying on the Keller edge-diffracted ray cone of half-angle β_0 as in Fig. 58b. In order to extend the above I_ℓ and M_ℓ to include observation points which are not restricted to lie on the Keller cone of edge-diffracted rays, one must introduce an additional angle (of diffraction), β, into the expressions in (215) and (216) for I_ℓ and M_ℓ. The angle β is shown in Fig. 58c.* It is noted that β and β_0 are in general different (except on the Keller edge-diffracted ray cone which is defined by $\beta = \beta_0$). A generalization of I_ℓ and M_ℓ to include information on β as well as β_0 can be performed heuristically, and in a manner consistent with reciprocity, by replacing the factor $\sin\beta_0$ in the previous definition of I_ℓ and M_ℓ with the new factor $\sqrt{\sin\beta_0 \sin\beta}$. With this change the previous I_ℓ and M_ℓ of (215) and (216) are replaced by the new edge currents $\tilde{I}_\ell$ and $\tilde{M}_\ell$, as follows:

$$\tilde{I}_\ell(Q) = \left(-\frac{1}{Z_0}\sqrt{\frac{8\pi}{k}}\, e^{-j\pi/4}\right) \frac{[\hat{\mathbf{e}} \cdot \mathbf{E}^i(Q)]}{\sqrt{\sin\beta_0 \sin\beta}} \tilde{D}_{es}^k(\phi, \phi'; \beta_0, \beta) \qquad (217)$$

and

$$\tilde{M}_\ell(Q) = \left(-\frac{1}{Y_0}\sqrt{\frac{8\pi}{k}}\, e^{-j\pi/4}\right) \frac{[\hat{\mathbf{e}} \cdot \mathbf{H}^i(Q)]}{\sqrt{\sin\beta_0 \sin\beta}} \tilde{D}_{eh}^k(\phi, \phi'; \beta_0, \beta) \qquad (218)$$

Furthermore,

$$\mathbf{H}^i(Q) = \hat{\mathbf{s}}^i \times Y_0 \mathbf{E}^i(Q)$$

---

*Note: $\beta = \beta_0$ if $Q' = Q$. Also, β_0 at Q may not be the same as β_0 at Q'. In general β does not have to equal β_0 for any point Q on the line current.

The above relation is true since $\mathbf{E}^i$ is assumed to be a ray optical field. It is observed that $D_{es,eh}^k$ in the old definitions of I_ℓ and M_ℓ contains the factor $(\sin\beta_0)^{-1}$ [see (67) and (68)] which must also be replaced by $(\sqrt{\sin\beta_0 \sin\beta})^{-1}$. Thus, one includes $\tilde{D}_{es,eh}^k$ in the new definitions for $\tilde{I}_\ell$ and $\tilde{M}_\ell$ of (217) and (218); the $\tilde{D}_{es,eh}^k$ contains $(\sqrt{\sin\beta_0 \sin\beta})^{-1}$ instead of just $(\sin\beta_0)^{-1}$ as in $D_{es,eh}^k$. In particular, $\tilde{D}_{es,eh}^k$ is simply related to $D_{es,eh}^k$ as follows:

$$\tilde{D}_{es,eh}^k(\phi,\phi';\beta_0,\beta) \equiv \frac{\sin\beta_0\, D_{es,eh}^k(\phi,\phi';\beta_0)}{\sqrt{\sin\beta_0 \sin\beta}} \tag{219}$$

The electric and magnetic fields $\mathbf{E}_c^d$ and $\mathbf{H}_c^d$ radiated by the equivalent edge currents of (217) and (218) are found via the radiation integrals [12]; thus,

$$\mathbf{E}_c^d(P) \approx \frac{jKZ_0}{4\pi} \oint_{\mathcal{L}} [\hat{\mathbf{R}} \times \hat{\mathbf{R}} \times \tilde{I}_\ell(\ell')\hat{e}' + Y_0 \hat{\mathbf{R}} \times \tilde{M}_\ell(\ell')\hat{e}'] \frac{e^{-jkR}}{R} d\ell' \tag{220}$$

and

$$\mathbf{H}_c^d(P) \approx \frac{-jk}{4\pi} \oint_{\mathcal{L}} [\hat{\mathbf{R}} \times \tilde{I}_\ell(\ell')\hat{e}' - Y_0 \hat{\mathbf{R}} \times \hat{\mathbf{R}} \times \tilde{M}_\ell(\ell')\hat{e}'] \frac{e^{-jkR}}{R} d\ell' \tag{221}$$

The integrations in (220) and (221) are over a closed path $\mathcal{L}$ formed by a line of diffraction as in Fig. 59, i.e., by a ring type of edge discontinuity on the surface of the scatterer from which the edge-diffracted rays are produced. If the contour $\mathcal{L}$ is not closed, then the end points of the path $\mathcal{L}$ will contribute to the integrals in (220) and (221); these end point contributions may, in some cases, result from incorrectly truncating the currents $\tilde{I}_\ell$ and $\tilde{M}_\ell$ on those portions of the edge which are shadowed by the surface of the scatterer for a certain range of aspects. In the latter case, however, the effect of the rays launched on the surface (i.e., surface rays) which

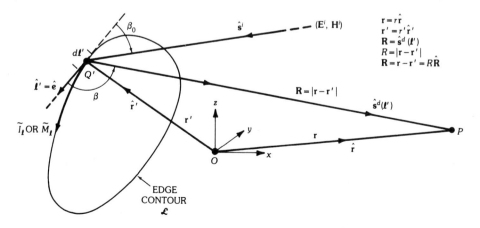

Fig. 59. Geometry associated with the radiation by the equivalent edge currents I_ℓ and M_ℓ on $\mathcal{L}$.

then undergo edge diffraction must also be taken into consideration so that $\bar{I}_\ell$ and $\bar{M}_\ell$ are not truncated, and spurious diffraction effects arising from such improper truncation can thereby be avoided. For a point P in the far zone of the scatterer one may employ the usual approximations in (220) and (221), namely,

$$\hat{\mathbf{R}} \cong \hat{\mathbf{r}}$$

$$\frac{e^{-jkR}}{R} \cong \frac{e^{-jkr}}{r} e^{+jk\hat{\mathbf{r}}\cdot\mathbf{r}'}$$

and

$$\mathbf{H}_c^d(P) \sim Y_0 \hat{\mathbf{r}} \times \mathbf{E}_c^d(P)$$

Here, $\hat{\mathbf{r}}$ denotes the radiation direction. The quantities $\hat{\mathbf{r}}$, $\hat{\mathbf{R}}$, $\hat{\ell}'$, β_0, and β are shown in Fig. 59. Far from the caustic directions, it can be shown that the dominant contribution to each of the integrals in (220) and (221) occurs from a few isolated stationary points corresponding to points of edge diffraction on $\mathscr{L}$ for which the law of edge diffraction ($\beta = \beta_0$) holds true, and in that case (220) and (221), in general, reduce to the GTD result given by

$$\mathbf{E}_c^d(P)\bigg|_{\substack{\text{far}\\\text{from}\\\text{caustics}}} \sim \sum_{p=1}^{N} \mathbf{E}^i(Q_p) \cdot \bar{\bar{\mathbf{D}}}_e^k(Q_p) \sqrt{\frac{\varrho_{ep}}{s_p^d(\varrho_{ep} + s_p^d)}} e^{-jks_p^d} \qquad (222)$$

and

$$\mathbf{H}_c^d(P)\bigg|_{\substack{\text{far}\\\text{from}\\\text{caustics}}} \sim Y_0 \sum_{p=1}^{N} \hat{\mathbf{s}}_p^d \times \left[\mathbf{E}^i(Q_p) \cdot \bar{\bar{\mathbf{D}}}_e^k(Q_p) \sqrt{\frac{\varrho_{ep}}{s_p^d(\varrho_{ep} + s_p^d)}} e^{-jks_p^d} \right] \qquad (223)$$

The subscript p in (222) and (223) denotes the pth edge diffraction point on $\mathscr{L}$ for which $\beta = \beta_0$; away from the caustic direction there are N such points ($p = 1, 2, 3, \ldots, N$), where N is a finite number (usually no larger than 4). Within the caustic region the entire ring $\mathscr{L}$ generally contributes to the integrals in (220) and (221); these integrals can be evaluated numerically (or analytically in some special cases). It is clear that the ECM integrals in (220) and (221) should be replaced by the GTD results of (222) and (223), respectively, when the observation point P moves sufficiently far from the caustic of the edge-diffracted rays where the numerical integration of (220) and (221) becomes inefficient and unnecessary.

6. The Physical Theory of Diffraction and Its Modifications

The physical theory of diffraction (PTD), which constitutes an extension of physical optics (PO), was developed originally by Ufimtsev [10, 11] for analyzing the high-frequency scattering from conducting surfaces. In particular, the PTD refines the PO field approximation just as the GTD refines the GO field approximation. In the PO technique the currents induced on the surface of a scatterer are approximated according to GO as described earlier in Section 3; however, it is clear that the GO approximation for the currents would be accurate only on the portion of the scatterer which is strongly illuminated by the source, whereas it would be totally inaccurate in the shadowed portion of a smooth convex surface, where, the GO yields a zero value for the surface current. The GO approximation for the surface current is also erroneous within the transition regions adjacent to the shadow boundary which divides a smooth convex surface into its illuminated (or lit) and shadowed portions; furthermore, it is expected to be inaccurate in the regions on the surface where a discontinuity (such as an edge) could exist. In the PTD approach, the GO current approximation is improved by including a correction which Ufimtsev refers to as a "nonuniform" component of the current. This nonuniform component of the current in the PTD formulation is supposed to include the effects not accounted for in the GO approximation to the surface current. In his original work Ufimtsev [10] considers the effects of the nonuniform component of the current only for conducting surfaces with edge type discontinuities; in addition, he neglects the effects of surface rays on smooth, convex bodies containing edges. Ufimtsev also does not give any expressions for the nonuniform component of the current nor does he explicitly integrate over these currents to obtain the fields radiated by them; instead, he obtains those fields via indirect considerations that are much simpler. The latter considerations involve an asymptotic high-frequency analysis of the canonical problem of the diffraction of an electromagnetic plane wave by a perfectly conducting wedge. As a result the fields of the nonuniform component of the current, as originally obtained by Ufimtsev, exhibit a ray optical character and hence become infinite at and near caustics, even though, unlike the GTD, they maintain their validity at and near the optical shadow boundaries except for grazing angles of incidence. Thus Ufimtsev introduced caustic matching functions in an *ad hoc* manner into the fields radiated by the nonuniform component of the current to correct for the infinite field behavior along the edge diffracted–ray caustic directions. That *ad hoc* procedure, which employs caustic matching functions, can be circumvented via an application of the ECM, which employs equivalent line currents deduced from Ufimtsev's expressions for the fields that are radiated far from the caustic by the nonuniform component of the current, rather than from GTD as done in Section 5; of course, these fields of nonuniform component of the current emanate from a line of discontinuity (such as an edge) on the surface of the scatterer. The latter modification of Ufimtsev's work based on the concept of the ECM was also suggested in [9], and it will be employed in the present development to modify the original version of the PTD.

The PTD field will be presented here as the superposition of the PO field and the field of the ECM which employs Ufimtsev's diffraction correction to PO instead of the GTD for deducing the equivalent line currents. In those situations where the PO and ECM type integrals occurring in this PTD approach for edged bodies can be evaluated asymptotically, one generally recovers the GTD solution; of course, if that asymptotic evaluation is performed in a uniform manner, then the corresponding uniform GTD (i.e., UTD or UAT) solution which remains valid even within the GO shadow boundary transition regions may be recovered from the

PTD. The GTD (and UTD or UAT) constitutes a far more efficient and physically appealing solution than the PTD, which in general requires an integration over the GO currents and also over the Ufimtsev-based equivalent line currents. Nevertheless, the PTD is very useful for estimating the fields in regions where there is a confluence of the transition regions associated with ray caustics and shadow boundaries, respectively; it is noted that the GTD, UTD, and UAT all fail in this special situation. The PTD also automatically remains valid in the neighborhood of the caustics of GO rays, since it contains the PO solution which provides a good estimate for the dominant fields near the caustics of GO rays associated with edge bodies. The PTD also remains valid in the neighborhood of the caustics of edge-diffracted rays. However, if the diffracted-ray caustic lies far from the GO shadow boundaries, then it is more efficient to employ the ECM of Section 5, which employs GTD based equivalent line currents, rather than to use the PTD (which involves the PO surface integral).

PTD for Edged Bodies

A Canonical Problem in the PTD Formulation—First, consider the 3-D canonical problem of the diffraction of an obliquely incident electromagnetic plane wave by a perfectly conducting wedge. A PO solution to this problem can be expressed formally in terms of the PO integral (over the GO currents) that is evaluated only on the illuminated faces of the wedge. Thus, the total PO electric field $\mathbf{E}^{PO}$ external to the wedge becomes

$$\mathbf{E}^{PO} = \mathbf{E}^i + \mathbf{E}^s_{PO} \tag{224}$$

as in (30), where $\mathbf{E}^i$ is the incident field in the absence of the wedge, and the scattered field $\mathbf{E}^s_{PO}$ is given by the integral in (30). For the present it is assumed that the observation point lies outside the GO shadow boundary transition regions. An asymptotic high-frequency evaluation of the integral for $\mathbf{E}^s_{PO}$ far from the GO shadow boundaries yields

$$\mathbf{E}^s_{PO} \sim -\mathbf{E}^i(1 - U_i) + \mathbf{E}^r U_r + \mathbf{E}^d_{PO} \tag{225}$$

Note that the step functions U_i and U_r have been defined previously under "GTD for Edges" in Section 4.

Combining (224) and (225) one obtains

$$\mathbf{E}^{PO} \sim \mathbf{E}^i U_i + \mathbf{E}^r U_r + \mathbf{E}^d_{PO} \tag{226}$$

where $\mathbf{E}^d_{PO}$ is viewed as the edge-diffracted field, as predicted by PO, at an observation point P; it can be expressed as in (57) by

$$\mathbf{E}^d_{PO}(P) = \mathbf{E}^i(Q) \cdot \bar{\bar{\mathbf{D}}}^{PO}_e(Q) \sqrt{\frac{\varrho_e}{s^d(\varrho_e + s^d)}} \, e^{-jks^d} \tag{227}$$

The PO dyadic edge-diffraction coefficient $\bar{\bar{\mathbf{D}}}^{PO}_e$ at the point of diffraction Q on the edge is of the form

Techniques for High-Frequency Problems

$$\bar{\bar{D}}_e^{PO} = -\hat{\beta}_0'\hat{\beta}_0 D_{es}^{PO} - \hat{\phi}'\hat{\phi} D_{eh}^{PO} \qquad (228)$$

which is similar to $\bar{\bar{D}}_e$ of (59c). The precise form of $D_{es,eh}^{PO}$ will be indicated shortly. The PO edge-diffracted field $\mathbf{E}_{PO}^d$ arises from the truncation of the GO current by the edge. On the other hand, a GTD solution to the same canonical problem yields [see (40) and (41) of Section 4]

$$\mathbf{E}_{GTD} \sim \mathbf{E}^i U_i + \mathbf{E}^r U_r + \mathbf{E}_k^d \qquad (229)$$

where

$$\mathbf{E}_k^d(P) = \mathbf{E}^i(Q) \cdot \bar{\bar{D}}_e^k(Q) \sqrt{\frac{\varrho_e}{s^d(\varrho_e + s^d)}} e^{-jks^d} \qquad (230)$$

$$\bar{\bar{D}}_e^k = -\hat{\beta}_0'\hat{\beta}_0 D_{es}^k - \hat{\phi}'\hat{\phi} D_{eh}^k \qquad (231)$$

The GTD result in (229) together with (230) and (231) constitutes the leading terms in the direct asymptotic approximation of the exact solution to this canonical problem [25, 26], whereas the corresponding asymptotic PO field expression in (226) together with (227) and (228) constitutes an asymptotic approximation of a PO based solution which in itself is approximate. Thus the diffraction correction to $\mathbf{E}^{PO}$ should now be readily evident from (226) and (229); it is denoted here by $\mathbf{E}_u^d$ and is simply obtained by subtracting (226) from (229). In particular,

$$\mathbf{E}_u^d = \mathbf{E}_{GTD} - \mathbf{E}^{PO} = \mathbf{E}_k^d - \mathbf{E}_{PO}^d \qquad (232)$$

or

$$\mathbf{E}_u^d(P) = \mathbf{E}^i(Q) \cdot \bar{\bar{D}}_e^u \sqrt{\frac{\varrho_e}{s^d(\varrho_e + s^d)}} e^{-jks^d} \qquad (233)$$

with

$$\bar{\bar{D}}_e^u = -\hat{\beta}_0'\hat{\beta}_0(D_{es}^k - D_{es}^{PO}) - \hat{\phi}'\hat{\phi}(D_{eh}^k - D_{eh}^{PO}) \qquad (234a)$$

or, more compactly,

$$\bar{\bar{D}}_e^u = \bar{\bar{D}}_e^k - \bar{\bar{D}}_e^{PO} \qquad (234b)$$

Ufimtsev's PTD ansatz is then essentially based on adding the required diffraction correction $\mathbf{E}_u^d$ to $\mathbf{E}^{PO}$ to arrive at an expression $\mathbf{E}_{PTD}$ for the electric field which constitutes a refinement over PO; thus

$$\mathbf{E}_{PTD} = \mathbf{E}^{PO} + \mathbf{E}_u^d \qquad (235a)$$

or

$$\mathbf{E}_{PTD} = (\mathbf{E}^i + \mathbf{E}_{PO}^s) + \mathbf{E}_u^d \qquad (235b)$$

It is important to note the $\mathbf{E}_{PO}^s$ of (235b) in the PTD formulation is left formally as an integral over the GO surface current approximation; this is in contrast to $\mathbf{E}_{GTD}$ of (229), which employs rays and requires no integration over the currents. Although the $\mathbf{E}_u^d$ above was developed initially for observation points sufficiently far from the GO shadow boundaries, it is easily verified that the singularities in the nonuniform edge-diffracted fields $\mathbf{E}_k^d$ and $\mathbf{E}_{PO}^d$, which occur at the GO shadow boundaries, exactly cancel each other in (232), thereby making $\mathbf{E}_u^d$ bounded, continuous, and valid at these boundaries (and their associated transition regions).

PTD for 3-D Edged Bodies—A PTD analysis of the problem of EM radiation from an antenna in the presence of perfectly conducting edge bodies, which are otherwise smooth and convex, can be developed directly from the canonical PTD wedge solution of (234) due to the local nature of $\mathbf{E}_u^d$. Thus the PTD electric field $\mathbf{E}_{PTD}$ for edged bodies can be expressed as in (235a) and (235b), namely,

$$\mathbf{E}_{PTD}(P) \sim \mathbf{E}^{PO}(P) + \mathbf{E}_u^d(P) \qquad (236a)$$

where

$$\mathbf{E}^{PO}(P) = \mathbf{E}^i(P) + \frac{jkZ_0}{4\pi} \iint_{S_{lit}} \hat{\mathbf{R}} \times \hat{\mathbf{R}} \times [2\hat{\mathbf{n}}' \times \mathbf{H}^i] \frac{e^{-jkR}}{R} dS' \qquad (236b)$$

as in (30), and

$$\mathbf{E}_u^d(P) = \sum_{p=1}^{N} \mathbf{E}^i(Q_p) \cdot \bar{\bar{\mathbf{D}}}_e^u(Q_p) \sqrt{\frac{\varrho_{ep}}{s_p^d(\varrho_{ep} + s_p^d)}} e^{-jks_p^d} \qquad (236c)$$

as in (233), except that the above expression in (236c) accounts for all the N edge-diffraction points which give rise to N diffracted rays that reach the point P. These ($p = 1, 2, \ldots, N$) points of edge diffraction obey Keller's law of edge diffraction so that the field $\mathbf{E}_u^d$ propagates along the same paths as the GTD edge-diffracted rays, i.e., the ray field $\mathbf{E}_u^d$ is associated with the Keller cone of edge-diffracted rays as in Figs. 1 and 12. It is important to note once again that $\mathbf{E}_u^d$ represents only a correction to the edge-diffracted field as predicted by PO. Since $\mathbf{E}_u^d$ is a ray optical type field, it becomes singular at the caustics of edge-diffracted rays; therefore a more useful form of $\mathbf{E}_u^d$ which circumvents this problem is available from (220) via the concept of the ECM. Thus, more generally, $\mathbf{E}_u^d$ is given by

$$\mathbf{E}_u^d \cong \frac{jkZ_0}{4\pi} \oint_L [\hat{\mathbf{R}} \times \hat{\mathbf{R}} \times \tilde{I}_\ell^u(\ell')\hat{\ell}' + Y_0 \hat{\mathbf{R}} \times \tilde{M}_\ell^u(\ell')\hat{\ell}'] \frac{e^{-jkR}}{R} d\ell' \qquad (237)$$

in which the equivalent Ufimtsev type electric and magnetic edge currents $\tilde{I}_\ell^u$ and $\tilde{M}_\ell^u$ on the edge contour are now defined as in (217) and (218), but with $\bar{D}_{es}^u$ and $\bar{D}_{eh}^u$ replacing $\bar{D}_{es}^k$ and $\bar{D}_{eh}^k$, respectively, in those equations. Hence,

$$\bar{I}_\ell^u(\ell') = \left(-\frac{1}{Z_0}\sqrt{\frac{8\pi}{k}}\, e^{-j\pi/4}\right) \frac{[\hat{e}\cdot\mathbf{E}^i(\ell')]}{\sqrt{\sin\beta_0\sin\beta}}\, \tilde{D}_{es}^u(\phi,\phi';\beta_0,\beta) \qquad (238)$$

and

$$\bar{M}_\ell^u(\ell') = \left(-\frac{1}{Y_0}\sqrt{\frac{8\pi}{k}}\, e^{-j\pi/4}\right) \frac{[\hat{e}\cdot\mathbf{H}^i(\ell')]}{\sqrt{\sin\beta_0\sin\beta}}\, \tilde{D}_{eh}^u(\phi,\phi';\beta_0,\beta) \qquad (239)$$

It is clear from (219) that $\tilde{D}_{es,eh}^u$ are related to $D_{es,eh}^u$ via

$$\tilde{D}_{es,eh}^u(\phi,\phi';\beta_0,\beta) \equiv \frac{\sin\beta_0\, D_{es,eh}^u(\phi,\phi';\beta_0)}{\sqrt{\sin\beta_0\sin\beta}} \qquad (240)$$

The contour $\mathscr{L}$ in (237) is shown to be closed; however, if it is nonclosed, then the discussion under (221) applies also to this case. The result in (237) remains valid along the diffracted-ray caustic directions even if these occur within GO shadow boundary transition regions, since $\bar{I}_\ell^u$ and $\bar{M}_\ell^u$ are valid there. On the other hand, recall that $\bar{I}_\ell$ and $\bar{M}_\ell$ of (217) and (218) are not valid near the GO shadow boundaries. Away from the caustic directions it is more efficient to replace (237) by its asymptotic approximation given previously in (236c).

It only remains to give explicit expressions for $D_{es}^u(\phi,\phi';\beta_0)$ and $D_{eh}^u(\phi,\phi';\beta_0)$, which occur in $\bar{\bar{D}}_e^u$ of (234a) and also in the definition of the equivalent Ufimtsev type edge currents of (238) and (239), to complete the PTD solution. From (234a) and (234b) it is clear that

$$D_{es,eh}^u(\phi,\phi';\beta_0) \equiv D_{es,eh}^k(\phi,\phi';\beta_0) - D_{es,eh}^{PO}(\phi,\phi';\beta_0) \qquad (241)$$

The $D_{es,eh}^{PO}$ is defined as follows:

$$D_{es,eh}^{PO}(\phi,\phi';\beta_0) = \begin{cases} -\dfrac{e^{-j\pi/4}}{2\sqrt{2\pi k}}\dfrac{1}{\sin\beta_0}\left[\tan\left(\dfrac{\phi-\phi'}{2}\right) \mp \tan\left(\dfrac{\phi+\phi'}{2}\right)\right] \\ \quad \text{if } 0 \leq \phi' \leq (\Omega-\pi) \text{ (or only } \phi=0 \text{ face is illuminated)} \\[6pt] -\dfrac{e^{-j\pi/4}}{2\sqrt{2\pi k}}\dfrac{1}{\sin\beta_0}\left[\mp\tan\left(\dfrac{\phi+\phi'}{2}\right) \mp \tan\left[\dfrac{2\Omega-(\phi+\phi')}{2}\right]\right] \\ \quad \text{if } (\Omega-\pi) \leq \phi' \leq \pi \text{ (or both faces are illuminated)} \\[6pt] -\dfrac{e^{-j\pi/4}}{2\sqrt{2\pi k}}\dfrac{1}{\sin\beta_0}\left[-\tan\left(\dfrac{\phi-\phi'}{2}\right) \mp \tan\left[\dfrac{2\Omega-(\phi+\phi')}{2}\right]\right] \\ \quad \text{if } \pi \leq \phi' \leq \Omega \text{ (or only } \phi=\Omega \text{ face is illuminated)} \end{cases} \qquad (242a)$$

with

$$\Omega = n\pi \qquad (242b)$$

(see Fig. 19). Note that $n = 2$ for a half-plane.

The $D_{es,eh}^k$ was defined earlier in (68). It is clear from (219), (240), and (234) that

$$\bar{\bar{D}}_{es,eh}^{PO}(\phi, \phi'; \beta_0, \beta) \equiv \frac{\sin \beta_0 \, D_{es,eh}^{PO}(\phi, \phi'; \beta_0)}{\sqrt{\sin \beta_0 \sin \beta}} \tag{243}$$

and

$$\bar{\bar{D}}_{es,eh}^u(\phi, \phi'; \beta_0, \beta) = \bar{\bar{D}}_{es,eh}^k(\phi, \phi'; \beta_0, \beta) - \bar{\bar{D}}_{es,eh}^{PO}(\phi, \phi'; \beta_0, \beta) \tag{244}$$

PTD for 2-D Edged Bodies—The PTD analysis of the problem of the radiation by a line source in the presence of a perfectly conducting 2-D structure with edges is similar to that in (236a), (236b), and (236c) except that (236b) becomes the following in the 2-D case [see (36)]:

$$\mathbf{E}^{PO}(P) = \mathbf{E}^i(P) + \frac{kZ_0}{4} \int_{C_{\text{lit}}} [\hat{\mathbf{R}} \times \hat{\mathbf{R}} \times (2\hat{\mathbf{n}}' \times \mathbf{H}^i)] H_0^{(2)}(kR) \, d\ell' \tag{245}$$

and (236c) reduces in the 2-D case to

$$\mathbf{E}_u^d(P) \cong \sum_{p=1}^N \mathbf{E}^i(Q_p) \cdot \bar{\bar{\mathbf{D}}}_e^u(Q_p) \frac{e^{-jks_p^d}}{\sqrt{s_p^d}} \tag{246}$$

It is noted that there is no diffracted-ray caustic in the 2-D case; hence there is no need for any integration as in (237) for calculating $\mathbf{E}_u^d$.

Results based on an application of the PTD for analyzing some simple antenna problems are illustrated in Figs. 60 through 63. Fig. 60 shows a comparison of the PTD and UTD based far-zone radiation pattern calculations for a 2-D parabolic reflector excited by a magnetic line source at the focus; furthermore, an independent formally exact moment method solution to this problem is also included in that figure for comparison. Outside the diffracted-ray caustic regions and exterior to the regions of confluence of diffracted-ray caustics with the GO shadow boundaries and/or GO ray caustics, a uniform asymptotic evaluation of the PO integral in the PTD solution generally allows the PTD solution to yield the leading terms of the UTD solution.

Finally, it may be mentioned that effects of slope edge diffraction can be incorporated in the equivalent currents of the ECM of Section 5; such a modification can also be similarly employed to include a slope diffraction correction to the $\mathbf{E}_u^d$ term of the PTD. It is noted that $\mathbf{E}^{PO}$ of the PTD inherently contains partial information on the slope diffraction effects just as it does on the ordinary edge diffraction [via $\mathbf{E}_{PO}^d$; see (226)]. Since these slope diffraction effects are not treated in this section dealing with the PTD, the results of this section are therefore valid if those effects are negligible; otherwise, their effects must be included.

Acknowledgments

The author wishes to thank Dr. Ayhan Altintas of the Ohio State University ElectroScience Laboratory for his review of this chapter and for his helpful comments.

The work in this chapter was supported in part by Contract No. ONR N00014-78-C-0049, between the Office of Naval Research and the Ohio State University Research Foundation.

7. References

[1] J. B. Keller, "Geometrical theory of diffraction," *J. Opt. Soc. Am.*, vol. 52, pp. 116–130, 1962.

[2] J. B. Keller, "A geometrical theory of diffraction," in *Calculus of Variations and Its Applications*, ed. by L. M. Graves, New York: McGraw-Hill Book Co., 1958, pp. 27–52.

[3] B. R. Levy and J. B. Keller, "Diffraction by a smooth object," *Commun. Pure Appl. Math.*, vol. 12, pp. 159–209, 1959.

[4] R. G. Kouyumjian, "The geometrical theory of diffraction and its applications," in *Numerical and Asymptotic Techniques in Electromagnetics*, ed. by R. Mittra, New York: Springer-Verlag, 1975.

[5] R. G. Kouyoumjian, P. H. Pathak, and W. D. Burnside, "A uniform GTD for the diffraction by edges, vertices, and convex surfaces," in *Theoretical Methods for Determining the Interaction of Electromagnetic Waves with Structures*, ed. by J. K. Skwirzynski, Amsterdam, the Netherlands: Sijthoff and Noordhoff, 1981.

[6] S. W. Lee and G. A. Deschamps, "A uniform asymptotic theory of EM diffraction by a curved wedge," *IEEE Trans. Antennas Propag.*, vol. AP-24, pp. 25–34, January 1976. Also see D. S. Ahluwalia, R. M. Lewis, and J. Boersma, "Uniform asymptotic theory of diffraction by a plane screen," *SIAM J. Appl. Math.*, vol. 16, pp. 783–807, 1968.

[7] C. E. Ryan, Jr., and L. Peters, Jr., "Evaluation of edge diffracted fields including equivalent currents for caustic regions," *IEEE Trans. Antennas Propag.*, vol. AP-7, pp. 292–299, 1969.

[8] W. D. Burnside and L. Peters, Jr., "Axial RCS of finite cones by the equivalent current concept with higher-order diffraction," *Radio Sci.*, vol. 7, no. 10, pp. 943–948, October 1972.

[9] E. F. Knott and T. B. A. Senior, "Comparison of three high-frequency diffraction techniques," *Proc. IEEE*, vol. 62, pp. 1468–1474, 1974.

[10] P. Ya. Ufimtsev, "Method of edge waves in the physical theory of diffraction" (from the Russian "Method Krayevykh voin v fizicheskoy teorii difraktsii," *Izd-Vo Sov. Radio*, pp. 1–243, 1962), translation prepared by the U.S. Air Force Foreign Technoloy Division, Wright-Patterson AFB, Ohio; released for public distribution September 7, 1971.

[11] S. W. Lee, "Comparison of uniform asymptotic theory and Ufimtsev's theory of EM edge diffraction," *IEEE Trans. Antennas Propag.*, vol. AP-25, no. 2, pp. 162–170, March 1977.

[12] S. Silver, *Microwave Antenna Theory and Design*, Boston: Boston Technical Publishers, Inc., 1964.

[13] S. Choudhary and L. B. Felsen, "Analysis of Gaussian beam propagation and diffraction by inhomogeneous wave tracking," *Proc. IEEE*, vol. 62, pp. 1530–1541, 1974.

[14] W. D. Burnside, C. L. Yu, and R. J. Marhefka, "A technique to combine the geometrical theory of diffraction and the moment method," *IEEE Trans. Antennas Propag.*, vol. AP-23, pp. 551–558, July 1975.

[15] G. A. Thiele and T. H. Newhouse, "A hybrid technique for combining moment methods with the geometrical theory of diffraction," *IEEE Trans. Antennas Propag.*, vol. AP-23, no. 1, January 1975.

[16] L. B. Felsen and A. H. Kamel, "Hybrid ray-mode formulation of parallel plane waveguide green's functions," *IEEE Trans. Antennas Propag.*, vol. AP-29, no. 4, pp. 637–649, July 1981.

[17] R. K. Luneberg, *Mathematical Theory of Optics*, Providence: Brown Univ., 1964.

[18] M. Kline, "An asymptotic solution of Maxwell's equation," *Commun. Pure Appl. Math.*, vol. 4, pp. 225–263, 1951.

[19] R. G. Kouyoumjian and P. H. Pathak, "The dyadic diffraction coefficient for a curved edge," *Rep. 3001-3*, ElectroScience Laboratory, Dept. Electr. Eng., Ohio State Univ., Columbus. Prepared under Grant NGR 36-008-144 for NASA, Langley Research Center, Hampton, Va. (see appendix III), August 1973.

[20] G. A. Deschamps, "Ray techniques in electromagnetics," *Proc. IEEE*, vol. 60, pp. 1022–1035, September 1972.

[21] S. W. Lee, "Electromagnetic reflection from a conducting surface: geometrical optics solution," *IEEE Trans. Antennas Propag.*, vol. AP-23, pp. 184–191, 1975.

[22] S. W. Lee, M. S. Sheshadari, V. Jamnejad, and R. Mittra, "Refraction at a curved dielectric interface: geometrical optics solution," *IEEE Trans. Microwave Theory Tech.*, MTT-30, no. 1, pp. 12–19, January 1982.

[23] S. W. Lee, P. Cramer, Jr., K. Woo, and Y. Rahmat-Samii, "Diffraction by an arbitrary subreflector: GTD solution," *IEEE Trans. Antennas Propag.*, vol. AP-27, pp. 305–316, 1979.

[24] R. G. Kouyoumjian, "Asymptotic high-frequency methods," *Proc. IEEE*, vol. 53, pp. 864–876, August 1965.

[25] P. H. Pathak and R. G. Kouyoumjian, "The dyadic diffraction coefficient for a perfectly conducting wedge," *Rep. 2183-4*, ElectroScience Lab., Dept. Elec. Eng., Ohio State Univ., Columbus. Prepared under Contract AF 19(628)-5929 for AF Cambridge Res. Labs. (AFCRL-69-0546), also *ASTIA Doc. AD 707 827*, June 5, 1970.

[26] R. G. Kouyoumjian and P. H. Pathak, "A uniform geometrical theory of diffraction for an edge in a perfectly conducting surface," *Proc. IEEE*, vol. 62, pp. 1448–1461, November 1974.

[27] F. A. Sikta, W. D. Burnside, T. T. Chu, and L. Peters, Jr., "First-order equivalent current and corner-diffraction scattering from flat-plate structures," *IEEE Trans. Antennas Propag.*, vol. 31, no. 4, pp. 584–589, July 1983.

[28] P. H. Pathak, "An asymptotic result for the scattering of a plane wave by a smooth convex cylinder," *Radio Sci.*, vol. 14, no. 3, pp. 419–435, May–June 1979.

[29] P. H. Pathak, W. D. Burnside, and R. J. Marhefka, "A uniform GTD analysis of the diffraction of electromagnetic waves by a smooth convex surface," *IEEE Trans. Antennas Propag.*, vol. AP-28, no. 5, pp. 631–642, September 1980.

[30] M. Abramowitz and I. A. Stegun, eds., *Handbook of Mathematical Functions* (Applied Mathematics Series 55). Washington, D.C.: National Bureau of Standards, p. 478, 1964.

[31] P. H. Pathak, N. Wang, W. D. Burnside, and R. G. Kouyoumjian, "Uniform GTD solution for the radiation from sources on a smooth convex surface," *IEEE Trans. Antennas Propag.*, vol. AP-29, no. 4, pp. 609–621, July 1981.

[32] W. D. Burnside, "Analysis of on-aircraft antenna patterns," PhD dissertation, Ohio State Univ., Columbus, 1972.

[33] N. Wang and W. D. Burnside, "An efficient-geodesic path solution for prolate spheroids," *Quarterly Report 711305-2*, The Ohio State Univ. ElectroScience Lab., Dept. Elec. Eng., July 1979.

[34] S. W. Lee, "Mutual admittance of slots on a cone: solution by ray technique," *IEEE Trans. Antennas Propag.*, vol. AP-26, no. 6, pp. 768–773, November 1978.

[35] J. R. Wait and W. E. Mientka, "Calculated patterns of slotted elliptic-cylinder antennas," *Appl. Sci. Res.*, Sec. B, vol. 7, pp. 449–462, 1959.

[36] P. C. Bargeliotes, A. T. Villeneuve, and W. H. Kummer, "Pattern synthesis of conformal arrays," Radar Microwave Lab., Aerospace Groups, Hughes Aircraft Co., Culver City, Calif., 1975.

[37] P. H. Pathak and R. G. Kouyoumjian, "The radiation from apertures in curved surfaces," *Rep. 3001-2*, ElectroScience Lab., Dept. Elec. Eng., Ohio State Univ. Prepared under Grant NGR 36-008-144 for NASA Langley Research Center, Hampton, Va., December 1972.

[38] P. H. Pathak, "Uniform GTD solutions for a class of problems associated with the diffraction by smooth convex surfaces," in volume 1 of the Ohio State University short course notes on *The Modern Geometrical Theory of Diffraction*, June 1983.

[39] P. H. Pathak and N. N. Wang, "Ray analysis of mutual coupling between antennas on a convex surface," *IEEE Trans. Antennas Propag.*, vol. AP-29, no. 6, pp. 911–922, November 1981.

[40] S. W. Lee and S. Safavi-Naini, "Approximate asymptotic solution of surface field due to a magnetic dipole on a cylinder," *IEEE Trans. Antennas Propag.*, vol. AP-26, no. 4, pp. 593–598, July 1978.

[41] R. F. Harrington, *Time Harmonic Electromagnetic Fields*, New York: McGraw-Hill Book Co., 1961.

[42] J. H. Richmond, "A reaction theorem and its application to antenna impedance calculations," *IRE Trans. Antennas Propag.*, vol. AP-9, no. 6, pp. 515–520, November 1961.

[43] K. E. Golden, G. E. Stewart, and D. C. Pridmore-Brown, "Approximation techniques for the mutual admittances of slot antennas on metallic cones," *IEEE Trans. Antennas Propag.*, vol. AP-22, pp. 43–48, 1974.

Appendixes

CONTENTS

A. Physical Constants, International Units, Conversion of Units, and Metric Prefixes — A-3
B. The Frequency Spectrum — B-1
C. Electromagnetic Properties of Materials — C-1
D. Vector Analysis — D-1
E. VSWR Versus Reflection Coefficient and Mismatch Loss — E-1
F. Decibels Versus Voltage and Power Ratios — F-1

Appendix A

Physical Constants, International Units, Conversion of Units, and Metric Prefixes

Yi-Lin Chen
*University of Illinois**

Physical Constants

Quantity	Symbol	Value
Speed of light in vacuum	c	$2.997925 \times 10^8 \, \text{ms}^{-1}$
Electron charge	e	$1.602192 \times 10^{-19} \, \text{C}$
Electron rest mass	m_e	$9.109558 \times 10^{-31} \, \text{kg}$
Boltzmann constant	k	$1.380622 \times 10^{-23} \, \text{JK}^{-1}$
Dielectric constant in vacuum	ϵ_0	$8.854185 \times 10^{-12} \, \text{Fm}^{-1}$
		$\cong (36\pi \times 10^9)^{-1} \, \text{Fm}^{-1}$
Permeability in vacuum	μ_0	$4\pi \times 10^{-7} \, \text{Hm}^{-1}$

International System of Units (SI Units): Basic Units

Quantity	Symbol	Units
Length	ℓ	meters (m)
Mass	m	kilograms (kg)
Time	t	seconds (s)
Electric current	I	amperes (A)
Temperature	T	kelvins (K)
Luminous intensity	I	candelas (cd)

---

*On leave from the Chinese Aeronautical Laboratory, Beijing, China, during 1983.

Derived Units in Electromagnetics

Quantity	Symbol	Units
Electric-field strength	$\mathbf{E}$	volts per meter (V/m)
Magnetic-field strength	$\mathbf{H}$	amperes per meter (A/m)
Electric-flux density	$\mathbf{D}$	coulombs per meter squared (C/m^2)
Magnetic-flux density	$\mathbf{B}$	teslas (T) = Wb/m^2
Electric-current density	$\mathbf{J}$	amperes per meter squared (A/m^2)
Magnetic-current density	$\mathbf{K}$	volts per meter squared (V/m^2)
Electric-charge density	ϱ	coulombs per meter cubed (C/m^3)
Magnetic-charge density	ϱ_m	webers per meter cubed (Wb/m^3)
Voltage	V	volts (V)
Electric current	I	amperes (A)
Dielectric constant (permittivity)	ϵ	farads/meter (F/m)
Permeability	μ	henrys/meter (H/m)
Conductivity	σ	siemens per meter (S/m) = ℧/m
Resistance	R	ohms (Ω)
Inductance	L	henrys (H)
Capacitance	C	farads (F)
Impedance	Z	ohms (Ω)
Admittance	Y	siemens (S) or mhos (℧)
Power	P	watts (W)
Energy	W	joules (J)
Radiation intensity	I	watts per steradian (W/sr)
Frequency	f	hertz (Hz)
Angular frequency	ω	radians per second (rad/s)
Wavelength	λ	meters (m)
Wave number	k	1 per meter (m^{-1})
Phase shift constant	β	radians per meter (rad/m)
Attenuation factor	α	nepers per meter (Np/m)

Conversions of Units

Quantity	Symbol	SI Unit	Equivalent Number of	
			CGS Electromagnetic Unit	CGS Electrostatic Unit
Electric charge	q	coulombs	10^{-1} abcoulomb	3×10^{9} statcoulombs
Current	I	amperes	10^{-1} abampere	3×10^{9} statampheres
Volume current density	$\mathbf{J}$	amperes/meter^2	10^{-5} abampere/centimeter^2	3×10^{5} statampheres/centimeter^2
Voltage	V	volts	10^{8} abvolts	$\frac{1}{3} \times 10^{-2}$ statvolt
Electric-field intensity	$\mathbf{E}$	volts/meter	10^{6} abvolts/cm	$\frac{1}{3} \times 10^{-4}$ statvolt/centimeter
Electric-flux density	$\mathbf{D}$	coulombs/meter^2	$4\pi \times 10^{-5}$ abcoulomb/centimeter^2	$12\pi \times 10^{5}$ statcoulombs/centimeter^2
Magnetic-field intensity	$\mathbf{H}$	amperes/meter	$4\pi \times 10^{-3}$ oersted	$12\pi \times 10^{7}$ oersteds
Magnetic-flux intensity	$\mathbf{B}$	webers/meter^2	10^{4} gausses	$\frac{1}{3} \times 10^{-6}$ gauss
Permittivity	ϵ	farads/meter	$4\pi \times 10^{-11}$ abfarad/centimeter	$36\pi \times 10^{9}$ statfarads/centimeter
Permeability	μ	henrys/meter	$\frac{1}{4\pi} \times 10^{7}$ gauss/oersted	$\frac{1}{36\pi} \times 10^{-13}$ gauss/oersted
Magnetic flux	Φ	webers	10^{8} gilberts	$\frac{1}{3} \times 10^{-2}$ gilbert
Resistance	R	ohms	10^{9} abohms	$\frac{1}{9} \times 10^{-11}$ statohm
Inductance	L	henrys	10^{9} abhenrys	$\frac{1}{9} \times 10^{-11}$ stathenry
Capacitance	C	farads	10^{-9} abfarad	9×10^{11} statfarads
Conductivity	σ	siemens/meter	10^{-11} absiemen/centimeter	9×10^{9} statsiemens/centimeter
Work	W	joules	10^{7} ergs	10^{7} ergs
Power	P	watts	10^{7} ergs/second	10^{7} ergs/second

Conversion of Length Units

Meters	Centimeters	Inches	Feet	Miles
1	100	39.37	3.281	6.214×10^{-4}
0.01	1	0.3937	3.281×10^{-2}	
0.0254	2.540	1	8.333×10^{-2}	
0.3048	30.48	12	1	1.894×10^{-4}
1609			5279	1

Metric Prefixes and Symbols*

Multiplication Factor	Prefix	Symbol
10^{18}	exa	E
10^{15}	peta	P
10^{12}	tera	T
10^{9}	giga	G
10^{6}	mega	M
10^{3}	kilo	k
10^{2}	hecto	h
10	deka	da
10^{-1}	deci	d
10^{-2}	centi	c
10^{-3}	milli	m
10^{-6}	micro	μ
10^{-9}	nano	n
10^{-12}	pico	p
10^{-15}	femto	f
10^{-18}	atto	a

*From *IEEE Standard Dictionary of Electrical and Electronics Terms*, p. 682, The Institute of Electrical and Electronics Engineers, Inc., 1984.

Appendix B

The Frequency Spectrum

Li-Yin Chen
University of Illinois*

The wavelength of an electromagnetic wave in free space is $\lambda_0 = c/f$.

$$\lambda_0 = \frac{300\,000}{f(\text{kHz})}\,\text{m} = \frac{300}{f(\text{MHz})}\,\text{m} = \frac{30}{f(\text{GHz})}\,\text{cm}$$

$$= \frac{9.843 \times 10^5}{f(\text{kHz})}\,\text{ft} = \frac{9.843 \times 10^2}{f(\text{MHz})}\,\text{ft} = \frac{11.81}{f(\text{GHz})}\,\text{in}$$

The wave number of an electromagnetic wave in free space: $k_0 = \omega\sqrt{\mu_0 \epsilon_0} = 2\pi f/c$.

$$k_0 = f(\text{Hz}) \times 2.0944 \times 10^{-8}\,\text{m}^{-1} = f(\text{kHz}) \times 2.0944 \times 10^{-5}\,\text{m}^{-1}$$
$$= f(\text{MHz}) \times 2.0944 \times 10^{-2}\,\text{m}^{-1} = f(\text{GHz}) \times 20.944\,\text{m}^{-1}$$
$$= f(\text{Hz}) \times 6.383 \times 10^{-9}\,\text{ft}^{-1} = f(\text{kHz}) \times 6.383 \times 10^{-6}\,\text{ft}^{-1}$$
$$= f(\text{MHz}) \times 6.383 \times 10^{-3}\,\text{ft}^{-1} = f(\text{GHz}) \times 6.383\,\text{ft}^{-1}$$
$$= f(\text{GHz}) \times 0.532\,\text{in}^{-1}$$

---

*On leave from the Chinese Aeronautical Laboratory, Beijing, China, during 1983.

Nomenclature of Frequency Bands

Adjectival Designation	Frequency Range	Metric Subdivision	Wavelength Range
elf: Extremely low frequency	30 to 300 Hz	Megametric waves	10 000 to 1000 km
vf: Voice frequency	300 to 3000 Hz		1000 to 100 km
vlf: Very low frequency	3 to 30 kHz	Myriametric waves	100 to 10 km
lf: Low frequency	30 to 300 kHz	Kilometric waves	10 to 1 km
mf: Medium frequency	300 to 3000 kHz	Hectrometric waves	1000 to 100 m
hf: High frequency	3 to 30 MHz	Decametric waves	100 to 10 m
vhf: Very high frequency	30 to 300 MHz	Metric waves	10 to 1 m
uhf: Ultrahigh frequency	300 to 3000 MHz	Decimetric waves	100 to 10 cm
shf: Superhigh frequency	3 to 30 GHz	Centimetric waves	10 to 1 cm
ehf: Extremely high frequency	30 to 300 GHz	Millimetric waves	10 to 1 mm
	300 to 3000 GHz	Decimillimetric waves	1 to 0.1 mm

Standard Radar-Frequency Letter Bands*

Band Designation	Nominal Frequency Range
hf	3–30 MHz
vhf	30–300 MHz
uhf	300–1000 MHz
L	1000–2000 MHz
S	2000–4000 MHz
C	4000–8000 MHz
X	8000–12 000 MHz
K_u	12.0–18 GHz
K	18–27 GHz
K_a	27–40 GHz
Millimeter	40–300 GHz

*Reprinted from ANSI/IEEE Std. 100-1984, *IEEE Standard Dictionary of Electrical and Electronics Terms*, © 1984 by The Institute of Electrical and Electronics Engineers, Inc., by permission of the IEEE Standards Department.

Television Channel Frequencies*

Channel Number[†]	Band (MHz)	Channel Number[†]	Band (MHz)	Channel Number[†]	Band (MHz)
2	54–60	29	560–566	57	728–734
3	60–66	30	566–572	58	734–740
4	66–72	31	572–578	59	740–746
5	76–82	32	578–584	60	746–752
6	82–88	33	584–590	61	752–758
7	174–180	34	590–596	62	758–764
8	180–186	35	596–602	63	764–770
9	186–192	36	602–608	64	770–776
10	192–198	37	608–614	65	776–782
11	198–204	38	614–620	66	782–788
12	204–210	39	620–626	67	788–794
13	210–216	40	626–632	68	794–800
14	470–476	41	632–638	69	800–806
15	476–482	42	638–644	70	806–812
16	482–488	43	644–650	71	812–818
17	488–494	44	650–656	72	818–824
18	494–500	45	656–662	73	824–830
19	500–506	46	662–668	74	830–836
20	506–512	47	668–674	75	836–842
21	512–518	48	674–680	76	842–848
22	518–524	49	680–686	77	848–854
23	524–530	50	686–692	78	854–860
24	530–536	51	692–698	79	860–866
25	536–542	52	698–704	80	866–872
26	542–548	53	704–710	81	872–878
27	548–554	54	710–716	82	878–884
28	554–560	55	716–722	83	884–890
		56	722–728		

Note: The carrier frequency for the video portion is the lower frequency plus 1.25 MHz. The audio carrier frequency is the upper frequency minus 0.25 MHz. All channels have a 6-MHz bandwidth. For example, channel 2 video carrier is at 55.25 MHz and the audio carrier is at 59.75 MHz.

[†]Channels 2 through 13 are vhf; channels 14 through 83 are uhf. Channels 70 through 83 were withdrawn and reassigned to tv translator stations until licenses expire.

Appendix C

Electromagnetic Properties of Materials

Yi-Lin Chen
University of Illinois[*]

Resistivities and Skin Depth of Metals and Alloys

Material	Resistivity* ($\mu\Omega$-cm)	Skin Depth[†] (μm at 1 GHz)
Aluminum	2.62	2.576
Brass	7.5	4.3586
(66% Cu, 34% Zn)		
Copper	1.7241	2.0898
Gold	2.44	2.4861
Iron	9.71	4.9594
Nickel	6.9	4.1807
Silver	1.62	2.0257
Steel	13–22	5.7384–7.465
(0.4–0.5% C, balance Fe)		
Steel, stainless	90	15.0988
(0.1% C, 18% Cr, 8% Ni, balance Fe)		
Tin	11.4	5.3737
Titanium	47.8	11.0036

*In solid form at 20°C; resistivity = (conductivity)$^{-1}$.
[†]Skin depth $\delta = (\pi\mu\sigma f)^{-1/2} = (20\pi)^{-1}[\sigma f(\text{GHz})]^{-1/2}$ m. The δs in the column are calculated at $f = 1$ GHz. For other frequencies, multiply them by $[f(\text{GHz})]^{-1/2}$.

---

[*]On leave from the Chinese Aeronautical Laboratory, Beijing, China, during 1983.

Characteristics of Insulating Materials*

Material Composition	T (°C)	Dielectric Constant[†] at (Frequency in Hertz)				Dissipation Factor[†] at (Frequency in Hertz)				Dielectric Strength in Volts/Mil at 25°C	DC Volume Resistivity in Ohm-cm at 25°C	Thermal Expansion (Linear) in Parts/°C	Softening Point in °C	Moisture Absorption in Percent
		10^4	10^6	3×10^9	2.5×10^{10}	10^4	10^6	3×10^9	2.5×10^{10}					
Ceramics:														
Aluminum oxide	25	8.80	8.80	8.79	—	0.00033	0.00030	0.0010	—	—	10^{12}–10^{13}	—	1400–1430	0.1
Barium titanate[‡]	26	1143	—	600	100	0.0105	—	0.30	0.60	75	10^{12}–10^{14}	—	—	0.1
Calcium titanate	25	167.7	167.7	165	—	0.0002	—	0.0023	—	100	—	—	1510	<0.1
Magnesium oxide	25	9.65	9.65	—	—	<0.0003	<0.0003	—	—	—	—	—	—	—
Magnesium silicate	25	5.97	5.96	5.90	—	0.0005	0.0004	0.0012	—	—	>10^{14}	9.2×10^{-6}	1350	0.1–1
Magnesium titanate	25	13.9	13.9	13.8	13.7	0.0004	0.0005	0.0017	0.0065	—	—	—	—	—
Oxides of aluminum, silicon, magnesium, calcium, barium	24	6.04	—	5.90	—	0.0011	—	0.0024	—	—	—	7.7×10^{-6}	1325	—
Porcelain (dry process)	25	5.08	5.04	—	—	0.0075	0.0078	—	—	—	—	—	—	—
Steatite 410	25	5.77	5.77	5.7	—	0.0007	0.0006	0.00089	—	—	—	—	—	—
Strontium titanate	25	232	232	—	—	0.0002	0.0001	—	—	100	10^{12}–10^{14}	—	1510	0.1
Titanium dioxide (rutile)	26	100	100	—	—	0.0003	0.00025	—	—	—	—	—	1667	—
Glasses:														
Iron-sealing glass	24	8.30	8.20	7.99	7.84	0.0005	0.0009	0.00199	0.0112	—	10^{10} at 250°	132×10^{-7}	484	poor
Soda-borosilicate	25	4.84	4.84	4.82	4.65	0.0036	0.0030	0.0054	0.0090	—	7×10^7 at 250°	50×10^{-7}	693	—
100% silicon dioxide (fused quartz)	25	3.78	3.78	3.78	3.78	0.0001	0.0002	0.00006	0.00025	410 (0.25")	>10^{18}	5.7×10^{-7}	1667	—
Plastics:														
Alkyd resin	25	4.76	4.55	4.50	—	0.0149	0.0138	0.0108	—	—	—	11–17×10^{-5}	—	2.3
Cellulose acetate-butyrate, plasticized	26	3.30	3.08	2.91	—	0.018	0.017	0.028	—	250–400 (0.125")	—	—	60–121	2.3
Cresylic acid–formaldehyde, 50% α-cellulose	25	4.51	3.85	3.43	3.21	0.036	0.055	0.051	0.038	1020 (0.033")	3×10^{12}	3×10^{-5}	>125	1.2
Cross-linked polystyrene	25	2.58	2.58	2.58	—	0.0016	0.0020	0.00019	—	—	—	—	—	—
Epoxy resin (Araldite CN-501)	25	3.62	3.35	3.09	—	0.019	0.034	0.027	—	405 (0.125")	>3.8×10^7	4.77×10^{-5}	109 (distortion)	0.14
Epoxy resin (Epon resin RN-48)	25	3.52	3.32	3.04	—	0.0142	0.0264	0.021	—	—	—	—	—	—
Foamed polystyrene, 0.25% filler	25	1.03	—	1.03	1.03	<0.0002	—	0.0001	—	—	—	—	85	low
Melamine–formaldehyde, α-cellulose	24	7.00	6.0	4.93	—	0.041	0.085	0.103	—	300–400	—	—	99 (stable)	0.4–0.6
Melamine–formaldehyde, 55% filler	26	5.75	5.5	—	—	0.0115	0.020	—	—	—	—	1.7×10^{-5}	—	0.6
Phenol–formaldehyde (Bakelite BM 120)	25	4.36	3.95	3.70	3.55	0.0280	0.0380	0.0438	0.0390	300 (0.125")	10^{11}	30–40×10^{-6}	<135 (distortion)	<0.6
Phenol–formaldehyde, 50% paper laminate	26	4.60	4.04	3.57	—	0.034	0.057	0.060	—	—	—	—	—	—
Phenol–formaldehyde, 65% mica, 4% lubricants	24	4.78	4.72	4.71	—	0.0082	0.0115	0.0126	—	—	—	—	—	—
Polycarbonate	—	2.96	—	—	—	0.010	—	—	—	—	2×10^{16}	7×10^{-5}	135 (deflection)	—
Polychlorotrifluoroethylene	25	2.42	2.32	2.29	2.28	0.0082	0.0028	0.0028	0.0053	364 (0.125")	10^{13}	—	—	—

Material	Temp										
Polyethylene	25	2.26	2.26	2.26	2.26	<0.0002	0.0002	0.00031	0.0006	1200 (0.033")	10^{17}	19×10^{-5} (varies)	95–105 (distortion)	0.03
Polyethylene-terephthalate	—	2.98	—	—	—	—	—	—	—	—	—	—	—	—
Polyethylmethacrylate	22	2.55	2.52	2.51	2.5	0.016	—	0.0075	0.0083	4000 (0.002")	—	—	—	low
Polyhexamethylene-adipamide (nylon)	25	3.14	3.0	2.84	2.73	0.0090	0.0200	0.0117	0.0105	400 (0.125")	8×10^{14}	10.3×10^{-5}	60 (distortion) 65 (distortion)	1.5
Polyimide	—	3.4	—	—	—	0.0218	—	—	—	570	—	—	—	—
Polyisobutylene	25	2.23	2.23	2.23	—	0.003	—	—	—	—	—	—	—	low
Polymer of 95% vinyl-acetate, 5% vinyl-acetate	20	2.90	2.8	2.74	—	0.0001	0.0003	0.00047	—	600 (0.010")	—	—	25 (distortion)	—
Polymethyl methacrylate	27	2.76	—	2.60	—	0.0150	0.0080	0.0059	—	—	—	—	—	0.3–0.6
						0.0140		0.0057			>5×10^{16}	$8–9 \times 10^{-5}$	70–75 (distortion)	
Polyphenylene oxide	—	2.55	—	2.55	—	0.0007	—	0.0011	—	990 (0.030")				
Polypropylene	—	2.55	—	—	—	<0.0005	—	—	—	500 (0.125")	10^{17}	5.3×10^{-5}	195 (deflection)	—
										650 (0.125")	6×10^{16}	$6–8.5 \times 10^{-5}$	99–116 (deflection)	
Polystyrene	25	2.56	2.55	2.55	2.54	0.00007	<0.0001	0.00033	0.0012	500–700 (0.125")	10^{18}	$6–8 \times 10^{-5}$	82 (distortion)	0.05
Polytetrafluoroethylene (Teflon)	22	2.1	2.1	2.1	2.08	<0.0002	<0.0002	0.00015	0.0006	1000–2000 (0.005"–0.012")	10^{17}	9.0×10^{-5}	66 (distortion) (stable to 300)	0.00
Polyvinylcyclohexane	24	2.25	2.25	2.25	2.25	<0.0002	<0.0002	0.00018	—	—	—	—	—	—
Polyvinyl formal	26	2.92	2.80	2.76	2.7	0.019	0.013	0.0113	0.0115	860 (0.034")	>5×10^{16}	7.7×10^{-5}	190	1.3
Polyvinylidene fluoride	—	6.6	—	—	—	0.17	0.050	0.0555	—	260 (0.125")	2×10^{14}	12×10^{-5}	148 (deflection)	—
Urea-formaldehyde, cellulose	27	5.65	5.1	4.57	—	0.027			—	375 (0.085")	—	2.6×10^{-5}	152 (distortion)	2
Urethane elastomer	—	6.5–7.1								450–500 (0.125")	2×10^{11}	$10–20 \times 10^{-5}$		
Vinylidene–vinyl chloride copolymer	23	3.18	2.82	2.71	—	0.057	0.0180	0.0072	—	300 (0.125")	$10^{14}–10^{16}$	15.8×10^{-5}	150	<0.1
100% aniline-formaldehyde (Dilectene-100)	25	3.58	3.50	3.44	—	0.0061	0.0033	0.0026	—	810 (0.068")	10^{16}	5.4×10^{-5}	125	0.06–0.08
100% phenol-formaldehyde	24	5.4	4.4	3.64	—	0.060	0.077	0.052	—	277 (0.125")	—	$8.3–13 \times 10^{-5}$	50 (distortion)	0.42
100% polyvinyl-chloride	20	2.88	2.85	2.84	—	0.0160	0.0081	0.0055	—	400 (0.125")	10^{14}	6.9×10^{-5}	54 (distortion)	0.05–0.15
Organic Liquids:														
Aviation gasoline (100 octane)	25	1.94	1.94	1.92	—	—	0.0001	0.0014	—	—	—	—	—	—
Benzene (pure, dried)	25	2.28	2.28	2.28	2.28	<0.0001	<0.0001	<0.0001	<0.0001	—	—	—	—	—
Carbon tetrachloride	25	2.17	2.17	2.17	—	<0.00004	<0.0002	0.0004	—	—	—	—	—	—
Ethyl alcohol (absolute)	25	24.5	23.7	6.5	—	0.090	0.062	0.250	—	—	—	—	—	—
Ethylene glycol	25	41	41	12	—	0.030	0.045	1.00	—	—	—	—	—	—
Jet fuel (JP-3)	25	2.08	2.08	2.04	—	0.0001	—	0.0055	—	—	—	—	—	—
Methyl alcohol (absolute analytical grade)	25	31	31.0	23.9	—	0.20	0.038	0.64	—	—	—	—	—	—
Methyl or ethyl siloxane polymer (1000 cs)	22	2.78	—	2.74	—	<0.0003	—	0.0096	—	—	—	—	—	—
Monomeric styrene	22	2.40	2.40	2.40	—	<0.0003	—	0.0020	—	300 (0.100")	3×10^{12}	—	—	0.06
Transil oil	26	2.22	2.20	2.18	—	<0.0005	0.0048	0.0028	—	300 (0.100")	—	—	−40 (pour point)	—
Vaseline	25	2.16	2.16	2.16	—	<0.0001	<0.0004	0.00066	—	—	—	—	—	—

Characteristics of Insulating Materials* (cont'd.)

Material Composition	T(°C)	Dielectric Constant† at (Frequency in Hertz)				Dissipation Factor† at (Frequency in Hertz)				Dielectric Strength in Volts/Mil at 25°C	DC Volume Resistivity in Ohm-cm at 25°C	Thermal Expansion (Linear) in Parts/°C	Softening Point in °C	Moisture Absorption in Percent
		10^4	10^6	3×10^9	2.5×10^{10}	10^4	10^6	3×10^9	2.5×10^{10}					
Waxes:														
Beeswax, yellow	23	2.53	2.45	2.39	—	0.0092	0.0090	0.0075	—	—	—	—	45–64 (melts)	—
Dichloronaphthalenes	23	2.98	2.93	2.89	—	0.0003	0.0017	0.0037	—	—	—	—	35–63 (melts)	nil
Polybutene	25	2.34	2.30	2.27	—	0.00133	0.00133	0.0009	—	—	—	—	—	—
Vegetable and mineral waxes	25	2.3	2.3	2.25	—	0.0004	0.0004	0.00046	—	—	—	—	57	—
Rubbers:														
Butyl rubber	25	2.35	2.35	2.35	—	0.0010	0.0010	0.0009	—	—	—	—	—	—
GR-S rubber	25	2.90	2.82	2.75	—	0.0120	0.0080	0.0057	—	870 (0.040")	2×10^{15}	—	—	—
Gutta-percha	25	2.53	2.47	2.40	—	0.0042	0.0120	0.0060	—	—	10^{15}	—	—	—
Hevea rubber (pale crepe)	25	2.4	2.4	2.15	—	0.0018	0.0050	0.0030	—	—	—	—	—	—
Hevea rubber, vulcanized (100 pts pale crepe, 6 pts sulfur)	27	2.74	2.42	2.36	—	0.0446	0.0180	0.0047	—	—	—	—	—	—
Neoprene rubber	24	6.26	4.5	4.00	4.0	0.038	0.090	0.034	0.025	300 (0.125")	8×10^{12}	—	—	nil
Organic polysulfide, fillers	23	110	30	16	13.6	0.39	0.28	0.22	0.10	—	—	—	—	—
Silicone-rubber compound	25	3.20	3.16	3.13	—	0.0030	0.0032	0.0097	—	—	—	—	—	—
Woods:‡														
Balsa wood	26	1.37	1.30	1.22	—	0.0120	0.0135	0.100	0.032	—	—	—	—	—
Douglas fir	25	1.93	1.88	1.82	1.78	0.026	0.033	0.027	—	—	—	—	—	—
Douglas fir, plywood	25	1.90	—	—	1.6	0.0230	—	—	0.0220	—	—	—	—	—
Mahogany	25	2.25	2.07	1.88	1.6	0.025	0.032	0.025	0.020	—	—	—	—	—
Yellow birch	25	2.70	2.47	2.13	1.87	0.029	0.040	0.033	0.026	—	—	—	—	—
Yellow poplar	25	1.75	—	1.50	1.4	0.019	—	0.015	0.017	—	—	—	—	—
Miscellaneous:														
Amber (fossil resin)	25	2.65	—	2.6	—	0.0056	—	0.0090	—	2300 (0.125")	Very high	—	200	—
DeKhotinsky cement	23	3.23	—	2.96	—	0.024	—	0.021	—	—	—	—	80–85	—
Gilsonite (99.9% natural bitumen)	26	2.58	2.56	—	—	0.0016	0.0011	—	—	—	—	—	155 (melts)	—
Shellac (natural XL)	28	3.47	3.10	2.86	—	0.031	0.030	0.0254	0.0013	—	10^{16}	—	80	low after baking
Mica, glass-bonded	25	7.39	—	—	—	0.0013	—	—	—	—	—	—	—	—
Mica, glass, titanium dioxide	24	9.0	—	—	—	0.0026	—	0.0040	—	—	—	—	—	—
Ruby mica	26	5.4	5.4	5.4	—	0.0003	0.0002	0.0003	—	3800–5600 (0.040")	5×10^{13}	9.8×10^{-5}	400	<0.5
Paper, royalgrey	25	2.99	2.77	2.70	—	0.038	0.066	0.056	—	202 (0.125")	—	—	—	—
Selenium (amorphous)	25	6.00	6.00	6.00	6.00	<0.0003	<0.0002	0.00018	—	—	—	—	—	—
Asbestos fiber–chrysotile paper	25	3.1	—	—	—	0.025	—	—	—	—	—	—	—	—
Sodium chloride (fresh crystals)	25	5.90	—	—	5.90	<0.0002	—	—	<0.0005	—	—	—	—	—

	Temp (°C)											
Soil, sandy dry	25	2.59	2.55	2.55	—	0.017	—	0.0062	—	—	—	—
Soil, loamy dry	25	2.53	2.48	2.44	—	0.018	—	0.0011	—	—	—	—
Ice (from pure distilled water)	−12	4.15	3.45	3.20	—	0.12	0.035	0.0009	—	—	—	—
Freshly fallen snow	−20	1.20	1.20	1.20	—	0.0215	—	0.00029	—	—	—	—
Hard-packed snow followed by light rain	−6	1.55	—	1.5	—	0.29	—	0.0009	—	—	—	—
Water (distilled)	25	78.2	78	76.7	34	0.040	0.005	0.157	0.2650	10^4	—	—

*Reproduced with permission of the publisher, Howard W. Sams & Company, Indianapolis, *Reference Data for Engineers: Radio, Electronics, Computer, and Communications*, 7th ed., by E. C. Jordan, ed., © 1985.

†The dissipation factor is defined as the ratio of the energy dissipated to the energy stored in the dielectric, or as the tangent of the loss angle. Dielectric constant and dissipation factor depend on electrical field strength.

‡Field perpendicular to grain.

Properties of Soft Magnetic Metals*

Name	Composition (%)	Permeability Initial	Permeability Maximum	Coercivity H_c (A/m)	Retentivity B_r (T)	B_{max} (T)	Resistivity (μΩ-cm)
Ingot iron	99.8 Fe	150	5 000	80	0.77	2.14	10
Low carbon steel	99.5 Fe	200	4 000	100	—	2.14	12
Silicon iron, unoriented	3 Si, bal Fe	270	8 000	60	—	2.01	47
Silicon iron, grain oriented	3 Si, bal Fe	1 400	50 000	7	1.20	2.01	50
4750 alloy	48 Ni, bal Fe	11 000	80 000	2	—	1.55	48
4-79 Permalloy	4 Mo, 79 Ni, bal Fe	40 000	200 000	1	—	0.80	58
Supermalloy	5 Mo, 80 Ni, bal Fe	80 000	450 000	0.4	—	0.78	65
2V-Permendur	2V, 49 Co, bal Fe	800	8 000	160	—	2.30	40
Supermendur	2V, 49 Co, bal Fe	—	100 000	16	2.00	2.30	26
Metglas† 2605SC	$Fe_{81}B_{13.5}Si_{3.5}C_2$	—	210 000	14	1.46	1.60	125
Metglas† 2605S-3	$Fe_{79}B_{16}Si_5$	—	30 000	8	0.30	1.58	125

*Reproduced with permission of the publisher, Howard W. Sams & Company, Indianapolis, *Reference Data for Engineers: Radio, Electronics, Computer, and Communications*, 7th ed., by E. C. Jordan, ed., © 1985.
†Metglas is Allied Corporation's registered trademark for amorphous alloys.

Appendix D

Vector Analysis

Yi-Lin Chen
*University of Illinois**

1. Change of Coordinate Systems

The transformations of the coordinate components of a vector **A** among the rectangular (x, y, z), cylindrical (θ, ϕ, z), and spherical (r, θ, ϕ) coordinates are given by the following relations (see Fig. 1):

$$A_x = A_\varrho \cos\phi - A_\phi \sin\phi = A_r \sin\theta \cos\phi + A_\theta \cos\theta \cos\phi - A_\phi \sin\phi$$
$$A_y = A_\varrho \sin\phi + A_\phi \cos\phi = A_r \sin\theta \sin\phi + A_\theta \cos\theta \sin\phi + A_\phi \cos\phi$$

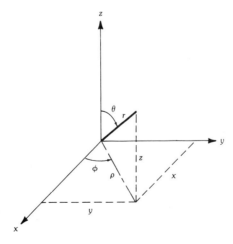

Fig. 1. Rectangular, cylindrical, and spherical coordinate systems.

---

*On leave from the Chinese Aeronautical Laboratory, Beijing, China, during 1983.

$$A_z = A_r \cos\theta - A_\theta \sin\theta$$
$$A_\varrho = A_x \cos\phi + A_y \sin\phi = A_r \sin\theta + A_\theta \cos\theta$$
$$A_\phi = -A_x \sin\phi + A_y \cos\phi$$
$$A_r = A_x \sin\theta \cos\phi + A_y \sin\theta \sin\phi + A_z \cos\theta = A_\varrho \sin\theta + A_z \cos\theta$$
$$A_\theta = A_x \cos\theta \cos\phi + A_y \cos\theta \sin\phi - A_z \sin\theta = A_\varrho \cos\theta - A_z \sin\theta$$

Differential element of volume:

$$dV = dx\,dy\,dz = \varrho\,d\varrho\,d\phi\,dz = r^2 \sin\theta\,dr\,d\theta\,d\phi$$

Differential element of vector area:

$$\begin{aligned}d\mathbf{S} &= \hat{\mathbf{x}}\,dy\,dz + \hat{\mathbf{y}}\,dx\,dz + \hat{\mathbf{z}}\,dx\,dy \\ &= \hat{\boldsymbol{\varrho}}\varrho\,d\phi\,dz + \hat{\boldsymbol{\phi}}\,d\varrho\,dz + \hat{\mathbf{z}}\varrho\,d\varrho\,d\phi \\ &= \hat{\mathbf{r}}r^2 \sin\theta\,d\theta\,d\phi + \hat{\boldsymbol{\theta}}r\sin\theta\,dr\,d\phi + \hat{\boldsymbol{\phi}}r\,dr\,d\theta\end{aligned}$$

Differential element of vector length:

$$\begin{aligned}d\boldsymbol{\ell} &= \hat{\mathbf{x}}\,dx + \hat{\mathbf{y}}\,dy + \hat{\mathbf{z}}\,dz \\ &= \hat{\boldsymbol{\varrho}}\,d\varrho + \hat{\boldsymbol{\phi}}\varrho\,d\phi + \hat{\mathbf{z}}\,dz \\ &= \hat{\mathbf{r}}\,dr + \hat{\boldsymbol{\theta}}r\,d\theta + \hat{\boldsymbol{\phi}}r\sin\theta\,d\phi\end{aligned}$$

2. ∇ Operator

In rectangular coordinates (x, y, z):

$$\nabla\Phi = \left(\hat{\mathbf{x}}\frac{\partial}{\partial x} + \hat{\mathbf{y}}\frac{\partial}{\partial y} + \hat{\mathbf{z}}\frac{\partial}{\partial z}\right)\Phi$$

$$\nabla\cdot\mathbf{A} = \frac{\partial A_x}{\partial x} + \frac{\partial A_y}{\partial y} + \frac{\partial A_z}{\partial z}$$

$$\nabla\times\mathbf{A} = \begin{vmatrix} \hat{\mathbf{x}} & \hat{\mathbf{y}} & \hat{\mathbf{z}} \\ \dfrac{\partial}{\partial x} & \dfrac{\partial}{\partial y} & \dfrac{\partial}{\partial z} \\ A_x & A_y & A_z \end{vmatrix}$$

$$\nabla^2\Phi = \nabla\cdot\nabla\Phi = \left(\frac{\partial^2}{\partial x^2} + \frac{\partial^2}{\partial y^2} + \frac{\partial^2}{\partial z^2}\right)\Phi$$

$$\nabla^2\mathbf{A} = \hat{\mathbf{x}}\nabla^2 A_x + \hat{\mathbf{y}}\nabla^2 A_y + \hat{\mathbf{z}}\nabla^2 A_z$$

In cylindrical coordinates (ϱ, ϕ, z):

$$\nabla\Phi = \left(\hat{\boldsymbol{\varrho}}\frac{\partial}{\partial \varrho} + \hat{\boldsymbol{\phi}}\frac{\partial}{\varrho\,\partial\phi} + \hat{\mathbf{z}}\frac{\partial}{\partial z}\right)\Phi$$

$$\nabla \cdot \mathbf{A} = \frac{1}{\varrho} \frac{\partial}{\partial \varrho} (\varrho A_\varrho) + \frac{1}{\varrho} \frac{\partial A_\phi}{\partial \phi} + \frac{\partial A_z}{\partial z}$$

$$\nabla \times \mathbf{A} = \frac{1}{\varrho} \begin{vmatrix} \hat{\varrho} & \varrho\hat{\phi} & \hat{z} \\ \frac{\partial}{\partial \varrho} & \frac{\partial}{\partial \phi} & \frac{\partial}{\partial z} \\ A_\varrho & \varrho A_\phi & A_z \end{vmatrix}$$

$$= \hat{\varrho}\left(\frac{1}{\varrho}\frac{\partial A_z}{\partial \phi} - \frac{\partial A_\phi}{\partial z}\right) + \hat{\phi}\left(\frac{\partial A_\varrho}{\partial z} - \frac{\partial A_z}{\partial \varrho}\right) + \hat{z}\left[\frac{1}{\varrho}\frac{\partial}{\partial \phi}(\varrho A_\phi) - \frac{1}{\varrho}\frac{\partial A_\varrho}{\partial \phi}\right]$$

$$\nabla^2 \Phi = \frac{1}{\varrho} \frac{\partial}{\partial \varrho}\left(\varrho \frac{\partial \Phi}{\partial \varrho}\right) + \frac{1}{\varrho^2} \frac{\partial^2 \Phi}{\partial \phi^2} + \frac{\partial^2 \Phi}{\partial z^2}$$

$$\nabla^2 \mathbf{A} = \nabla \nabla \cdot \mathbf{A} - \nabla \times \nabla \times \mathbf{A} \neq \hat{\varrho}\nabla^2 A_\varrho + \hat{\phi}\nabla^2 A_\phi + \hat{z}\nabla^2 A_z$$

In spherical coordinates (r, θ, ϕ):

$$\nabla \Phi = \left(\hat{r}\frac{\partial}{\partial r} + \hat{\theta}\frac{1}{r}\frac{\partial}{\partial \theta} + \hat{\phi}\frac{1}{r\sin\theta}\frac{\partial}{\partial \phi}\right)\Phi$$

$$\nabla \cdot \mathbf{A} = \frac{1}{r}\frac{\partial}{\partial r}(r^2 A_r) + \frac{1}{r\sin\theta}\frac{\partial}{\partial \theta}(A_\theta \sin\theta) + \frac{1}{r\sin\theta}\frac{\partial A_\phi}{\partial \phi}$$

$$\nabla \times \mathbf{A} = \frac{1}{r^2 \sin\theta}\begin{vmatrix} \hat{r} & r\hat{\theta} & (r\sin\theta)\hat{\phi} \\ \frac{\partial}{\partial r} & \frac{\partial}{\partial \theta} & \frac{\partial}{\partial \phi} \\ A_r & rA_\theta & (r\sin\theta)A_\phi \end{vmatrix}$$

$$= \hat{r}\frac{1}{r\sin\theta}\left[\frac{\partial}{\partial \theta}(A_\phi \sin\theta) - \frac{\partial A_\theta}{\partial \phi}\right] + \hat{\theta}\frac{1}{r}\left[\frac{1}{\sin\theta}\frac{\partial A_r}{\partial \phi} - \frac{\partial}{\partial r}(rA_\phi)\right]$$

$$+ \hat{\phi}\frac{1}{r}\left[\frac{\partial}{\partial r}(rA_\phi) - \frac{\partial A_r}{\partial \theta}\right]$$

$$\nabla^2 \Phi = \frac{1}{r^2}\frac{\partial}{\partial r}\left(r^2 \frac{\partial \Phi}{\partial r}\right) + \frac{1}{r^2 \sin\theta}\frac{\partial}{\partial \theta}\left(\sin\theta \frac{\partial \Phi}{\partial \theta}\right) + \frac{1}{r^2 \sin^2\theta}\frac{\partial^2 \Phi}{\partial \phi^2}$$

$$\nabla^2 \mathbf{A} = \nabla\nabla \cdot A - \nabla \times \nabla \times A \neq \hat{r}\nabla^2 A_r + \hat{\theta}\nabla^2 A_\theta + \hat{\phi}\nabla^2 A_\phi$$

3. Identities

$$\mathbf{a} \cdot \mathbf{b} \times \mathbf{c} = \mathbf{a} \times \mathbf{b} \cdot \mathbf{c} = \mathbf{b} \cdot \mathbf{c} \times \mathbf{a}$$

$$\mathbf{a} \times (\mathbf{b} \times \mathbf{c}) = (\mathbf{a} \cdot \mathbf{c})\mathbf{b} - (\mathbf{a} \cdot \mathbf{b})\mathbf{c}$$

$$(\mathbf{a} \times \mathbf{b}) \cdot (\mathbf{c} \times \mathbf{d}) = \mathbf{a} \cdot \mathbf{b} \times (\mathbf{c} \times \mathbf{d}) = \mathbf{a} \cdot [(\mathbf{b} \cdot \mathbf{d})\mathbf{c} - (\mathbf{b} \cdot \mathbf{c})\mathbf{d}] = (\mathbf{a} \cdot \mathbf{c})(\mathbf{b} \cdot \mathbf{d}) - (\mathbf{a} \cdot \mathbf{d})(\mathbf{b} \cdot \mathbf{c})$$

$$(\mathbf{a} \times \mathbf{b}) \times (\mathbf{c} \times \mathbf{d}) = (\mathbf{a} \times \mathbf{b} \cdot \mathbf{d})\mathbf{c} - (\mathbf{a} \times \mathbf{b} \cdot \mathbf{c})\mathbf{d}$$

$$\nabla(\Phi + \psi) = \nabla\Phi + \nabla\psi$$

$$\nabla(\Phi\psi) = \Phi\nabla\psi + \psi\nabla\Phi$$

$$\nabla \cdot (\mathbf{a} + \mathbf{b}) = \nabla \cdot \mathbf{a} + \nabla \cdot \mathbf{b}$$

$$\nabla \times (\mathbf{a}+\mathbf{b}) = \nabla \times \mathbf{a} + \nabla \times \mathbf{b}$$

$$\nabla \cdot (\Phi \mathbf{a}) = \mathbf{a} \cdot \nabla \Phi + \Phi \nabla \cdot \mathbf{a}$$

$$\nabla \times (\Phi \mathbf{a}) = \nabla \Phi \times \mathbf{a} + \Phi \nabla \times \mathbf{a}$$

$$\nabla (\mathbf{a} \cdot \mathbf{b}) = (\mathbf{a} \cdot \nabla)\mathbf{b} + (\mathbf{b} \cdot \nabla)\mathbf{a} + \mathbf{a} \times (\nabla \times \mathbf{b}) + \mathbf{b} \times (\nabla \times \mathbf{a})$$

$$\nabla \times (\mathbf{a} \times \mathbf{b}) = \mathbf{a} \nabla \cdot \mathbf{b} - \mathbf{b} \nabla \cdot \mathbf{a} + (\mathbf{b} \cdot \nabla)\mathbf{a} - (\mathbf{a} \cdot \nabla)\mathbf{b}$$

$$\nabla \cdot (\mathbf{a} \times \mathbf{b}) = \mathbf{b} \cdot \nabla \times \mathbf{a} - \mathbf{a} \cdot \nabla \times \mathbf{b}$$

$$\nabla \times \nabla \times \mathbf{a} = \nabla \nabla \cdot \mathbf{a} - \nabla^2 \mathbf{a}$$

$$\nabla \times \nabla \Phi \equiv 0$$

$$\nabla \cdot \nabla \times \mathbf{a} \equiv 0$$

$$\iiint_V \nabla \cdot \mathbf{a} \, dV = \oiint_S \mathbf{a} \cdot d\mathbf{S} \quad \text{(Gauss's theorem)}$$

$$\iint_S \nabla \times \mathbf{a} \cdot d\mathbf{S} = \oint_C \mathbf{a} \cdot d\boldsymbol{\ell} \quad \text{(Stokes's theorem)}$$

Green's first and second identities:

$$\iiint_V (\nabla \psi \cdot \nabla \Phi + \Phi \nabla^2 \psi) \, dV = \oint_S \Phi \nabla \psi \cdot d\mathbf{S}$$

$$\iiint_V (\Phi \nabla^2 \psi - \psi \nabla^2 \Phi) \, dV = \oint_S (\Phi \nabla \psi - \psi \nabla \Phi) \cdot d\mathbf{S}$$

$$\iiint_V (\nabla \times \mathbf{A} \cdot \nabla \times \mathbf{B} - \mathbf{A} \cdot \nabla \times \nabla \times \mathbf{B}) \, dV = \iiint_V (\nabla \cdot \mathbf{A} \times \nabla \times \mathbf{B}) \, dV$$

$$= \oint_S \mathbf{A} \times \nabla \times \mathbf{B} \cdot d\mathbf{S}$$

$$\iiint_V (\mathbf{B} \cdot \nabla \times \nabla \times \mathbf{A} - \mathbf{A} \cdot \nabla \times \nabla \times \mathbf{B}) \, dV = \oint_S (\mathbf{A} \times \nabla \times \mathbf{B} - \mathbf{B} \times \nabla \times \mathbf{A}) \cdot d\mathbf{S}$$

Appendix E

VSWR Versus Reflection Coefficient and Mismatch Loss

Yi-Lin Chen
University of Illinois[*]

The following relations are used in the construction of the vswr table below.

$$\text{vswr} = \frac{1 + |\Gamma|}{1 - |\Gamma|}, \qquad |\Gamma| = \frac{\text{vswr} - 1}{\text{vswr} + 1}$$

$$\text{mismatch loss (dB)} = -10\log_{10}(1 - |\Gamma|^2)$$

VSWR Versus Reflection Coefficient (Γ) and Mismatch Loss

VSWR	$|\Gamma|$	Mismatch Loss (dB)	VSWR	$|\Gamma|$	Mismatch Loss (dB)
1.01	.0050	.0001	1.12	.0566	.0139
1.02	.0099	.0004	1.13	.0610	.0162
1.03	.0148	.0009	1.14	.0654	.0186
1.04	.0196	.0017	1.15	.0698	.0212
1.05	.0244	.0026	1.16	.0741	.0239
1.06	.0291	.0037	1.17	.0783	.0267
1.07	.0338	.0050	1.18	.0826	.0297
1.08	.0385	.0064	1.19	.0868	.0328
1.09	.0431	.0081	1.20	.0909	.0360
1.10	.0476	.0099	1.21	.0950	.0394
1.11	.0521	.0118	1.22	.0991	.0429

---

[*]On leave from the Chinese Aeronautical Laboratory, Beijing, China, during 1983.

VSWR Versus Reflection Coefficient (Γ) and Mismatch Loss (cont'd.)

VSWR	|Γ|	Mismatch Loss (dB)	VSWR	|Γ|	Mismatch Loss (dB)
1.23	.1031	.0464	1.73	.2674	.3222
1.24	.1071	.0501	1.74	.2701	.3289
1.25	.1111	.0540	1.75	.2727	.3357
1.26	.1150	.0579	1.76	.2754	.3425
1.27	.1189	.0619	1.77	.2780	.3493
1.28	.1228	.0660	1.78	.2806	.3561
1.29	.1266	.0702	1.79	.2832	.3630
1.30	.1304	.0745	1.80	.2857	.3698
1.31	.1342	.0789	1.81	.2883	.3767
1.32	.1379	.0834	1.82	.2908	.3837
1.33	.1416	.0880	1.83	.2933	.3906
1.34	.1453	.0927	1.84	.2958	.3976
1.35	.1489	.0974	1.85	.2982	.4046
1.36	.1525	.1023	1.86	.3007	.4116
1.37	.1561	.1072	1.87	.3031	.4186
1.38	.1597	.1121	1.88	.3056	.4257
1.39	.1632	.1172	1.89	.3080	.4327
1.40	.1667	.1223	1.90	.3103	.4398
1.41	.1701	.1275	1.91	.3127	.4469
1.42	.1736	.1328	1.92	.3151	.4540
1.43	.1770	.1382	1.93	.3174	.4612
1.44	.1803	.1436	1.94	.3197	.4683
1.45	.1837	.1490	1.95	.3220	.4755
1.46	.1870	.1546	1.96	.3243	.4827
1.47	.1903	.1602	1.97	.3266	.4899
1.48	.1935	.1658	1.98	.3289	.4971
1.49	.1968	.1715	1.99	.3311	.5043
1.50	.2000	.1773	2.00	.3333	.5115
1.51	.2032	.1831	2.05	.3443	.5479
1.52	.2063	.1890	2.10	.3548	.5844
1.53	.2095	.1949	2.15	.3651	.6212
1.54	.2126	.2009	2.20	.3750	.6582
1.55	.2157	.2069	2.25	.3846	.6952
1.56	.2188	.2130	2.30	.3939	.7324
1.57	.2218	.2191	2.35	.4030	.7696
1.58	.2248	.2252	2.40	.4118	.8069
1.59	.2278	.2314	2.45	.4203	.8441
1.60	.2308	.2377	2.50	.4286	.8814
1.61	.2337	.2440	2.55	.4366	.9186
1.62	.2366	.2503	2.60	.4444	.9557
1.63	.2395	.2566	2.65	.4521	.9928
1.64	.2424	.2630	2.70	.4595	1.0298
1.65	.2453	.2695	2.75	.4667	1.0667
1.66	.2481	.2760	2.80	.4737	1.1035
1.67	.2509	.2825	2.85	.4805	1.1402
1.68	.2537	.2890	2.90	.4872	1.1767
1.69	.2565	.2956	2.95	.4937	1.2131
1.70	.2593	.3022	3.00	.5000	1.2494
1.71	.2620	.3088	3.10	.5122	1.3215
1.72	.2647	.3155	3.20	.5238	1.3929

VSWR Versus Reflection Coefficient (Γ) and Mismatch Loss

VSWR	|Γ|	Mismatch Loss (dB)	VSWR	|Γ|	Mismatch Loss (dB)
3.30	.5349	1.4636	8.10	.7802	4.0754
3.40	.5455	1.5337	8.20	.7826	4.1170
3.50	.5556	1.6030	8.30	.7849	4.1583
3.60	.5652	1.6715	8.40	.7872	4.1992
3.70	.5745	1.7393	8.50	.7895	4.2397
3.80	.5833	1.8064	8.60	.7917	4.2798
3.90	.5918	1.8727	8.70	.7938	4.3196
4.00	.6000	1.9382	8.80	.7959	4.3591
4.10	.6078	2.0030	8.90	.7980	4.3982
4.20	.6154	2.0670	9.00	.8000	4.4370
4.30	.6226	2.1302	9.10	.8020	4.4754
4.40	.6296	2.1927	9.20	.8039	4.5135
4.50	.6364	2.2545	9.30	.8058	4.5513
4.60	.6429	2.3156	9.40	.8077	4.5888
4.70	.6491	2.3759	9.50	.8095	4.6260
4.80	.6552	2.4355	9.60	.8113	4.6628
4.90	.6610	2.4945	9.70	.8131	4.6994
5.00	.6667	2.5527	9.80	.8148	4.7356
5.10	.6721	2.6103	9.90	.8165	4.7716
5.20	.6774	2.6672	10.00	.8182	4.8073
5.30	.6825	2.7235	11.00	.8333	5.1491
5.40	.6875	2.7791	12.00	.8462	5.4665
5.50	.6923	2.8340	13.00	.8571	4.7625
5.60	.6970	2.8884	14.00	.8667	6.0399
5.70	.7015	2.9421	15.00	.8750	6.3009
5.80	.7059	2.9953	16.00	.8824	6.5472
5.90	.7101	3.0479	17.00	.8889	6.7804
6.00	.7143	3.0998	18.00	.8947	7.0017
6.10	.7183	3.1513	19.00	.9000	7.2125
6.20	.7222	3.2021	20.00	.9048	7.4135
6.30	.7260	3.2525	30.00	.9355	9.0354
6.40	.7297	3.3022	40.00	.9512	10.2145
6.50	.7333	3.3515	50.00	.9608	11.1411
6.60	.7368	3.4002	60.00	.9672	11.9045
6.70	.7403	3.4485	70.00	.9718	12.5536
6.80	.7436	3.4962	80.00	.9753	13.1182
6.90	.7468	3.5435	90.00	.9780	13.6178
7.00	.7500	3.5902	100.00	.9802	14.0658
7.10	.7531	3.6365	200.00	.9900	17.0330
7.20	.7561	3.6824	300.00	.9934	18.7795
7.30	.7590	3.7277	400.00	.9950	20.0217
7.40	.7619	3.7727	500.00	.9960	20.9865
7.50	.7647	3.8172	600.00	.9967	21.7754
7.60	.7674	3.8612	700.00	.9971	22.4428
7.70	.7701	3.9049	800.00	.9975	23.0212
7.80	.7727	3.9481	900.00	.9978	23.5315
7.90	.7753	3.9909	1000.00	.9980	23.9881
8.00	.7778	4.0334			

Appendix F

Decibels Versus Voltage and Power Ratios*

Yi-Lin Chen
University of Illinois[†]

The decibel chart below indicates decibels for any ratio of voltage or power up to 100 dB. For voltage ratios greater than 10 (or power ratios greater than 100) the ratio can be broken down into two products, the decibels found for each separately, the two results then added. For example, to convert a voltage ratio of 200:1 to dB, a 200:1 voltage ratio equals the product of 100:1 and 2:1. Now, 100:1 equals 40 dB; 21:1 equals 6 dB. Therefore a 200:1 voltage ratio equals 40 dB + 6 dB, or 46 dB.

$$dB = 20\log_{10}(\text{voltage ratio}) = 10\log_{10}(\text{power ratio})$$

---

*Reprinted with permission of *Microwave Journal*, from *The Microwave Engineer's Handbook and Buyer's Guide*, 1966 issue, © 1966 Horizon House–Microwave, Inc.
[†]On leave from the Chinese Aeronautical Laboratory, Beijing, China, during 1983.

Decibels Versus Voltage and Power Ratios

Voltage Ratio	Power Ratio	−dB +	Voltage Ratio	Power Ratio	Voltage Ratio	Power Ratio	−dB +	Voltage Ratio	Power Ratio	Voltage Ratio	Power Ratio	−dB +	Voltage Ratio	Power Ratio
1.0000	1.0000	0	1.000	1.000	.5309	.2818	5.5	1.884	3.548	.2818	.07943	11.0	3.548	12.59
.9886	.9772	.1	1.012	1.023	.5248	.2754	5.6	1.905	3.631	.2786	.07762	11.1	3.589	12.88
.9772	.9550	.2	1.023	1.047	.5188	.2692	5.7	1.928	3.715	.2754	.07586	11.2	3.631	13.18
.9661	.9333	.3	1.035	1.072	.5129	.2630	5.8	1.950	3.802	.2723	.07413	11.3	3.673	13.49
.9550	.9120	.4	1.047	1.096	.5070	.2570	5.9	1.972	3.890	.2692	.07244	11.4	3.715	13.80
.9441	.8913	.5	1.059	1.122	.5012	.2512	6.0	1.995	3.981	.2661	.07079	11.5	3.758	14.13
.9333	.8710	.6	1.072	1.148	.4955	.2455	6.1	2.018	4.074	.2630	.06918	11.6	3.802	14.45
.9226	.8511	.7	1.084	1.175	.4898	.2399	6.2	2.042	4.169	.2600	.06761	11.7	3.846	14.79
.9120	.8318	.8	1.096	1.202	.4842	.2344	6.3	2.065	4.266	.2570	.06607	11.8	3.890	15.14
.9016	.8128	.9	1.109	1.230	.4786	.2291	6.4	2.089	4.365	.2541	.06457	11.9	3.936	15.49
.8913	.7943	1.0	1.122	1.259	.4732	.2239	6.5	2.113	4.467	.2512	.06310	12.0	3.981	15.85
.8810	.7762	1.1	1.135	1.288	.4677	.2188	6.6	2.138	4.571	.2483	.06166	12.1	4.027	16.22
.8710	.7586	1.2	1.148	1.318	.4624	.2138	6.7	2.163	4.677	.2455	.06026	12.2	4.074	16.60
.8610	.7413	1.3	1.161	1.349	.4571	.2089	6.8	2.188	4.786	.2427	.05888	12.3	4.121	16.98
.8511	.7244	1.4	1.175	1.380	.4519	.2042	6.9	2.213	4.898	.2399	.05754	12.4	4.169	17.38
.8414	.7079	1.5	1.189	1.413	.4467	.1995	7.0	2.239	5.012	.2371	.05623	12.5	4.217	17.78
.8318	.6918	1.6	1.202	1.445	.4416	.1950	7.1	2.265	5.129	.2344	.05495	12.6	4.266	18.20
.8222	.6761	1.7	1.216	1.479	.4365	.1905	7.2	2.291	5.248	.2317	.05370	12.7	4.315	18.62
.8128	.6607	1.8	1.230	1.514	.4315	.1862	7.3	2.317	5.370	.2291	.05248	12.8	4.365	19.05
.8035	.6457	1.9	1.245	1.549	.4266	.1820	7.4	2.344	5.495	.2265	.05129	12.9	4.416	19.50
.7943	.6310	2.0	1.259	1.585	.4217	.1778	7.5	2.371	5.623	.2239	.05012	13.0	4.467	19.95
.7852	.6166	2.1	1.274	1.622	.4169	.1738	7.6	2.399	5.754	.2213	.04898	13.1	4.519	20.42
.7762	.6026	2.2	1.288	1.660	.4121	.1698	7.7	2.427	5.888	.2188	.04786	13.2	4.571	20.89
.7674	.5888	2.3	1.303	1.698	.4074	.1660	7.8	2.455	6.026	.2163	.04677	13.3	4.624	21.38
.7586	.5754	2.4	1.318	1.738	.4027	.1622	7.9	2.483	6.166	.2138	.04571	13.4	4.677	21.88

.7499	.5623	2.5	1.334	1.778	.3981	.1585	8.0	2.512	6.310	.2113	.04467	13.5	4.732	22.39
.7413	.5495	2.6	1.349	1.820	.3936	.1549	8.1	2.541	6.457	.2089	.04365	13.6	4.786	22.91
.7328	.5370	2.7	1.365	1.862	.3890	.1514	8.2	2.570	6.607	.2065	.04266	13.7	4.842	23.44
.7244	.5248	2.8	1.380	1.905	.3846	.1479	8.3	2.600	6.761	.2042	.04169	13.8	4.898	23.99
.7161	.5129	2.9	1.396	1.950	.3802	.1445	8.4	2.630	6.918	.2018	.04074	13.9	4.955	24.55
.7079	.5012	3.0	1.413	1.995	.3758	.1413	8.5	2.661	7.079	.1995	.03981	14.0	5.012	25.12
.6998	.4898	3.1	1.429	2.042	.3715	.1380	8.6	2.692	7.244	.1972	.03890	14.1	5.070	25.70
.6918	.4786	3.2	1.445	2.089	.3673	.1349	8.7	2.723	7.413	.1950	.03802	14.2	5.129	26.30
.6839	.4677	3.3	1.462	2.138	.3631	.1318	8.8	2.754	7.586	.1928	.03715	14.3	5.188	26.92
.6761	.4571	3.4	1.479	2.188	.3589	.1288	8.9	2.786	7.762	.1905	.03631	14.4	5.248	27.54
.6683	.4467	3.5	1.496	2.239	.3548	.1259	9.0	2.818	7.943	.1884	.03548	14.5	5.309	28.18
.6607	.4365	3.6	1.514	2.291	.3508	.1230	9.1	2.851	8.128	.1862	.03467	14.6	5.370	28.84
.6531	.4266	3.7	1.531	2.344	.3467	.1202	9.2	2.884	8.318	.1841	.03388	14.7	5.433	29.51
.6457	.4169	3.8	1.549	2.399	.3428	.1175	9.3	2.917	8.511	.1820	.03311	14.8	5.495	30.20
.6383	.4074	3.9	1.567	2.455	.3388	.1148	9.4	2.951	8.710	.1799	.03236	14.9	5.559	30.90
.6310	.3981	4.0	1.585	2.512	.3350	.1122	9.5	2.985	8.913	.1778	.03162	15.0	5.623	31.62
.6237	.3890	4.1	1.603	2.570	.3311	.1096	9.6	3.020	9.120	.1758	.03090	15.1	5.689	32.36
.6166	.3802	4.2	1.622	2.630	.3273	.1072	9.7	3.055	9.333	.1738	.03020	15.2	5.754	33.11
.6095	.3715	4.3	1.641	2.692	.3236	.1047	9.8	3.090	9.550	.1718	.02951	15.3	5.821	33.88
.6026	.3631	4.4	1.660	2.754	.3199	.1023	9.9	3.126	9.772	.1698	.02884	15.4	5.888	34.67
.5957	.3548	4.5	1.679	2.818	.3162	.1000	10.0	3.162	10.000	.1679	.02818	15.5	5.957	35.48
.5888	.3467	4.6	1.698	2.884	.3126	.09772	10.1	3.199	10.23	.1660	.02754	15.6	6.026	36.31
.5821	.3388	4.7	1.718	2.951	.3090	.09550	10.2	3.236	10.47	.1641	.02692	15.7	6.095	37.15
.5754	.3311	4.8	1.738	3.020	.3055	.09333	10.3	3.273	10.72	.1622	.02630	15.8	6.166	38.02
.5689	.3236	4.9	1.758	3.090	.3020	.09120	10.4	3.311	10.96	.1603	.02570	15.9	6.237	38.90
.5623	.3162	5.0	1.778	3.162	.2985	.08913	10.5	3.350	11.22	.1585	.02512	16.0	6.310	39.81
.5559	.3090	5.1	1.799	3.236	.2951	.08710	10.6	3.388	11.48	.1567	.02455	16.1	6.383	40.74
.5495	.3020	5.2	1.820	3.311	.2917	.08511	10.7	3.428	11.75	.1549	.02399	16.2	6.457	41.69
.5433	.2951	5.3	1.841	3.388	.2884	.08318	10.8	3.467	12.02	.1531	.02344	16.3	6.531	42.66
.5370	.2884	5.4	1.862	3.467	.2851	.08128	10.9	3.508	12.30	.1514	.02291	16.4	6.607	43.65

Decibels Versus Voltage and Power Ratios, (cont'd.)

Voltage Ratio	Power Ratio	−dB +	Voltage Ratio	Power Ratio	Voltage Ratio	Power Ratio	−dB +	Voltage Ratio	Power Ratio	Voltage Ratio	Power Ratio	−dB +	Voltage Ratio	Power Ratio
.1496	.02239	16.5	6.683	44.67	.1259	.01585	18.0	7.943	63.10	.1059	.01122	19.5	9.441	89.13
.1479	.02188	16.6	6.761	45.71	.1245	.01549	18.1	8.035	64.57	.1047	.01096	19.6	9.550	91.20
.1462	.02138	16.7	6.839	46.77	.1230	.01514	18.2	8.128	66.07	.1035	.01072	19.7	9.661	93.33
.1445	.02089	16.8	6.918	47.86	.1216	.01479	18.3	8.222	67.61	.1023	.01047	19.8	9.772	95.50
.1429	.02042	16.9	6.998	48.98	.1202	.01445	18.4	8.318	69.18	.1012	.01023	19.9	9.886	97.72
.1413	.01995	17.0	7.079	50.12	.1189	.01413	18.5	8.414	70.79	.1000	.01000	20.0	10.000	100.00
.1396	.01950	17.1	7.161	51.29	.1175	.01380	18.6	8.511	72.44					
.1380	.01905	17.2	7.244	52.48	.1161	.01349	18.7	8.610	74.13			30		10^3
.1365	.01862	17.3	7.328	53.70	.1148	.01318	18.8	8.710	75.86	10^{-2}	10^{-3}	40	10^2	10^4
.1349	.01820	17.4	7.413	54.95	.1135	.01288	18.9	8.811	77.62		10^{-4}	50		10^5
.1334	.01778	17.5	7.499	56.23	.1122	.01259	19.0	8.913	79.43		10^{-5}	60		10^6
.1318	.01738	17.6	7.586	57.54	.1109	.01230	19.1	9.016	81.28	10^{-3}	10^{-6}	70	10^3	10^7
.1303	.01698	17.7	7.674	58.88	.1096	.01202	19.2	9.120	83.18		10^{-7}	80		10^8
.1288	.01660	17.8	7.762	60.26	.1084	.01175	19.3	9.226	85.11	10^{-4}	10^{-8}	90	10^4	10^9
.1274	.01622	17.9	7.852	61.66	.1072	.01148	19.4	9.333	87.10	10^{-5}	10^{-9}			
											10^{-10}	100	10^5	10^{10}

Index

Index

Abbe sine condition
  with lenses, 16-19 to 16-23, 16-29, 16-56, 21-18
  radiation pattern of lenses obeying, 19-97 to 19-99
Aberrations with lenses, 16-12 to 16-19, 21-23. *See also* Coma lens aberrations
Absolute gain measurements, 32-42 to 32-49
Absolute models, 32-69
Absolute polarization measurements, 32-60 to 32-61
Absorbers with UTD solutions, 20-17
Absorption
  of ionospheric propagation, 29-24
  of satellite-earth propagation, 29-8
ACS (attitude control systems), 22-46
Active element patterns
  of periodic arrays, 13-6
  with phased arrays, 21-8
Active region
  with log-periodic antennas
    dipole, 9-17 to 9-21, 9-24
    mono arrays, 9-65
    phasing of, 9-56
    zigzag wire, 9-35 to 9-36
  with log-spiral antennas, 9-75
    conical, 9-83, 9-84, 9-98 to 9-99
Adcock rotating antennas systems, 25-4, 25-9 to 25-10
  for in-rotating null patterns, 25-11 to 25-12
Admittance
  of edge slot arrays, 12-26
  of medium, 1-18
  mutual, between slots, 4-91
  of ridged TE/TM waveguides, 28-47
  of thin-wire antennas, 7-7 to 7-8, 7-14 to 7-15
  of transmission lines, 9-47 to 9-49, 28-5
Admittance matrices, 3-61
  and symmetries, 3-65, 3-67 to 3-68
Admittivity and dielectric modeling, 32-72
Advanced microwave sounding unit, 22-25, 22-48
Advanced Multifrequency Scanning Radiometer (AMSR), 22-48, 22-50, 22-51 to 22-52
AF method. *See* Aperture field method
Airborne Radiation Pattern code, 20-4, 20-70 to 20-85
Air coaxial transmission lines, 28-16
Aircraft and aircraft antennas
  blade antennas, 30-12, 30-16 to 30-19
  EMP responses for, 30-4, 30-6, 30-17 to 30-31
  interferometers mounted on, 25-21 to 25-23
  military, simulation of, 20-78 to 20-88
  modeling of, 32-74 to 32-76, 32-79 to 32-86
  monopoles, 20-62 to 20-68, 20-70, 20-73 to 20-75, 20-87 to 20-88
  numerical solutions for, 20-70 to 20-90
  radomes on, 31-3 to 31-4
  simulation of, 20-30, 20-37 to 20-42, 20-63 to 20-65
  structure of, as antenna, 20-61
Air gap with leaky-wave antennas, 17-97 to 17-98
Airport radar, antennas for, 19-19, 19-56
Air striplines, 21-58, 21-60
  for power divider elements, 21-50 to 21-51
  with satellite antennas, 21-7
Air traffic control antennas, 19-19, 19-56
Alignment of antenna ranges, 32-33 to 32-34
Aluminum alloys for satellite antennas, 21-28 to 21-29
AM antennas
  directional feeders for, 26-37 to 26-47
  ground systems for, 26-37

AM antennas (*cont.*)
  horizontal plane field strength in, 26-33 to 26-34
  patterns for
    augmented, 26-36
    size determination of, 26-22 to 26-33
    standard, 26-35 to 26-36
    theoretical, 26-34 to 26-35
    two-tower, 26-22
  power for system losses in, 26-36 to 26-37
  sky-wave propagation with, 29-25 to 29-27
  standard reference, 26-3 to 26-22
Ammeters, 2-30 to 2-32
Amplitude meters, 2-30 to 2-32
Amplitude patterns, measurement of, 32-39 to 32-41
Amplitude quantization errors, 21-12
Amplitude source, ideal, 2-29 to 2-30
AMSR, 22-48, 22-50, 22-51 to 22-52
AMSU (advanced microwave sounding unit), 22-25, 22-50, 22-51 to 22-52
Analytical formulations of modeling codes, 3-89
Analytical validation of computer code, 3-76
Anechoic chambers for indoor ranges, 32-25 to 32-28
  evaluation of, 32-38 to 32-39
Angles with log-periodic antennas, 9-12, 9-14, 9-15, 9-23
Annular patches for microstrip antennas, 10-43 to 10-45
  characteristics of, 10-17
  resonant frequency for, 10-47, 10-48
Annular phased array applicator for medical applications, 24-16 to 24-18
Annular-sector patches for microstrip antennas
  characteristics of, 10-17
  resonant frequency of, 10-47, 10-49
Annular slot antennas, 30-13
Antarctica, brightness temperature image of, 22-15
Antenna arrays. *See* Arrays
Antenna ranges. *See* Ranges, antenna
Antenna sampling systems for feeder systems, 26-41 to 26-44
Antipodal fin lines, 28-55
Aperiodic arrays, 14-3 to 14-6
  linear arrays as, 11-41
  optically fed, 19-57 to 19-59
  probabilistic approach to, 14-8 to 14-35
  space-tapered, 14-6 to 14-8
Aperture antennas, 5-5
  and discrete arrays, 11-29 to 11-30
  optimization of, 11-74
Aperture blockage, 5-16
  with Cassegrain feed systems, 19-62
  in compact ranges, 32-29, 33-23
  with HIHAT antennas, 19-87
  and lens antennas, 16-5, 21-16
  with millimeter-wave antennas, 17-28
  and offset parabolic antennas, 15-80
    dual-reflector, 15-61
    with off-focus feeds, 15-58
  with phased array feeds, 19-60 to 19-61
  with reflector antennas, 15-17 to 15-18
  and space-fed arrays, 19-54 to 19-55
Aperture couplers power division elements, 21-51 to 21-52
Aperture distributions. *See* Apertures and aperture distributions
Aperture efficiency
  of dielectric-loaded horn antennas, 8-73
  and effective area, 5-27 to 5-28
  of geodesic antennas, 17-32
  with lenses, 16-19, 16-20
  with offset phased-array feeds, 19-61
  and parallel feed networks, 19-6
  with reflector antennas, 21-24 to 21-25
  and reflector surface errors, 15-105
Aperture field method, 5-5
  for pyramidal horn fields, 15-92 to 15-93
  with reflector antennas, 8-72, 15-7, 15-13 to 15-15
Aperture fields, 15-88
  for circular arrays, 21-92
  for circumferential slots, 21-92
  of horn antennas
    conical, 15-93
    corrugated, 8-53
    E-plane, 8-5 to 8-7
    H-plane, 8-20
    pyramidal, 8-34, 8-36
Aperture-matched horn antennas, 8-50, 8-64
  bandwidths of, 8-65 to 8-66
  radiated fields of, 8-58, 8-66 to 8-67
  vswr of, 8-68
Apertures and aperture distributions, 11-13. *See also* Circular apertures; Planar apertures; Tapered aperture distributions; Uniform aperture distributions

with Butler matrix, 19-8
on curved surfaces, 13-50 to 13-51
efficiency of. *See* Aperture efficiency
electric and magnetic fields in, 5-8
with feed systems
  radial transmission line, 19-28, 19-31, 19-33
  semiconstrained, 19-17
  unconstrained, 19-51
and gain, 5-26 to 5-27, 17-7, 17-8
illumination of, 13-30 to 13-32, 13-36 to 13-37
with lens antennas, 21-14, 21-16, 21-17
  phase distributions of, 16-7, 16-36
  power distributions of, 16-33, 16-36
of longitudinal array slots, 12-24
and parallel feed networks, 19-6
in perfectly conducting ground, 3-22
radiation from, 1-28
  circular, 5-20 to 5-25
  and equivalent currents, 5-8 to 5-10
  near-field, 5-24 to 5-26
  planar aperture distributions, 5-10 to 5-11
  and plane-wave spectra, 5-5 to 5-7
  rectangular aperture, 5-11 to 5-20
reflections of, 8-67
with reflector antennas, 15-15 to 15-23, 21-20
phase error of, 15-107 to 15-108
for satellite antennas, 21-5, 21-7, 21-14, 21-16, 21-17
size of
  and conical horn beamwidth, 8-61
  of E-plane horn antenna directivity, 8-16
  of lenses, 16-6
square, 5-32 to 5-33
with Taylor line source synthesis, 13-26
Aperture taper
  and radiometer beam efficiency, 22-32, 22-33
  with reflector antennas, 21-24
Apparent phase center, 8-75
Appleton-Hartee formula for ionospheric refractive index, 29-21
Applied excitation of periodic arrays, 13-6 to 13-7
Arabsat satellite antenna, 21-77
Arbitrary ray optical field, 4-8
Archimedean-spiral curves, 9-107 to 9-108, 9-110

Array blindness, 13-45 to 13-49
  with flared-notch antennas, 13-59
Array collimations, 13-7 to 13-10
Array geometry, linear transformations in, 11-23 to 11-25
Array pattern functions, 3-35, 11-8
  aperiodic, 14-4 to 14-5, 14-8 to 14-9, 14-15, 14-22, 14-35
  Dolph-Chebyshev, 11-17 to 11-18
  linear, 11-8, 11-9, 11-11 to 11-14
  with longitudinal slots, 12-17
  phased, 21-10
  transformation of, 11-23 to 11-48
  with UTD solutions, 20-6
Arrays, 3-33, 3-35 to 3-36, 3-38 to 3-43. *See also* Aperiodic arrays; Array theory; Broadside arrays; Circular arrays; Log-periodic arrays; Log-spiral antennas; Periodic arrays; Planar arrays; Slot arrays
  element patterns with, 14-25
  errors in, from phase quantization, 13-52 to 13-57
  horn antennas as elements in, 8-4
  large, design of, 12-28 to 12-34
  of microstrip antennas, for medical applications, 24-25
  millimeter-wave, 17-23 to 17-27
  phase control of, 13-62 to 13-64
  scan characteristics of, 14-24 to 14-25
  of tapered dielectric-rod antennas, 17-47
  thinning of, with aperiodic arrays, 14-3
Array theory
  directivity in, 11-63 to 11-76
  general formulation of, 11-5 to 11-8
  for linear arrays, 11-8 to 11-23
  linear transformations in, 11-23 to 11-48
  pattern synthesis in, 11-76 to 11-86
  for planar arrays, 11-48 to 11-58
  SNR in, 11-58 to 11-63, 11-70 to 11-76
Artificial-dielectric plates, 21-19
A-sandwich panels, 31-17 to 31-19
Aspect ratio
  of dielectric grating antennas, 17-77 to 17-78
    surface corrugations for, 17-64
  and leakage constants, 17-62, 17-63
  with log-periodic antennas, 9-12, 9-14
  and directivity, 9-21 to 9-24
Assistance with computer models, 3-68
Astigmatic lens aberrations, 16-15, 16-20, 21-13

Astigmatic ray tube, 4-8
Asymmetric aperture distributions, 21-17
Asymmetric strips, 17-84 to 17-87, 17-97 to 17-98
AT-536/ARN marker beacon antenna, 30-27 to 30-28
AT-1076 uhf antenna, 30-21
Atmosphere
 absorption by
  of millimeter-wave antennas, 17-5 to 17-6
  and satellite-earth propagation, 29-8
 noise from, 6-25, 29-50 to 29-51
 refraction of, and line-of-sight propagation, 29-32 to 29-33
 refractive index of, 29-30 to 29-32
 remote sensing of, 22-6 to 22-7, 22-13 to 22-14, 22-16, 22-21 to 22-22
 standard radio, 29-32
Attachment coefficients, 4-96
 for conducting cylinder, 21-97
 of microstrip line, 17-118
Attenuation and attenuation constants
 of atmosphere, 17-5 to 17-6, 29-8
 with biconical horns, 21-104
 for conducting cylinder, 21-97
 ground wave, 29-45
 with log-periodic antennas, 9-4 to 9-5
 with microstrip patch antennas, 17-110
 with periodic loads, 9-49 to 9-54
 from radiation, 7-12, 9-5
 by rain, 29-10 to 29-13
 of rf cables, 28-26
 of standard waveguides, 28-40 to 28-41
  circular, 1-49 to 1-50
  rectangular, 1-39, 1-43
  TE/TM, 28-37
 for transmission lines, 28-7
 with uniform leaky-wave antennas, 17-89, 17-92
Attitude control systems, 22-46
Augmented am antenna pattern, 26-36 to 26-37
Auroral blackout and ionospheric propagation, 29-25
Automated antenna ranges, 32-30 to 32-31
Axial current on thin-wire antennas, 7-5 to 7-6
Axial feed displacements, 15-49 to 15-55
Axial gain, near-field, 5-29 to 5-33
Axial power density, 5-32 to 5-33

Axial ratios
 with lens antenna feeds, 21-79, 21-80
 and polarization, 32-51
  with log-spiral antennas, 9-77 to 9-78, 9-90 to 9-93
  with polarization ellipse, 1-15 to 1-16
Axis slot on elliptical cylinder, 4-75, 4-77, 21-92, 21-95
Azimuth-over-elevation positioners, 32-12
Azimuthal symmetry, 1-33

Babinet principle, 2-13 to 2-16
 and EMP and slot antennas, 30-14 to 30-16
Backfire arrays, 9-64
Back lobes and horn antennas
 aperture-matched, 8-67
 corrugated, 8-55, 8-58
Baklanov-Tseng-Cheng design, 11-55 to 11-56
Balanced antennas
 dipole, 9-72 to 9-73
 log-periodic zigzag, 9-45
 slot, 9-72 to 9-73, 9-78 to 9-79
 spiral, 9-78
Balanced bifilar helix, 9-83
Balanced two- and four-wire transmission lines, 28-12, 28-13
Baluns
 for log-spiral antennas, 9-75
  conical, 9-95, 9-103, 9-105
 microstrip, 18-8
Bandwidth, 3-28 to 3-31
 of beam-forming networks, 18-21, 18-24, 18-26, 21-32
 and diode phase shifters, 13-62 to 13-64, 18-16
 of feed circuits
  array feeds, 13-61 to 13-62
  broadband array, 13-39 to 13-41
  parallel feed, 19-5 to 19-6
  space-fed beam-forming feeds, 18-18
  true-time-delay, 19-77 to 19-78
 of horn antennas
  aperture-matched, 8-64 to 8-65
  corrugated, 8-64
  hybrid-mode, 15-99
  millimeter-wave, 17-23
 of leaky-wave antennas, 17-98
 of lens antennas
  constrained, 16-55
  dielectric, 16-54 to 16-56

equal group delay, 16-45 to 16-46
millimeter-wave, 17-14
Rinehart-Luneberg, 19-44, 19-46 to
  19-49
zoning of, 16-39, 16-43, 16-54 to
  16-55, 21-16
of log-periodic dipole antennas, 9-17, 9-21
of log-spiral conical antennas, 9-98 to
  9-99
of longitudinal-shunt-slot array, 17-33
of microstrip antennas, 10-6, 10-46,
  17-106, 17-118, 17-122
  with circular polarization, 10-59, 10-61
  dipole, 17-118 to 17-120, 17-125 to
    17-127
  and impedance matching, 10-50 to
    10-52
of millimeter-wave antennas, 17-5
  biconical, 17-32
  dielectric grating, 17-74 to 17-75
  holographic, 17-130
  horn, 17-13
  lens, 17-14
  microstrip arrays, 17-106, 17-114,
    17-116
  tapered dielectric-rod, 17-39 to 17-40,
    17-44
of phased arrays, 13-19 to 13-20, 18-8,
  21-12
and Q-factor, 6-22, 10-6
of receiving antennas, 6-22, 6-24
of satellite antennas, 21-4 to 21-5, 21-7
of self-complementary antennas, 9-10
with side-mount tv antennas, 27-21
of subarrays, 13-32 to 13-35, 13-39
and substrate height, 17-107
and temperature sounder sensitivity,
  22-22
of TEM waveguides, 21-57
Bar-line feed network, 21-74, 21-78
Bar line transmission types, 21-58
Barn antennas, 21-84 to 21-85
Base impedance of am antennas, 26-8 to
  26-17
Basic Scattering Code, 20-4
  for antennas on noncurved surfaces,
    20-68
  for numerical simulations
    of aircraft, 20-70, 20-85 to 20-86, 20-90
    of ships, 20-90 to 20-97
Basis functions in modeling codes, 3-81,
  3-83

Batwing tv antennas, 27-23 to 27-24
Bayliss line source pattern synthesis,
  13-27 to 13-29
BCS (bistatic cross section), 2-22 to 2-23
BDF. *See* Beam deviation factor
Beacon antennas
  EMP responses for, 30-23, 30-27 to
    30-28
  with satellite antennas, 21-84
Beads with coaxial lines, 28-21 to 28-22
Beam-broadening factor
  with linear arrays, 11-19 to 11-21
  with periodic arrays, 13-17 to 13-19
  with phased arrays, 21-10
Beam deviation factor
  with offset antennas, 15-81, 21-24
  with scanned beams, 15-51, 15-53, 15-55
    to 15-59, 15-62
Beam dithering, 13-56
Beam efficiency
  in arrays, 11-74, 11-76
  for conical horn antennas, 8-62
  for microwave radiometers, 22-28 to
    22-30, 22-32 to 22-33
Beam-forming feed networks
  constrained, 19-3 to 19-15
  cylindrical arrays, 19-98 to 19-119
  optical transform, 19-91 to 19-98
  for phased arrays, 18-17 to 18-26
  for satellite antennas, 21-5 to 21-6, 21-29
    to 21-57
  semiconstrained, 19-15 to 19-49
  transmission lines for, 21-63
  unconstrained, 19-49 to 19-91
Beam-pointing errors, 18-27 to 18-28
Beam scanning, 14-25 to 14-26
  of arrays
    aperiodic, 14-32 to 14-35
    Dolph-Chebyshev, 11-52
    millimeter-wave, 17-25 to 17-26
  beam-forming feed networks for, 18-21,
    18-24, 21-27 to 21-34, 21-37 to
    21-49
  conical, 22-40 to 22-41
  with microwave radiometers, 22-33 to
    22-43
  with millimeter-wave antennas, 17-9,
    17-10, 17-24 to 17-26
    array, 17-25 to 17-26
    dielectric grating, 17-69 to 17-72
  with offset parabolic antennas, 15-51,
    15-53

Beam scanning (*cont.*)
  with satellite antennas, 21-5
    phased-array, 21-71
    polar orbiting, 22-35 to 22-38
Beam-shaping efficiency, 21-5 to 21-6
Beam squint
  with constrained feeds, 18-19, 18-21 to 18-22
  with offset antennas, 15-48
  with phase-steered arrays, 13-20
  with reflector satellite antennas, 21-19
Beam steering
  computer for, in phased array design, 18-3 to 18-4, 18-26
  with lens antennas, 17-22
  with Wheeler Lab approach, 19-107
Beam tilt with tv antennas, 27-7, 27-26
Beamwidth, 1-20. *See also* Half-power beamwidths
  of arrays
    aperiodic, 14-3, 14-12
    linear, 11-15
    periodic, 13-10 to 13-12, 13-17 to 13-19
    phased, 19-61, 21-11 to 21-12
    planar, 11-53 to 11-55
    subarrays, 13-32
  and Dolph-Chebyshev pattern synthesis, 13-24 to 13-25
  of feed circuits
    limited scan, 19-57
    series feed networks, 19-5
  and gain, with tv antennas, 27-7
  of horn antennas
    diagonal, 8-70
    stepped-horn, 21-82 to 21-84
  and lens F/D ratio, 21-16
  of log-periodic antennas
    planar, 9-9 to 9-10
    zigzag, 9-39 to 9-40, 9-45 to 9-46
  of millimeter-wave antennas
    dielectric grating, 17-69 to 17-70, 17-74, 17-81 to 17-82
    geodesic, 17-32
    integrated, 17-134
    lens, 17-20
    metal grating, 17-69
    microstrip dipole, 17-119
    pillbox, 17-29
    reflector, 17-11, 17-13, 17-14
    slotted-shunt-slot array, 17-33
    spiral, 17-28
    tapered rod, 17-47
    uniform-waveguide leaky-wave, 17-85, 17-90, 17-98
  of multibeam antennas, 21-58
  and optimum rectangular aperture patterns, 5-18 to 5-20
  of reflector antennas
    offset, 15-37, 15-82
    tapered-aperture, 15-16 to 15-17
  of subarrays, 13-32
  and Taylor line source pattern synthesis, 13-26 to 13-27
  of tv antennas, 27-7
    side-mount tv antennas, 27-20
  of uhf antennas, 27-27
Beamwidths scanned and gain loss with reflector antennas
  dual-reflector, 15-73
  offset parabolic, 15-55, 15-58, 15-62 to 15-66
Benelux Cross antenna, 14-17 to 14-18
Beyond-the-horizon transmission, 29-21 to 29-30
Bhaskara-I and -II satellites, 22-17
Biconical antennas, 3-24 to 3-26, 3-29
  coaxial dipole, 3-30
  horn, 3-30
  millimeter-wave, 17-32
  for satellite antennas, 21-89, 21-100 to 21-109
Bifilar helix, 9-83
Bifilar zigzag wire antennas, 9-35, 9-37
Bifocal lenses, 16-30 to 16-33
Bilateral fin lines, 28-55
Binomial linear arrays, 11-10 to 11-11
Bistatic cross section, 2-22 to 2-23
Bistatic radar, 2-36 to 2-39
Blackbody radiation, 22-3 to 22-4
Blade antennas
  EMP responses for, 30-17 to 30-19
  equivalent circuits for, 30-16
  equivalent receiving area of, 30-12
Blass matrices
  with beam-forming feed networks, 18-25, 19-10 to 19-11
  with space-fed subarray systems, 19-68
  for switch beam circuits, 21-31 to 21-33
Blass tilt traveling-wave arrays, 19-80
Blind angles with arrays, 14-25 to 14-27, 14-32 to 14-35
Blindness, array, 13-45 to 13-49
  with flared-notch antennas, 13-59

Index

Blind spots with horn antennas, 8-4
Blockage. *See* Aperture blockage
Block 5D satellite, 22-18
Body-stabilized satellites, 21-90
Boeing 707 aircraft and EMP, 30-4, 30-6
Boeing 737 aircraft
   simulation of, 20-64, 20-70 to 20-78
   slot antennas on, 20-85 to 20-86, 20-89, 20-90
Boeing 747 aircraft and EMP, 30-4, 30-6
Bombs, aircraft, simulation of, 20-79
Bone, human, phantom models for, 24-55
Booker-Gordon formula for scattering, 29-28
Boom length of log-periodic antennas
   dipole, 9-21, 9-23, 9-24, 9-26
   zigzag, 9-39
Bootlace lenses, 16-46 to 16-48
Boresight beam
   gain of, with reflector antennas, 15-5
   with offset parabolic antennas, 15-81
   of phased-array satellite antennas, 21-71
   and radomes, 31-3 to 31-4, 31-7 to 31-9, 31-27 to 31-29
Born-Rytov and computer solutions, 3-54
Boundary conditions, 1-8 to 1-9
   with corrugated horns, 8-50, 15-96 to 15-97
   and current distributions, 3-38 to 3-39
   and lens zoning, 16-39
   in longitudinal slots, 12-6
   numerical implementation of, 3-55
   numerical validation of, 3-79 to 3-80
   with patches, 10-16 to 10-17
   and perfect ground planes, 3-18 to 3-19
   for thin wires, 3-44 to 3-45
Bow-tie dipoles, 17-137 to 17-138
Bragg condition for diffraction, 29-28
Brain, human, phantom models for, 24-55
Bra-Ket notation and reciprocity, 2-16
Branch guide coupler power division elements, 21-51 to 21-52
Branch line couplers
   for array feeds, 13-61
   directional power division element, 21-50
Branch line guide for planar arrays, 12-21 to 12-23
Brewster angle phenomena
   modeling of, 32-79
   and radome design, 31-12 to 31-13

Brightness temperature
   of emitters, 2-40 to 2-41
   and humidity, 22-6 to 22-7
   of microwave radiometers, 22-28
   vs. physical temperature, 22-10
   and surface emissivity, 22-7 to 22-9, 22-15
Broadband array feeds, 13-39 to 13-41
Broadside arrays, 3-40, 11-11, 11-15
   maximum directivity of, 11-64 to 11-67
   radiation pattern of, 13-12 to 13-13
Broadside coupled transmission lines, 28-16, 28-17
Broadside radiation of microstrip arrays, 17-115
Broadwall millimeter-wave array antennas, 17-26
Broadwall shunt-slot radiators, 18-8
Broadwall slots
   for center-inclined arrays, 12-24
   longitudinal, 12-4
BSC. *See* Basic Scattering Code
Bulge, earth, and line-of-sight propagation 29-34
Butler matrices
   with arrays
     multimode circular, 25-18 to 25-19
     multiple-beam, 19-80
   with beam-forming feed networks, 18-23 to 18-24, 19-8 to 19-10, 19-12
   with cylindrical array feeds, 19-101 to 19-103
   as Fourier transformer, 19-95
   with reflector-lens limited-scan feed concept, 19-68
   with space-fed subarray systems, 19-68
   for switch beam circuits, 21-31, 21-33
   in transform feeds, 19-92 to 19-93

Cable effect with modeling, 32-84
Cabling with scanning ranges, 33-16
CAD (computer-aided design), 17-120 to 17-122
Calibration
   accuracy of, with microwave radiometers, 22-23
   of AMSU, 22-50
   of field probes, 24-54 to 24-57
   horn antennas as standard for, 8-3, 8-43, 8-69
   of microwave radiometers, 22-25 to 22-26, 22-48

Cameras, thermographic, 24-51
Cancer therapy. *See* Hyperthermia
Candelabras for multiple-antenna installations, 27-37
Capacitance of in-vivo probes, 24-33 to 24-34
Capacitor-plate hyperthermia applicators, 24-46 to 24-47
Capture area of receiving antenna, 6-6
Cardioid patterns
    with loop antennas, 25-6 to 25-7
    with slot antennas
        coaxial, 27-28
        waveguide, 27-30
    from slotted cylinder reflectors, 21-110
    for T T & C, 21-89 to 21-90
Carrier-to-noise ratio, 29-48
Cartesian coordinate systems, 15-115 to 15-120
Cassegrain feed systems
    near-field, 19-62 to 19-63
    for reflector antennas
        millimeter-wave, 17-9, 17-13 to 17-14
        satellite, 21-21 to 21-22
    with wide-angle systems, 19-84, 19-88
Cassegrain offset antennas
    cross polarization with, 15-69 to 15-70, 15-72 to 15-74
    dual-mode horn antennas for, 8-72
    parameters of, 15-67 to 15-68
    performance evaluation of, 15-69 to 15-72
    scan performance of, 15-71 to 15-73, 15-78 to 15-79
Caustic distances, 4-10
    edge-diffracted, 4-27
    and GO reflected field, 4-18
Caustics, 4-6, 8-76
    with diffracted rays, 4-5, 4-6
        surface-diffracted, 4-83
    of diffracted waves, 4-96 to 4-102
    and GO representation, 4-16
    and lens antennas, 17-16
    matching functions for, 4-103
Cavity-backed radiators
    in panel fm antennas, 27-35 to 27-36
    in side-mount tv antennas, 27-20 to 27-23
Cavity-backed slots with log-periodic arrays, 9-68 to 9-71
Cavity model for microstrip antennas, 10-10 to 10-21
Centered broad wall slots, 12-4

Center-fed antennas
    dipole
        microstrip, 17-123 to 17-124
        transient response of, 7-22
    slot, for tv, 27-29 to 27-30
Center-fed arrays, 17-119 to 17-120
Center-fed dual series feeds, 18-20
Center-fed reflectarrays, 19-55
Center-inclined slot arrays, 12-24
Channel-diplexing, 22-46
Channel guide mode for leaky-wave antennas, 17-98
Characteristic impedance, 3-28 to 3-31
    of dipoles, 7-6, 9-28, 9-30
    of fin lines, 28-57, 28-58
    of holographic antennas, 17-130
    of log-periodic zigzag antennas, 9-42 to 9-44, 9-48
    of microstrip arrays, 17-109 to 17-111
    of slot lines, 28-53 to 28-54
    of transmission lines, 28-7
        biconical, 3-24 to 3-25
        coaxial, 28-20
        microstrip planar, 28-30 to 28-31
        TEM, 28-10 to 28-18
        triplate stripline, 28-22, 28-28 to 28-29
        two-wire, 7-39, 28-11, 28-19
        waveguide planar, 28-33
    of waveguides
        rectangular, 2-28
        ridged, 24-14 to 24-16
        TE/TM, 28-36, 28-37
Chebyshev polynomials, 11-15 to 11-17. *See also* Dolph-Chebyshev arrays
Check cases for validating computer code, 3-80
Chi-square distribution with aperiodic arrays, 14-10 to 14-12
Chokes
    with coaxial sleeve antennas, 7-31 to 7-33, 7-36
    with microstrip antennas, 10-68
Circles, 15-25 to 15-26
    and intersection curve, 15-29
Circuit characteristics
    measurements for, 32-4
    models for, 10-26 to 10-28
Circular apertures
    Bayliss line source pattern synthesis for, 13-29
    distribution of
        axial field, 5-32
        with reflector antennas, 15-17, 15-23

efficiency of, 5-29, 5-30
radiation patterns of, 5-20 to 5-24
near-field, 5-25
Circular arrays
for direction-finding antennas, 25-17 to 25-20
of probes, with transmission line feeds, 19-26 to 19-30
sections of, 13-8 to 13-10
transformations with, 11-33 to 11-36, 11-41 to 11-47
for T T & C, 21-89 to 21-100
Circular coaxial lines, 28-20 to 28-22
Circular cones
geometry for, 4-60
intersection of reflector surface with, 15-26 to 15-29
Circular corrugated horns, 15-97, 15-99
Circular cylinders
and intersection curve, 15-30
radiation patterns of, 4-75, 4-77 to 4-79
Circular-disk microstrip patches
characteristics of, 10-16, 10-25
with CP microstrip antennas, 10-61, 10-62
elements for, 13-60
principal-plane pattern of, 10-22
resonant frequency of, 10-47
Circularity, of multiple antennas, 27-38 to 27-40
Circular log-periodic antenna, 9-3, 9-4
Circular loop antennas
for geophysical applications, 23-19 to 23-21
as magnetic-field probes, 24-40 to 24-43
radiation resistance of, 6-19
thin-wire, 7-42
Circular open-ended waveguides
radiators for, with phased arrays, 18-6
with reflector antennas, 21-74
for reflector feeds, 15-90 to 15-93
Circular patches, 10-41 to 10-44
Circular-phase fields, 9-109
Circular pillbox feeds, 19-19, 19-21 to 19-22
Circular polarization, 1-15 to 1-16, 1-20, 1-28 to 1-29
of arrays, 11-7
with constrained lenses, 16-43
conversion of linear to, 13-60 to 13-61
with dual-mode converters, 21-78, 21-80
with fm antennas, 27-33 to 27-34
gain measurements for, 32-49 to 32-51
with horn antennas
biconical, 21-107 to 21-108
corrugated, 8-51, 8-64
diagonal, 8-70
with image-guide-fed slot array antennas, 17-50
of log-spiral conical antennas, 9-90 to 9-91, 9-99
measurements of, 32-52, 32-58, 32-60
with microstrip antennas, 10-57 to 10-63, 10-69 to 10-70
microstrip radiator types for, 13-60
with phased arrays, 18-6
for power dividers, 21-78, 21-80
with reflector antennas
offset parabolic, 15-36, 15-48 to 15-50
for satellites, 21-19, 21-73, 21-74, 21-78 to 21-80
for satellite antennas, 21-7, 21-29, 21-89
reflector, 21-19, 21-73, 21-74, 21-78 to 21-80
for shipboard simulation, 20-91 to 20-93
with spiral antennas, 17-27
for T T & C, 21-89
with tv antennas, 27-8 to 27-9, 27-38 to 27-40
helical, 27-15 to 27-19
side-mount, 27-20 to 27-23
skewed-dipole, 27-11 to 27-12, 27-18
vee-dipole, 27-9 to 27-11
Circular-sector patches, 10-17, 10-44
resonant frequency of, 10-47, 10-48
Circular tapered dielectric-rod antennas, 17-37 to 17-38
Circular waveguides, 1-44 to 1-50
array of, with infinite array solutions, 13-48 to 13-49
for conical horn antennas, 8-46, 15-93
dielectric, 28-50 to 28-51
open-ended, 15-90 to 15-93, 15-105
TE/TM, 28-42 to 28-45
transmission type, 21-60
Circumferential slots
on cones, 4-92
on cylinders, 4-75, 4-77, 21-92 to 21-93, 21-96
CLAD (controlled liquid artificial dielectrics), 17-22
Classical antennas, 3-26 to 3-31
Cluttered environments, simulation of, 20-90 to 20-97
Cmn, explicit expressions for, 2-11

Coaxial cable. *See also* Sleeve antennas
  with implantable antennas, 24-26 to 24-28
  for in-vivo measurements, 24-32 to 24-35
  for log-periodic zigzag antennas, 9-44 to 9-45
  with log-spiral antennas, 9-75, 9-80
  with medical applications, 24-24 to 24-25
  for microstrip antennas, 10-51, 17-120
  with TEM waveguides, 21-57
Coaxial current loops for hyperthermia, 24-45 to 24-46
Coaxial dipole antennas, 3-30
Coaxial hybrids as baluns, 9-95
Coaxial-line corporate feed networks, 13-61
Coaxial line transmission type, 21-57, 21-60
  for TEM transmission lines, 28-23 to 28-25
    circular, 28-20 to 28-22
    impedance of, 28-15, 28-16, 28-18
Coaxial-probe feeds, 10-28, 10-29
Coaxial radiators
  dipole, for phased arrays, 18-7
  excited-disk, 18-8 to 18-9
Coaxial slot antennas, 27-28 to 27-29
COBRA modeling code, 3-88 to 3-91
Codes, computer modeling, 3-80 to 3-96
Collimations, array, 13-7 to 13-10
Collinear arrays, 3-43
Collocation moments method, 3-60
Column arrays
  collimation of, 13-8
  of dipoles, coupling between, 13-44
Coma lens aberrations, 16-15, 16-17, 21-13 to 21-14
  and Abbe sine condition, 16-19 to 16-21
  and millimeter-wave antennas, 17-19 to 17-21
  with Ruze lenses, 19-39
  with spherical thin lenses, 16-29
  and zoning, 16-38 to 16-41
Communication satellite antennas, 21-4 to 21-6
  beam-forming networks for, 21-29 to 21-57
  design of, 21-6 to 21-8, 21-68 to 21-80
  feed arrays for, 21-27 to 21-29
  multibeam, 21-57 to 21-68
  types of, 21-8 to 21-26

Compact antenna ranges, 32-28 to 32-30
  for field measurements, 33-22 to 33-24
Compensation theorem and computer solutions, 3-54
Complementary planes, 2-14 to 2-15
Completely overlapped space-fed subarray system, 19-68 to 19-76
Complex environments
  airborne antenna patterns, 20-70 to 20-90
  antennas for, 20-3 to 20-4
  far fields with, 20-53 to 20-54, 20-61 to 20-70
  numerical simulations
    for antennas, 20-5 to 20-7
    for environment, 20-7 to 20-53
  shipboard antenna patterns, 20-90 to 20-97
Complex feeds for reflectors, 15-94 to 15-99
Complex pattern function for arrays, 11-7
Complex polarization ratios, 32-52
Complex poles of antennas, 10-11
Complex power, 1-7
Component errors and phased-array performance, 18-26 to 18-28
Compound distributions for rectangular apertures, 5-16 to 5-17
Computation with computer models, 3-54 to 3-55, 3-68 to 3-72
Computer-aided design, 17-120 to 17-122
Computers and computer programs
  for beam-forming network topology, 21-57
  for beam steering, 18-3 to 18-4, 18-26
  for complex environment simulations, 20-3 to 20-4, 20-18
  for dipole-dipole arrays, 23-7
  for Gregorian feed systems, 19-65
  for integral equations, 3-52 to 3-55
    codes for, 3-80 to 3-96
    computation with, 3-68 to 3-72
    numerical implementation of, 3-55 to 3-68
    validation of, 3-72 to 3-80
  for ionospheric propagation, 29-25
  modeling with, 32-65
  for rf personnel dosimeters, 24-43
  for spherical scanning, 33-15
Computer storage and time
  with frequency-domain solutions, 3-61 to 3-62, 3-65

using NEC and TWTD modeling codes, 3-94
and symmetry, 3-65, 3-67
with time-domain solutions, 3-63 to 3-65
and unknowns, 3-57
Concentric lenses, 19-80
Concentric ring arrays, 19-57 to 19-58
Conceptualization for computer models, 3-53 to 3-54
Conductance
  in inclined narrow wall slot arrays, 12-25 to 12-26
  with microstrip antennas, 10-7 to 10-8
  of resonant-length slots, 12-14
Conducting cones, fields of, 1-10
Conducting wedges, fields of, 1-9 to 1-10
Conduction losses, 1-7
  of microstrip patch antennas, 17-111
  of transmission lines
    circular coaxial, 28-21
    coplanar waveguide planar, 28-34
    microstrip planar, 28-32
    triplate stripline, 28-28
  of waveguides, 1-39
    rectangular TE/TM, 28-38, 28-42
Conductivity of troposphere, 29-31
Conductors, modeling of, 32-70 to 32-71
Conductor surfaces, linear current density in, 12-3
C-141 aircraft, simulation of, 20-82 to 20-84
Cone-tip diffraction, 4-93 to 4-94
Cones
  circular, 4-60
  conducting, 1-10
  coupling coefficient of slots on, 4-92
  radial slots in, 4-78 to 4-80
  surface-ray paths on, 4-75 to 4-76
Conformal arrays, 13-49 to 13-52
  nonplanar, 13-41
  probabilistic approach to, 14-33 to 14-34
  for T T & C, 21-90
Conical cuts, 32-6
Conical horn antennas, 8-4. *See also* Corrugated horn antennas, conical
  design procedures for, 8-48 to 8-49
  directivity of, 8-46 to 8-47
  dual-mode, 8-71 to 8-72, 15-95, 15-100 to 15-101
  for earth coverage satellite antennas, 21-82
  for feeds for reflectors, 15-93 to 15-94, 15-99 to 15-101
  gain of, 8-48
  millimeter-wave, 17-23, 17-25
  phase center of, 8-76, 8-78 to 8-80
  radiated fields of, 8-46
Conical log-spiral antennas, 9-3, 9-4
  active region of, 9-83, 9-84, 9-98 to 9-99
  axial ratios with, 9-90 to 9-93
  current wave of, 9-83 to 9-84
  directivity of, 9-87 to 9-90
  feeds for, 9-95 to 9-96, 9-99 to 9-104, 9-111 to 9-112
  front-to-back ratios with, 9-89, 9-92
  half-power beamwidth of, 9-86, 9-87, 9-89, 9-97 to 9-98
  impedance of, 9-95 to 9-97, 9-99, 9-107
  phase center of, 9-83, 9-91 to 9-92, 9-94
  polarization of, 9-86, 9-90 to 9-91, 9-99
  propagation constant with, 9-80 to 9-84
  radiation fields for, 9-85 to 9-86, 9-89 to 9-92, 9-99 to 9-109
  transmission lines with, 9-95
  truncation with, 9-79, 9-84, 9-89 to 9-90, 9-95 to 9-97
Conical scanning, 22-38 to 22-43, 22-48
Conic-section-generated reflector antennas, 15-23 to 15-31
Constant-K lenses, 17-15 to 17-16, 17-19
Constituent waves with radomes, 31-5
Constrained feeds, 18-19 to 18-26, 21-34
  multimode element array technique, 19-12 to 19-15
  multiple-beam matrix, 19-8 to 19-12
  for overlapped subarrays, 13-36
  parallel feed networks, 19-5 to 19-6
  series feed networks, 19-3 to 19-5
  true time-delay, 19-7 to 19-8
Constrained lenses, 16-5 to 16-6, 16-41 to 16-48
  analog of dielectric lenses for, 16-49 to 16-51
  bandwidth of, 16-55
  wide-angle multiple-beam, 19-80 to 19-85
Constrained variables and radome performance, 31-4
Contiguous subarrays, 13-32 to 13-35
Continuity equations, 1-6. *See also* Discontinuities
  for electromagnetic fields, 3-6
  for field distributions, 5-18 to 5-19

Continuous metal strips for leaky-wave antennas, 17-84 to 17-86
Continuous scans, 22-35
 compared to step scans, 22-43
Contour beam reflectors, 15-5, 15-80 to 15-87
Contours and tv antenna strength, 27-6
Controlled liquid artificial dielectrics, 17-22
Convergence
 in large arrays, 12-28
 measures of, and modeling errors, 3-72 to 3-75
Convex cylinder, surface-ray paths on, 4-75 to 4-76
Convex surfaces
 GTD for, 4-28 to 4-30
 UTD for
  and mutual coupling, 4-84 to 4-96
  and radiation, 4-63 to 4-84
  and scattering, 4-50 to 4-63
Coordinate systems
 for antenna ranges, 32-8 to 32-14
 feed vs. reflector, 15-88 to 15-89
 for lenses, 16-7 to 16-9
 for numerical aircraft solutions, 20-76
 for probe antennas, 32-6 to 32-7
 with reflector antennas, 15-7
 transformations of, 15-115 to 15-120
Cophasal arrays
 Dolph-Chebyshev, 11-19
 end-fire, 11-69, 11-70
 uniform, 11-64 to 11-65
 circular, linear transformations with, 11-46 to 11-47
Coplanar waveguide transmission types, 21-59, 21-61
 applicators for medical applications, 24-24 to 24-25
 planar quasi-TEM, 28-33 to 28-34
 strip planar quasi-TEM, 28-35
Copolarization. *See* Reference polarization
Copper for satellite antenna feeds, 21-28 to 21-29
Corner-diffracted fields
 coefficient for, 4-44 to 4-48
 and UTD solutions, 20-12 to 20-13
Corporate feed networks, 19-5 to 19-6
 for arrays, 13-61
 for beam-forming, 18-24
Correction factors
 for propagation, 29-5

with reflector surface errors, 15-106 to 15-107, 15-113
Corrugated horn antennas, 8-4, 8-50 to 8-51
 circular, 15-97, 15-99
 conical, 8-59 to 8-63
  cross-polarized pattern for, 8-69
  millimeter-wave, 17-23
  radiated fields of, 8-64, 8-65
 efficiency of, 8-50
 millimeter-wave, 17-23
 pyramidal, 8-51 to 8-52
  aperture fields of, 8-53
  half-power beamwidth of, 8-55
  radiated fields of, 8-53 to 8-59
 for reflector antennas, 15-96, 15-102 to 15-104
 for satellite antennas, 21-82
Corrugations
 depth of, and EDC, 17-55 to 17-56
 with periodic dielectric antennas, 17-48
Cosine lenses, hyperbolic, 16-53 to 16-54
Cosine representations of far fields, 1-28 to 1-29
Cosinusoidal aperture distributions
 and corrugated horns, 8-53
 near-field reduction factors for, 5-30 to 5-32
Cosmos-243 satellite, 22-17
Cosmos-384 satellite, 22-17
Cos type patterns in reflector feeds, 15-99, 15-102 to 15-105
Coupler/phase-shifter variable power dividers, 21-43 to 21-45
Couplers
 for array feeds, 13-61
 for series feed networks, 19-4 to 19-5
Coupling, 6-9. *See also* Mutual coupling
 of beam-forming feed networks, 18-25, 18-26
 with dipole arrays, 13-44
 microstrip antenna to waveguide, 17-119 to 17-121
 and phase shift, 9-58
 of transmitting and receiving antennas, 20-68
 with UTD solutions, 20-6 to 20-7
  for curved surfaces, 20-18
 of waveguides for planar arrays, 12-21 to 12-22
Coupling coefficient, 4-92
 with aperiodic arrays, 14-24, 14-26

Coupling equation
  for planar scanning, 33-8 to 33-9
  for spherical scanning, 33-13
Coverage area. *See also* Footprints
  of satellite antennas, 21-4 to 21-6, 21-73, 21-77
    maps for, 15-80 to 15-81
  of tv antennas, 27-4 to 27-6
Creeping waves
  with conformal arrays, 13-50 to 13-51
  with UTD solutions
    for curved surfaces, 20-20, 20-22
    for noncurved surfaces, 20-45
Critical frequencies of ionospheric propagation, 29-23
Crops, remote sensing for, 22-15
Cross dipole antennas, 1-24
Crossed yagis phased arrays, 21-12
Cross-flux of fields, 2-16, 2-18
Crossover
  of beam-forming feed networks, 18-25
  with pillbox feeds, 19-22 to 19-23
Cross polarization, 1-23 to 1-24
  with arrays
    center-inclined broad wall slot, 12-24
    conformal, 13-50
    inclined narrow wall slot, 12-24
    microstrip patch resonator, 17-108
    phased, 18-8
  with beam-forming networks, 21-57
  with edge-slot array antennas, 17-27
  with horn antennas
    aperture-matched, 8-68
    corrugated, 8-64, 8-69
    millimeter-wave, 17-23
  and matched feeds, 15-99
  with microwave antennas, 33-20 to 33-21
  with microwave radiometers, 22-30, 22-45
  with millimeter-wave antennas
    horn, 17-23
    reflector, 17-9
  with probe antennas, 32-8
  with reflector antennas, 15-5 to 15-6
    dual-reflector, 15-69 to 15-70, 15-72 to 15-74
    millimeter-wave, 17-9
    offset parabolic, 15-42, 15-46 to 15-48
    satellite, 21-19 to 21-20, 21-23
  with satellite antennas, 6-26, 6-27, 21-5, 21-7
  of side lobes, slots for, 8-63

Cross sections
  for leaky-wave antennas, 17-88
  receiving. *See* Receiving cross section
  with scattering, 2-22 to 2-25
  of tapered dielectric-rod antennas, 17-38 to 17-40
  of transmission lines, 21-63
Cross-track scanning, 33-7 to 33-10
Cubic phase errors with lenses, 16-14, 16-16 to 16-19, 16-38
  and coma aberrations, 17-19
Current and current distribution. *See also* Magnetic current distributions; Surface current
  with arrays, 3-33, 3-35
  in biconical transmission lines, 3-24
  and boundary-condition mismatch, 3-80
  with cavities, 10-12
  with coaxial sleeve antennas, 7-33, 7-35 to 7-37
  density of
    with microstrip antennas, 10-12
    PO, 4-19 to 4-21
    for thin wires, 3-44 to 3-45
  with dipole antennas
    arrays, 13-43
    folded, 7-37 to 7-39
    linear, 7-11
    loaded, 7-9, 7-19
    short, 6-14 to 6-15
    unloaded linear, 7-12 to 7-14, 7-18 to 7-19
  edge, 10-55 to 10-57
  of far fields, 1-21 to 1-22
  and imperfect ground planes, 3-27, 3-48
  with Kirchhoff approximations, 2-21
  measurement of, 2-31, 33-5 to 33-6
  of perfect conductors, 3-38 to 3-39
  in rectangular waveguides, 2-28 to 2-29, 12-3 to 12-4
  of reflected waves, 7-14
  short-circuit, 2-32, 6-5
  sinusoidal, 3-14, 3-17
  with sleeve antennas, 7-23 to 7-33
  with straight wire antennas, 3-49 to 3-50
  with thin-wire antennas, 7-5 to 7-6
    loops, 7-42 to 7-48
  for traveling-wave antennas, 3-16
  for Tseng-Cheng pattern, 11-56
  for UTD antenna solutions, 20-5
Current elements, 26-8
Current reflection coefficient, 2-28

Current source, ideal, 2-29 to 2-30
Current waves
  with log-spiral antennas, 9-74 to 9-75
    conical, 9-83 to 9-84
  in transmission lines, 28-6, 28-7
Curvature of field with lenses, 16-15, 16-20
Curved surfaces
  effects of, on conformal arrays, 13-50
  for diffraction, 8-64
  UTD solutions with, 20-4 to 20-5, 20-18 to 20-37
Cutoff frequency of waveguides, 28-40 to 28-41
  circular dielectric, 1-49, 28-50
  rectangular, 1-42
Cutoff wavelength
  of transmission lines, circular coaxial TEM, 28-20
  of waveguides, 28-40 to 28-41
    circular TE/TM, 28-42, 28-44
    rectangular TE/TM, 28-38
    ridged, 24-14 to 24-15
Cylinders and cylindrical antennas
  circular, 4-75, 4-77 to 4-79
  conducting, 21-92 to 21-100
  current distribution of, 3-14, 3-16
  dipole, 3-28
  dipole arrays on, 13-50 to 13-52
  elliptical, 4-75, 4-77
  illumination of, 4-12 to 4-13
  for intersection curve, 15-30
  for multiple-antenna analyses, 27-37 to 27-38
  radiation pattern for, 4-62 to 4-63
  strips mounted on, 20-23 to 20-24
  thin-wire, 7-10 to 7-11
Cylinder-to-cylinder interactions, 20-37, 20-44 to 20-53
Cylindrical array feeds, 19-98 to 19-101
  matrix-fed, 19-101 to 19-119
Cylindrical scanning for field measurements, 33-10 to 33-12

Data transmission, millimeter-wave antennas for, 17-5
Decoupling with coaxial sleeve antennas, 7-31 to 7-32
Defense meteorological satellites, 22-14
Defocusing techniques for field measurements, 33-24 to 33-26
Deicing of tv antennas, 27-9, 27-26
Delay lenses, equal group, 16-45 to 16-46

Design procedures and parameters
  for antenna ranges, 32-14 to 32-19
  for arrays
    Dolph-Chebyshev, 11-15 to 11-23
    feed, for satellite antennas, 21-27 to 21-29
    large, 12-28 to 12-34
    of longitudinal slots, 12-10 to 12-24
    phased, 18-3 to 18-28
    space-tapered, 14-6 to 14-8
    of wall slots, 12-24 to 12-28
  for dielectric grating antennas, 17-62 to 17-69
  for dielectric lenses, 16-8 to 16-9
  for horn antennas
    conical, 8-48 to 8-49
    E-plane, 8-19 to 8-20
    H-plane, 8-33 to 8-34
    pyramidal, 8-43 to 8-45
  for log-periodic antennas
    dipole, 9-20 to 9-32
    zigzag, 9-37 to 9-46
  for offset parabolic reflectors, 15-80 to 15-84
  and radome performance, 31-4, 31-8 to 31-10
  for satellite antennas, 21-6 to 21-8, 21-20 to 21-26
    feed arrays for, 21-27 to 21-29
Detour parameters, 4-42
Diagnostics, medical. *See* Medical applications
Diagonal horn antennas, 8-70 to 8-71
Diameter of offset parabolic reflectors, 15-31
Diathermy applicators, 24-46 to 24-48
Dicke radiometers, 22-24 to 22-25
Dielectric antennas
  bifocal lens, 16-30 to 16-33
  in integrated antennas, 17-132
  periodic, 17-34 to 17-35, 17-48 to 17-82
Dielectric constants, 1-6 to 1-7, 10-5. *See also* Effective dielectric constants
  and Abbe sine condition, 16-20
  and cross sections, 17-39
  of microstrip antennas, 10-7, 10-23 to 10-24
  vs. thickness, with lenses, 16-12 to 16-13, 16-23
Dielectric grating antennas, 17-50 to 17-53, 17-80 to 17-82

Dielectric horn antennas, 17-23, 17-25
Dielectric lenses, 16-5, 21-16
    constrained analog of, 16-49 to 16-51
    design principles of, 16-7 to 16-9
    and Snell's law, 16-6
    taper-control, 16-33 to 16-38
    wide-angle, 16-19 to 16-33
    zoning of, 16-38 to 16-41
Dielectric loading
    with horn antennas, 8-73
        for satellite antennas, 21-82, 21-84 to 21-86
    with lenses for cylindrical array feeds, 19-111, 19-114
    with pillbox feeds, 19-21
    with waveguides, 13-48 to 13-49
        for medical applications, 24-9
Dielectric logging for rock conductivity, 23-25
Dielectric loss
    in constant-K lenses, 17-19
    tangent for, 1-6
    with transmission lines, 28-9
        coplanar waveguide planar quasi-TEM, 28-34
        microstrip planar quasi-TEM, 28-31
    with waveguides
        rectangular, 1-39
        TE/TM, 28-37
Dielectric matching sheets, 16-56
Dielectric materials
    with cavities, 10-21
    and feed reactance, with microstrip antennas, 10-33
    and phase shift, 9-64
    properties of, 28-10
Dielectric millimeter-wave antennas, 17-48 to 17-82
Dielectric probes
    open-ended coaxial cable as, 24-32 to 24-34
    short monopoles as, 24-30 to 24-32
Dielectric resonator antennas, 17-35 to 17-36
Dielectric-rod antennas, 17-34 to 17-48, 17-133 to 17-134
Dielectrics
    and modeling, 32-70 to 32-75
    and radiation patterns, 32-81 to 32-82
Dielectric sheets, plane wave propagation through, 31-10 to 31-18, 31-20 to 31-25
Dielectric-slab polarizers, 21-30
Dielectric substrate in microstrip antennas, 10-5, 10-6
Dielectric waveguide transmission types, 21-59, 21-61
    circular, 28-50 to 28-51
    for fiber optics, 28-47
    rectangular, 28-51, 28-53
Dielectric weight of lens antennas, 17-15
Dielguides, 8-73
Difference beams with beam-forming feed networks, 18-25 to 18-26
Differential formulations, compared to integral, 3-55 to 3-58, 3-82
Diffracted rays, 4-3 to 4-5
    caustic regions with, 4-5
Diffractions and diffracted fields. *See also* Edges and edge-diffracted fields; High-frequency techniques
    coefficients of
        with reflector antennas, 15-13
        for UTD solutions, 20-13, 20-43, 20-44
    and curved surfaces, 8-64
    diffracted-diffracted, 20-28, 20-29, 20-34
    diffracted-reflected
        for curved surfaces, 20-24, 20-29, 20-34
        for noncurved surfaces, 20-37, 20-45, 20-46, 20-56
    and Fresnel ellipsoid, 29-36
    horn antennas to reduce, 8-4, 8-50
    propagation by, 29-30, 29-36, 29-41 to 29-44
    and radomes, 31-8, 31-10
    with UTD solutions, 20-13, 20-16, 20-43, 20-44
        correction of, 4-105 to 4-106
        for curved surfaces, 20-26, 20-29, 20-32, 20-34
        for noncurved surfaces, 20-49 to 20-50, 20-52, 20-59 to 20-60
Digital communications and rain attenuation, 29-13
Digital phase shifters
    for periodic arrays, 13-63 to 13-64
    for phased arrays, 18-12 to 18-13, 18-15, 18-18
    quantization errors wtih, 21-12
    in true-time-delay systems, 19-80
Dimensions of standard waveguides, 28-40 to 28-41

Diode detectors for rf radiation, 24-37
Diode-loaded circular loops, 24-40 to 24-43
Diode phase shifters
   for periodic arrays, 13-62 to 13-64
   for phased arrays, 18-12 to 18-17
   variable, for beam-forming networks, 21-38, 21-40 to 21-42
Diodes. *See* Pin diodes; Schottky diodes; Varactor diodes
Diode switches for beam-forming networks, 21-7, 21-37 to 21-38
Diode variable power dividers, 21-46 to 21-48
Diplexer/circulators, 19-110 to 19-113
Diplexers
   with fixed beam-forming networks, 21-7, 21-53, 21-54
   orthomode transducers as, 22-46
Dipole antennas, 2-27, 3-30. *See also* Dipole arrays; Electric dipole antennas; Log-periodic dipole antennas; Short antennas, dipole
   balanced, 9-72 to 9-73
   bow-tie, with integrated antennas, 17-137 to 17-138
   coaxial, 3-30
   compared to monopole, 3-20
   coupling between, 20-68
   for direction-finding antennas, 25-9 to 25-10
   as E-field probes, 24-38 to 24-39
   and EMP, 30-9 to 30-14
   in feed guides, for circular polarization, 8-64
   folded, 3-29, 7-36 to 7-40
   with implantable antennas, 24-27 to 24-28
   linear, 3-13 to 3-14, 7-6, 7-11 to 7-23
   magnetic, 4-59, 30-9
   microstrip, 17-106, 17-118 to 17-119, 17-122 to 17-126
   millimeter-wave, 17-32 to 17-33
   near-field radiation pattern of, 20-69
   with side-mount tv antennas, 27-20 to 27-21
   skewed, 27-11 to 27-18, 27-34 to 27-35
   sleeve, 7-26 to 7-29
   small, 3-28, 30-9
   spherical, 30-14
   thick, 3-28
   for vehicular-mounted interferometers, 25-23
Dipole arrays, 13-43 to 13-44, 13-58
   blindness in, 13-47 to 13-48
   on cylinders, 13-50 to 13-52
   mutual coupling with, 13-41 to 13-42
Dipole-dipole arrays
   for earth resistivity, 23-6 to 23-7
   for E-field probes, 24-51
Dipole radiators for phased arrays, 18-6 to 18-7, 18-10
Direct-contact waveguide applicators, 24-9
Direct-current mode with microstrip antennas, 10-25
Direct-current resistivity of earth, 23-4 to 23-8
Direct-ray method with radome analysis, 31-20 to 31-24
Direction-finding antennas and systems, 25-3
   with am antennas, 26-37 to 26-47
   with conical log-spiral antennas, 9-109 to 9-110
   interferometers, 25-21 to 25-23
   multimode circular arrays, 25-17 to 25-20
   multiple-signal, 25-24 to 25-25
   rotating antenna patterns, 25-4 to 25-17
Directive gain, 1-25, 3-12, 5-26, 5-27
   of arrays, 3-37
   of biconical antennas, 3-25
   of lens satellite antennas, 21-74
   of millimeter-wave antennas
     dielectric grating, 17-73 to 17-74, 17-76
     geodesic, 17-32
     integrated, 17-135 to 17-136
     lens, 17-18
     reflector, 17-12
     tapered dielectric-rod, 17-41, 17-44, 17-46 to 17-47
   of receiving antennas, 6-7
Directivity, 1-29, 3-12. *See also* Directive gain
   and aperture area, 17-7, 17-8
   of arrays, 11-5, 11-6, 11-58 to 11-76, 11-78
     aperiodic, 14-3, 14-15, 14-17 to 14-19, 14-32
     backfire, 9-64
     Dolph-Chebyshev, 11-19 to 11-21
     millimeter-wave, 17-33
     multimode element, 19-12

periodic, 13-10 to 13-11, 13-54 to 13-55
planar, 9-10, 11-63 to 11-65
scanned in one plane, 13-12
uniform linear, 11-11 to 11-13
of batwing tv antennas, 27-24
of beam-forming feed networks, 18-25
of conical log-spiral antennas, 9-87 to 9-90
for dielectric grating antennas, 17-79
and Dolph-Chebyshev pattern synthesis, 13-24 to 13-25
with DuFort-Uyeda lenses, 19-49
of ferrite loop antennas, 6-18, 6-20
and gain, 1-25
of horn antennas
  conical, 8-46 to 8-47
  E-plane, 8-14 to 8-18
  H-plane, 8-29 to 8-33
  pyramidal, 8-37, 8-39 to 8-42
of lens antennas, 17-18
of log-periodic antennas, 9-10
  dipole arrays, 9-21 to 9-24, 9-27, 9-61, 9-64, 9-66
  zigzag, 9-39 to 9-40
measurement of, 32-39 to 32-41
  with modeling, 32-85
of microwave antennas, 33-20
of millimeter-wave antennas, 17-5, 17-33
  lens, 17-18
  reflector, 17-12 to 17-14
  tapered dielectric-rod, 17-38 to 17-39
of monopole on aircraft, 20-76
of offset parabolic reflectors, 15-36
of open-ended circular waveguide feeds, 15-105
of short dipoles, 6-10 to 6-11
of side-mount tv antennas, 27-22
with spherical scanning, 33-12
Disc-cone antennas, 3-30
  millimeter-wave, 17-32
Discontinuities and UTD solutions, 20-10, 20-12
  with curved surfaces, 20-23 to 20-24, 20-30
  with noncurved surfaces, 20-37, 20-48, 20-52 to 20-53
Discrete arrays and aperture antennas, 11-29 to 11-30
Disk radiators for phased arrays, 18-6
Dispersion characteristics
  of dielectric grating antenna, 17-66
  of dielectric rod, 17-39
  of rectangular dielectric waveguides, 28-54
Displaced rectangular aperture distributions, 5-17 to 5-18
Dissipation factor of triplate stripline TEM transmission lines, 28-28, 28-29
Distortion
  from beam-forming networks, 21-57
  of integrated antennas, 17-135
  and lateral feed displacement, 21-20
  from lens aberrations, 16-15, 16-20, 21-13
  with microstrip patch resonator arrays, 17-109
  with offset parabolic antennas, 15-49
  from pin diodes, 21-38
  from reflector surface errors, 15-105 to 15-114
  with satellite antennas, 21-5
  scanning, 11-36 to 11-38
  with tapered dielectric-rod antennas, 17-44
  thermal, 21-28 to 21-29, 21-57
  from transformations, 11-57
  of wavefronts, from radomes, 31-7, 31-9
Distribution function with aperiodic arrays, 14-6, 14-10 to 14-12
Divergence factor and ocean reflection coefficient, 29-39 to 29-40
DMS (defense meteorological satellites), 22-14
DMSP Block 5D satellite, 22-17
DNA Wideband Satellite Experiment, 29-19
Documentation of computer models, 3-68, 3-70
Dolph-Chebyshev arrays
  linear, 11-13 to 11-23
  planar, two-dimensional, 11-49 to 11-52
  probabilistic approach to, 14-8
Dolph-Chebyshev pattern synthesis, 13-24 to 13-25
Doppler direction-finding antennas, 25-16 to 25-17
Doppler effects and mobile communication, 29-30
Dosimeters, personnel, rf, 24-43
Double-braid rf cable, 28-23
Double-ridged waveguides
  for medical applications, 24-12 to 24-16
  TE/TM waveguides, 28-45 to 28-47, 28-49

Doubly tuned waveguide array elements, 13-57
Downlink signals for satellite tv, 6-26
Driven-ferrite variable polarizer power dividers, 21-43 to 21-44
Driving-point impedance
  with directional antenna feeder systems, 26-43 to 26-45
  of microstrip antennas, 10-20, 10-34, 10-53
Drought conditions, remote sensing of, 22-15
Dual-band microstrip elements, 10-63 to 10-67
Dual fields, 2-5 to 2-6
  with Maxwell equations, 3-9, 3-11
Dual-frequency array with infinite array solutions, 13-48 to 13-49
Dual-hybrid variable power dividers, 21-43
Dual-lens limited-scan feed concept, 19-66 to 19-67
Dual-mode converters, 21-80
Dual-mode ferrite phase shifters, 13-64, 18-14 to 18-15, 21-38 to 21-40, 21-42
Dual-mode horn antennas
  conical, 8-71 to 8-72,
    radiation pattern of, 15-95, 15-100 to 15-101
  for satellite antennas, 21-82, 21-84 to 21-89
Dual-offset reflector antennas, 15-6 to 15-15
Dual polarizations
  circular, 21-73
  microstrip arrays with, 17-109
  reflector antennas with, 15-5
  satellite, 21-73 to 21-74, 21-78 to 21-80
Dual-reflector antennas, 15-61 to 15-62
  cross-polarization with, 15-69 to 15-71
  lens antennas as, 16-6
  millimeter-wave antennas, 17-9 to 17-11
  parameters of, 15-67 to 15-68
  for satellite antennas, 21-21
  scan performance of, 15-71 to 15-73
  shaped reflectors for, 15-73 to 15-80
  surface errors with, 15-109 to 15-111
Dual substrates with microstrip dipoles, 17-129
Dual-toroid ferrite phase shifters, 21-40 to 21-41
DuFort optical technique, 19-73 to 19-76

DuFort-Uyeda lenses, 19-49 to 19-53
Dummy cables
  for conical log-spiral antennas, 9-95, 9-103
  with log-spiral antennas, 9-75, 9-80
Dyadic coefficients
  edge-diffraction, 4-27, 4-101 to 4-105
    uniform, 4-32 to 4-33
  for surface reflection and diffraction, 4-52
Dyadic transfer function, 4-30
Dynamic programming technique, 14-5

Earth
  bulge of, and line-of-sight propagation, 29-34
  modeling of, 32-76 to 32-77
  resistivity of, 23-4 to 23-8
Earth coverage satellite antennas, 21-80 to 21-81
  horns
    dielectric-loaded, 21-84
    multistepped dual-mode, 21-84 to 21-89
    stepped, 21-82 to 21-84
    shaped beam, 21-89
Earth receiving antennas, 6-25 to 6-32
Earth-substended angles for satellites, 21-3
Eccentricity
  of conic sections, 15-25 to 15-26
  and intersection curve, 15-30
Eccentric line TEM transmission lines, 28-15
ECM. *See* Equivalent current method
EDC. *See* Effective dielectric constant method
Edge-coupled transmission lines, 28-16, 28-17
Edge-of-coverage gain with satellite antennas, 21-5
Edges and edge-diffracted fields, 4-32 to 4-33, 4-38
  and boundaries, 15-12
  corner-diffracted, 4-45 to 4-47
  currents on, 4-100 to 4-101
    magnetic, 10-55 to 10-57
  and ECM, 4-97 to 4-102
  elements with, and conformal arrays, 13-52
  GTD for, 4-25 to 4-28, 15-69
  and horn antenna fields, 8-3 to 8-4, 8-50

and image-plane ranges, 32-23
and Maxwell equations, 1-9
PO for, 4-22
PTD for, 4-104 to 4-108
slope diffraction by, 4-38 to 4-40
UAT for, 4-40 to 4-43
UTD solutions for, 4-32 to 4-38, 20-10 to 20-12
  curved surfaces, 20-21 to 20-23
  noncurved surfaces, 20-37
Edge-slot array antennas, 17-26 to 17-27
Edge slots for slot arrays, 12-26 to 12-28
Edge tapers with reflector antennas, 15-36 to 15-38, 15-41 to 15-45
  and cross-polarization, 15-47
  with off-focus feeds, 15-53
  and surface errors, 15-109
  tapered-aperture, 15-16 to 15-18, 15-21, 15-23
Effective area of receiving antennas, 22-4
  and aperture efficiency, 5-27 to 5-29
Effective dielectric constants
  of fin lines, 28-56, 28-58
  for microstrip patch antennas, 24-19 to 24-21
  of transmission lines
    coplanar strip planar quasi-TEM, 28-35
    coplanar waveguide planar quasi-TEM, 28-33
Effective dielectric constant (EDC) method
  for dielectric grating antennas, 17-65, 17-68
  for phase constants, 17-53 to 17-57
Effective heights. *See also* Vector effective height
  of hf antennas, 30-30
  of L-band antennas, 30-19
  of uhf communication antennas, 30-21
Effective isotropically radiated power, 1-27
Effective length, 2-34
Effective loss tangent with cavities, 10-20
Effective phase center of log-spiral antennas, 9-91 to 9-92, 9-94
Effective radiated power
  of fm antennas, 27-4, 27-6
  of tv antennas, 27-3 to 27-5
Effective receiving area and EMP, 30-10

Effective relative dielectric constant, 10-7
Efficiency. *See also* Aperture efficiency; Beam efficiency; Radiation efficiency
  of arrays, 11-74, 11-76
    planar, 11-65
    spaced-tapered, 14-6 to 14-7
  beam-shaping, with satellite antennas, 21-5 to 21-6
  of corrugated horn antennas, 8-50, 8-62
  of dielectric grating antennas, 17-63
  and inductive loading, 6-13
  of integrated antennas, 17-135
  of loop antennas, 25-8
  of microstrip antennas, 10-6, 10-49 to 10-50
    dipole, 17-124
    and matching filters, 10-52
  of microstrip patch resonator arrays, 17-109 to 17-115
  of microwave radiometers, 22-28 to 22-30, 22-32 to 22-33
  polarization, 1-26, 2-36, 2-39, 32-42, 32-56
  and Q-factor, 6-22 to 6-23, 10-6
  of receiving antennas, 6-22 to 6-24
  of reflector antennas
    dual-reflector, 15-73
    offset parabolic, 15-42
    tapered-aperture, 15-16, 15-23
  of small transmitting antennas, 6-9 to 6-10
E-4 aircraft, antennas on, 30-29, 30-30
Ehf (extra high frequencies), lens antennas for, 16-5, 16-28
EIRP (effective isotropically radiated power), 1-27
Electrically rotating patterns
  high-gain, 25-14 to 25-16
  null, 25-11 to 25-12
Electrically scanning microwave radiometer, 22-10
  calibration of, 22-26
  as conical scanning device, 22-40
  modulating frequency of, 22-24
  as planar scanning device, 22-37
Electrical scanning, 21-5
Electrical well logging, 23-24 to 23-25
Electric dipole antennas, 3-13 to 3-14
  and EMP, 30-9
  far-field pattern of, 1-29, 4-60

Electric fields. *See also* Electric-field strength
  with arrays, 3-36
  and cavities, 10-12
  conical-wave, 4-99
  distribution of, 12-4, 12-6 to 12-10, 12-35
  and equivalent currents, 5-8 to 5-10
    edge, 4-101
  in far zone, 6-3 to 6-4
  GO for, 4-13, 4-17
  Green's dyads for, 3-46, 3-48
  GTD for, 4-23
    for convex surfaces, 4-28
    for edged bodies, 4-25
    for vertices, 4-30
  integral equations for, 3-44, 3-46 to 3-47, 3-51
  of isotropic medium, 4-9
  with plane-wave spectra, 1-32, 5-5 to 5-8
  PO, 4-19 to 4-21
  probes for, 24-37 to 24-41
    calibration of, 24-54 to 24-57
    implantable, 24-51 to 24-54
  PTD for, for edged bodies, 4-106 to 4-107
  of rectangular apertures, 5-11 to 5-12
  reflection coefficient for, 2-28
  scattering. *See* Scattering
  of short monopole, 4-65
  of small dipoles, 6-13
  of small loops, 6-18
  of spherical wave, 4-6 to 4-7
  of surface fields, 4-84
  of transmitting antenna, 6-3, 20-6
  UAT for, for edged bodies, 4-39
  in unbounded space, 2-8 to 2-9
  UTD for
    corner-diffracted, 4-44
    for edged bodies, 4-32, 4-37, 4-83
  vector for, 1-14, 7-46 to 7-47
  in waveguides, 12-4
Electric-field strength
  of hemispherical radiators, 26-31 to 26-32
  horizontal, with two towers, 26-30 to 26-34
  of isotropic radiators, 26-26
  of uniform hemispherical radiators, 26-27 to 26-29
  with UTD solutions, 20-6
Electric line dipole for UTD solutions, 20-19 to 20-20
Electric surface current, 2-26 to 2-27
Electric vector potential, 3-6 to 3-7, 3-22, 5-8 to 5-9
  of rectangular-aperture antennas, 3-22
Electrode arrays for resistivity measurements, 23-4 to 23-8
Electroformed nickel for satellite antennas, 21-28 to 21-29
Electromagnetic fields
  computer model for, 3-53
  equations for, 3-6 to 3-8
  integral vs. differential formulation of, 3-55 to 3-58
  models for
    materials for, 32-69 to 32-70
    theory of, 32-65 to 32-69
Electromagnetic pulses, 30-3
  and aircraft antennas, 30-17 to 30-31
  analyzing effects of, 30-8 to 30-16
  and mounting structures, 30-4 to 30-7
  and principal elements, 30-7 to 30-8
Electromagnetic radiation monitors, 24-37
Electromechanical measurements, 32-4
Electromechanical phase shifters, 21-38, 21-41 to 21-43
Electromechanical power dividers, 21-48 to 21-49
Electromechanical switches, 21-37 to 21-38
Electronically Scanning Airborne Intercept Radar Antenna, 19-55
Electronic scanning, 21-5, 22-36 to 22-37
Elements, array
  and EMP, 30-7 to 30-8
  for feed arrays for satellite antennas, 21-27
  interaction of, with aperiodic arrays, 14-3
  mutual impedance between, 11-5
  for periodic arrays, 13-57 to 13-60
    patterns of, 13-6
  in phased arrays, 21-12
  in planar arrays, 11-53
  spacing of, 3-36, 14-4 to 14-5
Elevated antenna ranges, 32-19 to 32-21
Elevated H Adcock arrays, 25-9 to 25-10
Elevation and noise in tvro antennas, 6-31
Elevation-over-azimuth positioners, 32-12
Elf. *See* Extremely low frequencies
Ellipses, 15-25 to 15-26
  and intersection curve, 15-27, 15-29 to 15-30
  polarization, 1-14 to 1-16

Index

Elliptical arrays, transformations with, 11-33 to 11-36, 11-48
   uniformly excited, 11-44 to 11-45
Elliptical cylinders, 4-75, 4-77
Elliptical loops, 6-19
Elliptical patches, 10-18
Elliptical polarization, 1-15 to 1-16
   gain measurements for, 32-49 to 32-51
   with log-spiral antennas, 9-74
EM Scattering modeling code, 3-88 to 3-91
Emissivity in state, 2-40
Emissivity of surfaces, 22-7 to 22-9
EMP. *See* Electromagnetic pulses
End-correction network for input admittance, 7-7
End-fire arrays, 3-41, 11-12, 11-15
   maximum directivity of, 11-65 to 11-71
   radiation pattern for, 11-76, 11-77
Engines, aircraft, model of, 20-37
Environments. *See also* Complex environments
   measurements for, 32-4
   numerical simulation of, 20-7 to 20-53
E-plane gain correction factor for biconical horns, 21-104
E-plane horn antennas, 8-4
   aperture fields of, 8-5 to 8-7
   design procedure for, 8-19 to 8-20
   directivity of, 8-14 to 8-18
   field intensity of, 8-13 to 8-14
   gain of, 8-18 to 8-19
   half-power beamwidth of, 8-14 to 8-15
   phase center of, 8-76 to 8-77
   radiated fields of, 8-7 to 8-10
   for reflector feeds, 15-92 to 15-93, 15-96 to 15-97
   universal curves of, 8-10, 8-13 to 8-14
E-plane radiation patterns, 1-28, 5-12
   of apertures
      circular, 5-22
      rectangular, 5-12
   of dielectric grating antennas, 17-74, 17-78
   of DuFort-Uyeda lenses, 19-51
   of horn antennas
      corrugated, 8-55 to 8-59
      E-plane, 8-9 to 8-12
      H-plane, 8-22 to 8-25
      pyramidal, 8-37 to 8-40
      square, 17-20
   for log-periodic antennas
      diode arrays, 9-64 to 9-66
      metal-arm, 9-11 to 9-12
      zigzag, 9-39 to 9-42, 9-45, 9-46
   for microstrip antennas, 10-24
   and UTD solutions, 20-13 to 20-16
Equal group delay lenses, 16-45 to 16-46
Equiangular spirals, 9-72, 9-110
Equivalent current method, 4-5, 4-96 to 4-102
   and surface-diffracted rays, 4-83
Equivalent currents for surface fields, 5-8
Equivalent cylinder techniques, 27-37 to 27-38
Equivalent radii for thin-wire antennas, 7-10 to 7-11
Equivalent receiving area
   for aircraft blade antenna, 30-12
   for annular slot antenna, 30-13
   for center-fed ellipsoid antenna, 30-10 to 30-12
   for spherical dipole antenna, 30-14
ERP. *See* Effective radiated power
Errors. *See also* Aberrations; Path length constraint and errors; Phase errors; Random errors; Systematic errors
   in antenna ranges, 32-34 to 32-35
   array, 13-52 to 13-57
   boresight, 31-3 to 31-4, 31-7 to 31-9, 31-27 to 31-29
   component, for phased arrays, 18-26 to 18-28
   with far-field measurements, 32-5
   with microwave radiometers, 22-23, 22-25 to 22-28
   modeling, 3-69
   pattern, 21-65
   quantization, 13-52 to 13-57, 21-12
   with scanning ranges, 33-15 to 33-18
ESAIRA (Electronically Scanning Airborne Intercept Radar Antenna), 19-55
ESMR. *See* Electrically scanning microwave radiometer
Eulerian angles, 15-116 to 15-117, 15-120
Evaluation of antenna ranges, 32-33 to 32-39
Even distributions in rectangular apertures, 5-14
Excitations
   array element, 11-23, 13-6 to 13-7
   linear transformations on, 11-39 to 11-41
   nonuniform, 11-29 to 11-30

Excitations (cont.)
  of overlapped subarrays, 13-37
  in planar arrays, 11-52 to 11-53
  with TEM transmission lines, 28-11, 28-19
Excited-patch radiators, 18-8 to 18-9
Experimentation
  and modeling errors, 3-69
  and validation of computer code, 3-72 to 3-76
External fields in longitudinal slots, 12-7
External networks and EMP, 30-8
External noise, 6-24 to 6-25, 29-50
External numerical validation of computer code, 3-77 to 3-78
External quality factor with cavities, 10-20
External validation of computer codes, 3-77 to 3-78
Extra high frequencies
  lens antennas for, 16-5
    aplanatic dielectric, 16-28
Extrapolation techniques for field measurements, 33-26 to 33-28
  of gain, 32-47 to 32-48
Extraterrestrial noise, 29-50 to 29-51
Extremely low frequencies
  with geophysical applications, 23-3
  ionospheric propagation of, 29-27 to 29-28

Fade coherence time and ionospheric scintillations, 29-18
Faraday rotation
  depolarization from, 29-16
  with phase shifters, 13-64
  variable power dividers using, 21-44 to 21-46
Far-field diagnostics for slot array design, 12-36 to 12-37
Far fields and far-field radiation patterns, 1-13, 1-17 to 1-18, 3-10, 3-22 to 3-24, 6-3 to 6-4. *See also* Radiation fields and patterns
  of antennas vs. structures, 20-53 to 20-54, 20-61 to 20-70
  of aperture antennas, 3-22 to 3-24, 15-21 to 15-24
    circular, 5-20 to 5-24
    rectangular, 3-22 to 3-24, 5-11 to 5-12, 5-14 to 5-15
    sinusoidal, 3-24
  of arrays, 11-5 to 11-8
    circular, 21-91 to 21-92, 21-94
    cylindrical, feeds for, 19-99 to 19-100, 19-104
    periodic, 13-5
    phased, 21-9 to 21-10
  calculations of, 1-21 to 1-23
  with cavities, 10-15, 10-19 to 10-20
  components of, 8-73, 8-75
  coordinates of, transformations with, 15-115, 15-117 to 15-120
  of dipole antennas
    folded, 7-41
    half-wave, 6-10 to 6-11
    log-periodic, 9-19 to 9-20
    short, 6-10 to 6-11, 20-8
    sleeve, 7-27 to 7-28
  of E-plane horn antennas, 8-7 to 8-10
  and equivalent currents, 5-10
  expressions for, 3-10
  Friis transmission formula for, 2-38 to 2-39
  integral representation for, 3-8 to 3-9
  with lens antennas, 16-6, 16-13 to 16-14
    coma-free, 16-22 to 16-23
    obeying Abbe sine condition, 19-97 to 19-99
    reflector, limited-scan feed concept, 19-70 to 19-71
  measurements of. *See* Measurements of radiation characteristics
  with microstrip antennas, 10-10, 10-22, 10-24
  with multiple element feeds, 19-13 to 19-16
  with plane-wave spectra, 5-7
  polarization of, 1-23 to 1-24
  power density in, 6-7
  with radial transmission line feeds, 19-24
  with radomes, 31-3
  with reflector antennas, 15-8 to 15-10, 15-13 to 15-14
    with axial feed displacement, 15-52 to 15-55
    with lateral feed displacement, 15-56 to 15-66
    offset parabolic, 15-34 to 15-35, 15-37 to 15-39, 15-43 to 15-45
    surface errors on, 15-110 to 15-112

with tapered apertures, 15-15 to 15-23
with reflector-lens limited-scan feed concept, 19-70 to 19-71
representation of, 1-18 to 1-21, 1-28 to 1-29
scattered, and PO, 4-20
with single-feed CP microstrip antennas, 10-58 to 10-59
for sleeve monopoles, 7-31
for slot array design, 12-36 to 12-37
of space-fed subarray systems, 19-69
from thin-wire antennas, 7-8, 7-15 to 7-17
  loops, 7-44 to 7-46
for triangular-aperture antennas, 3-24
with UTD solutions, 20-4 to 20-6, 20-13 to 20-14

Fat, human
  phantom models for, 24-55
  power absorption by, 24-8, 24-9, 24-47
F/D ratios. *See* Focus length to diameter ratio
FDTD modeling code, 3-88 to 3-91
Feed blockage with offset reflector, 21-19, 21-23. *See also* Aperture blockage
Feed circuits. *See also* Beam-forming feed networks; Cassegrain feed systems; Constrained feeds; Optical feeds; Semiconstrained feeds; Unconstrained feeds
  for am antennas, 26-37 to 26-47
  for arrays
    dipole, 9-15 to 9-18, 9-21, 9-27, 9-29 to 9-31, 9-61 to 9-63
    monopole, 9-66 to 9-67
    periodic, synthesis of, 13-61
    phased, design of, 18-3 to 18-4, 18-6, 18-17 to 18-26
    power dividers for, 13-61
    spaced-tapered, 14-6 to 14-7
    subarrays, 13-39 to 13-41
    of tapered dielectric-rod antennas, 17-47 to 17-48
  for conical horn antennas, 8-46
    log-spiral, 9-95 to 9-96, 9-99 to 9-104, 9-111 to 9-112
  cylindrical array, 19-98 to 19-119
  displacements of, with offset parabolic antennas, 15-49 to 15-61
  for log-periodic dipole arrays, 9-15 to 9-18, 9-21, 9-27, 9-29 to 9-31, 9-61

  for log-spiral antennas, 9-75 to 9-78, 9-80
    dipole arrays, 9-62 to 9-63
    mono arrays, 9-66 to 9-67
  for microstrip antennas, 10-5, 10-6, 10-28, 17-108 to 17-119
    and impedance matching, 10-51 to 10-52
  for millimeter-wave antennas, 17-12, 17-20, 17-47 to 17-48
  offset, 17-9, 19-54, 19-63 to 19-66
  reactance of, with microstrip antennas, 10-31 to 10-34
  for reflector antennas, 15-5 to 15-6, 15-13, 15-84 to 15-89
    complex, 15-94 to 15-99
    cos type patterns of, 15-99 to 15-105
    radiation patterns of, 15-89 to 15-94
    rays with, 15-11 to 15-12, 21-21 to 21-22
  for satellite antennas, 21-5 to 21-7, 21-27 to 21-29
  for subarrays, 13-39 to 13-41
Feed coordinates
  with reflector antennas, 15-7, 15-36, 15-88 to 15-89
  transformations with, 15-115, 15-117 to 15-120
Feed elements
  horn antennas as, 8-3
  location of, for feed arrays for satellite antennas, 21-27
Feed gaps with log-spiral antennas, 9-75 to 9-77
Feed patterns
  for reflector antennas
    dual-reflector, 15-77, 15-80
    offset parabolic, 15-34 to 15-36, 15-47
  transformation of, with taper-control lenses, 16-37
Feed tapers, 15-36 to 15-37, 15-39 to 15-40, 17-40
Feedthrough with lens antennas, 16-5
Feed transmission media for beam-forming networks, 21-56 to 21-63
Feed types of reflector antennas, 15-5 to 15-6
Fence guide transmission type, 21-62
Fermat's principle and rays, 4-8 to 4-9
Ferrite absorbers with UTD solutions, 20-17

Ferrite loop antennas, 6-18, 6-20 to 6-22
   radiation resistance of, 6-19
Ferrite phase shifters
   for arrays
      periodic, 13-62 to 13-64
      phased, 18-12, 18-14 to 18-17
   with planar scanning, 22-37
   variable, for beam-forming networks, 21-38 to 21-40, 21-42
Ferrite switches for beam-forming networks, 21-7, 21-37 to 21-38
Ferrite variable power dividers, 21-43 to 21-46
F-4 fighter aircraft, simulation of, 20-78 to 20-82
Fiber optics
   dielectric waveguides for, 28-47
   with implantable temperature probes, 24-50
Fictitious isotropic radiator, 2-35
Field curvature lens aberrations, 21-13
Field distributions
   for circular dielectric waveguides, 28-50
   measurements of, with radomes, 31-6 to 31-7
Field equations, electromagnetic, 3-6 to 3-8
Field equation pairs, 3-7
Field probes for evaluating antenna ranges, 32-33
Fields. *See also* Aperture fields; Electric fields; Incident fields; Magnetic fields; Reactive fields
   with arrays, 3-33, 3-35 to 3-36
   circular-phase, 9-109
   dual, 2-5 to 2-6, 3-9, 3-11
   and Huygen's principle, 2-18 to 2-19
   using integration to determine, 3-56 to 3-57
   and Kirchhoff approximation, 2-19 to 2-21
   in longitudinal slots, 12-7
   of perfect conductors, 3-39
   of periodic structures, 1-33 to 1-37
   plane-wave spectrum representations of, 1-31 to 1-33
   in radome-bounded regions, 31-5 to 31-6
   of rectangular-aperture antennas, 3-21 to 3-24
   with short dipole, 3-34
   surface, 4-84 to 4-89, 5-8
   tangential, 5-7 to 5-9, 7-5 to 7-6
   TE and TM, 1-30 to 1-31
   time-harmonic, 1-5 to 1-8, 3-6 to 3-8
   of transmission lines, 2-27 to 2-29
   transmitting, 2-19 to 2-21, 2-34 to 2-35
   of waveguides
      circular, 1-44 to 1-50
      circular TE/TM, 28-43
      rectangular, 1-37 to 1-44
      rectangular TE/TM, 28-39
Field strength of am antennas, 26-8
Figure of merit
   EIRP as, 1-27
   for horn antennas, 8-14
   for tvro antenna, 6-30 to 6-31
Filamentary electric current with cavities, 10-20 to 10-21
Filter type diplexers, 21-54
Fin line transmission types, 21-59, 21-61, 28-55 to 28-57
Finite sources, far fields due to, 1-32
Finite waveguide arrays and coupling, 13-44 to 13-45
Five-wire transmission lines, 28-13
Fixed beam-forming networks, 21-34, 21-37, 21-49 to 21-56
Fixed line-of-sight ranges, 32-9 to 32-11
Fixed-wire antennas, EMP responses for, 30-23 to 30-27
Flared-notch antennas, 13-59
Flaring with horn antennas
   conical, 8-46
   corrugated, 8-60, 8-64
   E-plane, 8-9
   H-plane, 8-20, 8-29 to 8-30
   and phase center, 8-76 to 8-77
Flat dielectric sheets, propagation through, 31-10 to 31-18
Flat plate array antennas, 17-26
Flat-plate geometry for simulation of environments, 20-8, 20-18
Flat-plate Luneburg lenses, 19-44, 19-46 to 19-47
Flat surface lenses, 16-9 to 16-11
Flexible coaxial transmission lines, 28-22
Flood predicting with remote sensing, 22-15
Floquet space harmonics
   and scattered fields, 1-34 to 1-35
   and UP structures, 9-32
Fluctuation of aperiodic array beams, 14-26 to 14-28

FM broadcast antennas, 27-3 to 27-8, 27-32 to 27-40
Focal length
  of conic sections, 15-24 to 15-26
  of offset parabolic reflectors, 15-31
  for reflector satellite antennas, 21-22 to 21-24
  with taper-control lenses, 16-36
Focal points. See Caustics
Focal region of reflector satellite antennas, 21-19
Fock radiation functions, 4-68 to 4-69, 4-73
  for mutual coupling, 4-95
  for surface fields, 4-88 to 4-89
  for surface-reflection functions, 4-52 to 4-53, 4-66
Fock-type Airy functions, 4-53, 4-69
Fock type transition functions, 4-50n
Focus length to diameter (F/D) ratio
  with lens antennas, 16-12
    and dielectric constants, 16-24
    planar surface, 16-56
    for satellites, 21-16
    taper-control, 16-37
    and zoning, 16-43
  with microwave radiometers, 22-30
  and pyramidal horn antenna dimensions, 8-45
  with reflector satellites
    and cross-polarization, 15-47
    with off-focus feeds, 15-51, 15-53, 15-55 to 15-56, 15-58, 15-62
    offset parabolic reflectors, 15-32 to 15-33, 15-37
    for satellites, 21-22 to 21-23
    and wide-angle scans, 19-34 to 19-35
Fog, scintillation due to, 29-36
Folded antennas
  dipole, 3-29, 7-36 to 7-40
  slot, and Babinet principle, 30-15 to 30-16
  unipole, 3-30
Folded Luneburg lenses, 19-49
Folded pillbox feeds, 19-17 to 19-23
Folded-T waveguide power division elements, 21-50 to 21-52
Footprints. See also Coverage area
  of microwave radiometers, 22-30 to 22-32
  of tvro signal, 6-26, 6-27
Formulation for computer models, 3-53 to 3-54, 3-81

Four-electrode arrays for earth resistivity, 23-4 to 23-8
Fourier transformers, 19-95
Four-port hybrid junctions, 19-5 to 19-6
Four-tower systems, 26-33
Four-wire transmission lines, 28-12
Fox phase shifter, 13-64
Fraunhofer region, 1-18
Free space, resistance of, 26-25 to 26-26
Free-space loss, 2-39, 29-5 to 29-6
  of satellite-earth path, 29-8
Free space vswr evaluation method for anechoic chambers, 32-38 to 32-39
Frequencies
  and active region
    with conical log-spiral antennas, 9-83
    of log-periodic dipole antennas, 9-17 to 9-19
  allocation of
    for satellite antennas, 21-6
    for tv, 6-28, 27-3
  and beamwidth, 9-87
  and computer time, 3-62
  dependence on, by reflector antennas, 17-21 to 17-13
  and image impedance, 9-55
  and input impedance, 10-9
  and noise temperature, 2-43
  and phase center, 9-92
  and reactance, 10-30, 10-65
  and space harmonics, 9-35
  of standard waveguides, 28-40 to 28-41
    circular dielectric, 1-49, 28-50
    rectangular, 1-42
Frequency-agile elements, 10-66 to 10-69
Frequency-domain
  compared to time-domain, 3-65 to 3-66, 3-82
  and EMP, 30-4
  method of moments in, 3-57 to 3-62, 3-64
Frequency-independent antennas, 9-3 to 9-12. See also Log-periodic dipole antennas; Log-periodic zigzag antennas; Log-spiral antennas; Periodic structures
  millimeter-wave, 17-27 to 17-28
Frequency-modulation broadcast antennas, 27-3 to 27-8, 27-32 to 27-40
Frequency-scanned array antennas, 17-26 to 17-27

Frequency selective surface diplexer, 21-53
Fresnel ellipsoid and line-of-sight propagation, 29-36
Fresnel fields, 1-18. *See also* Near fields
Fresnel integrals with near-field patterns, 5-25
Fresnel-Kirchhoff knife-edge diffraction, 29-42 to 29-43
Fresnel reflection coefficients, 3-27, 3-33, 3-48
Fresnel zone-plate lens antennas, 17-20
Friis transmission formulas, 6-8 to 6-9
 for far fields, 2-38 to 2-39
Front ends, integrated, 17-134
Front-to-back ratios
 for log-periodic dipole arrays, 9-60 to 9-61
 for log-spiral conical antennas, 9-89, 9-92
F-16 fighter aircraft, simulation of, 20-82, 20-85 to 20-88
Fuel tanks, aircraft, simulation of, 20-79, 20-82
Fundamental waves and periodic structures, 9-33 to 9-35, 9-55 to 9-56
Fuselages of aircrafts
 and EMP, 30-4, 30-6
 simulation of, 20-70, 20-78

Gain, 1-24 to 1-27, 3-12 to 3-13, 3-28 to 3-31. *See also* Directive gain
 aperture, 5-26 to 5-34
  circular, 5-23
  rectangular, 5-29
  uniform fields, 5-14
 of arrays, 3-36, 3-38 to 3-43, 13-10 to 13-11
  array columns, 13-54
  edge-slot, 17-27
  feed, for satellite antennas, 21-27
  multimode element, 19-12
  phased, 18-11, 19-59, 21-11 to 21-12
 axial near-field, 5-32
 and bandwidth, with tv antennas, 27-7
 with DuFort optical technique, 19-76
 and effective area, 5-27 to 5-28
 of horn antennas
  biconical, 21-105 to 21-106, 21-108 to 21-109
  conical, 8-48
  correction factors with, 21-104
  E-plane, 8-18 to 8-19
  pyramidal, 8-43 to 8-44, 15-98
 of lens antennas, 17-18
 and lens F/D ratio, 21-16
 with limited scan feeds, 19-57
 of log-periodic zigzag antennas, 9-45 to 9-47
 measurement of, 32-42 to 32-51
 of microwave antennas, 33-20 to 33-21
 of millimeter-wave antennas
  lens, 17-18
  maximum-gain surface-wave, 17-40 to 17-41
  reflector, 17-12 to 17-14
  tapered dielectric rod, 17-40 to 17-41, 17-47
 reduction factor for, near-field, 5-30 to 5-32
 of reflector antennas
  boresight, 15-81
  millimeter-wave, 17-12 to 17-14
  surface errors on, 15-106 to 15-107
 ripple of, with feed arrays, 21-27
 with satellite antennas, 21-5 to 21-7, 21-12
  lateral feed displacement with, 21-20
  tvro, 6-30 to 6-31
 of short dipole, 6-10 to 6-11
 of small loops, 6-13, 6-17
Gain-comparison gain measurements, 32-49 to 32-51
Gain-to-noise ratio of tvro antenna, 6-31
Gain-transfer gain measurements, 32-49 to 32-51
Galactic noise, 6-25, 29-50
Galerkin's method
 with methods of moments, 3-60
 as weight function, 3-83
Gallium arsenide
 in integrated antennas, 17-132, 17-137 to 17-138, 17-140
 in sensors with implantable temperature probes, 24-50
Gap capacitance and input admittance, 7-8, 7-15
Gaseous absorption
 and line-of-sight propagation, 29-34 to 29-35
 of millimeter waves, 17-5
 of satellite-earth propagation, 29-8, 29-10

Gaseous emissions, noise from, 29-50 to 29-51
GEMACS modeling code, 3-88 to 3-91
Generalized equation for standard reference am antennas, 26-20 to 26-22
Geodesic lens antennas, 17-29 to 17-32
  matrix-fed, 19-108 to 19-119
  for phased array feeds, 19-17, 19-41, 19-44
Geodesic paths, 20-19
Geometrical models, 32-69
Geometrical optics, 4-3
  and computer solutions, 3-54
  fields with, 4-8 to 4-9
    incident, 4-10 to 4-13
    reflected, 4-13 to 4-18
  for lens antennas, 16-6
  and PTD, 4-103
  with reflector antennas, 15-7, 15-13, 15-14
    dual-reflector, 15-69 to 15-73, 15-77, 15-80
  for simulation of environment, 20-10
  with UTD solutions
    for curved surfaces, 20-35 to 20-36
    for noncurved surfaces, 20-45 to 20-46
Geometrical theory of diffraction, 4-3, 4-23 to 4-24. *See also* Uniform geometrical theory of diffraction
  for apertures on curved surfaces, 13-50 to 13-51
  and computer solutions, 3-54
  for convex surfaces, 4-28 to 4-30
  and ECM, 4-97 to 4-98
  for edges, 4-25 to 4-28
  for pyramidal horn fields, 15-92
  with reflector antennas, 15-7 to 15-8, 15-11 to 15-14
    dual-reflector, 15-69 to 15-72, 15-76 to 15-77
  for vertices, 4-30 to 4-31
Geophysical applications, antennas for, 23-3, 23-23 to 23-25
  electrode arrays, 23-4 to 23-8
  grounded wire, 23-8 to 23-17
  loop, 23-17 to 23-23
Geostationary Operational Environmental Satellite, 22-34, 22-43
Geosynchronous satellite antennas, 6-26
  microwave radiometers for, 22-43 to 22-46
G for microstrip antennas, 10-34 to 10-41

Gimbaling
  angles of, and radome field measurements, 31-6
  with satellite antennas, 21-5
Glass plates for microstrip arrays, 17-108
GOES (Geostationary Operational Environmental Satellite), 22-34, 22-43
GO methods. *See* Geometrical optics
Goniometers, rotating, 25-15 to 25-16
Good conductors, modeling of, 32-70 to 32-71
Good dielectrics, modeling of, 32-70 to 32-75
Goubau antenna, 6-24
Gradient synthesis method, 21-65
Graphite epoxy material for satellite antenna feeds, 21-28 to 21-29, 22-46
Grating antennas, 17-34 to 17-35, 17-48 to 17-82
Grating lobes
  with arrays
    aperiodic, 14-3, 19-57
    blindness in, 13-47
    Dolph-Chebyshev, 11-18 to 11-19
    for feeding satellite antennas, 21-27
    periodic, 11-32, 13-10
    phased, 18-9 to 18-10, 21-10 to 21-12
    scanned in one plane, 13-13 to 13-14
    scanned in two planes, 13-15 to 13-16
    space-fed, 19-69
    and subarrays, 13-32, 13-35, 19-69
  with multiple element feeds, 19-15
  with radial transmission line feeds, 19-31
  and space harmonics, 1-35
Grating period
  and leakage constants, 17-62
  and radiation angle in leaky modes, 17-56 to 17-58
Great-circuit cuts, 32-6
Green's dyads, 3-46 to 3-48
Green's function
  for field propagation, 3-56
  in unbounded space, 2-6 to 2-11
  with UTD solutions, 20-4 to 20-5
Gregorian feed systems
  offset-fed, 19-63 to 19-66
  with reflector satellite antennas, 21-21 to 21-22
Gregorian reflector systems
  cross polarization in, 15-69 to 15-70
  gain loss in, 15-73
Gridded reflector satellite antennas, 21-21

Groove depths
  of dielectric grating antennas, 17-77 to 17-78
  and leakage constants, 17-60 to 17-61
Groove guide transmission type, 17-35, 21-59, 21-62
  for leaky-wave antennas, 17-82 to 17-92
Ground-based hf antennas, modeling of, 32-76 to 32-77
Grounded wire antennas, 23-8 to 23-17
Ground-plane antenna ranges, 32-19
Ground planes
  and dipole arrays, 13-42
  imperfect, 3-27, 3-32 to 3-33, 3-47 to 3-50
  infinite, in slot array design, 12-34 to 12-35
  with microstrip antennas, 10-5, 10-6, 10-23 to 10-24
    circular polarization, 10-59
  perfect, 3-18 to 3-20, 3-47 to 3-48
  in UTD solutions, 20-7
Ground-reflection antenna ranges, 32-21 to 32-23
  gain measurement on, 32-48 to 32-49
Ground systems for am antennas, 26-37, 26-38
Ground-wave attenuation factor, 29-45
Group delay and satellite-earth propagation, 29-15 to 29-16
GTD. *See* Geometrical theory of diffraction
Guided waves from radomes, 31-5
Guidelines with modeling codes, 3-87, 3-95
Gyros for satellite antennas, 22-46
Gysel hybrid ring power division element, 21-50, 21-51

HAAT. *See* Height above average terrain
Half-power beamwidths, 1-20, 1-29
  of aperture antennas, 8-67
    circular, 5-22 to 5-24
    rectangular, 5-13, 5-16 to 5-17, 5-20, 15-91
  of arrays
    aperiodic, 14-15, 14-18, 14-23, 14-27 to 14-32
    Dolph-Chebyshev, 11-19
    linear, 11-15
    scanned in one plane, 13-12 to 13-15
  of classical antennas, 3-28 to 3-31
  and directivity, 32-41

  of horn antennas
    corrugated, 8-55, 8-59, 15-102 to 15-104
    E-plane, 8-14 to 8-15
    H-plane, 8-29 to 8-30
  of log-periodic zigzag antennas, 9-39 to 9-46
  of log-spiral conical antennas, 9-86, 9-87, 9-89, 9-97 to 9-98, 9-104
  and microwave radiometer beam efficiency, 22-27
  of millimeter-wave antennas
    dielectric grating, 17-79 to 17-80
    maximum-gain surface-wave, 17-40 to 17-42
    tapered dielectric-rod, 17-39, 17-47
  of reflector antennas
    lateral feed displacement, 21-20
    offset parabolic, 15-37, 15-41
    tapered-aperture, 15-16 to 15-17, 15-23
  of waveguide feeds, open-ended
    circular, 15-105
    rectangular, 15-104
Half-wave antennas, 26-8
  dipole, 3-17, 6-14 to 6-15
    radiation pattern of, 6-11
    receiving current of, 7-18 to 7-19
Half-wave radomes, 31-17
Hallen antenna, 6-24
Hansen-Woodyard end-fire linear array, 11-15
HAPDAR radar, 19-56
Harmonic distortion from pin diodes, 21-38
Harvesting of crops, remote sensing for, 22-15
H/D (height-to-diameter ratio), 15-32 to 15-33, 15-37, 15-40
Health care. *See* Medical applications
Heat sinking with integrated antennas, 17-133
Heating patterns, phantoms for checking, 24-54 to 24-57. *See also* Hyperthermia
Height. *See also* Vector effective height
  effective, 30-19, 30-21, 30-30
  for grating antennas
    dielectric, 17-64
    metal, 17-68 to 17-69
  and leakage constants, 17-58 to 17-59
  of log-spiral conical antennas, 9-101
  offset, of parabolic reflectors, 15-31 to 15-32

Index

of substrates, 17-107, 17-109, 17-111, 17-124, 17-126
of test antennas in elevated antenna ranges, 32-20
Height above average terrain
   and fm antenna ERP, 27-4, 27-6
   and tv antenna ERP, 27-3 to 27-5
Height-to-diameter ratio, 15-32 to 15-33, 15-37, 15-40
Height-to-radius ratio
   and impedance of dipoles, 9-61 to 9-62
   of log-periodic antennas, and directivity, 9-27, 9-30
Helical antennas, 3-31
   circularly polarized tv, 27-15 to 27-19
Helical-coil hyperthermia applicators, 24-48 to 24-49
Helical geodesic surface-ray paths, 4-75 to 4-76
Helix phased arrays, 21-12, 21-68, 21-70 to 21-71
Hemispherical lenses, 19-81 to 19-82
Hemispherical radiation pattern of am antennas, 26-5
Hemispherical radiators, 26-9, 26-27 to 26-29, 26-32
Hemispherical reflector antennas
   high-resolution, 19-85, 19-87, 19-89 to 19-91
   with wide-angle systems, 19-80 to 19-81
HEMP. *See* Electromagnetic pulses
Hertzian source, characteristics of, 3-13 to 3-14
Hexagonal arrays, 11-30 to 11-31
H-fields. *See* Magnetic fields
H-guide transmission type, 21-59, 21-62
   modification of, for leaky-wave antennas, 17-93 to 17-94
High-attenuation rf cable, 28-25
High-delay rf cable, 28-25
Higher-order modes
   junction coupling and scattering, 12-35 to 12-36
   for leaky-wave antennas, 17-87 to 17-92, 17-98 to 17-103
   with microstrip antennas, 10-25
High frequencies, ionosphere propagation of, 29-21 to 29-25
High-frequency antennas
   fixed-wire, 30-23 to 30-27, 30-29 to 30-30
   modeling of, 32-64, 32-76 to 32-77

High-frequency limit, 9-3
   of log-spiral antennas, 9-75
High frequency techniques, 4-3 to 4-6
   equivalent current method, 4-96 to 4-102
   geometrical optics fields, 4-6 to 4-8
   geometrical theory of diffraction, 4-23 to 4-96
   physical optics field, 4-8 to 4-23
   physical theory of diffraction, 4-102 to 4-115
High-gain antennas
   direction finding, 25-14 to 25-16
   dual-reflector, 15-73
   lens, 16-5
   limited scan, 19-57
   microstrip dipole, 17-122 to 17-126
   millimeter-wave antennas, 17-9 to 17-27
   multimode element array, 19-12
   rotating, 25-10 to 25-11
High-resolution antennas, 19-85, 19-87, 19-89 to 19-91
High-temperature rf cable, 28-24
HIHAT (high-resolution hemispherical reflector antennas), 19-85, 19-87, 19-89, to 19-91
Holey plate experiment, 14-27 to 14-35
Hollow-tube waveguides, 28-35 to 28-36
Holographic antennas, 17-106, 17-129 to 17-131
Homology with millimeter-wave antennas, 17-14
Horizon, radio, 29-34
Horizontal field strength with two towers, 26-30 to 26-34
Horizontal optimization, 11-82 to 11-83
Horizontal polarization
   with leaky-wave antennas, 17-97
   with loop fm antennas, 27-33
   for microwave radiometers, 22-44 to 22-46
   and surface emissivity, 22-8
   with tv antennas, 27-23
Horizontal profiling for earth resistivity, 23-4
Horizontal scanners, 33-19
Horizontal stabilizers, simulation of, 20-82 to 20-84
Horizontal transmitting loop antennas, 23-17

Horn antennas, 3-31, 8-5. *See also* Aperture-matched horn antennas; Conical horn antennas; Corrugated horn antennas; Dielectric loading, with horn antennas; E-plane horn antennas; H-plane horn antennas; Pyramidal horn antennas
  as array elements, 8-4
  as calibration standard, 8-3
  millimeter-wave, 17-23 to 17-24
  multimode, 8-69 to 8-73, 15-95
  phase center, 8-73 to 8-84
  with satellite antennas, 21-7, 21-80, 21-82 to 21-84
Horn feeds, 15-5
Horn launchers, 17-46
Horn waveguides, 2-27
H-plane folded and septum T waveguide power division elements, 21-51 to 21-52
H-plane gain correction factor for biconical horns, 21-104
H-plane horn antennas, 8-4
  aperture fields of, 8-20
  design of, 8-23 to 8-34
  directivity of, 8-29 to 8-33
  half-power beamwidth of, 8-29 to 8-30
  phase center of, 8-76 to 8-77
  radiated fields of, 8-20 to 8-26, 8-28
  for reflector feeds, 15-92 to 15-93, 15-96 to 15-97
  universal curves of, 8-23, 8-27 to 8-29
H-plane radiation patterns, 1-28, 5-12
  of apertures
    circular, 5-22
    rectanguar, 5-12
  with cavity-backed slot arrays, 9-68 to 9-71
  of dielectric grating antennas, 17-73 to 17-74, 17-78 to 17-79
  of DuFort-Uyeda lenses, 19-52
  of horn antennas
    corrugated, 8-53 to 8-59
    E-plane, 8-9, 8-10
    H-plane, 8-23 to 8-26, 8-28
    pyramidal, 8-37 to 8-40
  for log-periodic antennas
    diode arrays, 9-63 to 9-66
    zigzag, 9-39 to 9-42, 9-45, 9-46
  with microstrip antennas, 10-25
  of monofilar zigzag antenna, 9-36
  and UTD solutions, 20-13 to 20-14

Hughes approaches with cylindrical array feeds
  matrix-fed Meyer geodesic lens, 19-108 to 19-119
  phased lens, 19-107 to 19-108
Human-made noise, 29-51 to 29-52
Humidity
  and brightness temperature, 22-10
  remote sensing of, 22-6 to 22-8
  and scintillations, 29-20
  sounders for, 22-14, 22-22
Huygens-Fresnel principle, 1-32, to 1-33
Huygen's principle and source, 2-18 to 2-20
Hybrid antennas with reflector antennas, 17-11
Hybrid beam-forming networks, 21-34
Hybrid-coupled phase shifters, 13-62 to 13-63, 18-13
Hybrid-mode horn feeds, 15-5, 15-97, 15-99
Hybrid-mode waveguides, 28-47, 28-50 to 28-58
Hybrid-ring coupler power division element, 21-50, 21-52
Hydrology, remote sensing for, 22-15 to 22-16
Hyperbolas, 15-25 to 15-26
Hyperbolic lenses, 16-10, 16-53 to 16-54
Hyperbolic reflectors, feeds with, 19-61 to 19-62
Hyperthermia for medical therapy, 24-5 to 24-7, 24-9
  applicators for, 24-43 to 24-49
  heating patterns with, 24-50 to 24-57
  microstrip antennas for, 24-18 to 24-25
  waveguide antennas for, 24-9 to 24-18
Hypodermic monopole radiators, 24-26 to 24-27

Ice, 22-9
  remote sensing of, 22-14 to 21-15
  and sidefire helical tv antennas, 27-26
Illuminating fields and imperfect ground planes, 3-48
Illumination control of apertures, 13-30 to 13-32
Image-guide-fed slot array antenna, 17-50
Image guides, 21-59, 21-61, 28-50 to 28-53

Image impedances
  for log-periodic mono arrays, 9-66
  with periodic loads, 9-50, 9-55
Image-plane antenna ranges, 32-11, 32-19, 32-22 to 32-24
Images
  and imperfect ground planes, 3-27, 3-32, 3-48
  and perfect ground planes, 3-20, 3-47
  theory of, 2-11 to 2-13
Imaging arrays, 17-137 to 17-141
Impedance. *See also* Characteristic impedance; Input impedance; Mean impedance; Wave impedance
  base, with am antennas, 26-8 to 26-17
  of bow-tie dipoles, 17-138
  of cavities, 10-20 to 10-21
  of complementary planar antennas, 2-15 to 2-16
  of dipoles, 9-56
    and height-to-radius ratio, 9-61 to 9-62
    skewed, 27-18
  driving point
    with directional antenna feeder systems, 26-43 to 26-45
    with microstrip antennas, 10-20, 10-34, 10-53
  and EMP, 30-4
  feeder, with log-periodic diode arrays, 9-65 to 9-66
  of ideal sources, 2-29 to 2-30
  of log-periodic planar antennas, 9-10
  of log-spiral conical antennas, 9-107
  for medical applications, 24-49 to 24-50
  of microstrip antennas, 10-26 to 10-44
  resonant, 10-28, 10-30 to 10-31
  of scale models, 32-64, 32-68, 32-76 to 32-77
  of short monopole probe, 24-31 to 24-32
  surface, 3-54, 32-8
  of transmission lines, 21-63, 28-5
    coplanar, 28-35
    triplate stripline TEM, 28-22, 28-28 to 28-29
    two-wire, 28-11, 28-19
  and UTD solutions, 20-5
  waveguide, 2-28
Impedance function of cavities, 10-11
Impedance-loaded antennas
  dipole, 7-19 to 7-20
  monopole, 7-20 to 7-21

Impedance matching
  with directional antenna feeder systems, 26-40 to 26-41
  and gain measurements, 32-42
  and lens antennas, 16-5
    transformers for, 21-19
  of microstrip antennas, 10-50 to 10-52
  with receiving antennas, 6-6 to 6-7
  of split-tee power dividers, 13-61
  of waveguide arrays, 13-57
Impedance matrices, 3-61
  and rotational symmetry, 3-67 to 3-68
Imperfect ground planes, 3-27, 3-32 to 3-33
  integral equations with, 3-47 to 3-50
Implantable antennas for cancer treatment, 24-25 to 24-29
  E-field probes, 24-51 to 24-54
  temperature probes, 24-50 to 24-51
Incident fields
  GO, 4-10 to 4-13
  reciprocity of, 2-32 to 2-33
  and UTD solutions, 20-8 to 20-9
Incident powers and gain, 1-24 to 1-25
Incident shadow boundary, 4-10 to 4-11, 4-17
Incident wavefronts, 4-15, 4-16
Inclined edge-slot radiators, 18-8
Inclined-slot array antennas, 17-26
Inclined slots, 12-4, 12-24
  narrow wall arrays, 12-25 to 12-28
Incremental conductance technique, 12-26, 12-34
Index of refraction. *See* Refractive index
Indoor antenna ranges, 32-26 to 32-30
Inductive loading of short dipole, 6-13
Inductive susceptance of microstrip antennas, 10-26
Infinite array solutions, 13-48 to 13-49
Infinite baluns, 9-103, 9-105
Infinite ground plane assumptions in slot array design, 12-34 to 12-35
Inhomogeneous lenses, 16-51 to 16-54
In-line power dividers for array feeds, 13-61
Input features in modeling codes, 3-84 to 3-85
Input impedance, 2-29
  of array elements, 11-5
  of biconical antennas, 3-26
  of coaxial sleeve antennas, 7-32 to 7-33, 7-36

Input impedance (*cont.*)
  of dipole antennas
    folded, 7-36, 7-39 to 7-41
    linear, 7-14 to 7-17
    microstrip, 17-125
    short, 6-11 to 6-12
    sleeve, 7-26 to 7-29
  of L-band antennas, 30-18
  of loaded microstrip elements, 10-54 to 10-55
  of log-periodic antennas
    dipole arrays, 9-20 to 9-21, 9-24, 9-27 to 9-30, 9-58, 9-62
    zigzag, 9-39, 9-42 to 9-44
  of log-spiral antennas, 9-78 to 9-79
    conical, 9-95 to 9-97, 9-99
  of loop antennas
    ferrite, 6-21 to 6-22
    small, 6-17 to 6-18
  of microstrip antennas, 10-9, 10-26 to 10-28, 10-31, 10-65
  of models, 32-85 to 32-86
  of open-ended coaxial cable, 24-35
  of self-complementary antennas, 9-5, 9-7
  of sleeve monopoles, 7-30 to 7-31, 7-34
  of thin-wire antennas, 7-7 to 7-8, 7-44, 7-47 to 7-48
  of transmission lines, 28-7
    two-wire, 7-40
  of uhf communication antennas, 30-21
  validation of computer code for, 3-74 to 3-75, 3-77
  of vhf antennas
    communication, 30-24
    localizer, 30-26
    marker beacon, 30-28
Input susceptance of linear thin-wire antenna, 7-15 to 7-17
Insertion delay phase and dielectric sheets, 31-14
Insertion loss
  and beam-forming network switches, 21-37
  of phase shifters, 18-15 to 18-17, 18-21, 21-41
  and power division elements, 21-51
    diode varible, 21-47
  of transmission lines, 21-63
  of waveguides, 21-7
Insertion-loss gain measurement method, 32-44 to 32-46

Instantaneous direction-finding patterns, 25-12 to 25-14
Instantaneous field of view, 22-30 to 22-32
Instrumentation for antenna ranges, 32-30 to 32-33
Insular waveguide transmission types, 21-59, 21-61
  for microstrip arrays, 17-117 to 17-118
Integral equations, 3-36, 3-38 to 3-39, 3-44 to 3-52
  codes for, 3-80 to 3-96
  compared to differential, 3-55 to 3-58, 3-82
  computation with, 3-68 to 3-72
  for far fields, 3-8 to 3-9
  numerical implementation of, 3-52 to 3-68
  for unbounded space, 2-9 to 2-11
  validation of, 3-72 to 3-80
Integrated antennas, 17-105 to 17-106, 17-131 to 17-141
Integrated optics, 28-47
Interaction terms and computational effort, 3-64
Interfaces, between media, 1-8 to 1-9
  reflections at, 3-32, 24-7, 31-12 to 31-13
Interference
  in antenna ranges, 32-19
  with satellite antennas, 21-5
Interferometers, 25-21 to 25-23
Intermodulation distortion from pin diodes, 21-38
Internal networks and EMP, 30-8
Internal noise, 6-24, 29-50
Internal validation of computer code, 3-78 to 3-80
Interpolation of measured patterns, 20-6
Interrupt ability of modeling codes, 3-86
Intersection curve of reflector surface, 15-27 to 15-30
Inverted strip guide transmission type, 21-62
  dielectric, 17-49 to 17-50
In-vivo measurement, antennas for, 24-30 to 24-35
Ionosondes, 29-24
Ionosphere
  propagation via, 29-21 to 29-30
    satellite-earth, 29-14 to 29-15
  scintillation caused by, 29-16 to 29-21

Index

Irrigation control, remote sensing for, 22-15
ISB (incident shadow boundary), 4-10 to 4-11, 4-17
Isolation
  with ferrite switches, 21-38
  for satellite antennas, 21-5 to 21-7
Isosceles beams, 19-11 to 19-12
Isosceles triangular lattices, 1-35 to 1-36
Isotropic radiators
  characteristics of, 3-28
  electric-field strength of, 26-26
  radiation intensity of, 1-27
  receiving cross section of, 2-35
Isotropic response for E-field probes, 24-39 to 24-40

Jamming signals, arrays for, 11-10
Johnson compact antenna range, 32-29 to 32-30
Junction coupling in slot array design, 12-36
Junction effect with sleeve antennas, 7-23 to 7-26

KC-135 aircraft, patterns for, 20-65 to 20-67
Keller cone, 4-33
Keller's diffraction coefficients, 15-13
Kirchhoff approximation, 2-19 to 2-21
  and physical optics approximation, 2-27
Knife-edge diffraction, 29-42 to 29-44
Kolmogroff turbulence, 29-20
Ku-band systems, 29-8

Land
  emissivity of, 22-10
  modeling of, 32-74 to 32-77
  remote sensing of, 22-15 to 22-16
Large arrays, design of, 12-28 to 12-34
LASMN3 modeling code, 3-88 to 3-91
Latching Faraday rotator variable power dividers, 21-44 to 21-46
Latching phase shifters, 13-63 to 13-64
Lateral feed displacement, 15-51 to 15-61, 21-20
Lateral width of dielectric grating antennas, 17-64
Lattices
  field representation of, 1-35
  for phased-array radiators, 18-9 to 18-10
  spacing of, for feed arrays, 21-27

Launching coefficients
  for conducting cylinder, 21-97
  surface-ray, 4-70
Law of edge diffraction, 4-33
L-band blade antennas, 30-17 to 30-19
Leakage constants
  for dielectric grating antennas, 17-65 to 17-69, 17-74 to 17-79, 18-82
  with leaky mode antennas, 17-53 to 17-63, 17-85, 17-87 to 17-90, 17-95 to 17-97
Leakage radiation in medical applications, 24-5, 24-50
Leaky modes with periodic dielectric antennas, 17-51 to 17-62
Leaky-wave antennas, 17-34 to 17-35
  uniform-waveguide, 17-82 to 17-103
Least squares moment method, 3-60
Left-hand circular polarization, 1-15, 1-20
  of arrays, 11-7
  dual-mode converters for, 21-78, 21-80
  far-field pattern for, 1-29
  for microstrip antennas, 10-59, 10-69
  with offset parabolic reflectors, 15-36, 15-48 to 15-50
  power dividers for, 21-78, 21-80
Left-hand elliptical polarization, 1-16, 9-74
Lenses and lens antennas, 16-5 to 16-6. *See also* Constrained lenses; Optical feeds; Semiconstrained feeds
  aberrations and tolerance criteria of, 16-12 to 16-19
  with analytic surfaces, 16-9 to 16-12
  in compact ranges, 32-29
  design principles of, 16-7 to 16-9
  with horn antennas
    conical, 8-46
    diagonal, 8-70
    H-plane, 8-23
  inhomogeneous, 16-51 to 16-54
  millimeter-wave, 17-14 to 17-22
  for satellite antennas, 21-7, 21-8, 21-12 to 21-19, 21-71 to 21-77
  surface mismatches of, 16-54 to 16-57
  taper-control, 16-33 to 16-38
  wide-angle, 16-19 to 16-33
  zoning of, 16-38 to 16-41
LEO (low earth orbiting) satellites, 22-35 to 22-43
Level curves, 14-12 to 14-14
Lightning and tv antennas, 27-8, 27-9

Limited-scan unconstrained feeds, 19-12, 19-52, 19-56 to 19-76
Lindenblad tv antenna, 27-12
Linear arrays, 11-8 to 11-23
  broadside, maximum directivity of, 11-64 to 11-68
  cophasal end-fire, 11-69, 11-70
  of longitudinal slots, 12-13 to 12-16
  millimeter-wave, 17-33
    uniform leaky-wave, 17-35
Linear dipole antennas, 3-13 to 3-14, 7-6, 7-11 to 7-23
Linear errors with lenses, 16-14
Linear polarization, 1-14 to 1-16, 1-20, 1-28 to 1-29
  conversion of, from circular, 13-60 to 13-61
  measurement of, 32-6, 32-8, 32-53, 32-59
  for microwave radiometers, 22-44 to 22-46
  with phased arrays, 18-6
  with satellite antennas, 21-7
    gridded reflector, 21-21
Linear transformations in arrays, 11-23 to 11-48
Linear wire antenna, polarization of, 1-24
Line-of-sight propagation, 29-30, 29-32 to 29-36
  with millimeter-wave antennas, 17-5
Line-of-sight ranges, 32-9 to 32-11
Line source illumination, 4-12 to 4-13
Line source patterns, 13-13
Link calculations for satellite tv, 6-27, 6-29 to 6-31
Liquid artificial dielectric, 17-22, 17-72
Lit zone, 4-75
  Fock parameter for, 4-73
  terms for, 4-71 to 4-72
Load current of receiving antennas, 6-3, 6-4
Loaded-line diode phase shifters, 13-62 to 13-63, 18-13
Loaded-lines, 9-46 to 9-62
Loaded loop antennas, 7-47 to 7-48
Loaded microstrip elements, 10-52 to 10-56
Loaded thin-wire antennas, 7-9
Lobes, 1-28. *See also* Grating lobes; Side lobes
  with log-spiral antennas, 9-75, 9-77
  shoulder, 32-16
Localizer antennas, 30-23, 30-25 to 30-26

Logarithmic spiral antennas, 17-28
Logarithmic spiral curve, 9-72
Log-periodic arrays, 9-3 to 9-4, 9-8 to 9-11
  periodic structure theory designs for, 9-62 to 9-71
  wire diameter of, 32-76
Log-periodic dipole antennas
  active regions in, 9-17 to 9-21, 9-24
  bandwidth of, 9-17, 9-21, 9-24
  circular, 9-3, 9-4
  design of, 9-20 to 9-32
  directivity of, 9-21 to 9-24, 9-27, 9-61, 9-66
  feeder circuits for, 9-16 to 9-18
  front-to-back ratios of, 9-60 to 9-61
  impedance of, 9-21, 9-24, 9-27 to 9-30, 9-58
  parameters for, 9-12, 9-14, 9-15, 9-24
  radiation fields for, and Q, 9-19 to 9-20, 9-62 to 9-66
  swr of, 9-58 to 9-59
  transposition of conductors in, 9-14 to 9-16
Log-periodic mono arrays, 9-64 to 9-67
Log-periodic zigzag antennas
  active region of, 9-35 to 9-36
  design procedures for, 9-37 to 9-46
  directivity of, 9-39 to 9-40
  half-power beamwidths for, 9-39 to 9-46
  radiation fields of, 9-39 to 9-40
Log-spiral antennas. *See also* Conical log-spiral antennas; Logarithmic spiral antennas
  feeds for, 9-75 to 9-78, 9-80
  geometry of, 9-8
  input impedance of, 9-78 to 9-79
  polarization of, 9-74, 9-77
  radiated fields of, 9-75
  truncation of, 9-72 to 9-73
Long-slot array antennas, 17-26
Longitude amplitude taper in antenna ranges, 32-15
Longitudinal broadwall slots, 12-4
Longitudinal currents in waveguides, 12-3
Longitudinal section electric mode, 8-73
Longitudinal slot arrays
  aperture distribution of, 12-24
  design equations for, 12-10 to 12-13
  E-field distribution in, 12-6 to 12-10, 12-35
  mutual coupling with, 12-30 to 12-31
  nonresonantly spaced, 12-16 to 12-20
  planar, 12-20 to 12-24

resonantly spaced, 12-13 to 12-16
shunt-slot antennas, 17-33
Look angles for tvro, 6-29
Loop antennas
coaxial current, 24-45 to 24-46
ferrite, 6-18 to 6-22
fm, 27-33 to 27-34
for geophysical applications, 23-17 to 23-23
as magnetic-field probes, 24-40 to 24-43
rotating systems, 25-4 to 25-9
small, 3-16, 3-19, 6-13, 6-17 to 6-19, 7-46 to 7-47
spaced, 25-9 to 25-10
thin-wire, 7-40 to 7-48
vector effective area of, 6-5 to 6-6
Loop mutual impedance, 26-12 to 26-13
Looper radiators for medical applications, 24-20 to 24-22
Lorentz reciprocity theorem, 2-16 to 2-18
and planar scanning, 33-7
Losses. *See also* Conduction losses; Path losses
with conical horn antennas, 8-46
free space, 2-39, 29-5 to 29-6, 29-8
with microwave radiometers, 22-32 to 22-33
system, power for, with am antennas, 26-36 to 26-37
tangents for, 1-6
Lossy dielectrics, modeling of, 32-71 to 32-75
Low capacitance rf cable, 28-25
Low earth orbiting satellites, 22-35 to 22-43
Low frequencies, surface propagation by, 29-44
Low-frequency limit, 9-3
of log-periodic planar antennas, 9-8 to 9-9
of log-spiral antennas, 9-75
Low-frequency techniques, 3-5
for classical antennas, 3-13 to 3-36
for EMP analysis, 30-9 to 30-14
integral equations for. *See* Integral equations
theory with, 3-6 to 3-13
Low-frequency trailing-wire antennas, 30-31
Low-noise amplifiers for satellite tv, 6-26
Low-noise detectors, 17-136
Low-profile arrays. *See* Conformal arrays

Low-Q magnetic-field probes, 24-43, 24-44
Lowest usable frequency of ionospheric propagation, 29-25
LSE (longitudinal section electric) mode, 8-73
L-section power dividing circuits, 26-39 to 26-40
Ludwig's definition for polarization, 1-23 to 1-24
Lumped impedance for thin-wire antennas, 7-9
Luneburg lenses, 16-23
for feeding arrays, 19-18, 19-44, 19-46 to 19-49
as inhomogeneous lenses, 16-51 to 16-52
vs. two-layer lenses, 17-18n
with wide-angle multiple-beam, 19-80

Magic-T power division elements, 21-50 to 21-52
Magnetic cores. *See* Ferrite loop antennas
Magnetic current distributions
and Kirchhoff approximation, 2-20 to 2-21
for microstrip antennas, 10-22 to 10-25
for dual-band elements of, 10-66
edge of, in, 10-55 to 10-57
Magnetic dipoles
and EMP, 30-9
radiation pattern of, 4-59
Magnetic fields
with arrays, 3-36
conical-wave, 4-99
and edge conditions, 1-9, 4-27 to 4-28, 4-101
and equivalent currents, 4-101, 5-8 to 5-10
in far zone, 6-3 to 6-4
of GO reflected ray, 4-14
Green's dyads for, 3-46, 3-48
GTD for, 4-23
for edge conditions, 4-27 to 4-28
integral equation for, 3-46 to 3-47, 3-51
with plane-wave spectra, 1-32, 5-7 to 5-8
PO, 4-19 to 4-20
probes for, 24-37 to 24-38, 24-40 to 24-44
calibration of, 24-54 to 24-57
reflection coefficient of, 2-28
of small dipoles, 6-13
of small loops, 6-18

Magnetic fields (*cont.*)
  surface, 4-84
    with cavities, 10-12 to 10-13
      for physical optics approximation, 2-26 to 2-27
    tangential, 1-32, 5-7
    with UTD, 4-83, 4-84, 4-95
    for vertices, 4-31
Magnetic loss tangent, 1-6
Magnetic vector potential, 3-6 to 3-7
  with arrays, 3-33, 3-35, 11-6
Magnetometers, 23-23 to 23-24
Magnetrode for hyperthermia, 24-45
Magnitude pattern for linear arrays, 11-9, 11-11
Main beam, 1-20
  with aperiodic arrays, 14-24, 14-26 to 14-34
  in array visible region, 11-31 to 11-32
  of corrugated horns, 8-55
  directivity of, 1-29
  with rectangular apertures, 5-12 to 5-14, 5-19
  and space harmonics, 1-35
Main line guide for planar arrays, 12-21 to 12-23
Main Scattering Program modeling code, 3-88 to 3-91
Maintenance of computer models, 3-68
Mangin Mirror with wide-angle multiple-beam systems, 19-85, 19-89
Mapping transformation, 11-25
Marine observation satellite (MOS-1), 22-47
Mariner-2 satellite, 22-17, 22-18
Marker beacon antennas, 30-23, 30-27 to 30-28
Masts, shipboard, simulation of, 20-94 to 20-96
Matched feeds and cross-polarization, 15-99
Matched four-port hybrid junctions, 19-5 to 19-6
Matched loads calibration method, 22-25 to 22-27
Matching. *See* Impedance matching
Matching filters, 10-52
Matching transformers for lenses, 21-19
Mathematical models, 32-65
Matrices. *See also* Admittance matrices; Blass matrices; Butler matrices
  impedance, 3-61, 3-67 to 3-68
  scattering, 12-31, 33-7
  transformation, 15-117, 15-119
Matrix-fed cylindrical arrays, 19-101 to 19-106
  with conventional lens approach, 19-107 to 19-108
  with Meyer geodesic lens, 19-108 to 19-119
Maximum effective height
  of short dipoles, 6-12, 6-14 to 6-15
  of small loops, 6-18
Maximum-gain antennas, 17-40 to 17-41
Maximum receiving cross section
  of short dipoles, 6-12 to 6-15
  of small loops, 6-18
Maximum usable frequency of ionospheric propagation, 29-22 to 29-23
Maxson tilt traveling-wave arrays, 19-80
Maxwell equations, 1-5 to 1-7
  for electromagnetic fields, 3-6
  using MOM to solve, 3-55
  in unbounded space, 1-9
Maxwell fish-eye lenses, 16-51 to 16-53
Meanderline polarizers, 21-29
Mean effective permeability of ferrite loops, 6-20 to 6-21
Mean impedance
  of log-periodic antennas
    dipole, 9-27
    zigzag, 9-46
  with self-complementary antennas, 9-10 to 9-11, 9-13
  of spiral-log antennas, 9-78
Mean resistance of log-periodic antennas
  dipole, 9-28
  zigzag, 9-40
Mean spacing parameter for log-periodic dipole antennas, 9-28
Measurement of radiation characteristics, 32-3 to 32-5, 32-39 to 32-63, 33-3 to 33-5
  compact ranges for, 33-22 to 33-24
  current distributions for, 33-5 to 33-6
  cylindrical scanning for, 33-10 to 33-12
  defocusing techniques for, 33-24 to 33-26
  errors in, 33-15 to 33-17
  extrapolation techniques for, 33-26 to 33-28
  interpolation of, 20-6

Index

modeling for, 32-74 to 32-86
planar scanning for, 33-7 to 33-10
plane-wave synthesis for, 33-21 to 33-22
and radiation cuts, 32-6 to 32-8
ranges for, 32-8 to 32-39, 33-15 to 33-21
for slot array design, 12-26 to 12-37
spherical scanning for, 33-12 to 33-15
Mechanical scanning
  conical, 22-39 to 22-40
  with millimeter-wave antennas, 17-9, 17-10, 17-24 to 17-26
  with satellite antennas, 21-5
Mechanical switches with satellite antennas, 21-7
Medical applications, 24-5 to 24-9
  characterization of antennas used in, 24-49 to 24-57
  hyperthermia applicators for, 24-43 to 24-49
  implantable antennas for, 24-25 to 24-29
  in-vivo antennas for, 24-30 to 24-35
  microstrip antennas for, 24-18 to 24-25
  monitoring rf radiation in, 24-36 to 24-43
  waveguide- and radiation-type antennas for, 24-9 to 24-18
Medium frequencies, propagation of
  ionospheric, 29-25 to 29-27
  surface, 29-44
Medium-gain millimeter-wave antennas, 17-9 to 17-27
Medium propagation loss, 29-6
Menzel's antennas, 17-102 to 17-104
Metal grating antennas, 17-49, 17-51, 17-68 to 17-69
  bandwidth with, 17-74
Metal waveguides, 17-117 to 17-118
Meteorology and remote sensing, 22-13 to 22-15
Meteors, ionization trails from, 29-30
Meteor satellite, 22-17
Meters for receiving antennas, 2-30 to 2-32
Method of moments
  for complex environment simulations, 20-4
  for coupling, 20-68
  frequency-domain, 3-57 to 3-62, 3-64
  with Maxwell's equations, 3-55
  radomes, 31-27 to 31-29
  with reflector antennas, 15-7
  time-domain, 3-62 to 3-64

Meyer lenses for feeds, 19-19, 19-32 to 19-37
  matrix-fed, 19-108 to 19-119
MFIE modeling code, 3-88 to 3-91
Microbolometers, 17-133, 17-137 to 17-138, 17-141
Microstrip antennas, 10-5 to 10-6
  applications with, 10-56 to 10-70
  arrays of, for medical applications, 24-25
  circular polarization with, 10-57 to 10-63
  coupling of, to waveguides, 17-199 to 17-121
  efficiency of, 10-49 to 10-50
  elements for
    dual-band, 10-63 to 10-66
    frequency-agile, 10-66 to 10-69
    loaded, 10-52 to 10-56
    polarization-agile, 10-69 to 10-70
  feeds for, 10-28, 10-51
  impedance of, 10-28 to 10-44
  leaky modes with, 17-35
  loop radiators, 24-20 to 24-22
  matching of, 10-50 to 10-52
  for medical applications, 24-18 to 24-20
  models of
    circuit, 10-26 to 10-28
    physical, 10-7 to 10-21
  radiation patterns of, 10-21 to 10-26
  resonant frequency of, 10-44 to 10-49
  slot, for medical applications, 24-22 to 24-24
Microstrip array antennas, 17-25
Microstrip baluns, 18-8
Microstrip dipole antennas, 17-106, 17-118 to 17-119, 17-122 to 17-126
Microstrip excited-patch radiators, 18-8 to 18-9
Microstrip lines, 13-58, 21-59, 21-60
  for leaky-wave antennas, 17-82 to 17-83, 17-98 to 17-103
  for planar transmission lines, 28-30 to 28-33
Microstrip patch resonator arrays, 17-106 to 17-118
Microstrip radiator types, 13-60
  dipole, for phased arrays, 18-7
Microstrip resonator millimeter-wave antennas, 17-103 to 17-106
  with thick substrates, 17-122 to 17-129
  with thin substrates, 17-107 to 17-122

Microstrip techniques for integrated antennas, 17-131, 17-134 to 17-137
Microstrip waveguides for microstrip arrays, 17-117 to 17-118
Microwave-absorbing material for indoor ranges, 32-25
Microwave antennas
 constrained lenses for, 16-42
 measurements for, 33-20
Microwave diode switches, 21-38
Microwave fading, 29-40 to 29-41
Microwave radiation
 antennas to monitor, 24-36 to 24-43
 transfer of, 22-4 to 22-7
Microwave radiometers and radiometry
 antenna requirements, 22-16 to 22-46
 fundamentals of, 22-3 to 22-9, 22-22 to 22-23
 future of, 22-47 to 22-52
 for remote sensing, 22-9 to 22-16
 spacecraft constraints with, 22-46 to 22-47
Microwave sounder units, 22-14
 modulation frequency of, 22-24 to 22-25
 onboard reference targets for, 22-26
 for planar scan, 22-37
 as step scan type, 22-43
 on TIROS-N, 22-37
Microwave transmissions, 29-32
Military aircraft, simulation of, 20-78 to 20-88
Military applications, antennas for, 17-6 to 17-7, 17-15
Millimeter-wave antennas, 17-5 to 17-8
 fan-shaped beam, 17-28 to 17-32
 high-gain and medium-gain, 17-9 to 17-27
 holographic, 17-129 to 17-131
 integrated, 17-131 to 17-141
 microstrip
  monolithic, 17-134 to 17-137
  with thick substrates, 17-122 to 17-129
  with thin substrates, 17-107 to 17-122
 near-millimeter-wave imaging array, 17-137 to 17-141
 omnidirectional, 17-32 to 17-34
 periodic dielectric, 17-48 to 17-82
 printed-circuit, 17-103 to 17-131
 spiral, 17-27 to 17-28
 tapered-dielectric rod, 17-36 to 17-48, 17-133 to 17-134
 uniform-waveguide, leaky-wave, 17-82 to 17-103
Milimeter-wave Sounders, 22-47, 22-50 to 22-51
Mine communications, 23-3
 grounded wire antennas for, 23-8 to 23-17
 loop antennas for, 23-17 to 23-23
Minimax synthesis method, 21-65
Mismatches
 with lenses, 16-54 to 16-55
 and signal-to-noise ratio, 6-25
 transmission line and antenna, 1-24
Missile racks, aircraft, simulation of, 20-79 to 20-80, 20-82
Mixer diodes for integrated antennas, 17-133 to 17-134, 17-138 to 17-139, 17-141
MMIC (monolithic microwave integrated circuit technology), 17-12
Modal field distributions of loaded elements, 10-53
Modal transverse field distributions
 in circular waveguides, 1-47 to 1-48
 in rectangular waveguides, 1-40 to 1-41
Mode converters, odd/even, 21-54 to 21-56
Mode distributions of patches, 10-16 to 10-19
Modeling. *See also* Scaling
 for antenna ranges, 32-63 to 32-86
 with computers
  codes for, 3-80 to 3-96
  computation with, 3-68 to 3-72
  validation of, 3-72 to 3-80
Modes, patch, 10-7 to 10-21, 10-24 to 10-26, 10-35 to 10-37
Mode voltage in common waveguides, 12-13, 12-16
Modulating radiometers, 22-24
MOM. *See* Method of moments
Moment Method Code for antenna currents, 20-91
Monitors
 for directional antenna feeder systems, 26-41 to 26-43
 of rf radiation, 24-36 to 24-43
Monofilar zigzag wire antennas, 9-34, 9-36
Monolithic microstrip antennas, 17-134 to 17-137
Monolithic microwave integrated circuit (MMIC) technology, 17-12

Monopole antennas
  on aircraft, 20-62 to 20-68
    Boeing 737, 20-70, 20-73 to 20-75
    F-16 fighter, 20-87 to 20-88
  bandwidth of, 6-24
  hypodermic, 24-26 to 24-27
  images with, 3-20
  impedance-loaded, 7-20 to 7-21
  for in-vivo measurements, 24-30 to 24-32
  millimeter-wave, 17-32
  radiation of, on convex surface, 4-64 to 4-66, 4-71, 4-83 to 4-85
  short, 4-94, 6-12
  sleeve, 7-29 to 7-31
Monopulse pyramidal horn antennas, 8-72 to 8-75
MOS (marine observation satellite), 22-47
Mounting structures, 30-4 to 30-7
Movable line-of-sight ranges, 32-9 to 32-11
MSU. *See* Microwave sounder units
Multiarm self-complementary antennas, 9-6 to 9-8
Multibeam applications, lens antennas for, 16-6, 17-21
  wide-angle dielectric, 16-19
Multibeam satellite antennas, 15-5, 15-80 to 15-87, 21-57 to 21-58, 21-62, 21-64 to 21-67
  zoning in, 16-38
Multidetectors with integrated antennas, 17-137
Multifocal bootlace lenses, 16-46 to 16-48
Multifrequency Imaging Microwave Radiometer (MIMR), 22-47, 22-48 to 22-25
Multimode circular arrays, 25-17 to 25-20
Multimode element array beam feed technique, 19-12 to 19-15
Multimode feeds, 9-110, 9-112
  radial transmission line, 19-23 to 19-32
Multimode generators, Butler matrix as, 19-10
Multimode horn antennas, 8-69 to 8-73
  for reflector antennas, 15-95
Multipath propagation, 29-30, 29-37 to 29-41
Multipath reflections with scanning ranges, 33-16, 33-18
Multiple-access phased arrays, 21-68 to 21-71
Multiple-amplitude component polarization measurement method, 32-61 to 32-62
Multiple-antenna installations, 27-36 to 27-40
Multiple-beam antennas, 15-80
  reflector antennas with, 15-5
Multiple-beam constrained lenses, 19-79 to 19-85
Multiple-beam-forming networks, 19-23
Multiple-beam matrix feeds, 19-8 to 19-12
Multiple-feed circular polarization, 10-61 to 10-63
Multiple quarter-wavelength impedance matching transformers, 21-19
Multiple reflections
  with compact ranges, 33-27
  with thin-wire transmitting antennas, 7-13
Multiple-signal direction finding, 25-24 to 25-25
Multiport impedance parameters, 10-34, 10-44
Multiprobe launcher polarizers, 21-30
Multistepped dual-mode horns, 21-82, 21-84 to 21-89
Muscle, human
  phantom models for, 24-55
  power absorption by, 24-8, 24-9, 24-24, 24-47
Mutual admittance between slots, 4-91
Mutual base impedance of am antennas, 26-12 to 26-17
Mutual conductance with microstrip antennas, 10-9
Mutual coupling, 4-49
  and array transformations, 11-23
  of arrays
    aperiodic, 14-20 to 14-27
    circular, on cylinder, 21-100
    conformal, 13-49 to 13-52
    inclined narrow wall slot, 12-25
    large, 12-28 to 12-33
    periodic, 13-7, 13-41 to 13-46
    phased, 21-8 to 21-10, 21-12
    planar, of edge slots, 12-27 to 12-28
    scanned in one plane, 13-12
    slots in, 12-5
  of tapered dielectric-rod antennas, 17-47
  of convex surfaces, UTD for, 4-84 to 4-96
  with integrated antennas, 17-137
  with microstrip antennas, 10-7

Mutual coupling (*cont.*)
  of millimeter-wave antennas
    dielectric grating, 17-79
    dipoles, 17-128 to 17-129
    tapered dielectric-rod, 17-47
  nulls from, 13-45 to 13-46
  and open-periodic structures, 9-56
  and phase shift, 9-58
  with reflector antennas, 15-6
    offset parabolic, 15-83 to 15-84
  of slots
    longitudinal, 12-15, 12-18
    radiating, 12-35 to 12-36
Mutual impedance, 6-8 to 6-9
  with am antennas, 26-12 to 26-13
  with aperiodic arrays, 14-3, 14-20 to 14-21
  of array elements, 11-5
  and dipole spacing, 9-58
  of microstrip antennas, 10-34
  of monopoles, 4-94
  and radiation resistance, 9-56, 14-3

Narrowband conical horn antennas, 8-61 to 8-62
Narrow wall coupler waveguide power division elements, 21-51 to 21-52
Narrow wall slots, inclined, arrays of, 12-25 to 12-28
National Oceanic and Atmospheric Administration (NOAA)—series satellites, 22-51
Natural modes for dipole transient response, 7-21 to 7-22
Navigation systems, ionospheric propagation for, 29-27
Near-degenerate modes, 10-28, 10-29, 10-60
Near fields and near-field radiation patterns, 1-16 to 1-18
  of aperture antennas, 5-24 to 5-26, 15-21 to 15-24
    axial gain of, 5-29 to 5-33
    with Cassegrain feed systems, 19-62 to 19-63
  of dipole arrays, 13-44
  of dipoles, 20-69
  experimental validation of computer code for, 3-76
  gain reduction factor for, 5-30 to 5-32
  measurement of. *See* Measurement of radiation characteristics
  in medical applications, 24-5 to 24-8
  for monopole on aircraft, 20-62 to 20-65
  in periodic structure theory, 9-33
  with pyramidal horns, 15-92
  with reflector antennas, 15-8, 15-21 to 15-23
  transformation of, to far-field, 20-61 to 20-62
  with UTD solutions, 20-4, 20-6, 20-13 to 20-14
  in zigzag wire antennas, 9-33
Near-millimeter-wave imaging arrays, 17-137 to 17-141
NEC (Numerical Electromagnetic Code) modeling code, 3-87 to 3-95
Needle radiators, 24-26 to 24-27
Nickel for satellite antenna feeds, 21-28 to 21-29
Nimbus-5 satellite, 22-17
  planar scanning on, 22-37
Nimbus-6 satellite, 22-17 to 22-18
Nimbus-7 satellite, 22-17
  SMMR on, 22-22
NNBW (null-to-null beamwidths), 22-28
Nodal curves with frequency-agile elements, 10-68 to 10-69
Nodal planes with circular polarization, 10-62
Noise and noise power
  with arrays, 11-59
  emitters of, 2-40 to 2-42
  and propagation, 29-47 to 29-52
  and receiving antennas, 6-24 to 6-25
  at terminal of receiver, 2-42 to 2-43
  at tvro antenna, 6-26, 6-30
Noise temperature, 2-39 to 2-43
  effective, 22-27
  and figure of merit, 1-27, 6-31
  and microwave radiometers, 22-22 to 22-23
Noncentral chi-square distribution for array design, 14-10 to 14-12
Nonconverged solutions and computer solutions, 3-54
Nondirectional antennas for T T & C, 21-89
Nondirective couplers, 19-4 to 19-5
Nonoverlapping switch beam-forming networks, 21-30 to 21-31
Nonperiodic configuration with multibeam antennas, 21-66
Nonradiative dielectric guides, 17-93 to 17-98
Nonreciprocal ferrite phase shifters, 13-63, 18-14 to 18-15, 21-38 to 21-39

Nonresonantly spaced longitudinal slots, 12-16 to 12-20
Nontrue time-delay feeds, 19-54 to 19-56
Nonuniform component of current, 4-103
Nonuniform excitation of array elements, 11-29 to 11-30, 11-47 to 11-48
Normal congruence of rays, 4-6
Normalized aperture distributions
 circular, 5-21 to 5-22
 rectangular, 5-19 to 5-20
Normalized intensity of far shields, 1-20 to 1-21
Normalized modal cutoff frequencies
 for circular waveguides, 1-46, 1-49
 for rectangular waveguides, 1-42
Normalized pattern functions, 11-11 to 11-13
Normalized resonant modes of microstrip antennas, 10-13, 10-15
 frequency of, 10-46
Norton's theorem for receiving antenna loads, 6-3, 6-4
Nose section of aircraft, simulation of, 20-78
N-port analogy, 3-64, 3-85
NRD guides, 17-35, 17-82 to 17-83, 17-92 to 17-98
Nuclear detonations. *See* Electromagnetic pulses
Null-to-null beamwidths, 22-28
Nulls
 with apertures
  circular, 5-22 to 5-23
  rectangular, 5-13 to 5-15, 5-19
 in Bayliss line source pattern synthesis, 13-28
 electrically rotating, 25-11 to 25-12
 filling in of, with tv antennas, 27-7
 with linear arrays, 11-15, 11-40
 with log-spiral conical antennas, 9-102
 with loop antennas, 25-5 to 25-6
 with microstrip antennas, 10-25 to 10-26
 with multiple element feeds, 19-13
 from mutual coupling, 13-45 to 13-46
 with offset reflector antennas, 15-37, 15-41
 prescribed, arrays with, 11-9 to 11-10
 with two-tower antennas, 26-22 to 26-24
Numerical Electromagnetic Code, 3-87 to 3-95
Numerical implementation for electromagnetic field problems, 3-53 to 3-68

Numerical modeling errors, 3-69 to 3-71
Numerical simulations for antennas, 20-5 to 20-7
Numerical solutions
 for aircraft antennas, 20-70 to 20-90
 for ship antennas, 20-90 to 20-97
Numerical treatment in modeling codes, 3-81, 3-83 to 3-84, 3-90
Numerical validation of computer code, 3-76 to 3-80

Obstacles, scattering by, 2-21 to 2-27
Ocean
 emissivity of, 22-10 to 22-11
 modeling of, 32-79
 reflection coefficient over, 29-38 to 29-40
 remote sensing of, 22-14 to 22-15
 and remote sensing of humidity, 22-6 to 22-7
 simulation of, 20-91
Odd distribution in rectangular apertures, 5-15
Odd/even converters, 21-7, 21-54 to 21-56
Odd-mode amplitude control, 19-14 to 19-15
Off-focus feeds with reflectors, 15-49 to 15-61
Offset distance with reflector antennas, 21-23
Offset feeds
 for Gregorian feed systems, 19-63 to 19-66
 for reflectarrays, 19-54
 for reflector antennas, 17-9
Offset height of offset reflectors, 15-31 to 15-32
Offset long-slot array antennas, 17-26
Offset parabolic reflectors
 edge and feed tapers with, 15-36 to 15-37
 feed patterns for, 15-34 to 15-36
 geometrical parameters for, 15-31 to 15-34
 off-focus feeds for, 15-49 to 15-61
 on-focus feeds for, 15-37 to 15-49
Offset phased array feeds reflector, 19-61 to 19-62
Offset reflector antennas, 15-6, 15-23. *See also* Cassegrain offset antennas

Offset reflector antennas (*cont.*)
  dual-reflector, 15-6 to 15-15, 15-73, 17-11
  satellite, 21-19 to 21-26
Ohmic losses with remote-sensing microwave radiometers, 22-32
Omega navigation systems, 29-27
Omnidirectional antennas
  dielectric grating, 17-80 to 17-82
  millimeter-wave, 17-32 to 17-34
  for satellite T T & C, 21-90
  tv, 27-5
On-axis gain of horn antennas, 8-3 to 8-4
One-parameter model for reflector antennas, 15-18 to 15-21
One plane, arrays scanned in, 13-12 to 13-15
On-focus feeds with reflectors, 15-37 to 15-49
Open-circuit voltage, 2-32
  with UTD solutions, 20-7
  and vector effective height, 6-5
Open-ended coaxial cable, 24-32 to 24-35
Open-ended waveguides
  arrays of, 13-44
  circular
    for reflector feeds, 15-90 to 15-93
    with reflector satellite antennas, 21-74
    half-power beamwidths of, 15-105
  dielectric-loaded, for medical applications, 24-9
  radiators of, for phased arrays, 18-6 to 18-7
  rectangular
    electric-field distribution of, 5-14 to 5-15
    half-power beamwidths of, 15-104
    for reflector feeds, 15-89 to 15-92
  TEM, for medical applications, 24-9 to 24-13
Open-periodic structures, 9-55 to 9-56
Open-wire transmission lines, 28-11
  two-wire, impedance of, 28-12, 28-13
Optical devices as Fourier transformers, 19-95 to 19-98
Optical feeds. *See also* Unconstrained feeds
  for aperiodic arrays, 19-57 to 19-59
  corporate, for beam-forming feed networks, 18-22
  transform, 19-91 to 19-98
Optics fields. *See* Geometrical optics; Physical optics

Optimization array problems, 11-81 to 11-86
Optimum pattern distributions, 5-18 to 5-20
Optimum working frequency and ionospheric propagation, 29-25
Orbits of satellite antennas, 6-26
Orthogonal beams, pattern synthesis with, 13-22 to 13-23
Orthogonal polarization, 21-7
Orthomode junctions, 21-7
Orthomode transducers, 22-45
Out-of-band characteristics and EMP, 30-8
Outdoor antenna ranges, 32-19 to 32-24, 33-28
Outer boundaries of radiating near-field regions, 1-18
Output features in modeling codes, 3-86
Overfeeding of power with am antennas, 26-36 to 26-37
Overlapped subarrays, 13-35 to 13-39
  space-fed system, 19-68 to 19-76
Overlapping switch beam-forming networks, 21-31 to 21-32
Overreach propagation, 29-41
Oversized waveguide transmission type, 21-60
Owf (optimum working frequency), 29-25
Oxygen
  absorption by, 29-35
  remote sensing of, 22-12 to 22-13
  temperature sounders for, 22-16, 22-21

Pancake coils, 24-47 to 24-48
Panel fm antennas, 27-35 to 27-36
Parabolas, 15-25 to 15-26
Parabolic patches, 10-18
Parabolic pillbox feeds, 19-19 to 19-20
Parabolic reflector antennas, 3-31. *See also* Offset parabolic reflectors
  far-field formulas for, 15-15 to 15-23
  near-field radiation pattern of, 4-109
Paraboloidal lenses, 21-18
Paraboloidal surfaces and intersection curve, 15-29 to 15-30
Paraboloid reflectors, 15-76
  with compact antenna ranges, 32-28 to 32-29
  phased array feeds with, 19-58 to 19-61
Parallel feed networks, 18-19, 18-21 to 18-22, 18-26, 19-5 to 19-6
Parallel plate optics. *See* Semiconstrained feeds

Index

Parallel polarization, 31-11 to 31-15
Parallel-resonant power dividing circuits, 26-39 to 26-40
Parasitic arrays, 3-42
Parasitic reflectors, 9-64 to 9-65
Partial gain, 1-26
Partial time-delay systems, 19-52, 19-68
Partial zoning, lenses with, 16-28
Passive components for periodic arrays, 13-60 to 13-62
Patches and patch antennas. *See also* Microstrip antennas
   elements for, 10-5 to 10-7, 13-60
   medical applications of, 24-18 to 24-20
   parameters for, 10-16 to 10-17
   resonant frequency of, 10-47 to 10-49
Patch radiators for phased arrays, 18-6
Path length constraint and errors with lens antennas, 16-7 to 16-10, 16-21, 21-18
   bootlace, 16-47 to 16-48
   constrained, 16-41 to 16-51, 19-82 to 19-83
   geodesic, 19-115, 19-119
   microwave, 19-43
   and phase errors, 16-13
   with pillbox feeds, 19-22
   spherical cap, 16-25
   taper-control, 16-34 to 16-35
Path losses
   from ionospheric propagation, 29-25
   and line-of-sight propagation, 29-34 to 29-35
   with offset parabolic reflectors, 15-36 to 15-37
   of tvro downlink, 6-31
Pattern cuts and radiation patterns, 32-6 to 32-8
Pattern error with multibeam antennas, 21-65
Pattern footprint of tvro signal, 6-26, 6-27
Pattern functions. *See* Array pattern functions
Patterns. *See also* Far fields; Fields; Near fields; Radiation fields and patterns
   of arrays
      aperiodic, 14-30 to 14-32
      multiplication of, 3-36, 11-8
      periodic, 13-12 to 13-29
      synthesis of, 11-76 to 11-86
   distortion of. *See* Distortion
   with microstrip antennas, 10-21 to 10-26
   periodic, 13-20 to 13-29
Peak gains, 1-26 to 1-27

Pekeris functions, 4-52 to 4-54
Pencil-beam reflector antennas, 15-5 to 15-6
   AF method with, 15-13
   with direction-finding antennas, 25-20
   far-field formulas for, 15-15 to 15-23
   for satellites, 21-5
Perfect conductors, 3-36, 3-38 to 3-39, 3-44, 3-46
   time-domain analyses with, 3-51
Perfect ground planes, 3-18 to 3-20
   Green's functions for, 3-47 to 3-48
Periodic arrays, 13-5 to 13-11. *See also* Phased arrays
   linear transformations with, 11-31 to 11-33
   organization of, 13-23 to 13-29
   patterns of, 13-12 to 13-23
   practical, 13-30 to 13-64
Periodic configuration with multibeam antennas, 21-66
Periodic dielectric antennas, 17-34 to 17-35, 17-48 to 17-82
Periodic structures, 1-33 to 1-37
   theory of, 9-32 to 9-37
      log-periodic designs based on, 9-62 to 9-71
      and periodically loaded lines, 9-37 to 9-46, 9-46 to 9-62
Period of surface corrugations, 17-64
Permeability, 1-6
   of core in ferrite loop, 6-20
Permittivity
   of corrugation regions, 17-53 to 17-54
   and dielectric modeling, 32-72 to 32-73
   of human tissue, 24-12, 24-30 to 24-35, 24-47
   and leakage constants, 17-61 to 17-63
   of microstrip patch antennas, 17-109
   of open-ended coaxial cable, 24-33
   of substrates, 17-114, 17-124, 17-127
   of troposphere, 29-31
Perpendicular polarization, 31-11 to 31-12, 31-14 to 31-20
Personnel dosimeters, rf, 24-43
Phantoms for heating patterns of antennas, 24-54 to 24-57
Phase-amplitude polarization measurement methods, 32-57 to 32-60
Phase angle of ocean surface, 29-38
Phase center
   with horn antennas, 8-73 to 8-75
      conical, 8-61, 8-76, 8-78 to 8-80

Phase center (*cont.*)
    E-plane, 8-76 to 8-77
    H-plane, 8-76 to 8-77
    of log-periodic dipole antennas, 9-24
    of log-spiral conical antennas, 9-83, 9-91 to 9-92, 9-94
    technique to measure, 8-80 to 8-84
    testing for, 32-9
Phase constants
    for dielectric grating antennas, 17-68, 17-71
    with leaky mode antennas, 17-51, 17-53 to 17-62
    for leaky-wave antennas, 17-84 to 17-85, 17-89 to 17-92, 17-95 to 17-96, 17-100
    of TE/TM waveguides, 28-37
    for transmission lines, 28-7
Phase constraint with dielectric lenses, 16-21
Phase control, array, 13-62 to 13-64
Phased arrays. *See also* Beam-forming feed networks; Periodic arrays
    bandwidth of, 13-19 to 13-20
    conical scanning by, 22-39
    design of, 18-3 to 18-5
        and component errors, 18-26 to 18-28
        feed network selection in, 18-17 to 18-26
        phase shifter selection in, 18-12 to 18-17
        radiator selection in, 18-6 to 18-12
    feeds for, 19-58 to 19-62
    horn antennas in, 8-3
    with integrated antennas, 17-134 to 17-137
    for medical applications, 24-16 to 24-18
    for satellite antennas, 21-7 to 21-12, 21-68 to 21-71
Phase delay
    with equal group delay lenses, 16-45
    with series feed networks, 19-3
Phased lens approach, 19-107 to 19-108
Phase differences
    and direction-finding antennas, 25-3
    with interferometers, 25-21 to 25-22
Phase errors
    in antenna ranges, 32-16 to 32-18
    of array feeds, 13-61
    with arrays
        aperiodic, 14-32
        microstrip patch resonator, 17-109
        periodic, 13-55
        phased, 18-26 to 18-28, 21-12
    with beam-forming feed networks, 18-25
    with horn antennas
        conical, 8-46
        H-plane, 8-23
        pyramidal, 8-39 to 8-40
    with lenses, 16-12 to 16-20, 16-33, 21-13 to 21-14
        bootlace, 16-47 to 16-48
        constrained, 19-82
        equal group delay, 16-46
        geodesic, 19-115
        spherical, 16-24 to 16-25, 16-29
        surface tolerance, 21-18
        zone constrained, 16-43 to 16-45
    with offset parabolic antennas, 15-49, 15-55
    with pillbox feeds, 19-21
    from reflector surface errors, 15-107 to 15-108
    with satellite antennas, 21-5
Phase fronts with reflector antennas, 15-75 to 15-76, 15-80
Phase pattern functions, 11-7
Phase progression with longitudinal slots, 12-17 to 12-18
Phase quantization and array errors, 13-52 to 13-57, 21-12
Phase shift
    and mutual coupling, 9-58
    with periodic loads, 9-49 to 9-54
Phase-shifted aperture distributions, 5-17 to 5-18
Phase shifters
    with arrays
        periodic, 13-20, 13-60 to 13-64
        phased, 18-3 to 18-4, 18-6, 18-12 to 18-17
    with beam-forming networks
        scanned, 21-7, 21-38 to 21-43
        switched, 21-32
    with Butler matrix, 19-9
    with feed systems
        broadband array, 13-40 to 13-41
        directional antenna, 26-40 to 26-43, 26-45 to 26-47
        for gain ripple, 21-27
        parallel, 19-6
        series, 19-3 to 19-4
    with planar scanning, 22-37
    quantization errors with, 21-12

with subarrays
   contiguous, 13-32 to 13-35
   space-fed, 19-68
  in true-time-delay systems, 19-7, 19-80
  with Wheeler Lab approach, 19-105, 19-107
Phase squint with subarrays, 13-35
Phase steering, 13-7 to 13-8, 13-10
Phase taper in antenna ranges, 32-15 to 32-19
Phase term for aperture fields, 8-5
Phase velocity of TEM transmission lines, 28-10, 28-22
Phasors, Maxwell equations for, 1-5 to 1-7
Physically rotating antenna systems, 25-4 to 25-11
Physical models
  errors with, 3-69 to 3-71
  for microstrip antennas, 10-7 to 10-21
Physical optics and physical optics method, 4-5, 4-18 to 4-23
  and computer solutions, 3-54
  and PTD, 4-103
  with reflector antennas, 15-7 to 15-10, 15-14
    dual-reflector, 15-69, 15-72, 15-75 to 15-77
  for scattering, 2-25 to 2-27
Physical theory of diffraction, 4-5, 4-102 to 4-103
  for edged bodies, 4-104 to 4-108
Pillbox antennas, 17-28 to 17-30
Pillbox feeds, 18-22, 19-17 to 19-23
Pinched-guide polarizers, 21-30
Pin diodes
  for beam-forming network switches, 21-38
  for dielectric grating antennas, 17-71 to 17-72
  with integrated antennas, 17-135
  with microstrip antennas, 10-69
  in phase shifters, 13-62, 18-13
    variable, 21-40
  with variable power dividers, 21-46 to 21-47
Pin polarizers, 21-30
Pitch angle of log-periodic antennas, 9-39
Planar antennas
  impedance of, 2-15 to 2-16
  log-periodic antennas, 9-10 to 9-12
    radiated fields of, 9-8 to 9-9
    truncation with, 9-8
  log-spiral antennas, 9-72
  with polar orbiting satellites, 22-35 to 22-38
  power radiated by, 5-26 to 5-27
Planar apertures
  feed elements with, 15-88
  and plane-wave spectra, 5-5
  radiation patterns of, 5-10 to 5-26
  for reflector antennas, 15-14
Planar arrays, 11-48 to 11-58
  directivity of, 11-63 to 11-65, 13-11
  millimeter-wave antennas, 17-26
  mutual coupling in, 12-32 to 12-33
  optimization of, 11-63 to 11-65
  periodic, transformations with, 11-26 to 11-29
  rectangular, patterns of, 13-15 to 13-17
  slot, 12-4 to 12-5
    edge, mutual coupling of, 12-27 to 12-28
    longitudinal, 12-4, 12-20 to 12-24
  triangular, grating lobes of, 13-15 to 13-17
  of vertical dipoles, geometry of, 11-85
  waveguide, 17-26
Planar curves
  conic sections as, 15-23 to 15-24
  for reflector surfaces, 15-27
Planar lenses, 21-18
Planar quartz substrates, 17-137
Planar scanning for field measurements,

Planar transmission lines, 28-30 to 28-35
Plane symmetry and computer time and storage, 3-65, 3-67
Plane-wave illumination
  for distance parameters, 4-56
  and GO incident fields, 4-11 to 4-12
Plane wave propagation
  and dielectric sheets, 31-20 to 31-25
  flat, 31-10 to 31-18
Plane waves
  polarization of, 1-13 to 1-16
  propagation direction of, 1-16
  reciprocity of, 2-33 to 2-34
  from spherical wavefronts, 16-6, 16-9 to 16-10
  synthesis of, for field measurements, 33-21 to 33-22
Plane-wave spectra, 5-5 to 5-8
  representation of, 1-31 to 1-33
Plano-convex lenses, 16-20 to 16-21

Plate-scattered fields, 20-23
PO. *See* Physical optics
Pockington's equation for dipole arrays, 13-43
Poincare sphere, 32-54 to 32-56
Point-current source, 3-13 to 3-14
Point matching
  with methods of moments, 3-60
  weight function with, 3-83
Point-source illumination
  for distance parameter, with UTD, 4-56
  and GO incident field, 4-11
Polar cap absorptions and ionospheric propagation, 29-25
Polar coordinates with lens antennas, 16-8 to 16-9
Polarization, 1-14 to 1-16, 1-19 to 1-21, 1-28 to 1-29, 3-28 to 3-31. *See also* Circular polarization; Cross polarization; Horizontal polarization; Linear polarization; Vertical polarization
  for am antennas, 26-3
  of arrays, 11-7
    phased, 18-6, 21-12
  and diagonal horns, 8-70
  for earth coverage satellite antennas, 21-82
  efficiency of, 1-26, 2-36, 32-56
    and gain measurements, 32-42
    and mismatch, 2-39
  and Faraday rotation, 29-16
  and Fresnel reflection coefficients, 3-27
  and gain, 1-25 to 1-26
  of leaky-wave antennas, 17-90, 17-94, 17-97
  of log-periodic antennas, 9-8
  with log-spiral antennas, 9-74, 9-77
    conical, 9-86, 9-90 to 9-91, 9-99
  measurements of, 32-51 to 32-64
  of microstrip antennas, 10-57 to 10-63, 10-69 to 10-70
  for microwave radiometers, 22-43 to 22-45
  parallel, 31-11 to 31-15
  perpendicular, 31-11 to 31-12, 31-14 to 31-20
  positioners for, 32-12, 32-14
  and radome interface reflections, 31-11 to 31-13
  and rain, 29-14
  reference, 1-23

  of reflector antennas, 15-5 to 15-6, 15-89, 21-19 to 21-20, 21-73
  for satellite antennas, 21-6 to 21-7, 21-80, 21-82
  and surface emissivity, 22-7 to 22-8
Polarization-agile elements, 10-69 to 10-70
Polarization ellipses, 1-14 to 1-16
Polarization-matching factor, 2-36, 6-7 to 6-8
Polarization pattern, 32-6
  polarization measurement method, 32-62 to 32-63
Polarization-transfer methods, 32-56 to 32-57
Polarizers
  for satellite antennas, 21-7, 21-29, 21-30, 21-74, 21-79 to 21-80
  waveguide, 13-60 to 13-61
Polar orbiting satellites, 22-35 to 22-43
Poles of cavities, 10-11
Polystyrene antennas, 17-44 to 17-46
Polytetrafluoroethylene bulb for implantable antennas, 24-27, 24-29
Porous pots with earth resistivity measurements, 23-8
Positioning systems for antenna ranges, 32-8 to 32-14, 32-30 to 32-32
Potter horns
  earth coverage satellite antennas, 21-82
  for reflector antennas, 15-95, 15-100 to 15-101
Power
  accepted by antennas, 1-24, 1-25, 22-4
  complex, 1-7
  conservation of, numerical validation of, 3-79
  incident to antennas, 1-24, 1-25
  input to antenna, 6-6
  overfeeding of, with am antennas, 26-36 to 26-37
  received by antenna, 2-38, 32-42, 33-22
  reflected, 2-29
  reflection coefficient for, 1-24
  sources of, for millimeter-wave antennas, 17-11
  time-averaged, 1-7 to 1-8, 3-11
    for modal fields, 1-46
    in waveguides, 1-39
  transfer ratio for, 6-8 to 6-9
  transformation of, with taper-control lenses, 16-37
  with transmission lines, 28-8, 28-10

Power amplifiers, integration of, 17-136
Power conservation law, 16-33 to 16-34
Power density
    with arrays, 11-59
    with bistatic radar, 2-37 to 2-38
    near-field, 5-29 to 5-33
    of plane waves, 1-14
    of receiving antenna, 6-7
Power detectors, integration of, 17-136
Power dividers
    for array feeds, 13-61
    for beam-forming networks
        scanned, 21-7, 21-43 to 21-49
        switched, 21-32 to 21-33
    for directional antenna feeder systems, 26-39 to 26-40
    fixed beam-forming networks, 21-49 to 21-53
    for gain ripple with feed arrays, 21-27
    LHCP and RHCP, 21-80
    quantiziation errors with, 21-12
Power flow integration method for am antennas, 26-22 to 26-33
Power gain with multibeam antennas, 21-64
Power handling capability
    measurements for, 32-4
    of transmission lines, 21-63
        circular coaxial TEM, 28-21
        microstrip planar quasi-TEM, 28-32 to 28-33
        rectangular TE/TM, 28-39, 28-42
        triplate stripline TEM, 28-28 to 28-29
        two-wire TEM, 28-19
    of waveguides
        circular TE/TM, 28-44
Power law coefficients for rain attenuation, 29-12
Power lines, noise from, 29-51
Power radiated. *See* Radiated power
Power rating
    of rf cables, 28-27
    of standard waveguides, 28-40 to 28-41
    of tv antennas, 27-7 to 27-8
Power ratio with bistatic radar, 2-36 to 2-39
Power transmittance, 2-29
    of A-sandwich panels, 31-18 to 31-19
    and dielectric sheets, 31-14 to 31-16
    efficiency of, 2-43
    with radomes, 31-3, 31-7
Poynting theorem, 1-7 to 1-8
Poynting vector for radiated power, 3-11

P-percent level curves, 14-12 to 14-14
Precipitation distributions, remote sensing of, 22-14
Prescribed nulls, 11-9 to 11-10
Pressurized air with transmission lines, 28-10
Principal polarization. *See* Reference polarization
Principle of stationary phase, 4-6 to 4-7
Printed circuits
    dipoles, 13-58
        bandwidth of, 17-122 to 17-127
    and integrated antennas, 17-130 to 17-131
    millimeter-wave, 17-25, 17-103 to 17-131
    waveguides, 21-57
        ridged, 28-57 to 28-58
        TEM, 21-57
    zigzag, 24-39, 24-40
Probabilistic approach to aperiodic arrays, 14-5, 14-8 to 14-35
Probability mean, 11-81
Probes
    for alignment, 32-33
    E-field, 24-37 to 24-41
        calibration of, 24-54 to 24-57
        implantable, 24-51 to 24-54
        in-vivo, 24-30 to 24-35, 24-49 to 24-57
    magnetic field, 24-37 to 24-38, 24-40 to 24-44
        calibration of, 24-54 to 24-57
    and polarization, 32-6 to 32-8
    for rf radiation detection, 24-37 to 24-43
    temperature, 24-40 to 24-51
Prolate spheroid
    geometry of, 4-82
    rectangular slot in, 4-79 to 4-82
Propagation, 29-5 to 29-6
    in computer models, 3-55 to 3-56
    ionospheric, 29-21 to 29-30
    with millimeter-wave antennas, 17-5
    noise, 29-47 to 29-52
    and polarization, 1-13 to 1-16
    satellite-earth, 29-7 to 29-21
    tropospheric and surface, 29-40 to 29-47
Propagation constants
    with conical log-spiral antennas, 9-80 to 9-84
    for dielectric grating antennas, 17-65
    of microstrip patch resonator arrays, 17-110
    for plane waves, 29-6

Propagation constants (*cont.*)
   for transmission lines, 28-7
   and UP structures, 9-32
Protruding dielectric waveguide arrays, 13-48 to 13-49
PTD. *See* Physical theory of diffraction
Ptfe bulb for implantable antennas, 24-27, 24-29
Pulse rf cable, 28-24 to 28-25
Purcell-type array antennas, 17-27
Push-pull power dividing circuits, 26-39
Pyramidal horn antennas, 8-4
   aperture and radiated fields of, 8-34 to 8-40
   corrugated, 8-51 to 8-59
   design procedure for, 8-43 to 8-45
   directivity of, 8-37 to 8-42
   for feeds for reflectors, 15-92 to 15-93, 15-96 to 15-98
   gain of, 8-43 to 8-44
   as surface-wave launcher, 17-46
Pyramidal log-periodic antenna, 9-4

Q. *See* Quality factor
Quadratic phase errors
   in antenna ranges, 32-16 to 32-18
   and conical horn antennas, 8-46
   with lenses, 16-3, 16-14, 16-16 to 16-18, 16-33
Quadratic ray pencil, 4-8
Quadrature hybrids for circular polarization, 10-61
Quadrature power dividing circuits, 26-39 to 26-40
Quadrifocal bootlace lenses, 16-47 to 16-48
Quality factor
   of arrays, 11-61 to 11-62
     dipoles in, 9-61 to 9-63
     directivity in, 11-67 to 11-68, 11-76
     planar, 11-65
     and SNR, 11-70 to 11-72
   of cavities, 10-11 to 10-13, 10-20, 10-21
   of dipoles
     in arrays, 9-61 to 9-63
     and phase shift, 9-58
   of ferrite loop, 6-22
   with log-periodic structures, 9-55 to 9-56
     dipole arrays, radiation fields in, 9-62 to 9-63
   of microstrip antennas, 10-6, 10-9, 10-49, 10-53
   of receiving antennas, 6-22 to 6-23
   and stopband width, 9-49
Quantization errors with phased arrays, 13-52 to 13-57, 21-12
Quarter-wave matching layers for lenses, 21-19
Quarter-wave vertical antennas, 26-9
Quartic errors with lenses, 16-16, 16-18
Quartz substrate with integrated antennas, 17-137

Radar cross section, 2-23 to 2-24
   measurements of, 32-64
Radar equation, 2-36 to 2-39
Radar requirements and phased array design, 18-4 to 18-5
Radial slots in cone, 4-78 to 4-80
Radial transmission line feeds, 19-23 to 19-32
Radiated power, 1-25, 3-11 to 3-12, 5-26 to 5-27, 6-6 to 6-7
   with arrays, 11-60
   of biconical antennas, 3-25
   and cavities, 10-20
   and directive gain, 5-26
   of fm antennas, 27-4, 27-6
   into free space, 2-29
   of space harmonics, 17-52
   of tv antennas, 27-3 to 27-5
Radiating edges, 10-24
Radiating elements
   for feed arrays, 21-27 to 21-28
   for phased array design, 18-3 to 18-12
Radiating near-field region, 1-18
Radiation angle and grating period in leaky modes, 17-56 to 17-58
Radiation conditions and Maxwell equations, 1-9
Radiation efficiency, 1-25, 3-12
   of microstrip dipole antennas, 17-119, 17-124 to 17-127
   of microstrip patch resonator arrays, 17-108 to 17-109
   of receiving antennas, 6-22 to 6-24
Radiation fields and patterns, 1-11 to 1-13, 1-20 to 1-21. *See also* Apertures and aperture distributions, radiation from; Far fields, Near fields; Patterns; Reactive fields
   of am antennas, 26-5 to 26-9, 26-17, 26-22 to 26-33
   standard reference, 26-3 to 26-8

with arrays, 3-36, 3-37
   aperiodic, 14-15, 14-17, 14-19 to 14-21
   circular, on cylinder, 21-100
   conformal, 13-52
   Dolph-Chebyshev, 11-19 to 11-20
   end-fire, 11-76, 11-77
   periodic, 13-5, 13-8 to 13-10, 13-12 to 13-17, 13-45 to 13-48
   subarrays, 13-30 to 13-32, 13-35 to 13-39
of biconical antennas, 3-26, 21-107, 21-109
blackbody, 22-3 to 22-4
with circular loop antennas, 23-21
of convex surfaces, 4-63 to 4-84
of dipole antennas
   short, 20-8 to 20-9
   skewed, 27-14 to 27-18
with dual fields, 2-5 to 2-6, 3-9, 3-11
with DuFort optical technique, 19-76, 19-78
of feed systems
   radial transmission line, 19-31 to 19-32
   reflector-lens limited-scan concept, 19-72 to 19-73
   for reflectors, 15-11 to 15-12, 15-86 to 15-94
   semiconstrained, 19-49, 19-51 to 19-53
   simple, 15-89 to 15-94
with grounded wire antennas, 23-9 to 23-16
with ground planes
   imperfect, 3-27, 3-33
   perfect, 3-22
of horn antennas, 8-3
   aperture-matched, 8-58, 8-66 to 8-67
   biconical, 21-107, 21-109
   corrugated, 8-53 to 8-59, 8-64, 8-65
   dielectric-loaded, 21-84 to 21-86
   dual-mode, 15-95, 15-100 to 15-101
   E-plane, 8-7 to 8-10
   E-plane-flared biconical, 21-107
   H-plane, 8-20 to 8-26, 8-28
   H-plane-flared biconical, 21-107
   pyramidal, 8-34 to 8-39, 15-92 to 15-93
and Huygen's principle, 2-18 to 2-19
and image theory, 2-11 to 2-13
of implantable antennas, 24-27 to 24-29
of lenses
   constrained, 19-82, 19-86 to 19-87
   effect of aperture amplitude distributions on, 21-14

   microwave, 19-45
   modified Meyers, 19-115 to 19-118
   pillbox, 17-28
   for satellites, 21-73, 21-75 to 21-77
   spherical thin, 16-30
   Teflon sphere, 17-188
of log-periodic antennas
   dipole, 9-17, 9-18, 9-62 to 9-63
   dipole arrays, 9-64 to 9-66
   planar, 9-8 to 9-9
   zigzag antennas, 9-39 to 9-40
of log-spiral antennas, 9-75
   conical, 9-85 to 9-86, 9-89 to 9-92, 9-99 to 9-109
of loop antennas, 25-6 to 25-7
measurements for. *See* Measurement of radiation characteristics
of microstrip antennas, 10-10
of millimeter-wave antennas
   dielectric grating, 17-69 to 17-70, 17-72 to 17-81
   dipoles, 17-118 to 17-119, 17-127 to 17-128
   holographic, 17-132
   leaky-wave, 17-86
   spiral, 17-28 to 17-29
   tapered dielectric-rod, 17-41 to 17-43
with models, 32-64
of monofilar zigzag wire antennas, 9-34, 9-36
of monopole on aircraft, 20-64, 20-74 to 20-75, 20-86 to 20-88
of multibeam antennas, 21-64
and polarization, 1-20 to 1-21, 1-25 to 1-26
with radomes, 31-25 to 31-27
reciprocity of, 2-32 to 2-33
with reflector antennas, 15-8 to 15-10, 21-20 to 21-21
   random surface distortion on, 21-23
   slotted cylinder, 21-110 to 21-111
with satellite antennas, 21-5
   lens, 21-73, 21-75 to 21-77
   multibeam, 21-64
   slotted cylinder reflector, 21-110 to 21-111
of side-mount tv antennas, 27-20, 27-22
for slot antennas, 27-27
   axial, 21-95, 21-100
   circumferential, 21-92 to 21-93, 21-96
of small loops, 6-13, 6-17

Radiation fields and patterns (cont.)
  of subarrays, 13-30 to 13-32, 13-35 to 13-39
  for TEM waveguides, 24-11
  in tissue, 24-51 to 24-54
  of transmission lines, 21-63
    radial feeds for, 19-31 to 19-32
    TEM, 28-11
  with traveling-wave antennas, 3-16, 3-18
    slot, 27-27
  and UTD solutions, 20-5
    for convex surfaces, 4-63 to 4-84
    with curved surface, 20-18, 20-20, 20-25 to 20-26, 20-31 to 20-36
    for noncurved surfaces, 20-51
  of vee-dipole antennas, 27-11
  of waveguide-slot antennas, 27-30 to 27-31
Radiation resistance, 3-12
  of ferrite loop, 6-21 to 6-22
  of microstrip patch antennas, 17-107
  and mutual impedance, 9-56
  of short dipole, 6-11 to 6-12, 6-14 to 6-15
  of small loops, 6-19, 25-7 to 25-8
  of vertical antenna, 6-16
Radiation sphere, 1-19, 1-21, 32-6 to 32-7
Radiation-type antennas for medical applications, 24-9 to 24-18
Radiative transfer, microwave, 22-4 to 22-7
Radio-astronomy
  feeds for, 15-94
  and microwave remote sensing, 22-9
  millimeter-wave antennas for, 17-14, 17-16
Radio-frequency cables, list of, 28-23 to 28-25
Radio-frequency power absorption, 24-30 to 24-35
Radio-frequency radiation, monitoring of, 24-36 to 24-43
  personnel dosimeters for, 24-43
Radio horizon, 29-34
Radius vector with conical log-spiral antennas, 9-79, 9-98, 9-100
Radome Antenna and RF Circuitry, 19-54 to 19-55
Radomes, 31-3 to 31-5
  on aircraft, numerical solutions for, 20-74
  and boresight error, 31-27
  design of, 31-18 to 31-20
  and flat dielectric sheets, 31-10 to 31-18, 31-20 to 31-25
  materials for, 31-29 to 31-30
  modeling of, 32-82 to 32-83
  and moment method, 31-27 to 31-29
  with omnidirectional dielectric grating antennas, 17-81 to 17-82
  patterns with, 31-25 to 31-27
  physical effects of, 31-5 to 31-10
Rain
  line-of-sight propagation path loss from, 29-35
  and millimeter-wave antennas, 17-5
  and noise temperature, 29-49, 29-50
  remote sensing of, 22-12, 22-14
  and satellite-earth propagation
    attenuation of, 29-10 to 29-13
    depolarization of, 29-14
Random errors
  with arrays
    periodic, 13-53 to 13-57
    phased, 18-26 to 18-28
  with microwave radiometers, 22-23
  with reflector surfaces, 15-105 to 15-114, 21-23 to 21-24
Random numbers with aperiodic array design, 14-18 to 14-19
Range equation, 2-38
Ranges, antenna
  design criteria for, 32-14 to 32-19
  errors with, 33-15 to 33-18
  evaluation of, 32-33 to 32-39
  for field measurements, 33-15 to 33-24
  indoor, 32-25 to 32-30
  instrumentation, 32-30 to 32-33
  outdoor, 32-19 to 32-25
  positioners and coordinate systems for, 32-8 to 32-14
RAR (Reflect Array Radar), 19-55
RARF (Radome Antenna and RF Circuitry), 19-54 to 19-55
Rayleigh approximation and computer solutions, 3-54
Rayleigh criterion for smoothness of antenna ranges, 22-30 to 22-32, 32-22
Rayleigh-Jeans approximation with blackbody radiation, 22-3
Rays, 4-6 to 4-8
  caustics of
    distance of, 4-8
    reflected and transmitted, 4-15
  construction of, with reflector antennas, 15-11 to 15-12

Index

paths of, with UTD solutions, 20-48 to 20-49, 20-51
RCS (radar cross-section), 2-23 to 2-24
  measurements of, 32-64
Reactive fields, 1-16 to 1-18
  of aperture and plane-wave spectra, 5-5
  coupling of, in antenna ranges, 32-14
Reactive loads with dual-band elements, 10-63 to 10-66
Realized gain, 1-25
Real poles of cavities, 10-11
Real space of arrays scanned in one plane, 13-14
Received power, 2-38, 32-42, 33-22
Receiving antennas
  bandwidth and efficiency of, 6-22 to 6-24
  effective area of, 22-4
  equivalent circuit of, 6-3
  ferrite loop, 6-18 to 6-22
  Friis transmission formula for, 6-8 to 6-9
  grounded-wave, 23-16 to 23-17
  impedance-matching factor of, 6-6 to 6-7
  linear dipole, 7-18 to 7-19
  loops, 23-22 to 23-24
  meters for, 2-30 to 2-32
  mutual impedance between, 6-9
  and noise, 2-42 to 2-43, 6-24 to 6-25
  polarization-matching factor of, 6-7 to 6-8
  power accepted by, 1-24, 1-25, 22-4
  receiving cross section of, 6-6
  reciprocity of, 2-32 to 2-36
  satellite earth stations, 6-25 to 6-32
  small. *See* Small receiving antennas
  thin-wire, 7-8 to 7-9
  thin-wire loop, 7-46 to 7-47
  vector effective height of, 6-3 to 6-6
Receiving cross section, 2-35 to 2-36
  of receiving antennas, 6-6
  of short dipoles, 6-12 to 6-15
  of small loops, 6-18
Receiving polarization, 32-53
Receiving systems for antenna ranges, 32-30 to 32-32
Receptacles and plugs with modeling, 32-83
Reciprocal bases in transformations, 11-25
Reciprocal ferrite phase shifters
  dual-mode, 18-14 to 18-15, 21-38 to 21-40
  for periodic arrays, 13-64
Reciprocity, 2-32 to 2-36
  numerical validation of, 3-79
  and reciprocity theorem, 2-16 to 2-18, 33-7
  with scattering, 2-24 to 2-25
Recording systems for antenna ranges, 32-30 to 32-32
Rectangular anechoic chambers, 32-25 to 32-26
Rectangular apertures
  antennas with, 3-21 to 3-24
  compound distributions with, 5-16 to 5-17
  directive gain of, 5-27
  displaced, phase-shifted distributions with, 5-17 to 5-18
  effective area of, 5-27 to 5-28
  efficiency of, 5-28 to 5-29
  gain of, 5-29
  near-field gain reduction factors with, 5-30 to 5-31
  near-field pattern with, 5-25 to 5-26
  optimum pattern distributions with, 5-18 to 5-20
  radiation fields of, 5-11 to 5-20, 5-27
  simple distributions with, 5-13 to 5-15
  and uniform aperture distribution, 5-11 to 5-13
Rectangular coaxial TEM transmission lines, 28-18
Rectangular conducting plane, RCS of, 2-23 to 2-24
Rectangular coordinates with lens antennas, 16-7 to 16-8
Rectangular lattices
  field representation for, 1-35
  for phased-array radiators, 18-9
Rectangular loops
  for mine communications, 23-20
  radiation resistance of, 6-19
Rectangular patch antennas, 10-41 to 10-45, 24-19
  characteristics of, 10-16, 10-21 to 10-26
  medical applications of, 24-19 to 24-20
  principle-plane patterns of, 10-22
  resonant frequency of, 10-46 to 10-47
Rectangular planar arrays, patterns of, 13-15 to 13-17
Rectangular slots
  in cones, mutual coupling of, 4-91 to 4-92
  in prolate spheroid, radiation pattern of, 4-79 to 4-82

Rectangular tapered dielectric-rod antennas, 17-37 to 17-38
Rectangular waveguides, 1-37 to 1-44, 21-60
  currents in, 12-3 to 12-4
  dielectric, 28-51 to 28-54
  impedance of, 2-28
  with infinite array solutions, 13-48 to 13-49
  mutual coupling with, 13-44 to 13-45
  open-ended
    feeds for, half-power beamwidths of, 15-104
    radiators of, for phased arrays, 18-6
    for reflectors, 15-89 to 15-92, 15-104
  radiation from, 1-28
  TE/TM, 28-38 to 28-42
  voltage and current in, 2-28 to 2-29
Rectangular XY-scanners, 33-19
Reduction factor for line-of-sight propagation, 29-36
Redundant computer operations and symmetry, 3-65
Reference polarization, 1-23
  of microwave antennas, 33-20
  with probe antennas, 32-8
Reflect Array Radar, 19-55
Reflectarrays, 19-54 to 19-55
Reflected fields. *See* Reflections and reflected fields
Reflected power, 2-29
Reflection boundaries and UTD, 4-34
Reflection coefficients
  approximation of, for imperfect grounds, 3-48
  with coaxial sleeve antennas, 7-36
  for dielectric grating antennas, 17-71, 17-81
  and imperfect ground planes, 3-27
  with linear dipole antennas, 7-23
  of ocean surface, 29-38 to 29-40
  with tapered dielectric-rod antennas, 17-44 to 17-46
  with transmission lines, 2-28, 28-7, 28-8
  with unloaded transmitting antennas, 7-13
  with UTD noncurved surface solutions, 20-43, 20-44
Reflections and reflected fields
  from aircraft wings, 20-78
  in antenna ranges, 32-34 to 32-37
  errors from, in scanning ranges, 33-16
  and Fresnel ellipsoid, 29-36
  GO, 4-13 to 4-18
  by human tissue, 24-7
  and imperfect ground planes, 3-27, 3-33
  lobes from, 19-5
  and multipath propagation, 29-37
  and plane boundaries, 31-10 to 31-12
  with radomes, 31-5, 31-6, 31-10 to 31-17
  reflected-diffracted, 20-29, 20-33, 20-37, 20-43, 20-45, 20-55
  reflected-reflected, 20-29, 20-33, 20-45
  in scattering problems, 2-13 to 2-14
  shadow boundary, 4-13
  space-fed arrays with, 21-34, 21-36
  space feed types, 18-17 to 18-19
  and spatial variations in antenna ranges, 32-19
  with UTD solutions, 20-8 to 20-9, 20-15
    for curved surfaces, 20-21, 20-23 to 20-25, 20-29, 20-31
    for noncurved surfaces, 20-37, 20-42, 20-44, 20-49, 20-53 to 20-54, 20-59 to 20-60
Reflectivity of surfaces, 22-8 to 22-9
Reflector antennas, 15-5
  for antenna ranges, 32-28 to 32-30
  basic formulations for, 15-6 to 15-15
  bifocal lens, 16-30 to 16-33
  contour beam, 15-80 to 15-84
  and coordinate transformations, 15-115 to 15-120
  diameter of, 21-24
  dual. *See* Dual-reflector antennas
  far-field formulas for, 15-15 to 15-23
  with feed, rotating antenna systems, 25-4
  feeds for, 15-84 to 15-105
  generated, 15-23 to 15-31
  with implantable antennas, 24-27 to 24-28
  millimeter-wave, 17-9 to 17-14
  with off-focus feeds, 15-49 to 15-61
  offset parabolic, 15-31 to 15-61
  with on-focus feeds, 15-37 to 15-49
  phased arrays, 21-12
  random surface errors on, 15-105 to 15-114
  satellite, 21-7, 21-8, 21-19 to 21-26, 21-73 to 21-74, 21-77 to 21-81
  surfaces for, 15-26 to 15-29
  types of, 15-5 to 15-6, 21-20 to 21-22

Reflector coordinates
  with reflector antennas, 15-7, 15-88 to 15-89
  transformations with, 15-115, 15-117 to 15-119
Reflector-lens limited-scan feed concept, 19-67 to 19-73
Refraction
  and line-of-sight propagation, 29-32 to 29-33
  and plane boundaries, 31-10 to 31-12
Refractive index, 29-7
  of atmosphere, 29-30 to 29-32
  and ionosphere, 29-14 to 29-15, 29-21 to 29-22
  lenses with varying, 16-51 to 16-54
  of short monopoles, 24-31
  of troposphere, 29-46
Regularization synthesis method, 21-65
Relabeling of bases in transformations, 11-25
Relative dielectric constant, 1-7
Relative permeability, 1-7
Remote sensing
  antenna requirements for, 22-16 to 22-46
  microwave radiometry for, 22-9 to 22-16
  sleeve dipole for, 7-26
Reradiated fields and imperfect ground planes, 3-48
Reradiative coupling in antenna ranges, 32-14
Resistance. *See also* Radiation resistance
  of free space, 26-25 to 26-26
  measurements of, electrode arrays for, 23-4 to 23-8
  surface, 1-43
Resonance
  in inclined narrow wall slot arrays, 12-26
  with planar arrays, 12-21 to 12-22
  in slot arrays, 12-14 to 12-15
  and thin-wire antennas, 7-10
  validation of computer code for, 3-73 to 3-76
Resonant arrays, 17-114, 17-118 to 17-119
Resonant frequency
  of am towers, 26-9 to 26-12
  for microstrip antennas, 10-9, 10-44 to 10-49
  of loaded elements, 10-53 to 10-54
Resonant impedance, 10-28, 10-30 to 10-31

Resonant loads with dual-band elements, 10-63 to 10-66
Resonant modes of cavities, 10-13, 10-15
Resonantly spaced longitudinal slots, 12-13 to 12-16
Resonator millimeter-wave antennas, 17-35 to 17-36
Rexolite with lenses, 16-22, 16-36 to 16-37
Rf. *See* Radio-frequency
RG-type cables, 28-22 to 28-25
Rhombic antennas, 3-31
  dielectric plate, 17-23, 17-25
Richmond formulation with radomes, 31-27 to 31-29
Ridged-waveguide antennas
  for medical applications, 24-12 to 24-16
  phased arrays for satellite antennas, 21-12
  TE/TM, 28-45 to 28-49
Ridge-loaded waveguide arrays, 13-48 to 13-49
Ridge waveguide feed transmissions, 21-56 to 21-58
Right-hand circular polarization, 1-15, 1-20
  of arrays, 11-7
  dual-mode converters for, 21-78, 21-80
  far-field pattern for, 1-29
  for microstrip antennas, 10-59, 10-69
  with offset parabolic reflectors, 15-36, 15-48 to 15-50
  power dividers for, 21-78, 21-80
Right-hand elliptical polarization, 1-16, 9-74
Rinehart-Luneburg lenses, 19-41 to 19-49
Ring focus
  with bifocal lenses, 16-32
  with DuFort-Uyeda lenses, 19-49
Ring-loaded slots, 8-64
Ripples
  with conformal arrays, 13-50 to 13-52
  with feed arrays, 21-27
Rock conductivity, dielectric logging for, 23-24 to 23-25
Rod antennas, effective receiving area of, 30-10
Roll-over-azimuth positioners, 32-12
Rotary-field phase shifters, 13-64
Rotary joints for beam-forming networks, 21-53, 21-55
Rotating antenna patterns, 25-4 to 25-11

Rotating reflector
  with conical scanning, 22-40
  with direction-finding antennas, 25-10
  with feed rotating antenna systems, 25-4 to 25-5
Rotating-source polarization measurement method, 32-63
Rotational symmetry
  and computer time and storage, 3-67 to 3-68
  dielectric grating antennas with, 17-80 to 17-81
Rotman and Turner line source lenses, 19-37 to 19-41
ROTSY modeling code, 3-88 to 3-91
Rounded-edge triplate striplines, 28-17
RSB (reflection shadow boundary), 4-13
R-3 singularity, integration involving, 2-9 to 2-10
R-2R lenses, 16-48 to 16-49
Run-time features in modeling codes, 3-86
Ruze lenses, 19-39

Saltwater, modeling of, 32-79, 32-80. *See also* Ocean
Sam-D Radar, 19-55
Sample directivity in aperiodic array design, 14-15, 14-17
Sample radiation patterns in aperiodic array design, 14-15, 14-17, 14-19 to 14-21
Sampling systems, 26-41 to 26-44
Sampling theorem, 15-10
SAR, 22-30
Satellite antennas and systems, 21-3
  communication. *See* Communication satellite antennas
  conformal array for, 13-49
  contour beam antennas for, 15-5, 15-80 to 15-87
  earth coverage, 21-80 to 21-89
  earth receiving antennas for, 6-25 to 6-32
  lenses for, 16-22
    millimeter-wave, 17-15, 17-19 to 17-20
    spherical-thin, 16-28
    taper-control, 16-36
    zoning of, 16-38
  limited scan antennas for, 19-56
  low earth orbiting 22-35 to 22-43
  for microwave remote sensing, 22-9 to 22-10
  millimeter-wave antennas for, 17-7, 17-11 to 17-12, 17-14
    lens, 17-15, 17-19 to 17-20
  modeling of, 32-86
  offset parabolic antennas for, 15-58
  polar orbiting, 22-34 to 22-42
  propagation for, 29-7 to 29-21
  reflector antennas for, 15-5
  testing of, 33-4
  tracking, telemetry, and command, 21-89 to 21-111
  weather, 22-14
S-band phased array satellite antennas, 21-68 to 21-71
Scalar conical horn antennas, 8-61
Scalar wave equation, 2-6 to 2-7
Scaling. *See also* Modeling
  with log-periodic dipole antennas, 9-12, 9-14
    and directivity, 9-20, 9-22 to 9-24
    and element length, 9-31 to 9-32
    and swr, 9-27, 9-28
  with log-periodic planar antennas, 9-9 to 9-10
  for millimeter-wave antennas, 17-7 to 17-9
  with modeling, 32-74 to 32-86
  with offset parabolic antennas, 15-81
  and principal-plane beamwidths, 9-9 to 9-10
SCAMS (scanning microwave spectrometer), 22-26, 22-37, 22-43
Scan angles vs. directivity, with arrays, 11-78
Scan characteristics, array, 14-24 to 14-25
Scanned beams
  beam-forming feed networks for, 18-21, 18-24, 21-27 to 21-34, 21-37 to 21-49
  with offset parabolic antennas, 15-51, 15-53
  with phased array satellite antennas, 21-71
Scanning. *See* Beam scanning
Scanning microwave spectrometers, 22-26, 22-37, 22-43
Scanning multichannel microwave radiometer, 22-22
  calibration of, 22-25 to 22-26
  compared to SSM/I, 22-48
  as conical scanning device, 22-40 to 22-42
  as continuous scan type, 22-43

modulating frequency of, 22-24
momentum compensation devices in, 22-46 to 22-47
onboard reference targets for, 22-26
as step scan type, 22-43
Scanning ranges. *See* Ranges, antenna
Scanning thermographic cameras, 24-51
Scan performance
   of dual-reflector antennas, 15-71 to 15-73, 15-78 to 15-79
   of zoned lenses, 16-28
Scattering and scattered propagation fields, 2-21 to 2-27, 4-49
   and Babinet principle, 2-13 to 2-15
   cross section of, 2-22 to 2-25
   and dielectric thickness, 10-6
   with dual fields, 2-5 to 2-6
   field representation of, 1-33 to 1-37
   ionospheric, 29-28 to 29-30
   in longitudinal slots, 12-7, 12-10 to 12-11
   losses from, with remote-sensing microwave radiometers, 22-32, 22-33
   plate-scattered fields, 20-23
   and PO, 4-20, 4-22
   by radomes, 20-74, 31-5
   with reflector antennas, 15-7, 15-11
   with shipboard antennas, 20-90 to 20-97
   in slot array design, 12-35 to 12-36
   tropospheric, 29-46 to 29-47
   for UTD solutions
     centers of, 20-4 to 20-6
     for convex surfaces, 4-49 to 4-63
     for curved surfaces, 20-18
     for noncurved surfaaces, 20-47, 20-49, 20-51 to 20-52, 20-57 to 20-60
Scattering matrix for planar arrays, 12-21
Scattering matrix planar scanning, 33-7
Schelkunoff's induction theorem, 31-25
Schiffman phase shifters, 13-62
Schlumberger array, 23-5 to 23-7
Schmidt corrector with wide-angle lenses, 16-23 to 16-28
Schottky diodes for integrated antennas, 17-133 to 17-134, 17-138 to 17-139, 17-141
Schumaun resonance and ionospheric propagation, 29-27
Scintillation
  from fog and turbulence, 29-36
  index for, 29-16 to 29-17
  and satellite-earth propagation, 29-16 to 29-21

SCS (scattering cross section), 2-24 to 2-25
Sea. *See* Ocean
Seasat satellite, 22-17
  SMMR on, 22-22
Second-order effects in slot array design, 12-34 to 12-36
Second-order scattering, 20-49
Security, transmission, with millimeter-wave antennas, 17-6 to 17-7
Seidel lens aberrations, 16-14, 21-13
Self-admittance with microstrip antennas, 10-7
Self-baluns, 9-16, 9-17
Self base impedance of am antennas, 26-8 to 26-12
Self-complementary antennas, 9-5 to 9-7
  log-spiral, 9-98 to 9-99
Self-conductance with microstrip antennas, 10-7
Self-resistance of am antennas, 26-8
Semicircular arrays, 11-78 to 11-80
Semicircular-rod directional couplers, 21-50
Semiconstrained feeds, 19-15 to 19-19
  DuFort-Uyeda lens, 19-49
  Meyer lens, 19-32 to 19-37
  pillbox, 19-19 to 19-23
  radial transmission line, 19-23 to 19-32
  Rinehart-Luneburg lens, 19-41 to 19-49
  Rotman and Turner line source microwave lens, 19-37 to 19-41
Sense of rotation with polarization ellipse, 1-16
Sensing systems, millimeter-wave antennas for, 17-5. *See also* Remote sensing
Sensitivity and reception, 6-10
  with planar arrays, 11-65
SEO-I and -II satellites, 22-17
Septum polarizers, 21-30
  with reflector satellite antennas, 21-74, 21-79 to 21-80
  tapered, 13-60 to 13-61
Serial shift registers, 18-3 to 18-4
Series couplers, 19-4 to 19-5
Series-fed arrays
  for microstrip dipole antennas, 17-119
  microstrip patch resonator, 17-113 to 17-116
Series feed networks, 18-9, 18-21, 18-26, 19-3 to 19-5

Series impedance of transmission lines, 28-5
Series resonant magnetic field probes, 24-43, 24-44
Series-resonant power dividing circuits, 26-39 to 26-40
Series slot radiators with phased arrays, 18-8
Shadow boundaries
 and GO, 4-10, 4-17
 and PO, 4-22
 and PTD, 4-103 to 4-104
 and UTD, 4-32, 4-34, 4-36, 4-74, 20-8, 20-12
Shadow regions, 4-5, 4-75
 and diffracted rays, 4-3, 4-108
 Fock parameter for, 4-69
 and GO incident field, 4-10
 terms for, 4-67
 and UTD, 4-52, 4-63, 4-65
  with curved surfaces, 20-20 to 20-21
Shaped-beam antennas
 in antenna ranges, 32-19
 horns, 21-5, 21-82, 21-87 to 21-89
Shaped lenses, 21-18
Shaped reflectors, 15-73, 15-75 to 15-80
Shaped tapered dielectric-rod antennas, 17-37 to 17-38
Shapes for simulation of environments, 20-8
Shaping techniques for lens antennas, 17-19
Sheleg method, 19-103 to 19-105
Shielded-wire transmission lines, 28-11
 two-wire, impedance of, 28-15
Shields
 with loops, 25-9
 with scanning ranges, 33-16
Shift registers, 18-3 to 18-4
Ships and shipboard antennas
 models for, 32-74, 32-77 to 32-79
 numerical solutions for, 20-90 to 20-97
 simulation of, 20-7 to 20-8
Short antennas
 dipole, 3-13 to 3-15, 6-10 to 6-13, 6-15
  array of, 3-38 to 3-39
  compared to small loop antenna, 3-16
  fields with, 3-34
  UTD solutions for, 20-8 to 20-15
 monopole
  impedance of, 6-12

 for in-vivo measurements, 24-30 to 24-32
 mutual impedance of, 4-94
Short-circuit current, 2-32
 and vector effective height, 6-5
Shorted patch microstrip elements, 13-60
Shoulder lobes and phase taper, 32-16
Shunt admittance of transmission lines, 9-47 to 9-49, 28-5
Shunt couplers, 19-4
Shunt-slot array antennas, 17-26
Shunt slot radiators, 18-8
Side firing log-periodic diode arrays, 9-63 to 9-65
Side lobes, 1-20
 and antenna ranges
  compact, 33-23
  measurement of, 33-17
  phase errors in, 32-19
 and aperture field distibutions, 21-14, 21-16
 with apertures
  circular, 5-22 to 5-23
  rectangular, 5-12 to 5-14, 5-16, 5-18 to 5-20
 with arrays, 11-15
  aperiodic, 14-4, 14-5, 14-8, 14-14 to 14-19, 14-23 to 14-32
  binomial, 11-11
  conformal, 13-52
  Dolph-Chebyshev, 11-13 to 11-14, 11-18, 11-21, 11-49 to 11-50
  edge-slot, 17-27
  feed, for satellite antennas, 21-27
  periodic, 9-44, 9-45, 13-10 to 13-11, 13-18
  phased, 18-3, 18-26 to 18-28, 19-60, 19-61
  from random errors, 13-53 to 13-55
  scanned in one plane, 13-12
 with beam-forming networks, 18-25, 21-57
 and beamwidth, 9-44, 9-45, 13-18
 with DuFort optical technique, 19-76
 with feed systems
  arrays for satellite antennas, 21-27
  constrained, 18-20
  Gregorian, 19-65
  parallel, 19-5
  transmission line, radial, 19-26, 19-31, 19-33

with horn antennas
  corrugated, 8-55
  multimode, 8-70
and lateral feed displacement, 21-20
with lens antennas, 16-5
  coma aberrations, 16-22 to 16-23, 16-41, 21-14
  distortion in, 16-17
  DuFort-Uyeda, 19-53
  millimeter-wave, 17-15, 17-19, 17-20
  obeying Abbe sine condition, 19-98
  power distribution of, 21-18
  quadratic errors in, 16-14
  taper-control, 16-33
  TEM, 21-16
and log-periodic zigzag antennas, 9-44, 9-45
measurement of, 33-17
and microwave radiometer beam efficiency, 22-28
with millimeter-wave antennas
  dielectric grating, 17-69, 17-75 to 17-77
  geodesic, 17-32
  lens, 17-15, 17-19, 17-20
  metal grating, 17-69
  microstrip patch resonator arrays, 17-115 to 17-116
  tapered dielectric-rod, 17-38 to 17-43, 17-47
with offset parabolic antennas, 15-33, 15-37, 15-41 to 15-42
with orthogonal beam synthesis, 13-22
with parabolic pillbox feeds, 19-20, 19-22
from radomes, 31-6
and reflections, 32-34 to 32-37
with reflector antennas
  millimeter-wave, 17-9
  satellite, 21-23
  surface errors on, 15-105 to 15-106, 15-109 to 15-110, 15-112 to 15-114, 21-26
  tapered-aperture, 15-16 to 15-17, 15-21 to 15-23
and reradiative coupling, 32-14
with satellite antennas, 21-5, 21-7
with Schiffman phase shifters, 13-62
and series feed networks, 19-5
slots for, 8-63
with space-fed arrays, 19-55
  and subarrays, 13-30, 13-32, 13-39
  with synthesized patterns, 13-23 to 13-29
  with waveguides, open-ended
    circular, 15-94 to 15-95
    rectangular, 15-91 to 15-92
Side-mount antennas
  circularly polarized tv, 27-20 to 27-23
  for multiple-antenna installations, 27-37
Sidefire helical tv antennas, 27-24 to 27-26
Sidewall inclined-slot array antennas, 17-26
Signal-power to noise-power ratio for tvro link, 6-30
Signal-to-noise ratio and mismatches, 6-25
  with arrays, 11-58 to 11-63, 11-70 to 11-76
  with satellite antennas, 6-31
Significant height and ocean reflection coefficient, 29-38 to 29-40
Simple distributions
  with circular apertures, 5-23
  with rectangular apertures, 5-13 to 5-15
Simple lenses, 16-9 to 16-12
Simulation
  of antennas, 20-5 to 20-7
  of environment, 20-7 to 20-53
Simultaneous multi-beam systems, 19-79 to 19-80
Single-dielectric microstrip lines, 28-16
Single-feed circular polarization, 10-57 to 10-61
Single-wire transmission lines, 28-13 to 28-15
Sinusoidal-aperture antennas, 3-24
Sinusoidal current distribution, 3-14, 3-17
Skewed dipole antennas
  fm, 27-34 to 27-35
  tv, 27-11 to 27-18
Skin-effect resistance, 1-43
Skylab satellite, 22-17
Slant antenna ranges, 32-24 to 32-25
Sleeve antennas
  coaxial sleeve, 7-29 to 7-37
  dipole, 7-26 to 7-29
  with implantable antennas, 24-26 to 24-27
  junction effect with, 7-23 to 7-26
  monopole, 7-29 to 7-31
Slope of lenses, 16-9
Slope diffraction, 4-38 to 4-40, 4-108

Slot antennas, 3-31
　on aircraft wing, 20-85 to 20-86, 20-90
　annular, 30-13
　Babinet principle for, 30-14 to 30-16
　balanced, 9-72 to 9-73
　effective receiving area of, 30-10
　impedance of, 2-15 to 2-16, 9-78 to 9-79
　microstrip, for medical applications, 24-22 to 24-24
　radiation pattern of, 4-61
　stripline, 13-58 to 13-59
　tv, 27-26 to 27-31
Slot arrays
　cavity-backed, 9-68 to 9-72
　on curved surfaces, 13-51 to 13-52
　millimeter-wave, 17-26
　waveguide-fed, design of, 12-3 to 12-6
　　aperture distribution in, 12-24
　　and center-inclined broad wall slots, 12-24 to 12-25
　　and E-field distribution, 12-6 to 12-10
　　and equations for slots, 12-10 to 12-13
　　far-field and near-field diagnostics for, 12-36 to 12-37
　　and inclined narrow wall slots, 12-25 to 12-28
　　for large arrays, 12-28 to 12-34
　　and nonresonantly spaced slots, 12-16 to 12-20
　　for planar arrays, 12-20 to 12-24
　　and resonantly spaced slots, 12-13 to 12-16
　　second-order effects in, 12-34 to 12-36
Slot line transmission types, 21-59, 21-61
Slot line waveguides, 28-53 to 28-57
Slot radiators
　for aperture antennas, 4-64
　with phased arrays, 18-8
Slots, 3-31
　axis, 4-75, 4-77, 21-92, 21-95
　in circular cylinder, 4-77 to 4-79
　mounted in plate-cylinder, 4-61
　for side lobes, 8-63
　in sphere, 4-77 to 4-79
　voltages in, 12-10 to 12-12
Slotted cylinder reflector antennas, 21-110 to 21-111
Slotted waveguide array antennas, 17-26, 17-33
Small dipole antennas, 3-28
　and EMP, 30-9

Small loop antennas, 3-16, 3-19, 6-13, 6-17 to 6-19, 7-46 to 7-47
Small receiving antennas, 6-9
　bandwidth of, 6-22 to 6-24
　short dipole, 6-10 to 6-16
　small loop, 3-16, 3-19, 6-13, 6-17 to 6-18, 7-46 to 7-47
SMMR. *See* Scanning multichannel microwave radiometer
Snell's law of refraction, 32-70
　and lenses, 21-18
　　constrained, 16-51
　　dielectric, 16-6, 16-8 to 16-9
　　taper-control, 16-34
　with rays, 4-8 to 4-9
Snow and snowpack
　depolarization by, 29-14
　probing of, 23-25
　remote sensing of, 22-7, 22-15 to 22-16
SNR. *See* Signal-to-noise ratio
Soil
　emissivity of, 22-9 to 22-10
　probing of, 23-25
　remote sensing of moisture in, 22-13, 22-15 to 22-16
Solar radiation and ionospheric propagation, 29-14, 29-25
　noise from, 29-50
Solid bodies
　integral equations for, 3-46
　time-domain solutions with, 3-63
Solid-state components
　in integrated antennas, 17-131
　for millimeter-wave antennas, 17-11
Solid thin-wire antennas, 7-10 to 7-11
Sommerfield treatment, 3-49 to 3-50
Sounders. *See also* Microwave sounder units; Temperature sounders
　humidity, 22-14, 22-22
Source field patterns with UTD solutions, 20-15
　for curved surfaces, 20-25, 20-29, 20-31
　for noncurved surfaces, 20-37 to 20-42, 20-48
Sources
　radiation from, 1-11 to 1-13
　for transmitting antennas, 2-29 to 2-30
Space applications, lens antennas for, 16-5 to 16-6
Space-combination with millimeter-wave antennas, 17-11

Space configuration with am antennas, 26-17 to 26-19
Spacecraft constraints for microwave radiometers, 22-46 to 22-47
Spaced loops for direction-finding antennas, 25-9 to 25-11
Spaced-tapered arrays, 14-6 to 14-8
Space factors, cylinder, 21-98 to 21-100
Space feeds
  for beam-forming feed networks, 18-17 to 18-19
  for arrays in, 21-34, 21-35 to 21-36
  for reflectarrays, 19-54 to 19-56
  for subarray systems, 13-26, 19-68 to 19-69, 19-70 to 19-76
Space harmonics
  for dielectric grating antennas, 17-72
  with periodic dielectric antennas, 17-51 to 17-53
  and periodic structure theory, 9-33 to 9-35, 9-55 to 9-56
  and scattered fields, 1-34 to 1-35
Space shuttle and remote sensing, 22-10
Space waves for leaky-wave antennas, 17-100 to 17-101
Spacing with arrays, 3-36
  aperiodic, 14-4 to 14-5
  with log-periodic dipole antennas, 9-28, 9-29, 9-31
S-parameter matrix, 10-34 to 10-37
Spatial coupling with UTD solutions, 20-6 to 20-7
Spatial diplexers, 21-54
Spatial resolution of microwave radiometers, 22-30 to 22-32
Spatial variations in antenna ranges, 32-19
Special sensor microwave/imager, 22-25
  momentum compensation devices in, 22-46 to 22-47
  polarization with, 22-48
  scanning by, 22-46 to 22-47
Special sensor microwave/temperature sounder, 22-14
Spectrum functions and aperture fields, 5-6
Specular reflections in indoor ranges, 32-27
Speed of phase propagation, 29-7
Sphere
  radiation, 1-19, 1-21, 32-6 to 32-7
  slot in, radiation pattern of, 4-78

Spherical aberrations with lenses, 16-15, 16-20
  spherical thin, 16-29
Spherical coordinates, 3-24
  positioning system for, 32-8 to 32-13
  transformations of, 15-115 to 15-120
Spherical dipole antennas, 30-14
Spherical lenses, 21-18. *See also* Luneburg lenses
  aberrations with, 21-13 to 21-14
  cap, 16-23 to 16-27
  Maxwell fish-eye, 16-52 to 16-53
  symmetrical, 17-15
  thin, 16-28 to 16-30
Spherical near-field test, 33-20
Spherical radiation pattern of am antennas, 26-4, 26-8 to 26-9
Spherical reflector systems
  with satellite antennas, 21-20 to 21-21
  with wide-angle multiple-beam systems, 19-80
Spherical scanning for field measurements, 33-12 to 33-15
Spherical surfaces
  and intersection curve, 15-29
  for lenses, 16-11 to 16-12
  for offset parabolic antennas, 15-56
Spherical-wave illumination
  for distance parameter, with UTD, 4-56
  and GO incident field, 4-11 to 4-12
Spherical waves
  and biconical antennas, 3-24 to 3-25
  transformation of, to plane waves, 16-6, 16-9 to 16-10
Spillover
  with DuFort optical technique, 19-76, 19-78
  and lens antennas, 16-5
  taper-control, 16-37
  with offset parabolic antennas, 15-37, 15-42, 15-47, 15-56
Spinning-diode circular polarization patterns, 10-60 to 10-62
Spinning geosynchronous satellites, 22-43
Spin-scan technique, efficiency of, 22-34
Spin-stabilized satellites, 21-90
Spiral angle with log-spiral antennas, 9-108
  and directivity, 9-89
Spiral antennas, 17-27 to 17-28. *See also* Log-spiral antennas
  balanced, 9-78

Spiral-phase fields, 9-109
Spiral-rate constant, 9-72, 9-102
Split-tee power dividers
　for array feeds, 13-61
　stripline, 21-50, 21-51
Spread-F irregularities
　and scintillations, 29-18 to 29-19
　and vhf ionospheric propagation, 29-29
Square apertures, 5-32 to 5-33
Square coaxial transmission lines, 28-18
Square log-periodic antennas, 9-3, 9-4
Square waveguides
　for circular polarization, 16-43
　normalized modal cutoff frequencies for, 1-42
SSB. *See* Surface shadow boundary
S-65-147 vhf antenna, 30-25 to 30-26
S-65-8262-2 uhf antenna, 30-22
SSM/I. *See* Special sensor microwave/imager
SSM/T (special sensor microwave/temperature sounder), 22-14
Stabilizers, aircraft, simulation of, 20-73, 20-78, 20-82 to 20-83
Staggered-slot array antennas, 17-26
Standard am antenna patterns, 26-35 to 26-36
Standard-gain horn, 8-43
Standard radio atmosphere, 29-32
Standard reference am antennas, 26-3 to 26-22
Standard waveguides, 28-40 to 28-41
Standing-wave fed arrays, 12-4, 12-13 to 12-16
Stationary phase point, 4-7
Steering
　with periodic arrays, 13-7 to 13-10
　with subarrays, 13-32 to 13-35
Step functions
　for convex surfaces, 4-28
　in scattering problem, 4-51
　for wedges, 4-25
Step scans compared to continuous scans, 22-42
Stepped-horn antennas, 21-82 to 21-84
Stepped-septum polarizer, 13-61
Stopbands
　with dielectric grating antennas, 17-71, 17-81
　with leaky-mode antennas, 17-52, 17-68
　for log-periodic dipole antennas, 9-61 to 9-62, 9-64 to 9-65
　with periodic loads, 9-49 to 9-50, 9-55, 9-58
Stored energy with cavities, 10-20
Straight lines and intersection curve, 15-27
Straight wire antennas, 3-49 to 3-50
Stratton-Chu formula, 2-21
Stray efficiency of microwave radiometers, 22-29
Strip antennas, impedance of, 2-15 to 2-16
Strip dielectric guide transmission type, 21-62
Strip edge-diffracted fields, 20-21 to 20-23
Stripline techniques, 17-106
　for arrays, 13-61
　　blindness in, 13-47
　asymmetric, 17-84 to 17-87, 17-97 to 17-98
　for beam-forming networks, 20-50 to 20-51, 21-37
　for hybrid phase shifters, 13-62
　for printed dipoles, 13-58
　for slot antennas, 13-58 to 13-59
　for transmission types, 21-58, 21-60
　　feeds for, 19-29 to 19-30
　　for satellite antennas, 21-7
　TEM, 28-11, 28-16
　for waveguides, 21-57 to 21-58
Structural integrity, measurements for, 32-4
Structural stopbands, 9-49 to 9-50, 9-55
　with log-periodic arrays, 9-64 to 9-65
Structure bandwidth of log-periodic antennas, 9-17
Structures
　effect of, on radiation patterns, 20-53 to 20-54, 20-61 to 20-70
　and EMP, 30-4 to 30-7
　for simulation of environment, 20-8
Stub-loaded transmission lines
　attentuation curves for, 9-52, 9-54
　dispersion curves for, 9-52, 9-54, 9-57
Subarrays
　aperture illumination control with, 13-30 to 13-32
　overlapped, 13-35 to 13-39, 19-68 to 19-76
　time-delayed, 13-32 to 13-35
　with unconstrained feeds, 19-52
Subdomain procedures, 3-60
Submarine communications, 23-16

Subreflectors
   of Cassegrain offset antenna, 15-67, 15-69
   with reflector antennas, 15-7
Subsectional moment methods, 3-60
Substrates
   for integrated antennas, 17-131
   for microstrip antennas, 10-5, 17-107, 17-109 to 17-111, 17-114, 17-122
   dipole, 17-124 to 17-129
Substructures for simulation of environments, 20-8
Sum beams with beam-forming feed networks, 18-25
Sum-hybrid mode with microstrip antennas, 10-60 to 10-61
Sun
   and ionospheric propagation, 29-14, 29-25
   noise from, 29-51
Superdirective arrays, 11-61, 11-68 to 11-70
   aperiodic, 14-3
Supergain antennas, 5-27
   arrays, 11-61
Superstrates with microstrip dipoles, 17-127 to 17-128
Superturnstile tv antennas, 27-23 to 27-24
Surface corrugations
   for dielectric grating antennas, 17-64
   with periodic dielectric antennas, 17-48
Surface current
   decays of, with corrugated horn antennas, 8-52 to 8-54
   measurement of, 33-5 to 33-6
   and PO, 2-26 to 2-27
   and PTD, 4-103
Surface-diffracted waves with UTD, 4-57
Surface emissivity and brightness temperature, 22-7 to 22-9, 22-15
Surface errors of reflector antennas, 15-105 to 15-114
   and gain, 17-12
Surface fields
   equivalent currents for, 5-8
   with UTD, 4-84 to 4-89
Surface impedances
   and computer solutions, 3-54
   and modeling, 32-68
Surface integration method, 31-24 to 31-25

Surface mismatches with dielectric lenses, 16-54 to 16-56, 21-19
Surface Patch modeling code, 3-88 to 3-91
Surface perturbations for dielectric grating antennas, 17-70 to 17-71
Surface propagation, 29-30 to 29-47
Surface-ray field
   in GTD, 4-30
   in UTD, 4-58, 4-84 to 4-86
Surface-ray launching coefficients, 4-70
Surface representation of lens antennas, 21-13
Surface resistance, waveguide, 1-43
Surface roughness and conduction loss, 17-111, 22-31
Surface shadow boundary
   and GO fields, 4-13, 4-17
   and UTD, 4-50 to 4-51, 4-56
   transition regions, 4-58 to 4-59
Surface tolerances
   with lenses, 21-18 to 21-19
   with reflector satellite antennas, 21-23, 21-24, 21-26
Surface-wave antennas, 17-34
   periodic dielectric, 17-48 to 17-82
   tapered-rod, 17-35 to 17-48
Surface-wave launchers, 17-46
Surface waves
   with integrated antennas, 17-137
   for leaky-wave antennas, 17-100
   propagation with, 29-44 to 29-46
   from substrates, 17-122, 17-129
Susceptance
   with microstrip antennas, 10-7, 10-9
   with thin-wire antennas, 7-14 to 7-17
Suspended substrate transmission types, 21-59, 21-60
Switch beam-forming networks, 21-30 to 21-32
Switched-line phase shifters, 13-62 to 13-63, 18-13
Switches for beam-forming networks, 21-7, 21-37 to 21-38
Switching networks with Wheeler Lab approach, 19-105 to 19-106
Switching speeds of phase shifters, 18-16
Swr
   and feed cables, for log-spiral antennas, 9-78, 9-80
   and image impedance, 9-55

Swr (*cont.*)
  and log-periodic antennas, 9-27, 9-28
    dipole array, 9-58 to 9-59
    dipole scale factor, 9-27, 9-28
  and Q, in dipole arrays, 9-61
Symmetrical directional couplers, 21-50, 21-51
Symmetric parabolic reflectors. *See* Offset parabolic reflectors
Symmetry
  and computer time and storage, 3-65, 3-67
  with Dolph-Chebyshev arrays, 11-52
Synchronous satellites, 21-3
Synthesis methods
  for array patterns, 11-42 to 11-43, 11-76 to 11-86
  with multibeam antennas, 21-65
  for plane waves, 33-21 to 33-22
Synthetic aperture radar (SAR), 22-30
Systematic errors
  with microwave radiometers, 22-23, 22-25 to 22-28
  and phased array design, 18-26 to 18-28
System losses, power for, 26-36 to 26-37

TACOL structure, 19-56
Tangential fields, 5-7 to 5-9
  with thin-wire antennas, 7-5 to 7-6
Taper-control lenses, 16-33 to 16-38
Tapered anechoic chambers, 32-26 to 32-28
Tapered aperture distributions
  with parallel feed networks, 19-6
  with radial transmission line feeds, 19-31, 19-33
  for reflector antennas, 15-15 to 15-23
  with series feed networks, 19-5
Tapered-arm log-periodic zigzag antennas, 9-37 to 9-39
Tapered dielectric-rod antennas, 17-34 to 17-48
  for integrated antennas, 17-133 to 17-134
Tapered septum polarizer, 13-60 to 13-61
Taperline baluns, 9-95
Taylor line source pattern synthesis, 13-25 to 13-27
TDRSS (Tracking and Data Relay Satellite System), 21-67 to 21-71
TE field. *See* Transverse electric fields
TE/TM. *See* Transverse-electric/transverse-magnetic waveguides

TEC (total electron content), 29-15 to 29-16
Tee-bars for multiple-antenna installations, 27-37
Tee power dividers, 13-61, 26-40 to 26-41, 27-13, 27-15, 27-17
Teflon sphere lens antennas, 17-17 to 17-18
Television broadcast antennas, 27-3 to 27-8
  batwing, 27-23 to 27-24
  circularly polarized, 27-8 to 27-9
  helical, 27-15 to 27-19
  side-mount, 27-20 to 27-23
  horizontally polarized, 27-23
  multiple, 27-37 to 27-40
  sidefire helical, 27-24 to 27-26
  skewed dipole, 27-11 to 27-15
  slot
    coaxial, 27-28 to 27-29
    traveling-wave, 27-26 to 27-27
    waveguide, 27-29 to 27-31
  uhf, 27-27 to 27-28
  vee dipole array, 27-9 to 27-11
  zigzag, 27-31 to 27-32
TEM. *See* Transverse-electromagnetic fields
Temperature inversion layers and scintillations, 29-20
Temperature probes, implantable, 24-50 to 24-51
Temperature sensitivity of microwave radiometers, 22-21 to 22-22, 22-23
Temperature sounders
  and AMSU-A, 22-51
  for oxygen band, 22-16, 22-20
  for remote sensing, 22-13 to 22-14
Theoretical am antenna pattern, 26-34 to 26-36
Theory of uniform-periodic structures. *See* Periodic structure theory
Therapy, medical. *See* Hyperthermia
Thermal considerations with satellite antennas, 22-46
Thermal distortion, 21-28 to 21-29, 21-57
Thermal noise, 2-39 to 2-43, 29-50
Thermal temperature and sea surface emissivity, 22-14
Thermal therapy. *See* Hyperthermia
Thermistor detectors with implantable temperature probes, 24-50

Index

Thermocoupler detectors for rf radiation, 24-37
Thermographic cameras, 24-51
Thevenin's theorem for receiving antenna load, 6-3, 6-4
Thick dipole, 3-28
Thickness of lenses, 16-10 to 16-13. *See also* Zoning of lenses
  constrained, 16-43 to 16-44
  spherical, 16-27 to 16-29
  and surface tolerance, 21-18
  taper-control, 16-36
Thin dipole, 3-28
Thin-wall radome design, 31-17
Thin-wire antennas, 7-5 to 7-11
  and computer solutions, 3-54
  integral equations for, 3-44 to 3-46
  log-periodic zigzag, 9-45 to 9-47
  loop, 7-42 to 7-48
  receiving, 7-18 to 7-19
  time-domain analyses of, 3-51 to 3-52
  transmitting, 7-12 to 7-18
Thin-Wire Time Domain modeling code, 3-87 to 3-95
Three-antenna measurement methods
  for gain, 32-44, 32-46
  for polarization, 32-60 to 32-61
Three-axis stabilized satellites, 22-43 to 22-44
Three-wire transmission lines, 28-12
Through-the-earth communications. *See* Geophysical applications
Thunderstorms, noise from, 29-50
TID (traveling ionospheric disturbances), 29-25
Tilt angle and polarization, 32-51 to 32-53
  with polarization ellipse, 1-16
  with polarization pattern method, 32-62
  with reflector antennas, 15-48, 15-69 to 15-70, 15-72 to 15-73
Time-averaged power, 1-7 to 1-8, 3-11
  for modal fields, 1-46
  in rectangular waveguides, 1-39
Time delay
  with constrained analog of dielectric lenses, 16-51
  with constrained lenses, 16-44 to 16-46
  feed systems with
    matrix feed for beam-forming feed networks, 18-23, 18-25
    partial, 19-52, 19-68
    true, 19-7 to 19-8, 19-54, 19-76 to 19-91
  lenses for, 16-45 to 16-46, 16-54
  offset beams with, 13-39 to 13-41
  steering with
    with periodic arrays, 13-7 to 13-8
    with subarrays, 13-32 to 13-35
  with subarrays
    and contiguous subarrays, 13-32 to 13-35
    wideband characteristics of, 13-32
Time-domain
  compared to frequency-domain, 3-65 to 3-66, 3-82
  and EMP, 30-4
  integral equations in, 3-50 to 3-52
  method of moments in, 3-62 to 3-64
Time-harmonic excitation
  for grounded wire antennas, 23-8 to 23-12
  with loop antennas, 23-17 to 23-20
Time-harmonic fields, 1-5 to 1-8, 3-6 to 3-8
Tin-hat Rinehart lenses, 19-44, 19-46 to 19-47
TIROS-N satellite, 22-18
  planar scanning on, 22-36 to 22-37
  sounder units on, 22-14
Tissue, human
  permittivity of, 24-12, 24-30 to 24-35, 24-47
  phantoms for, 24-54 to 24-57
  power absorption by, 24-6 to 24-8
  radiation patterns in, 24-51 to 24-54
TM field. *See* Transverse magnetic fields
TMI microwave radiometer, 22-48 to 22-49
T-network as power dividers and phase shifters, 13-61, 26-40 to 26-41, 27-13, 27-15, 27-17
Toeplitz matrix, 3-68
Tolerance criteria for lenses, 16-16 to 16-20
Top loading of short dipole, 6-13
Topology with beam-forming networks, 21-57
Top-wall hybrid junction power division elements, 21-51 to 21-52
Toroidal beams for T T & C, 21-89
Toroidal ferrite phase shifters, 18-14, 21-40 to 21-42
Torsion factor, 4-87
Total electron content and satellite-earth propagation, 29-15 to 29-16

Towers for am antennas. *See also* Two-tower antenna patterns
  height of, and electric-field strength, 26-7
  mutual base impedance between, 26-12 to 26-17
  self-base impedance of, 26-8 to 26-12
  vertical radiation pattern of, 26-3 to 26-9
Tracking, telemetry, and command for satellite antennas, 21-89 to 21-111
Tracking and Data Relay Satellite System, 21-67 to 21-71
Trailing-wire antennas, EMP responses for, 30-27, 30-31
Transfer, microwave radiative, 22-4 to 22-7
Transformations
  of coordinates, 15-115 to 15-120
  distortion from, 11-57
  linear, 11-23 to 11-48
  of near-field data to far-field, 20-61 to 20-62
  for scale models, 32-67, 32-69
Transform feeds, 19-91 to 19-98
Transient conditions, 3-51
Transient excitation
  with grounded-wire antennas, 23-12 to 23-16
  with loop antennas, 23-20 to 23-21
Transient response
  of dipole antennas, 7-21 to 7-23
  of thin-wire antennas, 7-10
Transionospheric satellite-earth propagation, 29-14 to 29-15
Transition function with UTD solutions, 20-11
Transition regions, 4-5
  and PTD, 4-104
  and UTD, 4-32, 4-58 to 4-59
Translational symmetry and computer time and storage, 3-67 to 3-68
Transmission lines, 2-27 to 2-28. *See also* Constrained feeds
  analytical validation of computer code for, 3-76
  for beam-forming networks, 21-56 to 21-63
  with constrained lenses, 16-41
  current in, with folded dipoles, 7-38 to 7-39
  equations for, 28-5 to 28-10
  impedance of, with folded dipoles, 7-41
  and input admittance computations, 7-7
  with log-spiral conical antennas, 9-95
  for medical applications, 24-24 to 24-25
  for microstrip antennas, 10-7 to 10-10
  and mismatches, 1-24
  periodically loaded, 9-46 to 9-53, 9-57, 9-59 to 9-60
  planar quasi-TEM, 28-30 to 28-35
  radiator, for feeds, 19-23 to 19-32
  resonator for, with dual-band elements, 10-63, 10-67
  TEM, 28-10 to 28-29
Transmission space-fed arrays, 21-34, 21-35
Transmission-type space feeds, 18-17 to 18-19
Transmitting antennas
  linear dipole, 7-12 to 7-18
  reciprocity of, with receiving antennas, 2-32 to 2-36
  sources for, 2-29 to 2-30
  thin-wire, 7-8
  loop, 7-42 to 7-44
Transmitting fields
  and effective length, 2-34 to 2-35
  and Kirchhoff approximation, 2-19 to 2-21
Transmitting systems for antenna ranges, 32-30 to 32-32
Transposition of conductors with log-periodic antennas, 9-14 to 9-16, 9-59 to 9-60
Transverse amplitude taper in antenna ranges, 32-14
Transverse current in waveguides, 12-3
Transverse electric fields
  and circular waveguides, 1-44 to 1-50
  in rectangular waveguides, 1-37 to 1-41, 13-44
  representation of, 1-30 to 1-32
Transverse electric/transverse magnetic waveguides, 21-56 to 21-57, 28-35 to 28-47
  for medical applications, 24-9 to 24-12
Transverse electromagnetic fields
  and chambers for calibration of field probes, 24-54 to 24-57
  lenses for, 21-16
  transmission lines for, 28-10
    balanced, 28-12, 28-13
    circular coaxial, 28-20 to 28-22

triplate stripline, 28-22 to 28-29
two-wire, 28-11, 28-19 to 28-20
and waveguide antennas
for beam-forming networks, 21-56 to 21-58
for medical applications, 24-9 to 24-13
Transverse magnetic fields
and circular waveguides, 1-44 to 1-50
in rectangular waveguides, 1-37 to 1-41
representation of, 1-30 to 1-32
Transverse modes for leaky-wave antennas, 17-86, 17-88
Transverse slots, mutual coupling with, 12-30 to 12-31
Trapezoidal-tooth log-periodic antennas, 9-37 to 9-38
Trapped image-guide antenna, 17-49, 21-62
Trapped inverted microstrip transmission type, 21-59, 21-61
Trapped-miner problem, 23-23 to 23-24
Trapped surface waves, 29-45
Traveling ionospheric disturbances, 29-25
Traveling-wave antennas, 3-16, 3-18
arrays
for microstrip antennas, 17-118 to 17-119
microstrip patch resonator, 17-103 to 17-106, 17-114 to 17-115
television
sidefire helical, 27-24 to 27-26
slot, 27-26 to 27-27
Traveling-wave-fed arrays, 12-5
linear array of edge slots, 12-26
longitudinal slots, 12-16 to 12-20
planar, 12-23 to 12-24
Triangular-aperture antennas, 3-24
Triangular lattices
field representation of, 1-35
for phased-array radiators, 18-9 to 18-10
Triangular patches, 10-18
Triangular planar arrays and grating lobes, 13-16 to 13-17
Triangular-tooth log-periodic antennas, 9-37 to 9-38
Trifocal bootlace lenses, 16-47
Triplate striplines, 28-22, 28-26 to 28-29
impedance of, 28-17
Tropical Rainfall Measurement Mission (TRMM), 22-48 to 22-49
Troposphere and tropospheric propagation, 29-30 to 29-47
permittivity and conductivity of, 29-31

scatter propagation by, 29-46 to 29-47
scintillations caused by, 29-20
True-time-delay feed systems, 19-7 to 19-8, 19-54, 19-76 to 19-91
Truncation
of aperiodic arrays, 14-20
with log-periodic antennas, 9-4 to 9-5, 9-8
of log-spiral antennas, 9-72 to 9-73
conical, 9-79, 9-84, 9-89 to 9-90, 9-95 to 9-97
with microstrip antennas, 10-23 to 10-24
and mutual coupling, 12-29, 12-34
with optical devices, 19-98
and scanning range errors, 33-17
T T & C (tracking, telemetry, and command), 21-89 to 21-111
Tumors. See Hyperthermia
Turbulence, scintillation due to, 29-36
Turnstile antennas, 3-29
phased arrays for satellite antennas, 21-12
with radial transmission line feeds, 19-24, 19-26 to 19-27
TV. See Television broadcast antennas
TVRO (television receiving only) antennas, 6-25 to 6-32
Twin-boom construction for log-periodic dipole antennas, 9-15 to 9-16
Twist reflector for millimeter-wave antennas, 17-11
Two-antenna gain measurement method, 32-43 to 32-44, 32-46
Two-dimensional arrays, 13-8, 13-9
Dolph-Chebyshev planar, 11-49 to 11-52
Two-dimensional lenses
aberrations in, 16-19
waveguide, 16-42 to 16-43
Two-dimensional multiple-beam matrices, 19-11 to 19-12
Two-layer lenses, 17-15 to 17-18
Two-layer pillbox, 19-17 to 19-23
Two-parameter model for reflector antennas, 15-15 to 15-18
Two planes, arrays scanned in, 13-15 to 13-17
Two-point calibration method, 22-25 to 22-26
Two-terminal stopbands, 9-49 to 9-50
with log-periodic diode arrays, 9-65
phase shift in, 9-58
Two-tower antenna patterns, 26-22 to 26-24

Two-tower antenna patterns (*cont.*)
  horizontal electric-field strength with, 26-30 to 26-34
  mutual impedance of, 26-12 to 26-13
Two-wire transmission lines, 28-12
  field configuration of, 28-19
  impedance of, 7-39, 28-11, 28-14, 28-15, 28-20
  validation of computer code for, 3-76
TWTD (Thin-Wire Time Domain) modeling code, 3-87 to 3-95

UAT (uniform asymptotic theory), 4-5
  for edges, 4-32, 4-40 to 4-43
Uhf antennas, 27-3, 27-27 to 27-28
  bandwidth and gain of, 27-7
  EMP responses for, 30-17, 30-20 to 30-21
  and ionospheric propagation, 29-28 to 29-30
  slot, 27-29 to 27-31
Unconstrained feeds, 18-23 to 18-24, 19-49 to 19-54, 21-34
  limited scan, 19-56 to 19-76
  wide field of view, 19-54 to 19-56, 19-76 to 19-91
Underground communications, 23-3
  grounded-wire antennas for, 23-8 to 23-17
  loop antennas for, 23-17 to 23-23
Uniform aperture distributions, 5-11 to 5-13
  far-zone characteristics of, 3-23
  near-field reduction factors for, 5-30 to 5-31
  phase distortion with, 15-109
  radiation pattern with, 21-17
Uniform broadside linear arrays, 11-15
Uniform circular arrays, 11-46 to 11-47
Uniform dielectric layers, 17-65
Uniform diffracted fields, 20-44
Uniform dyadic edge-diffraction coefficient, 4-32 to 4-33
Uniform end-fire linear array, 11-15
Uniform geometrical theory of diffraction, 4-5
  for complex environment simulations, 20-4
  with convex surfaces
    mutual coupling, 4-84 to 4-96
    radiation, 4-63 to 4-84
    scattering, 4-50 to 4-63
  for edges, 4-32 to 4-38
  for numerical solutions. *See* Numerical solutions
  for vertices, 4-43 to 4-48
Uniform hemispherical radiators
  characteristics of, 26-9
  electric-field strength of, 26-27 to 26-29
Uniform linear arrays, 11-11 to 11-13
Uniformly perturbed millimeter waveguides, 17-35
Uniform reflected fields, 20-43
Uniform spherical radiators, 26-8
Uniform structures. *See* Periodic structures
Uniform-waveguide leaky-wave antennas, 17-35, 17-82 to 17-103
Unilateral fin lines, 28-55
Universal curves
  of E-plane horn antennas, 8-10, 8-13 to 8-14
  of H-plane horn antennas, 8-23, 8-27 to 8-29
Unknowns, and computer storage and time, 3-57
Unloaded linear dipole antennas
  receiving, 7-18 to 7-19
  transmitting, 7-12 to 7-18
Updating of computer models, 3-68
Uplink transmitters for satellite tv, 6-25 to 6-26
Upper frequency limit, 9-73
  of log-spiral antennas, 9-75
Use assistance with computer models, 3-68
UTD. *See* Uniform geometrical theory of diffraction

Validation of computer code, 3-55, 3-72 to 3-80
Varactor diodes
  with microstrip antennas, 10-67, 10-70
  in phase shifters, 18-13, 21-40
Variable amplitude networks, 21-33 to 21-34
Variable offset long-slot arrays, 17-26
Variable phase networks, 21-33 to 21-34
Variable phase shifters
  for beam-forming networks
    scanned, 21-38 to 21-43
    switched, 21-32
  for gain ripple with feed arrays, 21-27
  with Wheeler Lab approach, 19-105, 19-107

Variable power dividers
  for beam-forming networks
    scanned 21-43 to 21-49
    switched, 21-32 to 21-33
  for gain ripple with feed arrays, 21-27
  quantization errors with, 21-12
Vector effective area, 6-5 to 6-6
Vector effective height, 6-3 to 6-6
  and power ratio, 6-8 to 6-9
  of short dipole, 6-12
  with UTD solutions, 20-6
Vector far-field patterns, 7-8
Vector wave equation, 2-7 to 2-8
Vee dipole tv antennas, 27-9 to 27-11
Vegetation and surface emissivity, 22-10, 22-15
Vehicular antennas, 32-9, 32-19
Vehicular-mounted interferometers, 25-21 to 25-23
Vertical antennas, radiation resistance of, 6-16
Vertical current element, 26-9
Vertical full-wave loops, 3-31
Vertical optimization, 11-82 to 11-83
Vertical polarization
  for am antennas, 26-3
  with leaky-wave antennas, 17-94
  for microwave radiometers, 22-44 to 22-46
  and surface emissivity, 22-8
  with surface-wave propagation, 29-44 to 29-45
Vertical radiation characteristics for am antennas, 26-3 to 26-8
Vertical sounding, 23-4, 23-18
Vertical stabilizers, simulation of, 20-73, 20-82, 20-84
Vertical transmitting loop antennas, 23-17
Vertices
  GTD for, 4-30 to 4-31
  UTD for, 4-43 to 4-48
Very high frequencies
  communication antennas for, and EMP, 30-17, 30-20 to 30-24
  and ionospheric propagation, 29-21, 29-28 to 29-30
  localizer antennas for, and EMP, 30-23, 30-25 to 30-26
  marker beacon antennas for, and EMP, 30-23, 30-27 to 30-28
Very low frequencies
  ionospheric propagation of, 29-27 to 29-28

  trailing-wire antennas for, and EMP, 30-27, 30-31
Viscometric thermometers, 24-50
Visible regions
  of linear arrays, 11-9
  and log-periodic arrays, 9-56
  with planar periodic arrays, 11-27
  single main beam in, 11-31 to 11-33
Vokurka compact antenna range, 32-30
Voltage-controlled variable power dividers, 21-46 to 21-47, 21-49
Voltage reflection coefficient, 2-28
  of surfaces, 22-8
  for transmission lines, 28-7
Voltages
  in biconical transmission line, 3-24
  diagrams of, with standard reference am antennas, 26-19 to 26-20
  mode, 12-13, 12-16
  in rectangular waveguide, 2-28 to 2-29
  slot, 12-10 to 12-12
  in transmission lines, 28-6 to 28-8
  with UTD solutions, 20-7
Voltage source, ideal, 2-29 to 2-30
Voltmeters for receiving antenna, 2-30 to 2-32
Vswr
  of corrugated horns, 15-97
  and couplers, 19-4 to 19-5
  and lens surface mismatch, 16-54 to 16-55
  with log-periodic antennas
    front-to-back ratios in, 9-60 to 9-61
    zigzag, 9-42 to 9-44
  and log-spiral conical antennas impedances, 9-97, 9-107
  with microstrip antennas, 17-120
  with phase shifters, 21-40 to 21-41
  for pyramidal horns, 8-68
  and stopbands, 17-52, 17-81
  with transmission lines, 28-8
  of tv broadcast antennas, 27-3

Wall currents in waveguides, 12-3, 12-4
Water
  modeling of, 32-74 to 32-75
  remote sensing of, 22-15
Water bolus for loop radiators in medical applications, 24-20, 24-22
Water vapor
  absorption by, 29-35
  noise from, 29-51
  remote sensing of, 22-12 to 22-14, 22-22

Watson-Watt direction-finding system, 25-12 to 25-14
Wavefronts
 aberrations in, and radomes, 31-7 to 31-9
 incident, 4-15, 4-16
 and lenses, 16-6, 16-9 to 16-10
 and rays, 4-6 to 4-8
Waveguide-fed slot arrays. *See* Slot arrays, waveguide-fed
Waveguide feeds
 for beam-forming networks, 21-56 to 21-58
 for reflector antennas, 15-5
Waveguide hybrid ring waveguide power division elements, 21-50 to 21-52
Waveguide lenses, 16-42 to 16-45, 21-16
 for satellite antennas, 21-71 to 21-77
Waveguide loss. *See also* Leaky-wave antennas; Surface wave antennas
 circular, 1-49
 rectangular, 1-39
 with reflector antennas, 17-11
Waveguide nonreciprocal ferrite phase shifters, 21-38 to 21-39
Waveguide phased arrays, 21-12
Waveguides, 2-27 to 2-29
 arrays of, 13-44 to 13-45, 13-57
 blindness in, 13-45, 13-47 to 13-49
 circular. *See* Circular waveguides
 coplanar, 28-30 to 28-35
 coupling of, to microstrip antennas, 17-119 to 17-121
 hybrid-mode, 28-47 to 28-58
 impedance of, 2-28 to 2-29
 for medical applications, 24-9 to 24-18
 for microstrip patch resonator arrays, 17-117 to 17-118
 open-ended, for reflector feeds, 15-89 to 15-93, 15-104 to 15-105
 polarizers for, 13-60 to 13-61
 power divider elements for, 21-50 to 21-52
 rectangular. *See* Rectangular waveguides
 ridged, 28-45 to 28-49
 standard, 28-40 to 28-41
 TE/TM, 28-35 to 28-47
 uniformly perturbed millimeter, 17-35
Waveguide slot radiators, 18-6, 18-10
Waveguide slot tv antennas, 27-29 to 27-31
Waveguide transmission lines, 21-37

Wave impedances, 1-6, 1-13 to 1-14, 1-18 to 1-19
 of hollow-tube waveguides, 28-36, 28-37
 and modeling, 32-68
Wavelength, 1-6 to 1-7. *See also* Cutoff wavelength
Wave number, 1-6 to 1-7
Weapon-locating radar, limited scan, 19-56
Weather forecasting, remote sensing for, 22-13 to 22-15
Wedge-diffraction and UTD solutions, 20-11, 20-17 to 20-18
 coefficient of, 20-46
Wedges, conducting, 1-10
Weight functions in modeling codes, 3-83 to 3-84
Well logging, 23-24 to 23-25
Wenner array for earth resistivity, 23-5, 23-7
Wheeler Lab approach to cylindrical array feeds, 19-105 to 19-107
Wide-angle antennas, 17-29 to 17-30
Wide-angle lenses
 dielectric, 16-19 to 16-33
 multiple-beam, 19-37, 19-80 to 19-85
Wide-angle scanning
 and F/D ratios, 19-34 to 19-35
 lens antennas for, 17-15
Wide antennas, leaky mode with, 17-53 to 17-62
Wideband conical horn antennas, 8-61 to 8-62
Wideband Satellite Experiment, 29-19
Wideband scanning with broadband array feeds, 13-40 to 13-41
Wide field of view unconstrained feeds, 19-52, 19-54 to 19-56, 19-76 to 19-91
Width
 of dielectric grating antennas, 17-64
 main beam, with rectangular apertures, 5-13
 of metal grating antennas, 17-68 to 17-69
 of microstrip antennas, 10-8
Wilkinson power dividers for array feeds, 13-61
Wind speed, remote sensing of, 22-14
Wind sway, effect of, 27-38
Wings, aircraft, simulation of, 20-37, 20-70 to 20-71, 20-78, 20-80
Winter anomaly period for ionospheric propagation, 29-25

Index

Wire-antennas, 7-5 to 7-11
  folded dipole, 7-37 to 7-40
  linear dipole, 7-11 to 7-23
  loop, 7-40 to 7-48
  sleeve, 7-23 to 7-37
Wire diameter and radiation patterns, 32-76, 32-79
Wire objects, computer storage and time for, analysis of, 3-57
Wire-outline log-periodic antennas, 9-13
Wire problems
  subdomain procedures for, 3-60
  time-domain, 3-63
Wullenweber arrays, 25-14 to 25-16

X-band horn antennas, 8-41 to 8-42, 8-44 to 8-45, 8-67
X-polarized antennas, 1-28
XY-scanners, 33-19

Yagi-Uda antennas, 32-76
Yardarm, shipboard, simulation of, 20-93 to 20-94
Yield of crops, forecasting of, 22-15
Y-polarized antennas, 1-28

Zernike cylindrical polynomials, 16-13
Zero-bias diodes, 24-43
Zeroes of the Airy function, 4-59
Zeroes of Bessel functions, 1-45
Zeroes of W, 4-89
Zigzag antennas. *See also* Log-periodic zigzag antennas
  dipoles, as E-field probes, 24-39 to 24-40
  tv, 27-31 to 27-32
  wire, 9-33 to 9-38
Zone-plate lens antennas, 17-20, 17-22
Zones, FCC allocation and assignment, 27-4
Zoning of lenses, 16-6, 16-28 to 16-30
  and bandwidth, 16-54 to 16-55, 21-16
  with constrained lenses, 16-43 to 16-45
  with dielectric lenses, 16-38 to 16-41
  with equal group delay lenses, 16-46
  with millimeter-wave antennas, 17-15, 17-19 to 17-21

DUE DATE